Springer Proceedings in Physics 62

Springer Proceedings in Physics

Managing Editor: H. K. V. Lotsch

44 *Optical Fiber Sensors*
Editors: H. J. Arditty, J. P. Dakin, and R. Th. Kersten

45 *Computer Simulation Studies in Condensed Matter Physics II: New Directions*
Editors: D. P. Landau, K. K. Mon, and H.-B. Schüttler

46 *Cellular Automata and Modeling of Complex Physical Systems*
Editors: P. Manneville, N. Boccara, G. Y. Vichniac, and R. Bidaux

47 *Number Theory and Physics*
Editors: J.-M. Luck, P. Moussa, and M. Waldschmidt

48 *Many-Atom Interactions in Solids*
Editors: R .M. Nieminen, M. J. Puska, and M. J. Manninen

49 *Ultrafast Phenomena in Spectroscopy*
Editors: E. Klose and B. Wilhelmi

50 *Magnetic Properties of Low-Dimensional Systems II: New Developments*
Editors: L. M. Falicov, F. Mejía-Lira, and J. L. Morán-López

51 *The Physics and Chemistry of Organic Superconductors*
Editors: G. Saito and S. Kagoshima

52 *Dynamics and Patterns in Complex Fluids: New Aspects of the Physics-Chemistry Interface*
Editors: A. Onuki and K. Kawasaki

53 *Computer Simulation Studies in Condensed Matter Physics III*
Editors: D. P. Landau, K. K. Mon, and H.-B. Schüttler

54 *Polycrystalline Semiconductors II*
Editors: J. H. Werner and H. P. Strunk

55 *Nonlinear Dynamics and Quantum Phenomena in Optical Systems*
Editors: R. Vilaseca and R. Corbalán

56 *Amorphous and Crystalline Silicon Carbide III, and Other Group IV – IV Materials*
Editors: G. L. Harris, M. G. Spencer, and C. Y.-W. Yang

57 *Evolutionary Trends in the Physical Sciences*
Editors: M. Suzuki and R. Kubo

58 *New Trends in Nuclear Collective Dynamics*
Editors: Y. Abe, H. Horiuchi, and K. Matsuyanagi

59 *Exotic Atoms in Condensed Matter*
Editors: G. Benedek and H. Schneuwly

60 *The Physics and Chemistry of Oxide Superconductors*
Editors: Y. Iye and H. Yasuoka

61 *Surface X-Ray and Neutron Scattering*
Editors: H. Zabel and I. K. Robinson

62 *Surface Science: Lectures on Basic Concepts and Applications*
Editors: F. A. Ponce and M. Cardona

63 *Coherent Raman Spectroscopy: Recent Advances*
Editors: G. Marowsky and V. V. Smirnov

64 *Superconducting Devices and Their Applications*
Editors: H. Koch and H. Lübbig

Volumes 1–43 are listed on the back inside cover

F. A. Ponce   M. Cardona (Eds.)

Surface Science

Lectures on Basic Concepts and Applications

Proceedings
of the Sixth Latin American Symposium
on Surface Physics (SLAFS-6),
Cusco, Peru, September 3–7, 1990

With 258 Figures

Springer-Verlag
Berlin Heidelberg New York
London Paris Tokyo
Hong Kong Barcelona
Budapest

Dr. Fernando A. Ponce
Xerox Corp., Palo Alto Research Center, 3333 Coyote Hill Road,
Palo Alto, CA 94304, USA

Professor Dr. Drs. h.c. Manuel Cardona
Max-Planck-Institut für Festkörperforschung, Heisenbergstrasse 1,
W-7000 Stuttgart 80, Fed. Rep. of Germany

ISBN-13:978-3-642-76378-6 e-ISBN-13:978-3-642-76376-2
DOI: 10.1007/978-3-642-76376-2

Softcover reprint of the hardcover 1st edition 1991

Typesetting: Camera ready by authors

54/3140-543210 – Printed on acid-free paper

Preface

This volume contains lectures and papers presented at the Sixth Latin American Symposium on Surface Physics (SLAFS) in Cusco, Peru, in September of 1990 and is dedicated to the memory of Nicolás Cabrera. For details about the SLAFS and biographical notes on Nicolás Cabrera the reader is referred to the introductory contributions. The volume is divided into nine parts (plus the introductory contributions) covering a broad range of chemical, physical and engineering aspects of two-dimensional solid systems such as surfaces, thin films, and artificial layer structures. Each part is headed by one or more invited lectures on topics of current interest, followed by a few shorter contributed papers which give a cross-sectional view of most regional activities in the field in Latin America. As such, they provide an inventory of manpower, know-how and equipment available in the area at this time.

The reader will be amazed by the wide spectrum of subjects covered. They range from state-of-the-art theoretical topics, some involving large scale computation, to sophisticated numerical treatment of experimental data obtained in the analytical characterization of practical materials. They further range from high-energy particle detection to the oxidation of barbed wire in Mesoamerican coastal areas, from surface magnetism to thin film photovoltaic cells, from high temperature superconductors to amorphous thin film semiconductor devices. The reader will also be surprised by the high degree of expertise and the quantity, quality, and sophistication of equipment available in what is generally considered a developing area. The reason for high development in the subject of this volume is to be sought, at least in part, in the considerable relevance of it to technologies and industries which are rapidly growing in Latin America, among others the chemical, petrochemical, and electronic industries. Industrial managers and scientists have been farsighted enough to devote a fraction of their analytical equipment and computational potential to basic studies in the related areas of surface science and thin films. Regular happenings, such as the SLAFS, have helped to forge a coherent community of competent scientists who speak the same language and know how to help each other in their endeavor.

The editors of this volume strongly recommend readers to attend one of the future SLAFS. It will be an unforgettable experience, as Cusco was for them.

Stuttgart
Palo Alto, CA
June 1991

Fernando Ponce
Manuel Cardona

Contents

Part IV Interaction Between Radiation and Surfaces

Part V Processes at Surfaces: Interface Formation

Part VI **Properties of Thin Films**

Part VII **Semiconductors**

Part I

Introductory Contributions

The Sixth Latin American Symposium on Surface Physics

F.A. Ponce[1] and *M. Cardona*[2]

[1]Xerox Palo Alto Research Center, Palo Alto, CA 94304, USA
[2]Max-Planck-Institut für Festkörperforschung, W-7000 Stuttgart, Fed. Rep. of Germany

«Yo en el medio, ustedes rodéandome. Yo hablando, ustedes escuchando. Vivimos, andamos. Esa es la felicidad.»
«Qué miserable debe ser la vida de los que no tienen, como nosotros, gentes que hablen, reflexionaba. Gracias a lo que cuentas, es como si lo que ha pasado volviera a pasar muchas veces. ... Escucha, no desperdicies estas historias, criatura.»
Mario Vargas Llosa, "El Hablador"

Abstract. This Symposium took place in Cusco, Peru, during the week of September 3-10, 1990. A historical perspective of the SLAFS, the organization of the meeting, the dedication to the memory of Prof. Nicolás Cabrera, and other features of this event are discussed in this introduction to the volume.

1. HISTORIC PERSPECTIVE

During the last decade, the Latin American Symposia on Surface Physics have become established as a forum for the discussion of recent research activities in surface science and related applications. SLAFS is an acronym which derives from its name in Spanish and Portuguese (in Spanish: Simposio Latino-Americano de Física de Superficies). The SLAFS series started as a colloquium organized by Norberto Majlis in 1980. Its original name was "Coloquio Latinoamericano de Física de Superficies" and was held at the Universidade Federal Fluminense, Niteroi, Brazil, during the week of 1-6 December 1980. The second meeting, SLAFS-2, was organized by José Luis Morán-López and Peter Halevi, and was held at the Universidad de Puebla, Puebla, Mexico, during 4-8 October 1982. San José, Costa Rica, was the host of SLAFS-3, held at the Universidad de Costa Rica, during the week of 24-28 September 1984, and was organized by Alejandro Sáenz y Neville Clark.

After three consecutive symposia in Central America, Venezuela hosted the SLAFS-4, held in Caracas, in 1986, and organized by Germán

Springer Proceedings in Physics, Volume 62
Surface Science Eds.: F.A. Ponce and M. Cardona

Castro[1]. Jairo Giraldo organized the SLAFS-5, in Bogotá, Colombia, on 11-15 July 1988[2].

Besides the high level of academic interaction, the SLAFSs have enjoyed a unique fraternity that can occur only in Latin America. For no other part of the world can enjoy a vast territory like that of Central and South America, and a rich diversity of nearly 20 nations which are united by a common Iberian culture. It is a unique experience to feel people from so many nations speaking their own language, which is either Spanish or Portuguese in the case of Brazil, both being similar to each other. It is also unique to be able to understand and identify with each other's songs, humor, and other cultural expressions.

The mixing of Latin America, and the migration of scientists to other parts of the world has precipitated the formation of a Latin American identity. "For the last decades we have seen thousands of Argentinians living in Ecuador, or Uruguayans living in Venezuela, or Chileans moving to Colombia, . . . masses of people moving around the continent. . . The people got together . . ., and there was a cultural exchange. For the first time you could hear people saying *I'm a Latin American*"[3], whether they were in Europe, the U. S., or an adoptive Latin American country. In addition to allowing scientists from Latin American nations to meet, the SLAFSs have also provided an opportunity for Latin Americans who live abroad to return home and find themselves again; and for colleagues from other cultures to come and meet the New World. Cultural affinity with Spain and its rapidly expanding science also provide a natural link to scientific and other human endeavor in the European community.

2. THE SLAFS-6

At the general assembly that took place at the end of the SLAFS-5 in Bogota, one of the authors (FP) was asked to organize the next symposium that was to be held two years later. Together with Manfred Horn and Victor Latorre, support from the Universidad Nacional del Cusco, Multiciencias, and the National Institute of Technology (INTINTEC) was obtained.

Cusco was chosen as the site based on its historical preeminence and its symbolic value as one of the most important cities of pre-Columbian America.

The purpose of the Symposium was determined to be twofold: (i) to serve as a forum for the communication of current research on Solid Surfaces and related fields being conducted in Latin America; and (ii) to stimulate collaborative research in this active area of science and technology between researchers in Latin America, United States, and Europe.

The attendance to SLAFS-6 set a new high record. As in previous symposia in the series, the core of this Symposium was constituted by invited talks delivered by leading specialists. 45 invited talks of 35 minutes each covered the topics presented in these proceedings. Poster sessions were held with 112 contributed papers in a similar order as the oral sessions. In keeping with the stated goal of fostering multilateral collaborations, time was allocated on a daily basis to ensure that informal discussions and interactions would take place. Two evening panel discussions on *Candent Problems in Surface Physics* and *International Cooperation and Funding* were held during the week. The Symposium ended with an assembly of all the participants, where the next site was confirmed. Blas Alascio, will organize the next SLAFS in Bariloche, Argentina, in November 1992. It was also decided to form the Latin American Society of Surface Science with open membership, starting with all those present in Cusco. The statutes of this new society will be prepared by the organizers of SLAFS-7 and shall be confirmed in Bariloche in 1992.

In addition to the intense scientific program, cultural enjoyment was available throughout the week, punctuated by a 5 hour talent show, given by all the participants, and weekend one-day-excursions to Machuppicchu and the Sacred Valley of the Incas.

On the following week, the Second Ibero-American Workshop on Surfaces and Interfaces (TIASI-2) was held in Yucay, one hour from Cusco. It consisted of extended lectures in the mornings, and workshops on contemporary problems in the afternoons. It was attended by 50 participants.

3. DEDICATION

The Symposium was dedicated to the memory of Prof. Nicolás Cabrera, a great Spanish physicist who contributed to the development of surface science in both North and South America . A description of the work and life of Prof. Cabrera is given in the following articles by his son, Blas Cabrera, and one of his students, Miguel José-Yacamán.

The Symposium was also part of the tricentennial celebration of the Universidad Nacional San Antonio Abad del Cusco.

4. ORGANIZATION OF THE SYMPOSIUM

The Symposium would not have been possible without the advice, help and support of many colleagues many of whom collaborated through some of the committees.

The Local Organizing Committee was chaired by M. Horn (UNI, Lima), and included H. Barrientos (UNSAAC, Cusco), F. Bartra (UNSAAC, Cusco), V. Latorre (Multiciencias,UNSAAC,Cusco), E. López-Carranza (SOPERFI, Lima), H. Nowak (UNMSM, Lima), L. M. Ponce de Rozas (UNSAAC, Cusco) and P. Zanabria (UNSAAC, Cusco).

The International Advisory Committee was composed of B. Alascio (Argentina), F. Alvarez (Brazil), F. Briones (Spain), G. Castro (Venezuela), G. F. Chiarotti (Italy), F. Claro (Chile), P. Echenique (Spain), F. García Moliner (Spain), J. Ferrón, (Argentina), R. Livi (Brazil), C. Ocal (Spain), A. Pantoja (Colombia), Y. Petroff (France), F. Sánchez-Sinencio (Mexico) and H. Verdún (USA).

The Organizing and Program Committee was chaired by F. A. Ponce (USA), and was composed of R. Barrera (Mexico), M. Cardona (Germany), A. Eguiluz (U.S.A.), J. Giraldo (Colombia), I. Hernández Calderon (Mexico), N. Majlis (Brazil) and M. José-Yacamán (Mexico).

5. ACKNOWLEDGMENTS

The support of several institutions is very gratefully acknowledged. The organizing institutions were the Faculty of Chemical, Physical and Mathematical Sciences of the Universidad Nacional del Cusco (FCQFM, UNSAAC), Cusco; Multiciencias, UNSAAC, Cusco; and the Peruvian Physical Society (SOPERFI), Lima.

Institutions sponsoring the event were the Centro Latinoamericano de Física (CLAF), Rio de Janeiro, Brazil; the Centro Internacional de Física (CIF), Bogotá, Colombia; CONCYTEC, Lima, Peru; the Faculty of Sciences, Universidad Nacional de Ingeniería, Lima, Peru; the Instituto de Cooperación Iberoamericana, Madrid, Spain; the Instituto de Física, UNAM, Mexico; the International Center for Theoretical Physics, Trieste, Italy; the International Union of Vacuum Science Techniques and Applications; Japan Electron Optics Co. (JEOL); the Ministry for Science and Education, Dr. Juan Rojo, Spain; the National Science Foundation, USA; the Organization of American States; and Xerox Corporation.

References

(1) G. R. Castro and M. Cardona (eds.): "Lectures on Surface Science," (Springer Verlag, Heidelberg, 1987). Proceedings of SLAFS-4.

(2) M. Cardona and J. Giraldo (eds.): "Thin Films and Small Particles," (World Scientific, Singapore, 1989). Proceedings of SLAFS-5.

(3) Isabel Allende, in "Allende of the Spirits", by Steve Kettman, Metro Guide, San Jose, California. November 16-21, 1989, p. 21-22.

Professor Nicolás Cabrera Sanchez (1913–1989): A Family Perspective of His Scientific Career

B. Cabrera

Physics Department, Stanford University, Stanford, CA 94305, USA

Abstract. The Sixth Latin American Symposium on Surface Physics is dedicated to the memory of Professor Nicolás Cabrera, a great Spanish physicist who pursued studies of surfaces throughout a rich career in the United States and Europe. He also contributed substantially to the development of his chosen field in Latin America. The following summary of his career is written from my family perspective.

1. Introduction

The life and career of my father were remarkable in that he found himself in totally different environments three different times during his lifetime. In each of these epochs he took advantage of the new environments to establish vigorous research in fundamental physics. He always found the positive aspects of each new situation even when support for research was weak, and he never paid attention to those who lamented that it was not possible to get anything done. He dismissed pessimistic views by insisting that it is always possible to establish interesting fundamental research, even with limited resources or in the face of obstacles. His life naturally divides into four epochs, beginning with his formative years in Spain (1913-1938), continuing through his development as a scientist in France (1938-1952), progressing through a mature epoch in the United States, and finally the return to Spain in his later career (1968-1989). A chronological list of his career is given in Section 7 and is followed by a selected list of his eight most important papers in Section 8.

2. Early Years in Spain (1913-1938)

Nicolás Cabrera Sanchez† was born in 1913 to the family of Blas Cabrera Felipe and Maria Sanchez Real. His father was a famous experimental physicist in Spain before the Spanish civil war. He was brought up in an intellectual atmosphere together with two older brothers, Blas Cabrera

† Throughout this paper the Hispanic style for names is often used. The next to last name comes from the first family name of the father and the last name comes from the first family name of the mother. The anglicized version is Nicolas Cabrera.

Springer Proceedings in Physics, Volume 62
Surface Science Eds.: F.A. Ponce and M. Cardona

Fig 1.
Nicolás Cabrera at age 22 had completed his undergraduate studies at the University of Madrid.

Sánchez, who became a medical doctor and Luis Cabrera Sánchez, who became an architect. When my father was nine or ten he recalls that Albert Einstein, while on a major tour of Europe, was visiting Spain and was hosted by my grandfather while in Madrid. During the visit, my father talked of an evening get-together at the family home which included a number of university scholars from both the sciences and humanities. In addition, Andrés Segovia had been invited and performed for the guests on his guitar. As my father told the story, he would always smile and recall that Einstein had requested a violin and performed at the gathering as well. This particular reunion was very special, but often the intellectual life of my grandfather was brought to their home. In this environment, my father's interests in the sciences and the humanities flourished. In fact, his first academic love was history. Only later did he decide on a scientific direction, first beginning his studies at the university in engineering and after one year transferring to the study of physics.

He obtained his undergraduate degree in the sciences at the University of Madrid in 1935. Fig 1 shows him as a young man of 22 having had a complete education in Madrid, but knowing very little of the outside world. After obtaining his undergraduate degree, he continued his study of physics at the Rockefeller Institute in Madrid which had been established under the direction of his father. My father's first scientific publication [1] was an experimental study of the magnetism of the rare earth elements together with his father, Blas Cabrera Felipe and Salvador Velayos. The magnetism of the rare earths formed the central theme of my grandfather's experimental studies and his compiled data formed the basis of an important early test of the then new quantum mechanics as calculated by J. H. van Vleck, the Harvard theorist. The time my father spent doing research in that famous laboratory taught him that he could perform scientific studies at an international level of significance.

The research in that superb scientific laboratory, as well as research in a number of other active laboratories in Spain, came to an abrupt end during the Spanish civil war. The careers of both my father and grandfather were disrupted when they and their family were forced to leave Spain towards the end of the war, and to travel as refugees to Paris in 1938. In 1941, when it became clear that my grandfather would not be allowed to return to Spain, he and my grandmother emigrated to Mexico where he joined the University of Mexico and helped to establish fundamental scientific research before his death in 1945.

3. The Years in France (1938-52)

My father remained in Paris throughout World War II. He met and married my mother, Carmen Navarro Clavero, in 1942. She had traveled to Paris from Madrid with her family, also as civil war refugees. Although at times they had lived within several blocks of one another in Madrid, they had never met in Spain. My father supported his family through a position at the International Bureau of Weights and Measures (BIPM) where he continued his scientific studies at the same time that he was completing his doctoral dissertation at the University of Paris under the famous physicists Louis de Broglie and Leon Brillouin. He obtained his Ph. D. in 1944 on a theoretical study of thermodynamic phase transitions [2].

Some of his work at the BIPM pertained to precision dimensional metrology. My father became interested in the effect surface oxides might have on metrology and he began to study metal oxides, in particular the surface oxides of aluminum. He wrote a series of publications, many with Hamon, while in Paris [3]. The famous physicist N. F. Mott, then at the University of Bristol in England, became interested in several of these papers and invited my father to Bristol as a Research Associate. The next three years produced the two most important scientific accomplishments of my father's career. The first paper published in Bristol was titled "Theory of Oxidation of Metals" and was written together with Mott [4]. It describes the first theory of oxide growth based on quantum mechanical principles. The second, entitled "The Growth of Crystals and the Equilibrium States of Their Surfaces", was written together with F. C. Frank and W. K. Burton [5]. It describes the first theory of crystal growth which includes the interaction between crystal growth and crystal dislocations. This publication also contains for the first time the idea that the surface of a crystal can melt at a temperature below the bulk melting temperature, a phase transition known as the roughening transition.

4. Years at the University of Virginia (1952-1968)

His publications from the research in Bristol attracted attention from a large number of physicists and led to several opportunities for my father to join a university. He decided to accept an invitation to join the Physics Department of the University of Virginia in the United States. In 1952, he took the family across the Atlantic and to Charlottesville. I was five years old and my sister Cristina was not yet a year old. A second daughter, Carmen, arrived in 1956.

While at the University of Virginia, my father continued his research. He became interested in the scattering of atoms, such as helium, from surfaces as a tool for studying the surface structure [6,7]. In addition, he also developed a talent for organization and he helped build the University of Virginia Physics Department into a first rate research institution. From 1962 through 1968, he was the Chairman of the Physics Department. During these years he made many contacts with scientists from Latin America and from Spain.

I remember the summer after we arrived in the United States, 1953, he took the whole family from Charlottesville to Mexico City by car to visit his brother Blas Cabrera Sánchez, who had emigrated to Mexico with the Spanish government in exile after the Spanish civil war. The trip was made by car. At that time, well before the interstate road system, the trip took fourteen days driving for eight hours each day. To make matters worse, my mother did not drive. When we arrived, we visited with the family of my uncle and with my grandmother. While in Mexico, my father made contacts with physicists at the University of Mexico and at the National Polytechnic Institute, and he maintained those contacts throughout his career. An account of these interactions is contained in the comments by Professor Miguel Jose Yacamán in these proceedings.

In 1963, at the invitation of Professor Gonzalo Castro Fariñias, my father took the family to Caracas, Venezuela where he spent a sabbatical. Many new contacts were made at the University of Venezuela.

During the sixteen years in the United States, our home was often the center of festivities for many Spanish and Latin American visitors, faculty and staff at the University of Virginia. There was always much singing and discussion about politics, history and, yes, about science and physics.

5. The Return to Spain (1968-1989)

In 1968, my father decided to leave Virginia, because he felt that he could still help to establish fundamental research in a Spanish-speaking country. He first went to Mexico, where he had a number of long term contacts. Then, the following year, he was offered the opportunity of returning to Spain to form the Physics Department at the newly established Autonomous University of Madrid (UAM).

For several years, tremendous progress was made by a large number of active researchers to establish the first international-caliber scientific

Fig 2.
Nicolás Cabrera at age 73 was a Professor Emeritus of the Autonomous University of Madrid.

investigation in physics in Spain since before the civil war. Then, in 1973, the Prime Minister of Spain, Carrero Blanco, was assassinated. When a new government was chosen, there was also a change in the Minister of Education. Suddenly, the support for the new University was withdrawn, and a very difficult period followed. Contracts were broken and about half of the department researchers left, mostly to foreign institutions. My father almost decided to return to the United States, but in the end, he decided to make the best of the hard situation and to maintain as much of the new research spirit as possible.

After the death of Franco in 1976, the situation improved slowly and continuously. My father continued his physics research throughout his career, in Spain concentrating mostly on studies of single atom scattering off of surfaces [8]. Today, there are excellent research groups at the Autonomous University of Madrid and at a number of other universities throughout Spain. My father was not the only one responsible for this resurgence, but there is no doubt that his contributions were significant. Fig 2 shows him at 73 as an Emeritus Professor at the Autonomous University of Madrid.

My father died in Madrid in 1989 after a protracted illness. He was 76 years old. His headstone reads Nicolás Cabrera Sánchez: "FISICO - Cientifico de Gran Humanidad" - roughly translated as "A PHYSICIST - A Scientist and A Caring Humanitarian". He treated everyone he met with respect and always maintained a gentlemanly presence which prompted respect in return.

6. Acknowledgements

It has been a great honor for me to participate at this international conference dedicated to my father. I wish to express the gratitude of my family to the Organizing Committee of the Sixth Latin American Symposium on Surface Physics. In particular, I wish to thank Fernando Ponce, the Chair of the Organizing Committee, and Miguel Jose Yacamán for proposing that the conference be dedicated to my father. I also thank the National University of San Antonio Abad in Cusco for the grand hospitality that they afforded all of us. We will all long remember the striking beauty of Cusco and the breathtaking adventure to Machupicchu.

7. Chronological Career Summary for Nicolás Cabrera

1935	Licenciado en Ciencias Físicas, Universidad de Madrid
1935-37	Profesor Auxiliar, Universidad de Madrid
1944	Doctor ès Sciences Physiques, Université de Paris
1938-52	Assistant Scientifique, Bureau International des Poids et Mesures, Sèvres
1947-50	Research Associate, H. H. Wills Physics Laboratory, University of Bristol
1952-54	Associate Professor of Physics, University of Virginia
1954-67	Professor of Physics, University of Virginia
1962-68	Chairman of Physics Department, University of Virginia
1963	Profesor de Física, Universidad Central de Venezuela
1967-72	Commonwealth Professor of Physics, University of Virginia
1968-69	Profesor de Física (Experto do la Unesco), Instituto Politécnico Nacional, México
1969-78	Catedrático contratado y Director de la División de Física, Universidad Autónoma de Madrid
1973-78	Director del Departamento de Física Fundamental Universidad Autónoma de Madrid
1973	Gran Cruz de Alfonso X el Sabio
1975-83	Catedrático Numerario, Cátedra de Física del Estado Sólido Universidad Autónoma de Madrid
1975-78	Director del Instituto de Física del Estado Sólido (CSIC)
1978-80	Decano de la Facultad de Ciencias Universidad Autónoma de Madrid
1982	Premio "Blas Cabrera Felipe", Universidad Menéndez Pelayo
1982	"Nicolas Cabrera Festschrift" Philsophical Magazine **45**, 221-356 (1982)
1983-89	Profesor Emérito, Universidad Autónoma de Madrid

8. List of Selected Publications

1. N. Cabrera, B. Cabrera y S. Velayos, "Constantes magnéticas de algunos sulfatos octohidratados de las tierras raras", Boletin Academia de Ciencias, Madrid **1**, 1 (1935).
2. N. Cabrera, "Perturbation des conditions aux limites", Cahiers de Physique Paris, **31-32**, 24 (1948).
3. N. Cabrera et Hamon, "Oxydation de l'aluminium",Comptes Rendus **224**, 1713 (1947); and **224**, 1958 (1947); **225**, 59 1947).
4. N. Cabrera and N. Mott, "Theory of oxidation of metals", Reports on Progress in Physics **12**, 163 (1949).
5. N. Cabrera, F. C. Frank and W. K. Burton, "The growth of crystals and the equilibrium structure of their surfaces",Philosophical Transactions of the Royal Society **243**, 299 (1951).
6. N. Cabrera, V. Celli and R. Manson, "Theory of surface scattering and detection of surface phonons",Physical Review Letters **42**, 346 (1969).
7. N. Cabrera, F. O. Goodman and J. Lui, "Model of the repulsive elastic scattering of atoms by solid surfaces", Journal of Chemical Physics **57**, 1968 (1972).
8. N. Cabrera and N. Garcia, "A new method for solving the scattering of waves from a periodic hard surface", Physical Review **B18**, 576 (1978).

Don Nicolás Cabrera in Mexico

M. José-Yacamán

Instituto de Física, Universidad Nacional Autónoma de México,
Apdo. Postal 20-364, 01000 México, D.F., Mexico

Abstract. Nicolás Cabrera had a strong impact in the development of surface science in Latin America. In this article, his contributions are described from the point of view of one of his students.

1. Introduction

I was very pleased when the organizer of this symposium, Dr. Fernando Ponce, decided to support my suggestion that the Sixth Latin American Symposium on Surface Physics be dedicated to Prof. Nicolás Cabrera. I was even more satisfied when he informed me that the organizing committee had unanimously agreed with the proposal.

Prof. Cabrera played a key role in the development of surface science in Latin America. His influence was very deep and it took many years before his effect was totally absorbed by the community. In my opinion, Nicolás Cabrera is one of the most outstanding scientists of the Iberoamerican Spanish-speaking community in this century. His achievements were indeed world-class and were made at the time when very few Spanish or Latin American scientists were making impact in the international community. The highlights of his scientific career are described by his son Blas in an article which is published in this volume. I will concentrate on the role that Cabrera played in the Latin American community, and particularly in Mexico.

2. Spanish Migration to America

The Spanish Civil War caused an internal disruption that had a great impact in the later life of that country. One of the consequences was the loss of a great number of high level intellectuals, who by their Republican conviction had to emigrate in order to protect their own lives. Latin America in general, and Mexico in particular, received a great influence from such migration. It could be said that this event produced an attitude that caused some of the best intellects of the Iberian Peninsula to move to America. The result was a notable change in the cultural panorama in almost all areas of knowledge in Latin America, physics not being an exception. The migration of physicists was perhaps not as numerous as in other fields, but critical people came who had a great effect in the development of science on this side of the Atlantic. Blas Cabrera, former rector of the Universidad Complutense of Madrid, was the most notable physicist from the decades of the 20's and 30's. Undoubtedly, he was a well known figure abroad and was internationally recognized for his studies on the magnetism of solids. He had also occupied very important positions within the Spanish scientific organizations. Blas Cabrera was forced to leave Spain, and arrived in Mexico to work in the Institute of Physics and in the Faculty of Science of the UNAM (Universidad Nacional Autónoma de México). He was already a man advanced in age and in failing health, with a deep resentment about the tragic situation that he had endured. Blas Cabrera passed away in 1945, only four years after his arrival in Mexico.

3. Nicolás Cabrera's Early Contributions

Nicolás Cabrera, son of Blas, inherited the same scientific interests and left to study Physics at the University of Bristol, in the H. H. Wills Laboratory. In this group, he worked with the brilliant Charles Frank, and published one of the classic papers of the century, the theory of crystal growth (Frank, Burton and Cabrera). This article appeared in the Proceedings of the Royal Society of London in 1949, and it is undoubtedly one of the keystones of the modern theory of materials. During those years, an atomic view of the structure and properties of materials was being developed. The article contains a large amount of information and is impressive for its depth and simplicity. This work is the start of the modern theory of the thermodynamics of solids and of the surface theory that Nicolás Cabrera would develop in the years to come. After his stay at Bristol, Cabrera worked at the department of Physics of the University of Virginia in the United States as professor and eventually became its chairman. During those years he became well known for his contributions to the understanding of thermodynamics of solids and surfaces. His work on surface morphology and roughness at equilibrium are well known.

4. Nicolás Cabrera and México

Don Nicolás, as we affectionately call him, was always interested in Mexico and in Latin America. During his chairmanship at the University of Virginia, he received many scientists that were trained in the techniques of nucleation and growth theory and in solid state physics. Many of the scientists in the Spanish speaking community knew that they had in Cabrera somebody from the front line of science that could guide them on current interesting scientific problems. He visited Mexico frequently to see his mother, who after the death of Blas Cabrera had decided to stay permanently in the country. During those visits, he maintained some contact and interest in the development of physics, with particular links to Manuel Sandoval Vallarta and with Marcos Moshinsky, who had been a student of Blas in the first courses on modern physics given at the UNAM.

Don Nicolás Cabrera's main contribution was perhaps when he came to Mexico in 1967 to teach a course on thermodynamics of solids. This visit was organized within the frame of the Latin American School of Physics which happened at the National Polytechnic Institute during that year. During the course he produced an excellent set of lectures that were recorded by some of the students and distributed among the solid state scientists of the time. Those notes on surface thermodynamics were probably the first introduction for many of us to the world of surfaces. In those notes,Don Nicolás started from the very general point of view and built in great depth, with a lot of respect for the intelligence of the audience. The students always felt challenged and they appreciated the special respect and concern that Don Nicolás manifested for them, something that had not been felt from other foreign lecturers.

His first visit was so successful that the authorities of the Instituto Politécnico Nacional decided to invite Don Nicolás as a UNESCO expert for a sabbatical during one year, which took place in 1969. This was a seminal moment for the development of the solid state physics in Mexico and then later in other parts of Latin America. When Don Nicolás came back to Mexico he spent part of the time at the Polytechnic Institute and part of the time at the Institute of Physics of the University of Mexico. At that time the Institute was trying to create a new group on solid state physics. The group was made up of young people and there was no clear leadership even in the fields that were thought highly important to develop. The situation in Mexico was very similar to the rest of Latin America in the late sixties, with very small budgets, lack of direction and experience, and absence of experts in the field to provide leadership. We had a lot of enthusiasm, however, and were willing to try new areas of research. When Don Nicolás came to Mexico, he taught courses, gave seminars, talked to his students, discussed important problems, etc. and he made a tremendous impact on many people. In my particular case, he influenced virtually all my future research, convincing me that surface physics would be one of the most interesting areas in the future.

He was a very kind person above all, and it was very important for many of us to know that one of the really famous names in the literature was present with us to discuss interesting problems. We were introduced to the magic of the paper by Frank, Burton and Cabrera, about the spiral growth of crystals generated by dislocations. I was deeply impressed by the beautiful physics that I saw in it. The paper taught me the important lesson that there were beautiful and interesting things in physics besides the then-popular subjects as elementary particles, structure of the universe, cosmology, etc., topics that attracted most of the young people. I think this paper convinced me that this area was something worth pursuing. On the other hand, having Don Nicolás around at the Institute convinced us that the authors of the papers corresponded to real names, and to real people with whom we could talk and discuss our problems. Cabrera also convinced the authorities at the Institute to invest in equipment and in people. He promoted further education of young scientists abroad, in areas related to surface science.

One of his students was Dr. Leonel Cota, who later went to England to obtain his Ph. D. He became the first doctorate in "hard" surface science techniques at the University of Warwick under the direction of Prof. Forty. A few years later, back in Mexico, Leonel started the development of experimental surface physics techniques, such as LEED and Auger, in his laboratory in Ensenada. Don Nicolás also collaborated with many other people. I clearly remember the experience of sending my first paper for publication in the Journal of Crystal Growth. I was very afraid at that time because it was a new experience for me and for most colleagues in my community. When the reviews came back they seemed to me to be disastrous. After reading the reviews, Don Nicolás told me that there was nothing unusual and explained to me the way things work in the scientific community. He helped me rearrange the paper to satisfy the referee requirements, giving me tips about theoretical interpretations. After so much help he refused to be considered coauthor of the final paper. The exercise gave me an enormous amount of confidence that has had a big effect on my scientific career. It also made me realize that one of the things we were most lacking in Latin America was confidence.

5. Return to Spain

Unfortunately for us, Don Nicolás returned to Spain to work on the scientific development of that country. However, the impact of his stay in Mexico lasted for many years and it was responsible for the start of surface science in Mexico. Through the Mexican efforts, many other groups in Latin American were initiated in this field. Nowadays, the Mexican Society of Vacuum and Surface Sciences has a membership of about two hundred members, including scientists from surface science groups from all over the country from Merida to Ensenada. Back in Spain, Don Nicolás maintained contact with many of us, overseeing the development of science in Mexico. He also had an important influence in Venezuela during his visit there. I visited him several times and received constant support and praise from him for the international recognition that our group in Mexico was receiving. I always think of Don Nicolás as a great man in all respects, with a tremendous intuition for science, a great human spirit, and a warm personality with a great sense of humor. I hope that the works presented in this volume represent a small tribute to the figure of one of the greatest Iberoamerican scientists of all times.

Part II

Electronic Structure of Surfaces

Dynamical Response of an Overlayer of Alkali-Metal Atoms Adsorbed on a Free-Electron Metal Surface

A.G. Eguiluz and J.A. Gaspar

Department of Physics, Montana State University,
Bozeman, MT 59717, USA

Abstract. We report a theoretical study of the spectrum of elementary excitations of an alkali-atom overlayer adsorbed on Al. This study is placed in the context of recent experiments via the evaluation of the loss-spectrum for near-specular electron backscattering, starting from the knowledge of the dynamical density-response function for an electron-gas model of the chemisorption interface. Detailed results are presented for the case of Na overlayers. For coverage < one monolayer the calculated loss spectrum shows the presence of a peak due to incoherent electron-hole pair excitation, combined with the kinematic factor which enters the scattering probability. For coverage > one monolayer the loss spectrum corresponds to plasmon-like excitation. For two-monolayer coverage an additional overlayer-induced collective loss is found. For thick Na overlayers this mode corresponds to the "dipole" surface plasmon long-predicted theoretically, and recently discovered experimentally. The transition from quasi two-dimensional response to three-dimensional response must be studied microscopically: macroscopic-response theory fails qualitatively for monolayer coverage.

1. Introduction

The subject of elementary excitations in metals has a long tradition. Pioneering calculations performed many years ago for the homogeneous, interacting electron gas,[1] established this model as a good starting point for the study of these excitations in systems with relatively slowly-varying electron density, such as sp-bonded metals.

A system which has attracted considerable attention for many years is that of overlayers of alkali-metal atoms chemisorbed on metal surfaces.[2-17] A major focus of this attention has been directed to the elucidation of the nature of the electronic excitations in the submonolayer and monolayer (ML) coverage regimes.[2-4,9,11-13] Aiming at an understanding of the physics of the adsorption and response of the overlayer at its simplest level, two experiments have been performed recently for the simplest possible chemisorption system, namely Na/Al(111).[3,4] The angle-resolved, high-resolution electron energy-loss (EELS) experiment of Heskett *et al.*[3] shows intriguing features. In particular, their loss spectrum undergoes a sharp jump at the completion of the first monolayer, with an apparent abrupt onset of the response of the electron gas, in the form of plasmon creation. This experiment was carried out for a low incident energy ($E_0 = 30$ eV).

The first purpose of this article is to discuss the physics of the response in the ML-coverage regime, and its manifestation in an EELS experiment, via a self-consistent quantum mechanical calculation of the loss spectrum for an electron-gas model of the Na/Al system. For coverage < 1ML we find that the overlayer does not display coherent behavior, *i.e.*, the loss function has no well-defined peak; however, for high E_0 the actual loss spectrum does show a peak. This loss originates as a combined effect of the spectral density for electron-hole-pair creation in the overlayer (the loss function), and the kinematic factor which multiplies the former to yield the scattering probability. The intensity of this loss decays rapidly as the incident energy E_0 is lowered. For coverage > 1ML the response of the electron gas becomes progressively more coherent (*i.e.*, the loss function develops a peak), and the loss spectrum becomes a direct mapping of the spectral features of the overlayer response. The spectrum is then the same for all E_0. These results reconcile the EELS data of Heskett *et al.*[3] with the expectation that the overlayer response should be a smooth function of coverage for monolayer coverage.

A related issue we pursue is the transition from the quasi-two-dimensional behavior embodied by the response for ~ 1ML, to the quasi-three-dimensional behavior represented by the response of two and more monolayers, and the applicability of macroscopic response theory. We find that, *unlike the case of a semi-infinite metal*, macroscopic response is not an adequate approximation in the small-wave vector limit for ML-coverage.

A further objective of the present paper is to report theoretical evidence for the existence of an additional surface plasmon in the overlayer for coverage ~ 2ML. This type of elementary excitation was predicted first on the basis of a hydrodynamic model [18-20] (it was also hinted at early on in microscopic models [21,22]). A rather large number of theoretical papers subsequently provided partial support for the possible existence of this collective mode.[23-27] The first unequivocal experimental confirmation of the existence of an additional surface plasmon has been given very recently by Tsuei *et al.*[28]. These authors performed EELS measurements for thick slabs of Na, and K, both in the specular direction, and away from it. They observed the presence of an additional collective mode for the cases of K and Na, in the expected energy range, and having the predicted dipolar character.[18-20]. We present calculated dispersion relations for both "regular" and additional surface plasmons as function of overlayer thickness. For thick slabs semi-quantitative agreement with experiment [28] is achieved with use of an effective electron mass. The effects of exchange and correlation on the dispersion curves are also discussed within density functional theory. They are estimated to be as important for finite wave vectors as the effects of crystallinity are for zero wave vector.

2. Theory of the Electron Energy-Loss Process

The energy-loss function describing the long-range linear coupling between an electron incident on the surface and the density fluctuations induced at the surface by that coupling is given by the equation [29]

$$P(\vec{q}_{\|};\omega) = -2\hbar e^2 \int dz \int dz' \, e^{-q_{\|}(z+z')} \, Im\chi(\vec{q}_{\|};\omega + i\eta | zz') \quad , \tag{1}$$

where the integrals run over the region of space occupied by the metal electrons, and $\vec{q}_{\|}$ is a two-dimensional wave vector in the plane of the surface (the plane xy). In Eq. (1) we have introduced the (imaginary part of the) density-response function for interacting electrons, $\chi(\vec{x},\vec{x}'|\omega)$, defined by the equation [30]

$$n_{ind}(\vec{x};\omega) = \int d^3x' \, \chi(\vec{x},\vec{x}'|\omega) \, U_{ext}(\vec{x}';\omega) \quad , \tag{2}$$

where n_{ind} is the electron number density induced by the external potential U_{ext} due to the incoming electron. The density response function is obtained by solving an integral equation of the Bethe-Salpeter form [30]

$$\chi = \chi^0 + \chi^0 V \chi \quad , \tag{3}$$

where $\chi^0(\vec{x},\vec{x}'|\omega)$ is the irreducible density-response function, *i.e.*, the response to the self-consistent field which is established in the electron system by the screening process.[30] This response function is computed, in the presence of the overlayer, from the one-electron energy eigenvalues and eigenfunctions of the Kohn-Sham equation in the local-density approximation (LDA) of density-functional theory.[31] The effective electron-electron interaction $V(\vec{x},\vec{x}')$ is given, also in LDA, by the equation

$$V(\vec{x},\vec{x}') = v(\vec{x}-\vec{x}') + \frac{dV_{axc}(\vec{x})}{dn(\vec{x})} \delta(\vec{x}-\vec{x}') \quad , \tag{4}$$

where $v(\vec{x}-\vec{x}')$ is the bare Coulomb interaction, and $V_{axc}(\vec{x})$ is the exchange-correlation potential, which is the many-electron contribution to the effective one-electron potential obtained self-consistently with the density $n(\vec{x})$ in the solution of the ground-state problem in LDA.

The long-range fields involved in the coupling described by the loss function favor small-angle deflections.[29] The incident electron is thus scattered mostly into a near-specular lobe; the *loss spectrum*, or scattering efficiency per unit frequency for a process in which the backscattered electron loses energy $\hbar\omega$, is given by the equation[29]

$$I(\omega) = \frac{2e^2 v_z^2}{\pi\hbar} |R_I|^2 \int d^2q_{\|} \frac{P(\vec{q}_{\|};\omega)}{[v_z^2 q_{\|}^2 + (\omega - \vec{v}_{\|} \cdot \vec{q}_{\|})^2]^2} \quad , \tag{5}$$

where v_z and $\vec{v}_{\|}$ are the components of the velocity of the incident electron along the surface normal and on the plane of the surface, respectively, and $|R_I|^2$ is the probability for elastic scattering. Note that the physics of the response enters the loss spectrum entirely via the loss function $P(\vec{q}_{\|};\omega)$; the kinematics of the loss process does so via the prefactor of the loss function.

With regard to the chemisorption model, for coverages up to 1ML we utilize the model proposed by Serena *et al.*,[15] which is a variation of the original uniform-background model of alkali chemisorption due to Lang.[16] The ionic backgrounds for both substrate and overlayer are replaced by jellium slabs; the width d of the jellium slab for the adsorbate is given by $d(\theta) = d_{ion} + (b - d_{ion})\theta$, where θ is the coverage, d_{ion} is twice the Na ionic radius, and b is the interplanar spacing for the most densely-packed lattice planes in bulk Na (the (110) planes). Monolayer coverage corresponds to $\theta = 1$ (for which the areal density of the overlayer is ~ 1/2 that of an Al(111) substrate). While the use of d_{ion} in this model is not supported by new (but not universally held [32]) evidence for the absence of an ionic low-coverage regime [8,14] the calculated loss spectrum is rather insensitive to the details of the formula for $d(\theta)$ for $\theta < 1$. The model does provide a very good account of the work function changes induced by alkali-atom adsorption. For $\theta > 1$ we simulate the overlayer growth according to $d(\theta) = b\theta$, which corresponds to keeping the density of the jellium slab for the second layer fixed; for $\theta = 2$ the second layer is complete. With this choice, the work function remains equal to its saturation value; by contrast, a variable-density model gives rise to a pronounced dip in the work function at the start of the second layer.

3. Numerical Results and Comparison with Experiment

The physics of the response recognizes distinct regimes, namely submonolayer-, monolayer-, and multilayer-coverage. It is convenient to consider them separately. The transition between 1ML- and 2ML-coverage is of considerable interest, since it illustrates the switch-over between quasi two-dimensional response and quasi three-dimensional response. In addition, it is of interest to discuss the applicability of a macroscopic model which is sometimes invoked in the analysis of experiments.

The calculation was performed for a substrate which is sufficiently thick (its thickness must be equivalent to at least that of a 45-layer Al slab bounded by a pair of (100) surfaces – this requirement dictates the use of a supercomputer) that spurious low-energy peaks, present for thinner slabs, are eliminated. (The presence of these spurious peaks would considerably cloud the physical interpretation of the results.) Since the loss function for the electron gas depends on $\vec{q}_{\|}$ only through its magnitude $q_{\|}$, Eq. (5) can be reduced to the form

$$I(\omega) = \int_0^{q_{\| max}} dq_{\|}\, K(q_{\|};\omega)\, P(q_{\|};\omega) \quad , \tag{6}$$

as done by Camley and Mills,[33] who give the explicit definition of the *kinematic factor* $K(q_{\|};\omega)$. As noted above, the long-range fields involved in the coupling described by the loss function favor small-angle deflections about the specular direction ($q_{\|} = 0$); the finite angle subtended by the spectrometer (of half-width $\Delta\Theta$) is taken into account via the definition $q_{\| max} = \Delta\Theta\sqrt{2mE_0}/\hbar$. In our calculations we have taken $\Delta\Theta = 1°$, which corresponds to the experimental setup of Ref. 3.

Submonolayer-coverage regime

Figure 1 shows $I(\omega)$ for Na/Al for $E_0 = 100$ eV. The presence of a low-energy loss starting with $\theta = 0.5$ is noteworthy. It occurs well below the surface plasmon frequency of sodium. With increasing coverage this loss shifts to higher energies, while gaining spectral weight. The main feature of Fig. 1 we want to emphasize at this point is that there is no abrupt jump in the peak position as the coverage goes through $\theta = 1$. We next discuss the physics of this loss.

In Fig. 2 we show the kinematic factor, the loss function, and their product for $\theta = 0.6$. Consider first the case E_0=100 eV, for a representative value of $q_{||}$. For this energy $q_{||\,max} = 0.09 Å^{-1}$ (for $\Delta\Theta = 1°$). It is apparent that the integrand of Eq. (6) (and $I(\omega)$ as well) develops a peak as a result of two competing effects: (*i*) the loss function drops off for small-energy transfers, but is quite constant for higher energies, and (*ii*) the kinematic factor favors small-energy transfers, and strongly suppresses large-energy transfers. Thus, the existence of an electron-gas peak in $I(\omega)$ for $\theta < 1$ and $E_0 = 100$ eV depends on "kinematic" effects; ***it is a "ghost peak"***, in the terminology adopted in the interpretation of similar loss structures in other physical systems.[34]

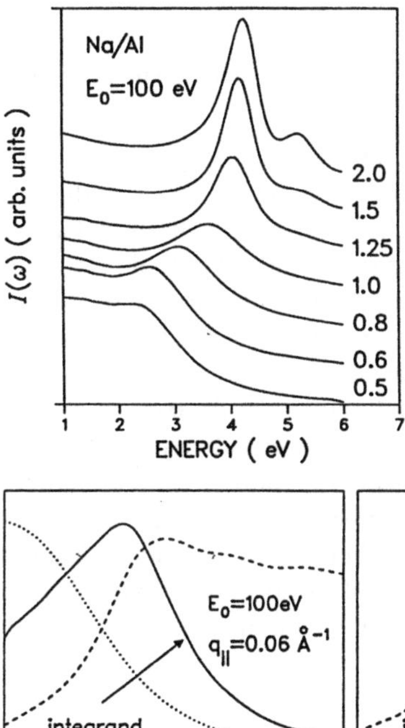

Fig. 1. Calculated loss spectrum for Na/Al for $E_0 = 100$ eV, for (Na coverage) θ ranging from 0.5 to 2 (half-ML to 2ML). Angle of incidence: 60°.

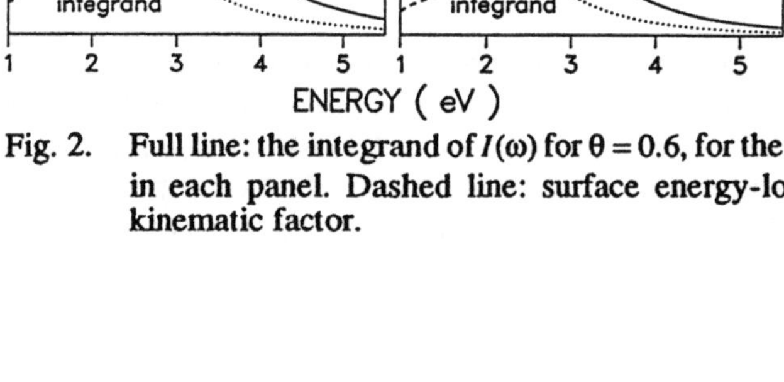

Fig. 2. Full line: the integrand of $I(\omega)$ for $\theta = 0.6$, for the values of E_0 and $q_{||}$ shown in each panel. Dashed line: surface energy-loss function. Dotted line: kinematic factor.

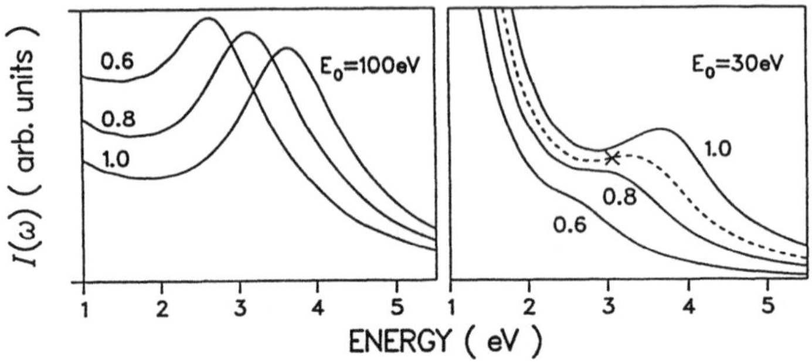

Fig. 3. $I(\omega)$ for E_0 = 100 eV (left panel), and for E_0 = 30 eV (right panel), for $\theta = 0.6, 0.8$, and 1. The dashed line is for $\theta = 0.89$, or 1ML as defined in the experiment of Ref. 3, the cross indicating the position of the loss measured for this coverage. Angle of incidence: 60°.

The above picture changes for a lower incident energy, such as the one employed in Ref. 3: As the right panel of Fig. 2 shows, for E_0 = 30 eV the kinematic factor succeeds in blurring the details of the loss function for small ω. In addition, since $q_{\|\max}$ scales as $\sqrt{E_0}$, the kinematic factor does not probe the larger-$q_\|$ part of the response (which favors the development of a peak for higher E_0). The net result is that for E_0 = 30 eV the peak-to-background ratio for the ghost peak is substantially smaller than for E_0 = 100 eV, as demonstrated in Fig. 3.

The above results lead us to propose that for $\theta < 1$ the electron-gas ghost-peak mechanism gives rise to a measurable loss for high E_0, but not for low E_0, which is consistent with the fact that the loss spectrum of Heskett *et al.*[3] shows a discontinuous change for ~ 1ML, *i.e.*, an electron-gas loss peak appears in their spectrum abruptly for a coverage for which it already is plasmon-like. This conclusion receives partial support from a higher-incident energy (E_0 = 100 eV) EELS experiment performed recently by Hohlfeld and Horn [4] on the same system, Na/Al(111). The loss reported in Ref. 4 is very broad, shifts little with coverage for $\theta < 1$, and goes smoothly through $\theta = 1$, all of which is consistent with the possibility that the loss includes (non-resolved) contributions from the ghost peak and the single-particle transition observed in Ref. 3. The fact that the experiment of Ref. 4 was not angle-resolved does not affect our conclusions; the fact that the energy-resolution was considerably lower than the one used in Ref. 3 leads us to suggest that the latter high-resolution experiment be performed for higher incident energy.

Monolayer-coverage regime

For $\theta > 1$ the electron-gas ghost peak discussed above evolves into a collective (plasmon) loss (Fig. 1). The physics of this changeover in the nature of the loss is best described with reference to Fig. 4, which shows the loss function for several coverages, and for wave vectors representative of the $q_\|$'s which contribute to $I(\omega)$. Clearly, for $\theta < 1$ the overlayer response is featureless (except for its low-energy

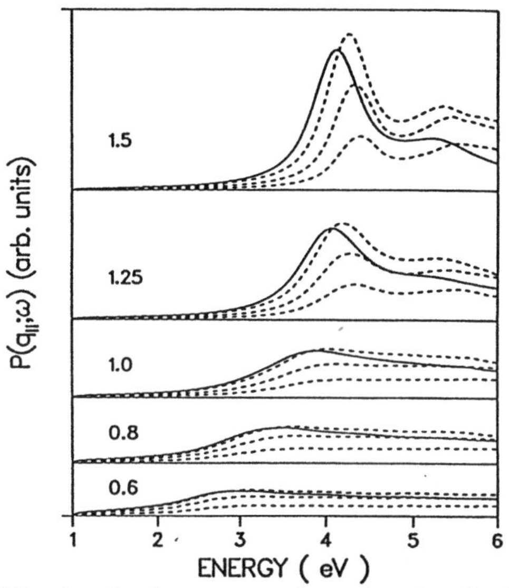

Fig. 4. Surface energy-loss function for Na/Al for representative wave vector transfers. Solid lines: $q_{||} = 0.09\,Å^{-1}$ ($= q_{||\max}$ for $E_0 = 100$ eV and $\Delta\Theta = 1°$). Dashed lines: $q_{||} = 0.02Å^{-1}, 0.035Å^{-1}, 0.05Å^{-1}$ respectively; for these three wave vectors the loss function increases monotonically with $q_{||}$.

drop-off); the same corresponds to incoherent electron-hole-pair excitation, which, combined with the kinematic factor, gives rise to the peak in the loss spectrum we discussed above. As the coverage is increased the response *gradually becomes coherent*, and the loss spectrum (Fig. 1) shows, for $\theta > 1$, a peak which maps the form of the loss function (Fig. 4). The spectrum is now basically the same for all incident energies E_0, unlike the situation discussed above for $\theta < 1$. These conclusions corroborate the assignment made by Heskett *et al.*[3] and Hohlfeld and Horn[4] of the experimental loss for $\theta > 1$ as being due to plasmon creation in the overlayer.

A subtle point behind these results is worth discussing. This has to do with the *upward* shift of the theoretical loss present in Fig.1 from a frequency well below $\omega_p(\mathrm{Na})/\sqrt{2}$ for $\theta = 0.5$ to a value close to this "classical" frequency for $\theta = 2$. (We note that the theoretical spectra have been calculated using an *ad hoc* effective electron mass equal to 1.13 m, for which $\omega_p(\mathrm{Na})/\sqrt{2} = 4$ eV.) This behavior, which agrees qualitatively with experiment,[3,4] is non-classical. It is a direct consequence of our microscopic, quantum mechanical treatment of the overlayer response. This statement is further substantiated by comparison with a macroscopic model of the response in which both substrate and overlayer are described by frequency-dependent dielectric constants. In that case the energy-loss function given by Eq. (1) reduces to the simple form [29]

$$P(q_{||};\omega) = \frac{2\hbar q_{||}}{\pi} Im\left(\frac{(-1)}{1+\varepsilon_{eff}(q_{||};\omega)}\right) \quad , \tag{7}$$

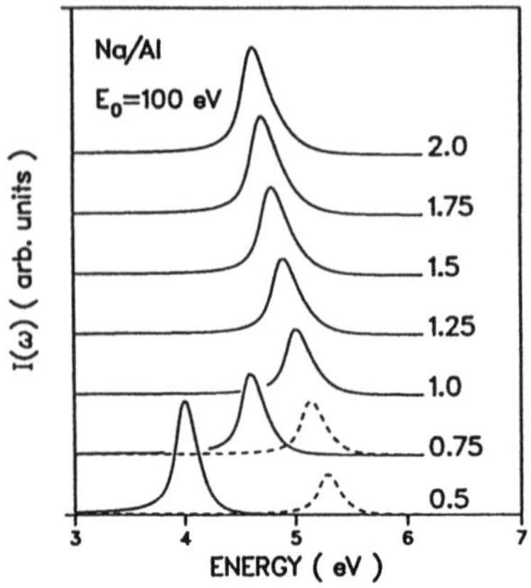

Fig. 5. Loss spectrum obtained for a macroscopic response model (see Section 3). Angle of incidence: 60°.

where ε_{eff} is the effective dielectric constant for the adsorbate-substrate system.[29] For a thick overlayer ε_{eff} equals the dielectric constant for the overlayer, and thus it becomes independent of $q_{||}$. Clearly, the only type of elementary excitation present in Eq. (7) *for all coverages* (including submonolayer coverage), is a delta function-like peak due to plasmon creation in the overlayer. Using Eq. (7) as the basis of a calculation of $I(\omega)$ yields the loss spectrum shown in Fig. 5, in which, for consistency, $d(\theta)$ was defined as done in the microscopic response calculation. The most striking difference between Figs. 1 and 5 is that in the latter case the loss shifts *downward* in energy as the second layer forms. (In fact, if the growth of the first layer is modelled in the same way as that of the second layer, *i.e.*, keeping the density fixed -- as is commonly done in macroscopic models -- the situation is even worse, as also illustrated in Fig. 5.) It is the mechanism of electron-hole pair excitation, inherent to the microscopic theory, which is responsible for this difference. The same gives rise to a low-frequency non-plasmon loss for submonolayer coverage which, with the formation of the second layer, becomes collective in nature, and shifts upward in energy, as it is found experimentally.[3,4]

Two and more monolayers

The spectrum for 2ML (Fig. 1) shows the presence of an additional loss, lying between the surface- and bulk-plasmon frequencies for Na. Because of its energy location, and because of its absence from the macroscopic, sharp-interface model outlined above, it is natural to expect that this mode should be the microscopic counterpart of the "additional" surface plasmon first obtained in hydrodynamic treatments of the response of surfaces with a diffuse electron density profile.[18-20]. In the early work an appealing physical picture was given:[19,20] The additional mode arises in hydrodynamics as a bound state of a Schrödinger equation in which the potential is proportional to the ground-state electron density profile at the surface; thus "binding" of the mode occurs for sufficiently-diffuse surfaces. The new mode was characterized as being dipolar in nature, the "usual" (or "regular") surface

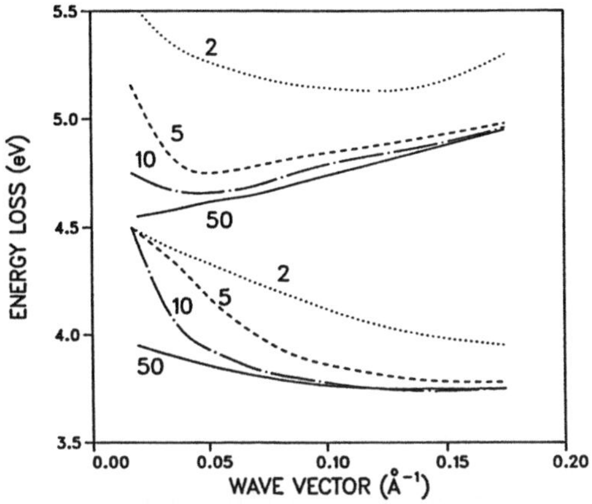

Fig. 6. Calculated dispersion curves for "regular" and "additional" surface plasmons for overlayers of growing number of layers (this number is indicated for each curve). The lower (higher) set of curves corresponds to the regular (additional) surface plasmon.

plasmon being referred to as a "monopole", terminology which makes reference to the zero or non-zero value, respectively, of the integral over all space of the density fluctuation associated with the mode for zero wave vector.[19,20] For sufficiently-diffuse interfaces higher multipoles (quadrupole, etc.) can also be "bound".

As noted in the Introduction, Tsuei *et al.*[28] were recently successful in observing the additional mode via EELS measurements performed for thick layers of Na and K. (The experiment performed on Al failed to show an additional mode, presumably due to insufficient spectral weight.) It is then of interest to inquire how the extra mode obtained here for 2ML evolves as more layers are added.

In Fig. 6 we show the dispersion relations for the "regular" and "additional" surface plasmons for several coverages, up to a film which is thick enough that it represents the response of semi-infinite Na. Note, in particular, that the dispersion curve for the additional plasmon for 2ML does evolve smoothly into that for the semi-infinite slab. An analysis of the charge fluctuation induced by the scattered electron [35] reveals that for the thick-film case the higher-frequency mode is indeed dipolar in character for $q_{||} = 0$. One then expects that this mode should be easier to observe in the impact regime (large wave vector tranfer) for which it can carry a net charge; this expectation is indeed verified experimentally.[28]

The downward trend of the dispersion relations of the regular and dipole surface plasmons as the number of layers is increased (Fig. 6) agrees qualitatively with the results of a new set of EELS experiments (E. W. Plummer, private communication.) However, this agreement does not appear to be universal, since similar data for Na show an opposite trend, namely the dispersion curves shift upward on going from a 2ML film to a thick film. Given the current interest in the properties of thin films, and in the process of film growth, this problem will be investigated further.

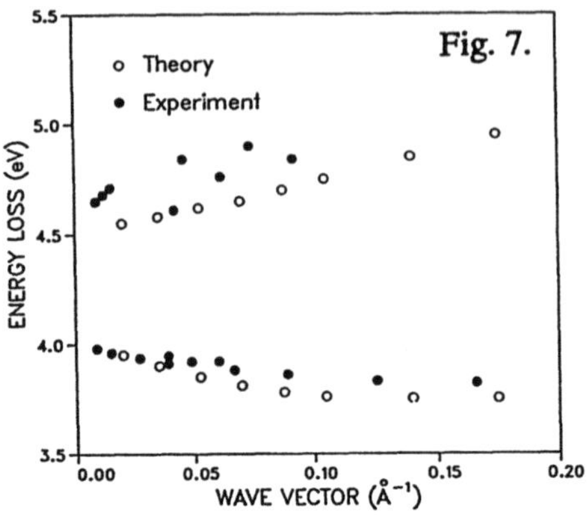

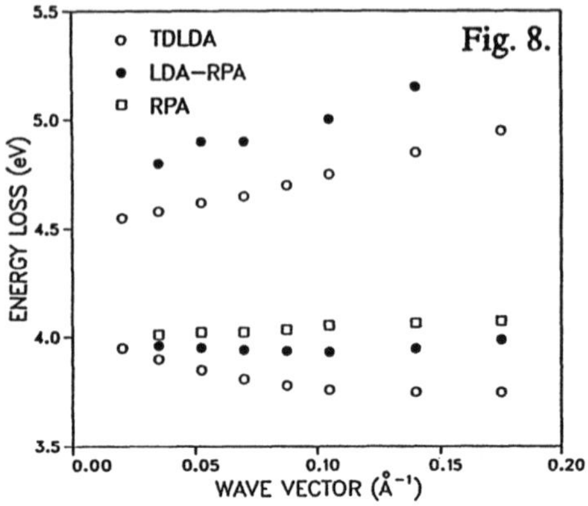

Fig. 7. Comparison of experimental (solid circles; Ref. 28) and theoretical (empty circles) dispersion curves for a thick sodium film.

Fig. 8. Effect of exchange and correlation (within density functional theory) on the calculated dispersion curves for a thick Na film. Empty circles: TDLDA. Solid circles: LDA-RPA. Squares: RPA.

In Fig. 7 we compare the theoretical dispersion curves for the thick-overlayer case with the data of Tsuei *et al.*[28] The excellent agreement observed for zero-wave vector for the regular mode is a consequence of our using an appropriate effective mass, as noted above. (We would like to note that we have been able to obtain the dispersion curves in the range of wave vectors for which the dipole mode was observed experimentally; by contrast, the theoretical dispersion curves obtained in Ref. 28 for Na correspond to larger wave vectors only.)

It is also of interest to assess the influence of exchange and correlation on the calculated dispersion curves. This is rather easily done within the context of density functional theory. In Fig. 8 we show three sets of results, which correspond to the following approximations: (*i*) density-functional theory in LDA is used both in the chemisorption ground-state and in the computation of the response, as outlined in Section 2 (TDLDA); (*ii*) exchange and correlation effects are completely neglected, which corresponds to the random-phase approximation (RPA); (*iii*) exchange and correlation are included in the ground state, but are neglected in the response calculation, *i.e.*, the second term in Eq. (4) is dropped (LDA-RPA). We note that for the full RPA calculation the dipole mode is absent (*i.e.* its spectral weight is vanishingly small). In addition, the spread in the calculated dispersion curves shown in Fig. 8 is comparable with the downward shift in the value of the surface plasmon frequency at zero wave vector introduced by our use of an effective mass of 1.13m. From these results we draw the conclusion that the effects of exchange a correlation on the dispersion curves are comparable to those of the crystal structure.

4. Summary

We have presented a novel picture of elementary excitations in an overlayer of Na atoms adsorbed on Al. The same leads to a smooth transition between incoherent (electron-hole pairs) and coherent (plasmon) response for ~ 1ML. For high E_0, and for $\theta < 1$, the kinematic factor, in conjunction with electron-hole pair excitation, give rise to a loss in the nature of a ghost peak.[34] For $\theta > 1$ the loss spectrum is indicative of coherent electron-gas response for all E_0. Our results are consistent with, and provide theoretical support for, the peak assignments made in the experiments of Heskett *et al.*[3] The dependence of the strength of the electron-gas loss for $\theta < 1$ on incident energy allows us to understand the apparent abrupt onset of the response of the electron gas in the data of Ref. 3 for 1ML coverage.

For a semi-infinite metal, macroscopic physics is recovered in the small wave vector limit. In particular, the loss spectrum is, to zeroth order, obtainable from a Drude-like model.[29] Our self-consistent, quantum-mechanical approach, in conjunction with the experimental findings,[3,4] reveal the breakdown of this correspondence between macroscopic and microscopic physics in the 1ML coverage regime. The physical reason for this breakdown is that for electrons in a confined geometry such as a monolayer film the mechanism of electron-hole pair excitation is important even for zero wave vector transfer (unlike the case of a semi-infinite metal).

We have reported theoretical evidence for the presence of a "higher multipole" in the loss spectrum for 2ML of Na/Al. This additional collective mode turns smoothly into the dipole mode recently observed experimentally for thick alkali-metal films.[28]

The theory presented here can be refined in two important respects. Inclusion of band-structure effects from first principles remains a major challenge, as is a rigorous treatment of dynamical many-electron effects at the surface. These effects are expected to play a role in *e.g.*, the one-electron transition observed in Ref. 3.

Acknowledgements

This work was supported by NSF Grant No. DMR 86-038920, by MONTS Grant No. 196504, and by the San Diego Supercomputer Center. J. A. G. acknowledges partial support from the Universidad de Sonora, Mexico.

References

1. See, for example, A. L. Fetter and J. D. Walecka, *Quantum Theory of Many-Particle Systems*, (McGraw-Hill, New York,1971).

2. A. U. MacRae *et al.*, Phys. Rev. Lett. **22**, 1048 (1969); S. Andersson and U. Jostell, Surf. Sci. **46**, 625 (1974); Faraday Discuss. Chem. Soc. **60**, 255 (1975); S. Å. Lindgren and L. Walldén, Phys. Rev. Lett. **59**, 3003 (1987); Phys. Rev. B **22**, 5967 (1980); J. Cousty, R. Riwan, and P. Soukiassian, J. Physique **46**, 109 (1985); S. Å. Lindgren and L. Walldén, Phys. Rev. Lett.

59, 3003 (1987); L. Walldén, Phys. Rev. Lett. **54**, 943 (1985); A. Hohlfeld, M. Sunjic, and K. Horn, J. Vac. Sci. Technol. **A5**, 679 (1987); D. Heskett *et al.*, Phys. Rev. B **36**, 1276 (1987).

3. D. Heskett, K. -H. Frank, K. Horn, E. E. Koch, H. -J. Freund, A. Baddorf, K. -D. Tsuei, and E. W. Plummer, Phys. Rev. B **37**, 10387 (1988).
4. A. Hohlfeld and K. Horn, Surf. Sci. **211/212**, 844 (1989).
5. B. Woratscheck, W. Sesselmann, J. Küppers, G. Ertl, and H. Haberland, Phys. Rev. Lett. **55**, 1231 (1985).
6. G. M. Lamble, R. S. Brooks, and D. A. King, Phys. Rev. Lett. **61**, 1112 (1988).
7. K. Horn *et al.*, Phys. Rev. Lett. **61**, 2488 (1988).
8. D. M. Riffe, G. K. Wertheim, and P. H. Citrin, Phys. Rev. Lett. **64**, 571 (1990).
9. J. W. Gadzuk, Phys. Rev. B **1**, 1267 (1970); D. M. Newns, Phys. Lett. **39A**, 341, (1972).
10. N. D. Lang and A. R. Williams, Phys. Rev. B **18**, 616 (1978); J. P. Muscat and I. P. Batra, Phys. Rev. B **34**, 2889 (1986); L. -A. Salmi, and M. Persson, Phys. Rev. B **39**, 6249 (1989); P. Nordlander and J. C. Tully, Surf. Sci. **211/212**, 207 (1989).
11. A. G. Eguiluz and D. A. Campbell, Phys. Rev. B **31**, 7572 (1985).
12. H. Ishida and M. Tsukada, Surf. Sci. **169**, 225 (1986).
13. R. Fuchs and W. Ekardt, J. Phys.: Condens. Matter **1**, 4081 (1989).
14. H. Ishida, Phys. Rev. B **39**, 5492 (1989); Phys. Rev. Lett. **63**, 1535 (1989); E. Wimmer *et al.*, Phys. Rev. B **28**, 3074 (1983).
15. P. A. Serena, J. M. Soler, N. García, and I. P. Batra, Phys. Rev B **36**, 3452 (1987).
16. N. D. Lang, Phys. Rev. B **4**, 4234 (1971).
17. B. N. J. Persson, and L. H. Dubois, Phys. Rev. B **39**, 6249 (1989); A. Liebsch, G. Hincelin, and T. López Ríos, Phys. Rev. B. **41**, 10463 (1990).
18. A. J. Bennett, Phys. Rev. B **1**, 203 (1970).
19. A. G. Eguiluz, S. C. Ying, and J. J. Quinn, Phys. Rev. B **11**, 2118 (1975).
20. A. G. Eguiluz and J. J. Quinn, Phys. Rev. B **14**, 1347 (1976); *ibid.* **12**, 4299 (1976); Phys. Lett. **53 A**, 151 (1976).
21. P. J. Feibelman, Phys. Rev. B **9**, 5077 (1974).
22. J. E. Inglesfield and E. Wikborg, J. Phys. C: Solid State Phys. **6** L158 (1975).
23. C. Schwartz and W. L. Schaich, Phys. Rev. B **26**, 7008 (1982).
24. K. Kempa and F. Forstmann, Surf. Sci. **129**, 516 (1983).
25. K. Kempa and R. R. Gerhardts, Solid State Commun. **53**, 579 (1985).
26. K. Kempa and W. L. Schaich, Solid State Commun. **61**, 357 (1987).
27. J. F. Dobson and G. H. Harris, J. Phys. C: Solid State Phys **21**, L729 (1988).

28. K. -D. Tsuei, E. W. Plummer, A. Liebsch, K. Kempa, and P. Bakshi, Phys. Rev. Lett. **64**, 44 (1990).

29. D. L. Mills, Surf. Sci. **48**, 59 (1975).

30. A. G. Eguiluz, Phys. Rev. Lett. **51**, 1907 (1983); A. G. Eguiluz, Phys. Rev. B **31**, 7472 (1985); Phys. Scr. **36**, 651 (1987).

31. For reviews of density-functional theory see, *e.g.*, articles in *Theory of the Inhomogeneous Electron Gas,* edited by S. Lundqvist and N. H. March (Plenum, New York,1983).

32. G. M. Lamble, R. S. Brooks, and D. A. King, Phys. Rev. Lett. **61**, 1112 (1988).

33. R. E. Camley and D. L. Mills, Phys. Rev. B **29**, 1695 (1984).

34. R. E. Palmer, J. F. Annett, and R. F. Willis, Phys. Rev. Lett. **58**, 2490 (1987).

35. J. A. Gaspar, and A. G. Eguiluz, to be published.

Self-Diffusion by Place Exchange on Smooth Surfaces

P.J. Feibelman and G.L. Kellogg

Sandia National Laboratories, Albuquerque, NM 87185, USA

Abstract. Theory and experiment show that a substitutional mode dominates self-diffusion on several fcc(001) metal surfaces–Al, for which the calculations were performed, and Pt and Ir, for which the unusual signature of substitutional diffusion has been observed–namely a preference for motion in [100] and [010] directions. The first principles electronic structure calculations for Al show that the dominance of substitutional over ordinary hopping diffusion stems from the fact that higher bond-order can be maintained in the substitutional process.

Diffusion of metal atoms on metal surfaces is a fundamental process in crystal growth and epitaxy. The usual picture of surface diffusion is that a migrating atom hops from one equilibrium site to another over saddles in the (static) atom-surface potential. This picture is consistent with direct Field Ion Microscopy (FIM) observations of single atom diffusion on a variety of transition metal surfaces [1]. There is one class of surfaces, however, for which observations suggest a more complex diffusion mode: On fcc(011) metal surfaces, specifically, on Pt, Ir, Ni and Al(011) which have grooves and ridges, the evidence suggests that what appears to be diffusion across a ridge actually occurs when an adatom replaces a ridge atom [2,3]. After that, it is the latter atom that continues the diffusion.

The apparent distinction between diffusion on ridged and on smooth surfaces tempts one to conclude that it is the low coordination of the atoms in a ridge that makes the replacement diffusion process favorable. This, presumably, is why replacement diffusion has not previously been considered in explaining adatom motion on smoother crystal faces such as fcc(001), for which the nominal coordination of surface atoms is high. Nevertheless, we have found compelling evidence, both theoretical and experimental, that self-diffusion by concerted displacement is the dominant diffusion mode at low temperatures on certain fcc(001) metal surfaces.

Calculations were performed for self-diffusion on Al(001) using the self-consistent scattering theory method, which has been the subject of several detailed publications [4]. The calculations imply that the energy required for an Al adatom to move from its equilibrium 4-fold hollow to a transition configuration at a 2-fold bridge is 0.65eV. This is roughly 0.2eV higher than the barrier suggested by Tung's FIM results for various Al surfaces [3]. Allowing for elastic surface relaxation in the 4-fold hollow as well as at the bridge does little to reduce to the calculated barrier height. This is because elastic energies are of the order of a few tens of meV in general, but also because the energy gains in equilibrium and at the bridge tend to cancel. Something quite different from ordinary hopping is therefore required to explain how Al can self-diffuse at relatively low temperatures.

We have considered a "concerted displacement" process in which the adatom replaces a surface layer Al, the latter emerging in a second neighbor four-fold hollow and becoming the new adatom. The transition state for this process is illustrated in Fig. 1. Atom A was the original

Springer Proceedings in Physics, Volume 62
Surface Science Eds.: F.A. Ponce and M. Cardona

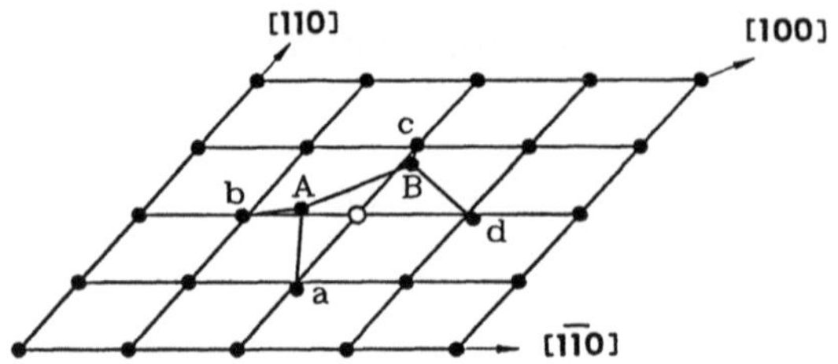

Fig. 1 Barrier configuration corresponding to the substitutional diffusion process described in the text. The open circle is the vacancy created when atom B emerged from the surface and which will be filled by atom A.

adatom. It is heading for the vacancy (shown as an open circle) left behind when atom B emerged from the surface. The interesting feature of the transition configuration shown in Fig. 1 is that each of the Al's above the surface can form three bonds. Indeed, with relatively small relaxations of the atoms labelled a, b, c, and d, each of the three bonds can take on a value close to twice the covalent radius of Al. This means that the bonding requirements of the trivalent Al atoms can be well-satisfied as compared to the transition state for ordinary hopping, where the Al at the bridge is two-fold bonded. The price for removing an Al from the surface layer is not as high as one may think, because even though a surface Al has eight nearest neighbors, it has only three valence electrons to bond to them. Thus each of the eight "metallic bonds" must be weak. In fact the calculations predict a barrier for the substitutional diffusion process that is less than 1/3 that for ordinary hopping, showing that the "chemical argument" for this process, based on valency and bond order, is indeed a strong one.

An important implication of the substitutional diffusion process illustrated in Fig. 1 is that diffusion is restricted to [100] and [010] directions at low temperatures. Substitutional diffusion in the [110] or [1$\bar{1}$0] directions is hindered on the fcc(001) plane by the unfavorable geometry for forming three bonds per diffusing atom. Thus, when substitutional is the dominant diffusion mode, an adatom will appear to move as though on a checkerboard - i.e., an adatom initially on a "black square" will only be found later on other "black squares." This is a strong prediction because it is just the opposite of what one would expect for ordinary hopping. In ordinary hopping, the preference is for motion over barriers at two-fold ("bridge") sites rather than one-fold ("atop") sites. (The argument is that metals that adopt the fcc crystal structure do so in order to maximize their number of nearest neighbors. Thus the barrier at the higher coordination bridge should be lower than at the lower coordination atop site.) This translates into a preference for diffusion in [110] or [1$\bar{1}$0] directions.

With this in mind, we have compiled a map of sites visited by a self-diffusing Pt adatom on the (001) plane of a Pt FIM tip. The orientation of the plane's axes has been determined by standard techniques. Figure 2(1) illustrates the pattern predicted by the replacement diffusion model, while 2(2) shows what we would observe if ordinary hopping were dominant. In each case, the lines join the sites (open dots) visited by the diffusing adatom, while the solid dots represent surface layer atom positions.

It is already of interest that the onset for self-diffusion on Pt(001) is found to be only 175K. This in itself is an indication that an extraordinary diffusion process is at work. The map of sites visited by a diffusing Pt, representing 300 observations, is shown in Fig. 3. It clearly indicates "checkerboard motion" as in Fig. 2(1), leaving us either with the unacceptable notion that ordinary hopping is occurring with the barrier at an atop site, or that self-diffusion is a concerted displacement

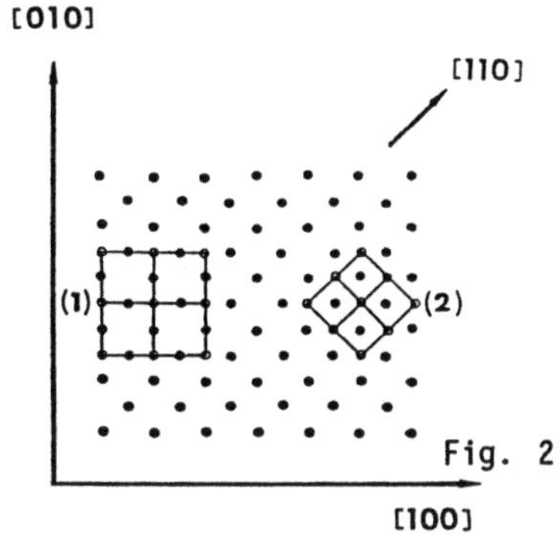

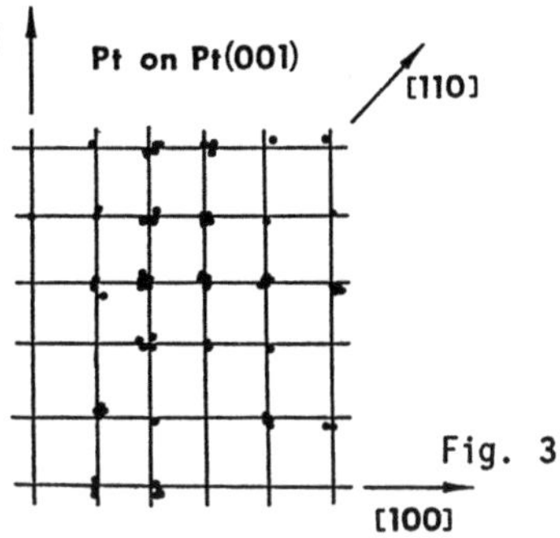

Fig. 2 A schematic of the fcc(001) plane. The solid circles represent substrate surface atoms and the open circles represent adatom binding sites. Pattern (1) shows the diffusion-accessible sites for migration confined to the [100] and [010] directions, i.e., motion along the surface diagonals. Pattern (2) shows the diffusion-accessible sites for bridge hopping.

Fig. 3 A map of diffusion-accessible sites for a Pt adatom diffusing on Pt(001) at 175K. The mesh forms a square pattern with sides parallel to the [100] and [010] directions of the substrate plane.

process as actually predicted for the case of Al. Similar results have also been found by Chen and Tsong [5] for self-diffusion on Ir(001).

We are now investigating inhomogeneous adsorption systems, in order to explore the systematics of diffusion by concerted displacement. Preliminary results indicate that Pd diffusion on Pt(001) proceeds via ordinary hopping, while Ni replaces a first layer Pt atom which then continues the diffusion process. Further information concerning our work is available in two recent publications [6,7].

Acknowledgements

This work performed at Sandia National Laboratories was supported by the U.S. Department of Energy under contract #DE-AC04-76DP00789.

References

1. See, e.g., D.W. Bassett, in Surface Mobilities on Solid Materials, ed. by Vu Thien Binh (Plenum, New York, 1983) p. 63; G. Ehrlich and K. Stolt, Ann. Rev. Phys. Chem. 31, 603 (1980); T.T. Tsong and P. Cowan, CRC Crit. Rev. St. Mater. Sci. 7, 289 (1978); G.L. Kellogg, T.T. Tsong and P.L. Cowan, Surf. Sci. 70, 485 (1978).
2. D.W. Bassett and P.R. Weber, Surf. Sci. 70, 520 (1978); R. Tung and W.R. Graham, Surf. Sci. 97, 73 (1980); J.D. Wrigley and G. Ehrlich, Phys. Rev. Lett. 44, 661 (1980).
3. R. Tung, Thesis, U. of Pennsylvania, 1980 (unpublished).
4. A.R. Williams, P.J. Feibelman, N.D. Lang, Phys. Rev. B36, 5433 (1982); P.J. Feibelman, Phys. Rev. B35, 2626 (1987), B38, 1849 (1988), B38, 7287 (1988), B39, 4866 (1989).
5. C. Chen and T.T. Tsong, Phys. Rev. Lett. 64, 3147 (1990).
6. G.L. Kellogg and P.J. Feibelman, Phys. Rev. Lett. 64, 3143 (1990).
7. P.J. Feibelman, Phys. Rev. Lett. 65, 729 (1990).

The European Synchrotron Radiation Facility in Grenoble

S. Ferrer

European Synchrotron Radiation Facility,
BP 220, F-38043 Grenoble, France

Abstract. After a short description of the basic concepts of synchrotron radiation, a general description of the characteristics of the ESRF is presented. The highest priority beamlines to be operative by mid 1994 are briefly described.

1. Basic concepts

A charged particle describing a circular orbit at relativistic speed emits electromagnetic radiation directed tangentially to the orbit along the forward direction. The synchrotron radiation is confined in a narrow cone with an **aperture angle** ($1/\gamma$) equal to the ratio between the rest energy and the kinetic energy of the particle [1,2]. For an electron at 6 GeV this gives 0.004 deg.

The spectral distribution of the radiation is characterized by the **critical energy** Ec (keV) = 2.2 E^3 (GeV) / R (m) = 0.7 B (T) E^2. E is the kinetic energy and B is the magnetic field that causes the particle to bend to an orbit with radius R. In practice, the available photon flux ranges from 0 to 2 or 3 times Ec. The two bottom curves of Figure 1 show the flux vs. photon energy for the ESRF corresponding to bending magnetic fields of 0.4 and 0.8 T. For the latter Ec = 19.2 keV and the flux at 60 keV is 100 times less than the maximum one.

The trajectory described by the electrons in the reality is not a geometric ideal curve but it is a beam with a finite size and divergence. This non-ideality is described by the **emittance** [3] of the orbit which is the product of the source size and divergence along a given direction. The **brilliance** of the radiation is the density of photons in phase space.

Springer Proceedings in Physics, Volume 62
Surface Science Eds.: F.A. Ponce and M. Cardona

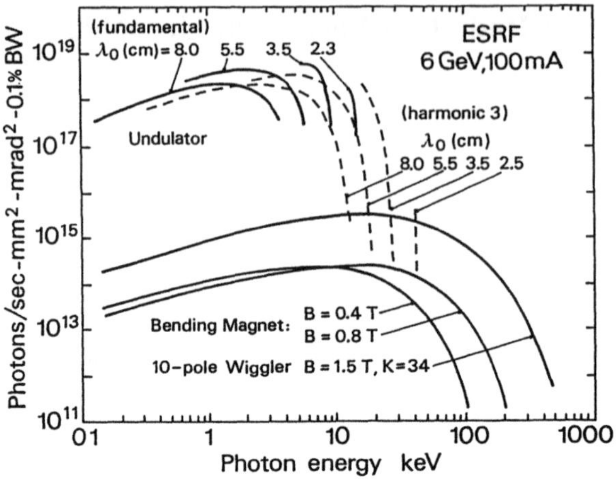

Fig. 1 Brilliance of some ESRF sources as a function of the photon energy.

A beam with high brilliance is narrow and parallel. It is easily seen that the emittance of the electron orbit and the brilliance of the corresponding synchrotron light are inversely proportional.

2. Present situation

The most recent generation of sources, which are all under construction or on the drawing boards, are characterized by two features: one is the very low emittance, generally $\leq$ 10 nm.rad in the horizontal plane; the other is the extensive use of insertion devices (wigglers and undulators) as radiation sources. The combined effect of the low-emittance, corresponding to a reduced source size and collimation, and of insertion devices, with improved radiation properties, results in an increase of many orders of magnitude in brilliance of the new sources with respect to existing ones.

The machines of this new generation can be divided in two groups, differing in the energy of the circulating electrons or positrons. One group of sources is planned to have an energy 1-2 GeV, optimized for the VUV and soft X-ray spectral ranges; the Advanced Light Source in Berkeley (USA), Elettra in Trieste (Italy), Bessy II in Berlin and

"Laboratorio Nacional de Luz Sincrotron" in Brazil belong to this category. Another type of source, optimized for the hard X-ray part of the spectrum, is represented by the Advanced Photon Source (APS) in Argonne (USA), with an energy of 7 GeV, by the Japanese 8 GeV project in the Kansai region, and by the European Synchrotron Radiation Facility (ESRF) with 6 GeV which is under construction in Grenoble.

3. The ESRF

The European Synchrotron that is being built in Grenoble will be a dedicated, high brilliance and high flux machine to serve the needs of the European scientific community for X-ray research in the 1 to ~ 100 keV energy range.

At the time of writing the project is supported by 12 European countries: France, Germany, Italy, UK, Spain, Switzerland, Belgium, Denmark, Finland, Norway, Sweden and The Netherlands.

We summarize now the main features of the machine.

Electrons (or positrons) will be accelerated up to 200 MeV with a linear accelerator. After injection into a booster synchrotron operating at 10 Hz they will be ramped to their final energy of 6 GeV. Then, they will be transfered to the storage ring that will have a circumference of 850 m. The stored current will be 100 mA. The electrons will be grouped in a number of bunches ranging from 1 to 992. The horizontal emittance of the electron beam will be 7 nm rad and the vertical one, one order of magnitude smaller. The beam lifetime will be at least 10 hours at normal operating conditions. Along the storage ring, previsions have been made for 29 straight sections to install insertion devices as radiation sources. In addition there will also be 27 bending magnet ports providing a fan of 4 mrad radiation of 19.2 keV critical energy.

4. Insertion Devices

In the new generation synchrotron sources, the emission of radiation takes place not only from the bending magnets, but mostly from the so-called insertion devices (undulators or wigglers) installed in straight sections of the storage ring. An insertion device consists generally of an array of magnets producing steady vertical magnetic fields, with a sinusoidal position dependence in the direction of the electron motion (see Fig. 2). The oscillatory field forces the electrons to follow a zig-zag trajectory, with many sharp curves, from which high intensity radiation is emitted. There are two basic advantages with respect to bending magnets: first, any magnetic field can be adopted in the device, independent of that adopted for the bending magnets of the ring; second, since many "wiggles" can be produced along the trajectory, the radiated flux is increased accordingly with respect to that for a series of bending magnets. Undulators are insertion devices with a small field, producing small angular deviations (compared with the characteristic emission angle $1/\gamma$) of the trajectory from a straight line. The emission cones from successive bends of the trajectory overlap, producing, by interference phenomena, a spectrum of lines with $\Delta E/E \sim 1/N$, where N is the number of magnetic periods in the device. High field devices, called wigglers, produce angular deviations much greater than $1/\gamma$, resulting in a broad emission band. Figure 1 shows

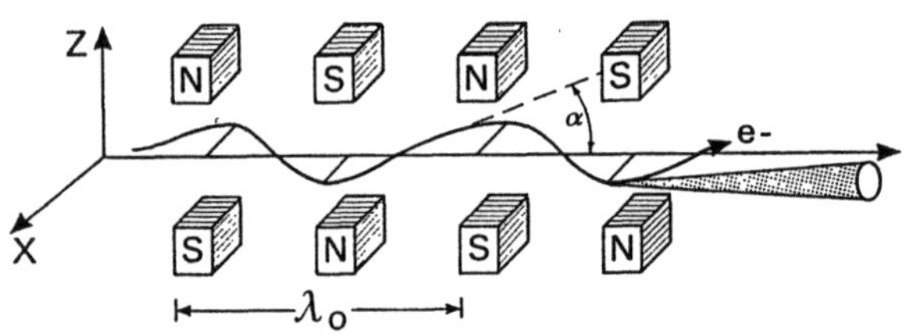

Fig. 2 Schematic view of the magnetic structure of an insertion device. Also shown are the electron trajectory, the maximum angle of deviation α and one radiation cone (of aperture $1/\gamma$) from a particular point on the trajectory.

calculated spectral characteristics for different insertion devices at the ESRF. λ_0 refers to the period of the magnetic field and K to the deviation angle of the trajectory in units of the characteristic angle $1/\gamma$.

For undulators of a given period λ_0, the energy of the fundamental and higher harmonics depends on the field, which can be tuned by varying the gap between the upper and lower set of magnetic poles (see Fig. 2). The curves in Fig. 1 show the tunability ranges of the first and third harmonics of various undulators. A remarkable increase in brilliance for insertion devices with respect to bending magnets is evident.

5. Scientific Programmes

It is foreseen that commissioning of instruments at ESRF will start in mid-1993. A first group of 7 beam lines will be operating for external users in mid-1994; within the following 12 months it is planned that 11 more beam lines will be commissioned; after that 3 new beam lines per year will be brought into operation, up to a total of 30 beam lines operating in 1999. Although the construction of such a large number of instruments is a major technical challenge, requiring the mobilization and collaboration of all Europe's expertise, the bulk of the instrumentation of the beam lines will be built within the budget of the facility. They will be operated and made available to users free of charge, on the basis of scientific merit. It is furthermore important to note that since there are 29 straight sections for insertion devices, plus 27 bending magnet ports, the 30 "public" beam lines are far from exhausting the potential of the machine, especially in so far as bending magnet sources are concerned (most of the 30 "public" beam lines will be on insertion devices). The ESRF is receiving and considering proposals from groups of users intending to build additional instruments at ESRF, with independently raised funding; they will in exchange receive privileged access to a large fraction of the available beam time.

The scientific interest of ESRF encompasses many fields of science, ranging from crystallography to surface science, from materials research to biology and medicine, from nuclear physics to earth sciences, etc. The choice of beam lines and their priority schedule is therefore a complex task, for which input from potential users from the entire European scientific community plays a crucial role. A priority list for the first 18 "public" beam lines was established in July 1989. The scientific goals for the first eight beam lines, for most of which the design phase is well under way, will be briefly described.

1. Microfocus Beam Line. This instrument will use the exceptional properties of high brilliance undulator radiation to produce a focal spot size of the order 10 μm x 10 μm. Further collimation of the beam by slits could allow reduction to 1 μ x 1 μ. Scientific uses include micro-crystal diffraction, small angle scattering and microtomography, with photon energies up to ~ 25 keV.

2. Materials Science Wiggler Beam Line. Using the intense wiggler radiation up to 40 keV, this beam line will provide tunable monochromatic radiation by a fixed-exit double crystal monochromator to allow high-momentum resolution crystallography studies in materials science, and other applications of diffraction. The experimental station will be equipped with a high stability 6-circle diffractometer.

3. Laue Protein Crystallography. A wiggler source similar to that of the previous beam line will be used to obtain white beam Laue diffraction patterns of macromolecular samples. This technique is one of the most promising for structural studies of biological molecules. The ESRF source will reduce exposure time, thus limiting radiation damage, and improve angular resolution. If supported by advances in position-sensitive detector technology, this instrument could make a very significant contribution not only to the solution of these very complex structures, but also to kinetic studies of the dynamics of

biological reactions. In addition, the white beam will be used for energy dispersive diffraction studies of materials under high hydrostatic pressure.

4. High-Brilliance Undulator Beam Line. The source for this beam line is optimized to produce the highest brilliance, at the expense of tunability. It will therefore operate in a narrow wavelength range, in the neighborhood of 1 Å, and will be used for real-time small angle scattering applications (with possible applications to muscle dynamics, interface dynamics, etc.) and for monochromatic protein crystallography.

5. High-Energy X-Ray Scattering Beam Line. A high field wiggler on ESRF can generate a critical energy in excess of 80 keV. At such energies the absorption corrections become very small and X-ray scattering can probe large volumes of the sample (as in neutron scattering). Studies of crystalline perfection and of the distribution of elastic strains in large crystals become possible. Compton scattering using such an intense source will become more important in the study of the electronic properties of solids.

6. Circular Polarisation Beam Line. Specially designed undulators may generate a magnetic field with vertical and horizontal components, forcing electrons to follow a helical trajectory and resulting in emission of radiation with more than 90% of circular or elliptic polarization for energies up to a few keV in the first harmonic. Absorption and photoemission studies with circularly polarized photons can provide very significant and novel information on magnetic materials, on surface magnetism, and on chiral molecules of importance in chemistry and biology.

7. Surface and Interface Diffraction Beam Line. Grazing incidence diffraction has proved to be a very powerful tool for the investigation

of long-range order, phase transitions, growth mechanisms and reconstruction phenomena in surface physics. The high brilliance up to 25 keV photon energy from an ESRF undulator will improve resolution, offer sensitivity to low-coverage and light-atom adsorbates, and allow access to buried interfaces in layered structures, in systems of electrochemical interest, etc.

8. Dispersive EXAFS. EXAFS is a technique for obtaining structural information on the short-range neighborhood of a given atom by measuring the absorption coefficient in an interval of a few hundreds of eV's above one of the atomic core absorption thresholds. In the dispersive mode, the whole spectrum in this region is collected simultaneously using dispersive optics with a position-sensitive linear detector. Detector read-out time limits to ~ 1 ms the acquisition time; this is, however, sufficient in some cases to follow slow processes that modify the atomic environment, for example in catalyst reactions.

The second generation of ESRF beam lines shall also cover fields of high interest, such as inelastic scattering, Mössbauer diffraction, magnetic scattering, X-ray topography, surface EXAFS and X-ray standing waves, and many others. This rapid and incomplete survey hopefully conveys the breadth of new and exciting fundamental research which will be possible at the ESRF. Furthermore, structural studies of systems relevant to the microelectronic, chemical and pharmaceutical industries also fall very much within the ESRF's overall objectives.

References

[1] Winick H. in: Synchrotron Radiation Research, Eds. H.Winick and S. Doriach (Plenum New York, 1980, 1)

[2] ESRF Foundation Phase Report (Grenoble), 1987

[3] Koch E.E., Eastman D.E. and Farge Y. in Handbook of Synchrotron Radiation, Vol. 1a, Ed. E.E. Koch (North Holland, Amsterdam, 1983)

Surface Normal Modes of a "Real" Electron Gas

*E.W. Plummer, G.M. Watson, and K.-D. Tsuei**

Department of Physics, University of Pennsylvania,
Philadelphia, PA 19104, USA
*Present address: Physics Department,
Brookhaven National Laboratory, Upton, NY 11973, USA

Abstract. The measured dielectric constants ε_1 and ε_2 for the simple metals are incorporated into the theoretical equation for the surface response function, which is derived from a calculation based upon a "jellium" metal. This procedure gives a first order estimation of the effect of the bulk band structure upon the energy, dissipation and dispersion of the surface modes, such as the surface plasmon and the multipole mode. When ε_2 is relatively small, the only noticeable effect is the increased damping for the surface plasmon for small $q_{||}$.

1. Introduction

In recent years, quite dramatic advances have been achieved in the understanding of normal modes of the electrons at the surface of simple metals [1-4]. This progress is due to two factors: (1) The availability of new angle-resolved inelastic electron scattering data taken from smooth films of alkali metals prepared in ultra high vacuum [3]; and (2) Theoretical calculations based upon a "jellium" model of the metal, which are in qualitative or semi-quantitative agreement with data [1-5]. The major successes of these studies of the normal modes on alkali surfaces have been: the verification of the negative surface plasmon dispersion at small momentum transfer (predicted by Feibelman in 1973 [9]); the identification of a new normal mode, called the "multipole mode" [2,3] (predicted by Bennett in 1970 [10,11]); and agreement between theory and experiment for the energy and dispersion of these modes [3]. This excellent agreement between theory based on a "jellium" representation of the solid and the experimental data for surface modes is quite surprising, given the dramatic failure of this type of theoretical treatment to explain the dispersion of the bulk plasmon [12]. The discrepancy between theory and experiment for the bulk plasmon is attributed to a breakdown in the current theories of Fermi liquids, due to the approximate treatment of exchange and correlation [12].

Even though the agreement between theory and experiment for the surface normal modes has been quite impressive, there are consistent discrepancies. The successes and failures of the theory can be seen in Figures 1 and 2 for the normal modes at a K surface. Figure 1 shows the measured and calculated dispersion for the surface plasmon (lower curve) and the multipole mode (top curve). In this case, the theoretical calculations use a Lang-Kohn ground state charge profile [13], and the dynamical calculation is done with a time-dependent local density scheme (TDLDA)[14]. The overall agreement for K and the other alkali metals is excellent, with the exception that the theory predicts a much larger negative dispersion at small

Springer Proceedings in Physics, Volume 62
Surface Science Eds.: F.A. Ponce and M. Cardona

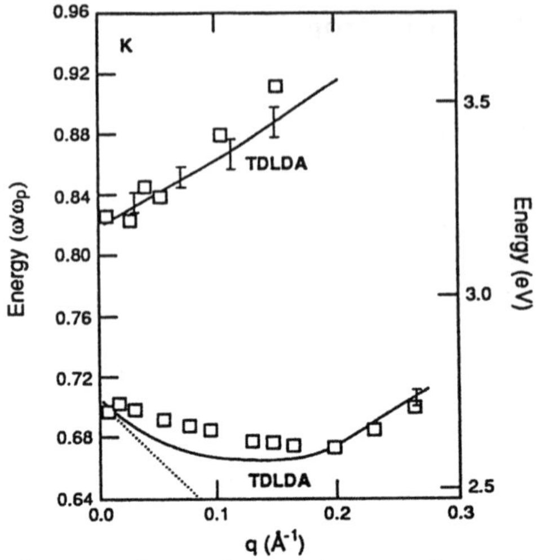

Figure 1. Comparison of theory and experiment (squares) for the dispersion of the surface plasmon (lower curve) and the multipole mode (upper curve) for K [2,3]. The dotted line shows the linear dispersion, at q=0, of the TDLDA calculation.

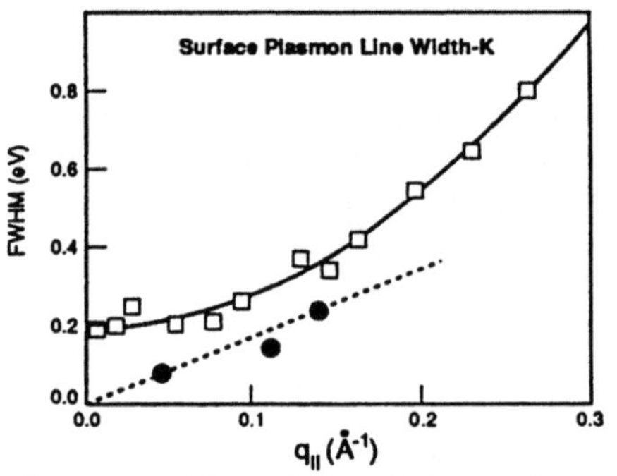

Figure 2. Line width of the surface plasmon peak as a function of momentum: experimental (squares) and theoretical for "jellium" (circles) [3].

q for the surface plasmon than is observed. The theoretical slope is three times larger than what is observed. This is a general behavior for all of the alkali metals that have been measured [3]. Figure 2 shows a comparison of the line widths observed in the experiment for K as a function of q compared to the theoretical calculation [3]. Here there is an obvious problem, since theory predicts that the inherent line width should be zero at q=0. Both the disagreement in the magnitude of the negative dispersion of the surface plasmon and the line width could be due to the effects of bulk or surface band structure or to the presence of surface roughness [3]. In this paper, the measured bulk dielectric response [15-18] for the alkali metals will be inserted into the theoretical equation derived for jellium to produce a first order estimate of the effect of the bulk band structure on the surface normal modes.

Before proceeding with the discussion of the effects of bulk band structure upon the surface normal modes, it is instructive to examine the analogous situation in the

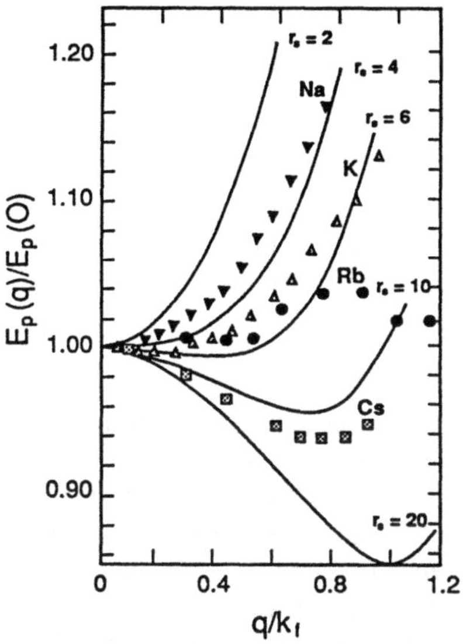

Figure 3. Dispersion of the volume plasmon normalized to the plasmon energy at q=0 [12]. The symbols represent the experimental data for Na (r_s=3.93), K (r_s=4.86), Rb (r_s=5.20) and Cs (r_s=5.62). The solid lines give the theoretical plasmon dispersion of Singwi et al. [19] for various values of r_s.

bulk. Figure 3 shows a comparison of the measured dispersion of the bulk plasmon mode, normalized to the q=0 value for Na, K, Rb and Cs, and the theoretical calculations, which are more sophisticated than the one used for the surface modes [12]. As r_s gets larger or the electron density becomes smaller, there is a dramatic disagreement between theory and experiment. The Cs data is best fit by a calculation with r_s=12, which is a metal with an effective density one eighth of the real Cs density. The theoretical curves use a local field correction, by Singwi [19], to a standard RPA calculation. The explanation that has been offered for this discrepancy between theory and experiment is that exchange and correlation effects are not properly treated by the current Fermi liquid theories [12].

At face value, it is perplexing that a simple RPA type calculation for jellium works so well for the surface region while it fails dramatically for the bulk. All of the response for the surface case is occurring in a region of large charge inhomogeneity, where it might be expected that the RPA would not work and any local correction for the exchange and correlation effects might also be inadequate. The answer to this puzzle seems to be contained in a suggestion put forward by Kempa [20]. The dispersion of the bulk plasmon is totally a property of the dynamics of an interacting electron gas system. In contrast, the ground state charge density profile is crucial in obtaining the surface plasmon dispersion and the energy of the multipole mode. For example, an infinite barrier model for K predicts that the surface plasmon dispersion is positive at small q and that there is no multipole mode [21]. Fortunately, the properties of the normal modes at a surface seem to depend more on the ground state charge density profile than they do on the dynamical processes. In the next section, an attempt will be made to evaluate the possible effects of the bulk band structure upon the properties of the surface normal modes on alkali metals.

2. Results

In the limit of small momentum transfer, it follows from standard dipole scattering theory [22, 23] that the probability for inelastic scattering is given by

$$P(\mathbf{k},\mathbf{k}') = A(\mathbf{k},\mathbf{k}') \,\mathrm{Im}\, g(q,\omega), \tag{1}$$

where **k** and **k'** are the wave vectors of the incident and reflected electron respectively, the parallel momentum transfer is $q = k_{\parallel} - k'_{\parallel}$, and the energy transfer is $\hbar\omega = \hbar(k^2 - k'^2)/2m$. The kinematic factor $A(\mathbf{k},\mathbf{k}')$ depends only on the scattering parameters, i.e., it is independent of the properties of the semi-infinite medium. The surface response function is $g(q,\omega)$ and $\mathrm{Im}\, g(q,\omega)$ is the *loss function*. The frequency of a normal mode is defined by the poles in the response function, i.e., the response diverges even for an infinitesimal field. A normal mode is an oscillation of the electron gas that once started sustains itself without the aid of a field. Theoretically, Persson and Zaremba have shown that the surface response function $g(q,\omega)$ can be written as [24]:

$$g(q,\omega) = \int dz\, e^{qz}\, \delta n(q,z,\omega), \tag{2}$$

where $g(q,\omega)$ determines the amplitude of the reflected component of an electrostatic potential interacting with the surface. The charge induced by the electromagnetic field is $\delta n(q,z,\omega)$. The response function g can be related to the $d(\omega)$ function, defined by [6]

$$d(\omega) = \int dz\, z\, \delta n(z,\omega)/\sigma(\omega) \tag{3}$$

with the total charge $\sigma(\omega)$ induced by the external field given as

$$\sigma(\omega) = \int dz\, \delta n(z,\omega) = [(\varepsilon(\omega)-1)/(\varepsilon(\omega)+1)]\, (E(\omega)/2\pi). \tag{4}$$

Expanding Equation 2 in the small q limit [6] gives

$$g(q,\omega) = \frac{(\varepsilon(\omega)-1)\,(1+qd(\omega))}{\varepsilon(\omega)+1 - (\varepsilon(\omega)-1)qd(\omega)}\,. \tag{5}$$

When q goes to zero, there is an obvious pole in the response function when $\varepsilon(\omega)+1=0$, which is the definition of the surface plasmon. Using the Drude function $\varepsilon(\omega) = 1 - \omega_p^2/\omega^2$ the energy of this normal mode is $\omega_{sp}=\omega_p/\sqrt{2}$. The energy of the surface plasmon is dictated by the bulk dielectric constant not by any surface property. If $d(\omega)$ is small and slowly varying with ω, the Drude dielectric response can be substituted into Equation 5 to obtain the equation for the dispersion of the surface plasmon in the small q limit:

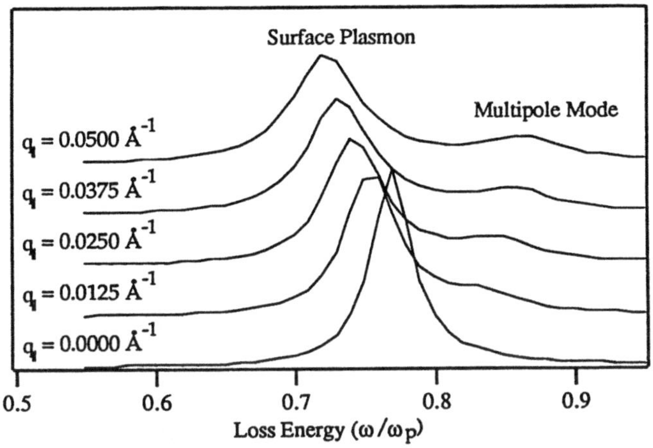

Figure 4. Calculated loss function Im {g(q,ω)} for K using the optical data of Palmer and Schnatterly [16].

$$\omega(q) = \omega_{sp}[1 - d(\omega_{sp})q/2\,]. \tag{6}$$

The Re d(ω) determines the surface plasmon dispersion and the Im d(ω) gives the width. The small q values shown in Figures 1 and 2 are obtained from this equation with the function d(ω) taken from the calculations by Liebsch [14].

A first order approximation to the surface response function for a "real" metal can be obtained by inserting the measured dielectric constants ($\varepsilon = \varepsilon_1 + i\varepsilon_2$) into Equation 5, while still using the d(ω) calculated for jellium by Liebsch [14]. In principle this approach will add the bulk band structure but will not include any effects of the surface band structure that should appear in d(ω). Figure 4 shows the Im g(q,ω) calculated for K using the dielectric constants measured by Palmer and Schnatterly [16]. Both the surface plasmon and the multipole mode can be seen in the theoretical curves. The surface plasmon has negative dispersion while the multipole mode exhibits a positive dispersion, in agreement with experiment and the more sophisticated theoretical calculations shown in Figure 1. In general, the shape of these curves agrees with the calculation by Liebsch using Equation 2 [3]. The one conspicuous difference is that the width of the surface plasmon mode does not go to zero at q=0, when the measured dielectric response is used. This is a consequence of the fact that ε_2 is not zero for a real metal, and the surface plasmon can decay, even at q=0.

Figure 5 compares the calculated loss function with the experimental data for Na (a), K (b) and Cs (c). Qualitatively, the agreement is very good. A large fraction of the width in the surface plasmon mode that could not be explained using the jellium model is accounted for when the real dielectric response in used. Table 1 compares the experimental data obtained from inelastic electron scattering with the jellium calculations and the new calculation using measured optical data for the line width and the dispersion at q=0. This inclusion of the measured dielectric constants significantly improves the agreement between theory and experiment for the

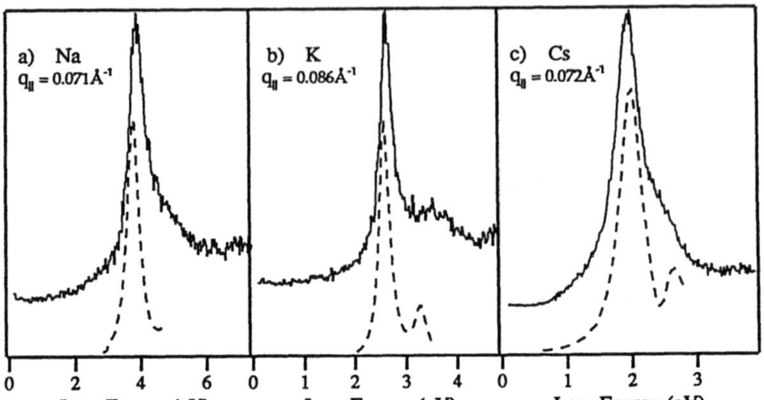

Figure 5. A comparison of the calculated loss function with the experimental data [3] for Na (a), K (b) and Cs (c). The value of the momentum q is shown in the figure. The optical data used is: a) Monin and Boutry [15], b) Palmer and Schntterly [16], and c) Smith [18].

Table 1: Physical Properties of the Surface Plasmon

Material	ω_p (eV) [9]	ω_{sp} (eV) [3]	$(\omega_p/\omega_{sp})^2$	d* (Å)[3]	$\Delta E_{q=0}$ (eV) [3]	ω_{MM}/ω_P	$\Delta E_{MM}/\omega_P$
Electron Scattering Data							
Li†	7.12	4.28	2.77	0.48	1.41		
Na	5.72	3.98	2.07	0.78	0.65	0.81	0.22
K	3.72	2.73	1.86	0.70	0.22	0.84	0.18
Cs	2.90	1.99	2.12	0.44	0.45	0.83	0.23
Jellium Theoretical Prediction[14]							
Na			2.00	1.81	0.0	0.79	0.03
K			2.00	2.10	0.0	0.82	0.03
Cs			2.00	2.22	0.0	0.84	0.04
Derived from Optical Data							
Na[15]	5.61	4.06	1.91	1.94	0.18	0.80	
Na[16]	5.40	3.95	1.87	1.66	0.14		
K[17]	3.80	2.83	1.80	2.80	0.11	0.83	0.03
K[15]	3.79	2.80	1.83	2.64	0.13	0.83	0.03
K[16]	3.80	2.86	1.77	2.93	0.13	0.83	0.03
Cs[18]	3.0	2.23	1.81	3.83	0.66	0.79	0.05

* $d = (2/\omega_{sp})$*(linear coeffecient of Plasmon Dispersion)

† For Li the bulk plasmon energy is from reference 25 while the data for the surface plasmon is from reference 26.

linewidth, but does nothing to change the theoretical dispersion. Also shown in Table 1 are the energies for the multipole modes (ω_{MM}) and the widths (ΔE_{MM}), where the width is obtained by inserting the $\varepsilon_1 = 1-1/\omega^2$ and $\varepsilon_2=0$ into Equation 5. The multipole mode can be identified as a pole in the d function. At q=0, Equation 5 predicts no multipole mode, the values shown in Table 1 for ω_{MM} and ΔE_{MM} are for $q=0.01 Å^{-1}$. As can be seen, the inclusion of ε_2 in Im $g(q,\omega)$, has no effect on the energy or width of the multipole mode. However, the theoretical width is consistently too small. This can be seen in Figure 5. The narrow width of the multipole plasmon for the optical data is due to the fact that interband transitions are not included in the calculation of $d(\omega)$.

Equation 5 can be evaluated analytically with $\varepsilon = \varepsilon_1+i\varepsilon_2$ and $d(\omega) = d_1(\omega) + id_2(\omega)$. The energy position of the surface plasmon peak will be determined by :

$$\varepsilon_1 + 1 = (\varepsilon_1-1)qd_1 - q\,\varepsilon_2\, d_2. \tag{7}$$

If it is assumed that ε_2, d_1, and d_2 are slowly varying in ω, then the approximate FWHM is given by:

$$\Delta E = \omega_{sp}(\varepsilon_2/2 + q\, d_2), \tag{8}$$

where the term, $\omega_{sp}qd_2$, is the lowest order correction due to the surface response. Equation 7 can be evaluated for the position of the surface plasmon peak if we assume that in the region of the surface plasmon energy the real part of the dielectric function can be fit to the following equation,

$$\varepsilon_1 = A - B/(\hbar\omega)^2. \tag{9}$$

Equations 7 and 9 give the following expression for the energy of the surface plasmon:

$$\omega = \omega_{sp}\,\{1 - (q/2)[d_1 - [\,\varepsilon_2 d_2 + (1-A)d_1\,]/(1+A)]\}\ . \tag{10}$$

Table 2 shows the values of A and B determined from the optical data, as well as the values of ε_2, d_1, and d_2 at the surface plasmon energy. Since A is so close to unity for all of the materials the last term in Equation 10 is not important and in general

Table 2: Data for evaluating Eqn. 8 and 10 ($\varepsilon_1=A - B/(\hbar\omega)^2$)

Material	A	B (eV^2)	ε_2 at ε_1 =-1	d_1 (Å)	d_2 (Å)	$\varepsilon_2 d_2$ (Å)
Na [15]	0.958	32.14	0.086	1.81	1.52	0.13
Na[16]	1.217	34.74	0.078	1.81	1.52	0.12
K[15]	1.200	17.28	0.10	2.10	1.03	0.10
K[16]	1.278	18.70	0.103	2.10	1.03	0.11
K[17]	0.939	15.36	0.084	2.10	1.03	0.09
Cs[18]	1.045	9.68	0.517	2.22	0.68	0.36

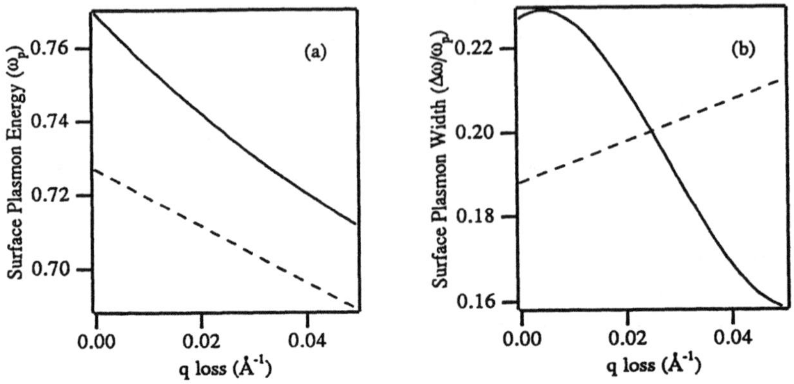

Fig. 6. Comparison of the numerical evaluation of Equation 5 for Cs (solid lines) using the optical data of Smith [18] with a) Equation 10 (dashed line) for the surface plasmon energy, and b) Equation 8 (dashed line) for the width of the surface plasmon.

$\varepsilon_2 d_2$ is always much smaller than d_1. This means that the dispersion is given by the term q $d_1/2$ for these materials. Numerical evaluation of Im g(q,ω), like that shown in Figure 4, confirms that the conclusions drawn from Equation 10 are correct. Equation 8, for the line width of the surface plasmon, agrees well with the numerical evaluation of Im g(q,ω) for Na and K. For Cs, however, the assumption that ε_2, d_1, and d_2 vary slowly with ω does not hold. Over the width of the surface plasmon peak, d_2 goes through a resonance and changes by a factor of more than 20! Figure 6 shows a comparission of the numerical evaluation of Im g(q,ω), for Cs, with a) Equation 10, and b) Equation 8. While the dispersion of surface plasmon is not significantly different from what is predicted by Equation 10. At q=0, both the energy, predicted by Equation 10, and the width, predicted by Equation 8, are to low. This is due solely to the fact that ε_2 is large and varies over the width of the surface plasmon. The resonance in d and changes in ε_2 result in a linewidth of the Cs surface plasmon which cannot be described by Equation 8.

The inclusion of ε_2 in the evaluation of Im g(q,ω) improves the agreement with the measured values of the line width. Differences in the values of the dielectric functions as measued by seperate groups, are a limiting factor in the comparision of Im g(q,ω) with the observed line widths. A careful measurement of ε_1 and ε_2 for smooth alkali films prepared in ultra high vacuum is necessary for further comparision of this model with experiment.

3. Conclusion

Calculations of the loss function including the measured values of the optical constants for the simple metals and the calculated d-functions for jellium produce the following conclusions:

1. The surface plasmon dispersion is not effected by the inclusion of ε_2 as long as the functional form of ε_1 is Drude-like and d_1 much larger than $\varepsilon_2 d_2$.

2. The width of the surface plasmon is, in general, increased by a term proportional to $\omega_{sp}(\varepsilon_2/2)$.
3. At q=0, the width and energy of the multipole mode is not effected by the inclusion of ε_1 and ε_2. Further understanding of the widths for the multipole mode requires the inclusion of interband transitions in the calculation of $d(\omega)$.

Any further improvement in the theoretical values of the surface plasmon (or multipole mode) dispersion or line width will require detailed calculations of the surface response functions $d(\omega)$ for "real" metals [27].

4. Acknowledgements

We would like to thank N. V. Smith for suggesting this calculation. This work was supported by the NSF under Grant No DMR-89-12666.

References

1. K.-D. Tsuei, E. W. Plummer and P. J. Feibelman, Phys. Rev. Lett. **63**, 2256 (1989).
2. K.-D. Tsuei, E. W. Plummer, A. Liebsch, K. Kempa and P. Bakshi, Phys. Rev. Lett. **64**, 44 (1990).
3. K. -D. Tsuei, E. W. Plummer, A. Liebsch, E. Pehlke, K. Kempa and P. Bakshi, to be published in Surface Science.
4. P. J. Feibelman and K.-D. Tsuei, Phys. Rev. B **41**, 8519 (1990).
5. Several good reviews of the theoretical models and predictions have been published [6-8].
6. P. J. Feibelman, Prog. Surf. Sci. **12**, 287 (1982), and references therein.
7. A. Liebsch, Physica Scripta **35**, 354 (1987).
8. B. B. Dasgupta and D. E. Beck, in *Electromagnetic Surface Modes* , edited by A. D. Boardman (Wiley, New York, 1982), p 77.
9. P. J. Feibelman, Phys. Rev. Lett. **30**, 975 (1973); Phys. Rev. B **9**, 5077 (1974).
10. A.J. Bennett, Phys. Rev. B **1**, 203 (1970).
11. A. G. Eguilúz, S. C. Ying, and J. J. Quinn, Phys. Rev. Lett. **58**, 2490 (1987).
12. A. vom Felde, F. Sprosser-Prou, and J. Fink, Phys. Rev. B **40**, 10181 (1989).
13. N. D. Lang and W. Kohn, Phys. Rev. B **1**, 4555 (1970); *ibid* **7**, 3541 (1973).
14. A. Liebsch, Phys. Rev. B **36**, 7378 (1987).
15. J. Monin and G.-A. Boutry, Phys. Rev. B **9**, 1309 (1974).
16. R. E. Palmer and S. E. Schnatterly, Phys. Rev. B **4**, 2329 (1971).
17. N. V. Smith, Phys. Rev **183**, 634 (1969).
18. N. V. Smith, Phys. Rev. B **2**, 2840 (1970).
19. K. S. Singwi, M. P. Tosi, R. H. Land, and A. Sjolander, Phys. Rev. **176**, 589 (1968).
20. K. Kempa, Private communication.

21. R. R. Gerhardts and K. Kempa, Phys. Rev. B **30**, 5704 (1984).
22. H. Ibach and D. L. Mills, *Electron Energy Loss Spectroscopy and Surface Vibrations* (Academic Press, New York, 1982).
23. B. N. J. Persson, Solid State Commun. **24**, 573 (1977).
24. B. N. J. Persson and E. Zaremba, Phys. Rev. B **30**, 5669 (1984).
25. C. Kunz, Phys. Lett. **15**, 312 (1965).
26. G. M. Watson and E. W. Plummer, to be published.
27. J. E. Inglesfield, Surface Science, **188**, L701 (1987).

On Plasmon Dispersion Measurements by EELS

M. Rocca, U. Valbusa, and F. Moresco

Centro CNR di Fisica delle Superfici e delle Basse Temperature,
Dipartimento di Fisica, Via Dodecaneso 33, I-16146, Genova, Italy

Abstract. The surface excitation spectrum of clean silver single crystal surfaces has been studied recently by angle resolved high resolution electron energy loss spectroscopy.

Some experimental aspects are elucidated with emphasis on the possible artifacts connected to the shape of the inelastic cross section or to elastic reflectivity variations which can complicate the interpretation of the data in this kind of measurement. Results will be presented for Ag(001).

1. Introduction

In spite of the importance of the surface excitation spectrum of clean metal surfaces, electron energy loss spectroscopy (EELS) has been applied only very recently to the investigation of the surface plasmons on clean single crystal surfaces of silver /1,2,3/ and of adsorbed thin films of simple metals /4,5/. These studies have opened up a promising new area of research in surface science. Interesting and unexpected effects were discovered relative to electronic excitations and to electron-surface interaction. Both surface plasmon frequency and dispersion were found to depend, in the case of silver, on crystal face and in the case of an (110) surface even on crystal azimuth /2/. Moreover the form of plasmon dispersion was found to be linear in the case of an (001) surface /3/ and quadratic for the other low Miller index surfaces /1,2/. The inelastic cross section displays clearly the lobular form predicted by dipole scattering theory but with the minimum clearly shifted off the expected position at vanishing transferred momentum.

In this contribution we will review some of these results and discuss some examples for Ag(001) and the reliability and the limits of the experimental data with respect to artifacts which can be induced in the spectra by the angular variation of the elastic and inelastic scattering cross section.

2. EELS measurement of surface plasmons

Surface plasmons can be excited by low energy electrons through dipolar interaction /6/. For dipole scattering the inelastic cross section is peaked around a parallel momentum transfer (surfing condition)

$$Q=\omega/V \qquad (1)$$

Springer Proceedings in Physics, Volume 62
Surface Science Eds.: F.A. Ponce and M. Cardona

where ω is the plasmon frequency and V the component of the velocity of the impinging electron parallel to the surface. The dipole intensity is relevant in a lobe of angular half width $\Delta\theta$

$$\Delta\theta = \hbar\omega/2E \qquad (2)$$

where E is the impinging energy of the electrons. The inelastic cross section vanishes near Q=0, so that two lobes will appear in an angular scan. Due to the angular acceptance α of the EEL spectrometer the spectra are integrated in Q space over a finite window ΔQ /7/. For spectrometers based on an electrostatic monochromator and analyzer with 127° geometry and rectangular slits, $\alpha \simeq 1.5^o$, ΔQ is large compared to the relevant plasmon momenta, except at low E and grazing scattering, θ_s, and incidence, θ_i, angles, where it can be reduced e.g. to $\Delta Q = \pm 0.0125$ A^{-1} for E= 10.5 eV, $\theta_s = 86.26^o$ and $Q = 0.05$ A^{-1}.

Under conditions where the energy loss ($\simeq$4 eV) is a large fraction of E, both scattering angle and energy loss, E_L, determine the transferred momentum in the spectra according to

$$\hbar Q = (2m)^{0.5}\left[(E)^{0.5}\sin\theta_i-(E-E_L)^{0.5}\sin\theta_s\right]. \qquad (3)$$

As usual the angle of incidence, rather than the angle of scattering, is varied in the experiment in order to achieve the desired parallel momentum transfer as illustrated in the inset of Fig.1.

The plasmon dispersion is obtained by plotting the energy loss as a function of Q according to eq. (3).

3. Surface plasmons on Ag(001)

EEL spectra taken for a Ag(001) surface display a single, very sharp surface plasmon loss at 3.69 eV for vanishing Q. As shown in Fig. 1 the loss disperses linearly upwards with Q with a slope of 1.5 eV/A /3,7/. At Q=0 the plasmon frequency coincides with the one reported by S. Suto et al. for the (111) surface /2/. They find however in accord with Contini and Layet a quadratic dispersion of the surface plasmon for both (111) and (110) surfaces. This indicates an astonishing qualitative difference of the behaviour of two phases of the same material which is so far unexplained. These findings are indicative of the realization, in the case of silver surfaces, of a two-dimensional electron gas largely decoupled from the bulk.

Jellium model calculations do predict for surface plasmons a linear dispersion, too. The slope should then however be negative as effectively found by Tsuei et al. for simple metal films /4/.According to Feibelman /8/ and Liebsch /9/ one should then have

$$\omega_s(Q) = \omega_s\,(1 - 0.5\,d(\omega_s)Q + O(Q^2)) \qquad (4)$$

where $d(\omega_s)$ is the centroid of the induced charge relative to

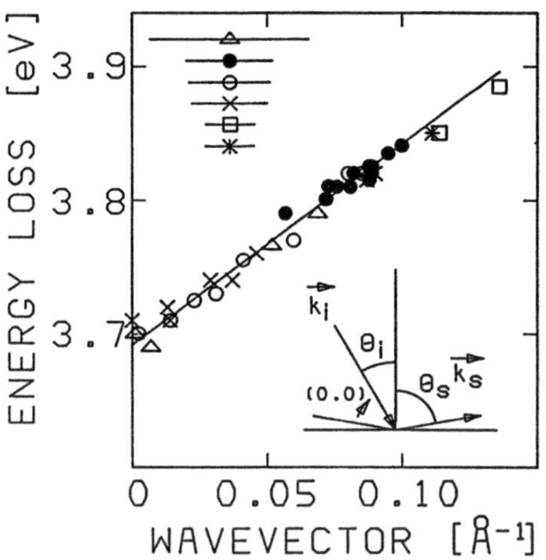

Fig.1 Plasmon dispersion for Ag(001) /3/. Different symbols refer to data collected at different impact energies or scattering geometries. The bars indicate ΔQ window.

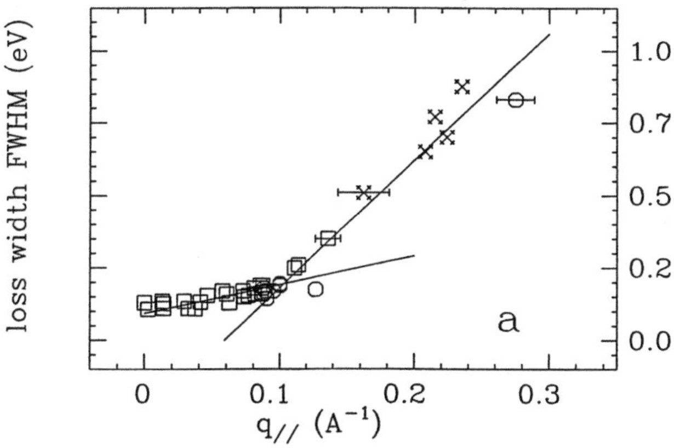

Fig.2 Loss width vs plasmon momentum for Ag(001) /7/. The bars indicate ΔQ.

the jellium edge. Although jellium is not suited to describe noble metals one can argue that our positive slope indicates a negative $d(\omega_s)$.

The plasmon loss on Ag(001) has a half width ΔE_L of 95 meV at room temperature, i.e. $\Delta E_L/E_L$=2.6%. The phonon contribution to the plasmon lifetime can be estimated from the shrinking of the loss width when cooling the sample with liquid nitrogen. The 10 meV shrinking observed from room temperature to 104 K corresponds to a 40 meV contribution of phonon-plasmon coupling to ΔE at room temperature /10/. As illustrated in Fig. 2 the plasmon lifetime decreases linearly till Q= 0.10 Å^{-1} due to the contribution of non - vertical intraband transitions as expected also for the jellium model.

Beyond this point the plasmon width explodes due to the opening of efficient damping channels associated with interband transitions involving the d bands.

4. Possible experimental artifacts and reliability of the data

In order to give the reader a feeling for the reliability of the experiment we will discuss some aspects which may complicate the interpretation of the loss spectra. These experimental effects are connected to variations of either elastic reflectivity or inelastic cross section and are due to the finite angular integration connected with the angular acceptance of the EELS.

4.1 Variations of inelastic cross section

In Fig 3 we report some spectra taken for the "ideal" conditions of grazing incidence and low impact energy recommended above to minimize ΔQ. The spectra were recorded for different angles of incidence corresponding thus to different Q values. As one can see, the loss peak position in the second spectrum is lower than in the neighboring ones. A strong anomaly appears therefore in loss dispersion as shown in Fig 4a.

The explanation of this effect is in the behaviour of the loss intensity shown in Fig. 4b. The two lobes are predicted by dipole scattering theory and are due to vanishing cross section at Q=0. An abrupt drop of intensity takes place however before the position of the maximum expected from eq. (1), in correspondence with the anomaly. This feature is connected to the appearence of the (-1-1) LEED beam which causes a sudden fall in specular elastic intensity to which the intensity of the loss is proportional. Due to integration over ΔQ the side wings contribute to the inelastic intensity,

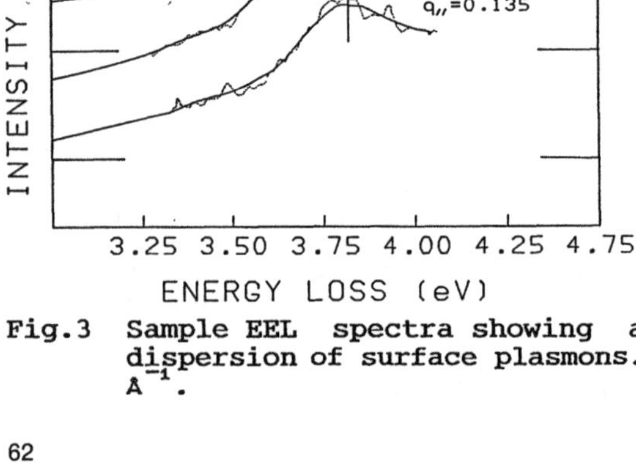

Fig.3 Sample EEL spectra showing an apparently anomalous dispersion of surface plasmons. Q values are given in $Å^{-1}$.

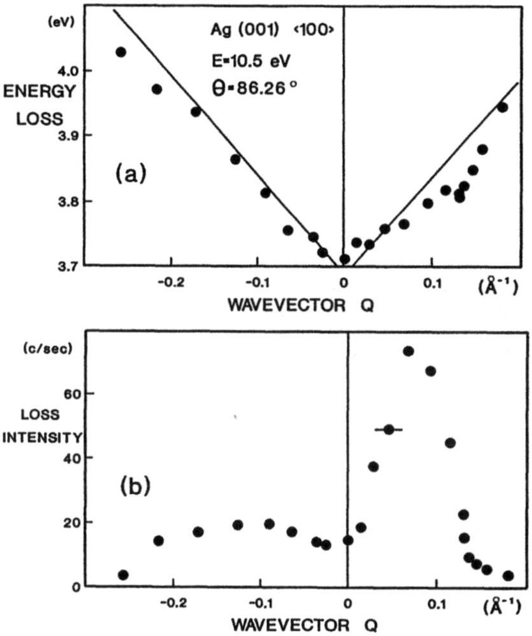

Fig.4 (a) Plasmon dispersion and (b) inelastic intensity vs.Q measured by EELS in the conditions discussed in the text.The full lines represent plasmon dispersion as in Fig.1.

shifting the apparent loss peak from its true position. The smoother behavior of inelastic cross section for negative Q values thus allows for more reliable measurements in spite of lower intensity.

A similar situation takes place on the other side of the maximum, near to Q=0, where the dipole scattering cross section vanishes. In this case larger frequencies than the true ones are mimicked as the right wing contributes to the inelastic intensity. The effect is less dramatic than before as ΔQ is smaller for more grazing angles and the intensity variation smoother. But ΔQ integration simulates in this case an unphysical quadratic dispersion of the plasmon loss. For other scattering conditions the minimum in cross section is fortunately displaced away from Q=0 /3/ thus allowing us to measure the correct plasmon frequency also for vanishing plasmon momentum.

When the data are strongly affected by inelastic cross section variations however the loss width is affected, too, thus permitting one to discriminate between valuable and useless data.

The off-specular shift of the minimum near Q=0 is interesting by itself. As recently demonstrated by Marvin and Toigo /11/ it is due to multiple inelastic interaction of the impinging electron with the surface and constitutes in fact a measure of the electron self-energy.

4.2. Effects associated with variations of reflectivity

In Fig. 5 another example of anomalous spectrum is shown. In this case two peaks instead of one are observed. The lower loss has the correct energy and can therefore be ascribed to the creation of a surface plasmon. The higher and broader loss at 4.38 eV, on the contrary, is mimicked by a variation of the elastic reflectivity as demonstrated by the apparent shift in loss energy that it undergoes when the primary energy is varied, showing that it is associated with the absolute energy of the outgoing electron rather than with the energy loss. Similar structures were already reported for EEL spectra of Ni(110) /13/ at very low impact energies and were associated with the Rydberg resonances in the specular channel. On Ag(001) we observe that such structures are present also at much higher absolute energies and can be as sharp as the surface plasmon loss and can eventually be superposed on it.

The effect is particularly troublesome beyond $Q=0.10$ $Å^{-1}$ where the plasmons are heavily damped. The plasmon loss can take place either on the way to the surface or after the reflection and the two processes interfere. Thus in that case the variation of elastic reflectivity inside the plasmon loss can cause a shift of the maximum of the loss structure.

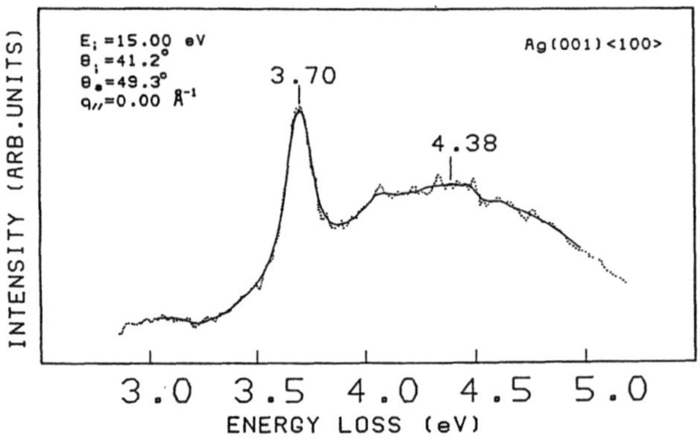

Fig.5 EEL spectrum showing the plasmon loss superposed on elastic reflectivity structures.

5. Conclusions

We have shown that the energy loss peaks in plasmon measurements can be shifted due to the angular acceptance of EELS and the behaviour of the inelastic cross section or by superposed structures caused by variations of elastic reflectivity. These effects are possibly responsible for at least part of the spread in the data points in Figs. 1 and 2. A careful investigation of the loss intensity and form of the peaks has therefore to be undertaken in order to interpret the data correctly.

7. References

1. R. Contini and J.M. Layet,Solid State Commun. 64, 1179 (1987)
2. S. Suto, K.D. Tsuei, E.W. Plummer, and E. Burstein, Phys. Rev. Lett. 63, 2590 (1989)
3. M. Rocca and U. Valbusa Phys. Rev. Lett. 63, 2398 (1990)
4. K.D. Tsuei, E.W. Plummer, and P.J. Feibelman, Phys. Rev. Lett. 63, 2256 (1990)
5. K.D. Tsuei, E. W. Plummer, A. Liebsch, K. Kempa, and P.Bakshi,Phys. Rev. Lett. 64, 44 (1990)
6. H. Ibach and D.L.Mills,*Electron Energy Loss Spectroscopy and Surface Vibrations* (Academic, New York, 1982)
7. M. Rocca,F.Biggio, and U. Valbusa, to appear in Phys Rev. B 42 (1990)
8. P.J. Feibelman, Prog. Surf. Sci. 12, 287 (1982)
9. A. Liebsch, Phys. Scr. 35, 354 (1987)
10. F. Moresco, M. Rocca and U. Valbusa, to be published
11. A. M. Marvin and F. Toigo, to be published
12. D. Rebenstorff, H. Ibach, J. Kirschner, Solid State Commun. 56, 885 (1985)

The Electronic Band Structure of Penrose Lattices: A Renormalization Approach

Chumin Wang[1] *and R.A. Barrio*[2]

[1]Instituto de Investigaciones en Materiales, Universidad Nacional Autónoma de México, Apdo. Postal 70-360, 04510 México, D.F., Mexico
[2]Instituto de Física, Universidad Nacional Autónoma de México, Apdo. Postal 20-364, 01000 México, D.F., Mexico

Abstract. The electronic band structure of two-dimensional Penrose lattices is analyzed by means of a novel renormalization method. This method allows one to calculate, in exact form, the local density of states on large clusters, with a tight-binding Hamiltonian. The results show the fractal structure of the spectrum. The method should be useful for the analysis of dynamical properties of Penrose lattices.

Recently developed technologies by which it is possible to construct artificial arrangements of atomic arrays of practically any form have produced an enormous interest in studying lattices or arrays of atoms which show quasi-crystalline order. Such is the case of quasi-periodic superlattices of two semiconducting compounds [1]. Similarly it is possible to produce two-dimensional and three-dimensional quasi-crystalline alloys [2]. The two-dimensional quasicrystals can be studied by models which contain a 5-fold symmetry, such as the Penrose tiles [3].

An interesting question is how the electronic properties of such systems differ from normal periodic networks. There have been numerous and detailed studies of electrons in Penrose lattices in recent years [4] and the band structure derived from those studies reveals a peculiar nesting of bands and gaps. However, the nature of the spectrum is not well agreed upon, probably because practically all the calculations are made in small clusters with different boundary conditions and the results depend dramatically on the cluster size [5].

It is therefore desirable to develop a method sufficiently efficient to manage a large number of atoms in a network. In this paper we present a renormalization approach, which takes advantage of the inflation properties of the Penrose tile to calculate the local electronic states in a large system. The method relies on a recipe to construct the Penrose tile by simple addition of two basic units (Robinson triangles [6]) and the use of Green's function techniques to investigate the density of states for an s-electron tight-binding Hamiltonian.

The construction of the Penrose tiles is analogous to the 1D Fibonacci chain system [7] and consists of steps of adding two tiles, which have been previously formed to obtain a new bigger tile $[T(n) = T(n-1) + T^*(n-2)]$. The star on the $T(n-2)$ indicates that one has to take the mirror image of the tile in order to guarantee the matching of sites. In Fig. 1a the first 4 generations produced by this method are shown.

Springer Proceedings in Physics, Volume 62
Surface Science Eds.: F.A. Ponce and M. Cardona

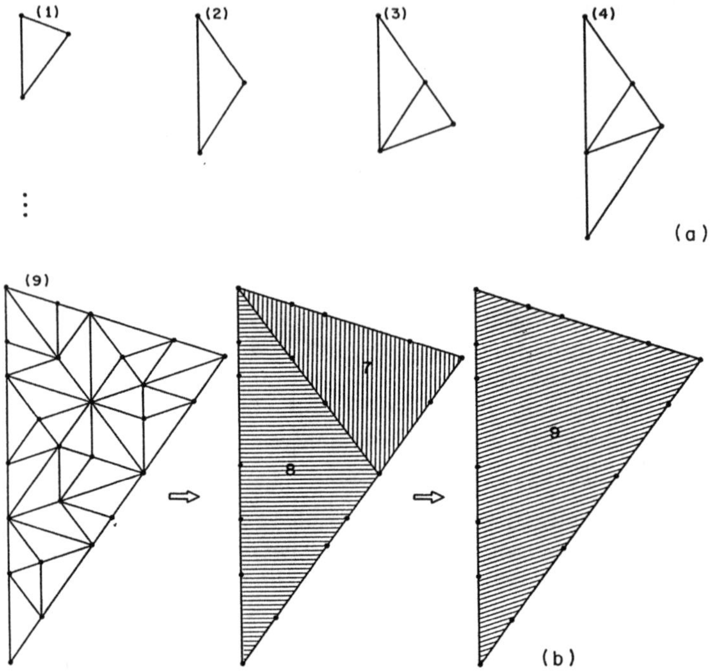

Fig. 1 (a) The first four generations of Penrose tiles are always of the type of two Robinson triangles. (b) A sketch representing the renormalization of the central sites of tiles 7 and 8 and the final result obtained renormalizing the border sites between the two tiles from the equations for Green's functions

The renormalization procedure consists in eliminating the coordinates of the *central* sites from the equations of motion for the Green's function in each generation. This allows us to handle only the atoms in the surface of each tile in order to construct the new generation. The procedure is exemplified in Fig. 1b for generation 9. The usual Penrose tile can be obtained by eliminating the shortest and the longest bonds of the Robinson network. The equations of motion are similar to those for the Fibonacci chain, which are written in full in [7]. A complete account of the algebra will be published elsewhere [8].

In order to illustrate the sort of calculations that one can carry out, we present the local density of states of two contiguous sites in the middle of a tile of generation 20 (3500 sites), for an electron tight-binding Hamiltonian of the form

$$H = \alpha \sum_{i} |i\rangle \langle i| + \beta \sum_{i \neq j} |i\rangle \langle j|$$

and using the vertex model (i.e. the atoms occupy the vertices of the tiles and the electrons hop only along the edges). Figure 2 shows the peculiar distribution of bands and gaps of the spectrum and the appearance of a peak at E = 0,

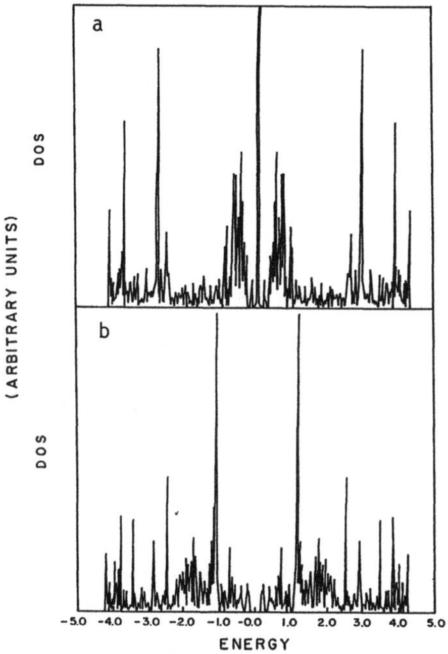

Fig. 2 Local density of states from a Penrose tile of generation 20. **(a)** and **(b)** correspond to two contiguous sites in the middle of the tile. The parameters used were $\alpha = 0$ and $\beta = -1$

only at selected sites. This behavior has been detected in other calculations [4] and apparently reflects a certain degree of spatial localization of this state. We have expanded the central part of the spectrum of big tiles in order to show the self-similarity of this spectrum, and a detailed account of this result will be published in the future [8].

We conclude with the following:

1) This method is efficient and exact, allowing one to calculate local quantities without much effort.

2) The method, as is, can be used to calculate properties measured between the surfaces of a system, such as conductivity in a network of LRC components [9], since Kirchhoff's laws can be mapped exactly onto models of isotropic phonons.

3) This method allows the study of samples of macroscopic size, without having to make approximate models for a real system.

4) The method can be extended to investigate the dynamic response of 2D quasi-crystalline systems by applying the method of partial summations used to investigate the Raman response of a Fibonacci superlattice [10].

Acknowledgements. Financial support from the Dirección General de Asuntos del Personal Académico de la UNAM under grant IN-01-02-89 is gratefully acknowledged.

References

1. J. Todd et al.: Phys. Rev. Lett. **57**, 1157 (1986)
2. D. Shechtman et al.: Phys. Rev. Lett. **53**, 1951 (1984)
3. R. Penrose: Bull. Inst. Math. Appl. **10**, 266 (1974)
4. M. Kohmoto, B. Sutherland: Phys. Rev. Lett. **56**, 2740 (1986); P. Ma, Y. Liu: Phys. Rev. B **39**, 9904 (1989)
5. F. Aguilera-Granja et al.: Phys. Rev. B **36**, 7342 (1987)
6. R. Robinson: University of California preprint (1975), unpublished
7. R.A. Barrio, Chumin Wang: In *Quasicrystal and Incommensurate Structures in Condensed Matter*, ed. by M. José Yacamán et al. (World Scientific, Singapore 1990), p. 448
8. Chumin Wang and R.A. Barrio: To be published
9. Chumin Wang, O. Navarro, R.A. Barrio: To be presented at the MRS 1990 Fall Meeting
10. Chumin Wang, R.A. Barrio: Phys. Rev. Lett. **61**, 191 (1988)

Part III

Atomic Arrangement at Surfaces and Interfaces

Spin vs Charge Asymmetry in the Dimers of the Si(100)-2×1 Surface

*E. Artacho**

Departamento de Física de la Materia Condensada, C-III,
Universidad Autónoma de Madrid, E-28049 Madrid, Spain
*Actual address: Department of Physics, University of California,
Berkeley, CA 94720, USA

Abstract. Spin polarization in the dimers is proposed instead of charge and geometric asymmetry for the Si(100)-2x1 surface. Current symmetric and asymmetric dimer models are considered. Core level shift and total energy calculations support this proposal. An energy gain of about 0.5 eV per surface atom is obtained by including spin correlations in the calculations. Electronic charge and spin densities, surface states bands and spin-spin correlations are presented. All these strongly suggest that the dimers buckling is much smaller than predicted by previous spin independent calculations.

1. Introduction

Since its discovery by Schlier and Farnsworth in 1959 [1], the 2x1 reconstruction of the (100) silicon surface has been extensively studied. At present it is generally accepted that the surface atoms move from their ideal positions to form rows of dimers. However, the situation concerning the actual geometry and electronic structure of the dimers formed is far from being clear. From the experimental point of view the results are to some extent contradictory. On one hand scanning tunneling microscopy (STM) measurements indicate that the dimers are symmetric away from defects and impurities [2]. On the other, grazing incidence X-ray diffraction measurements are interpreted in terms of asymmetric dimers [3]. Core level shifts of the Si $2p$ levels are consistent with the symmetric dimer configuration [4]. Recent polarization-dependent angle-resolved photoemission experiments [5] reveal the existence of a mirror plane symmetry normal to the dimers also indicating symmetric dimers. Other experiments are compatible with both symmetric and asymmetric dimer configurations as it has been already discussed by the authors [6].

There have been several spin independent theoretical calculations of the electronic and geometrical structure of the Si(100)-2x1 surface [7-12]. Most of the calculations indicate that the asymmetric dimer model is the more stable one although a recent calculation by Batra [12] finds an almost equal energy for the symmetric and asymmetric dimer configura-

Springer Proceedings in Physics, Volume 62
Surface Science Eds.: F.A. Ponce and M. Cardona

tions in agreement with previous calculations done by Pandey [10]. The surface band structure obtained in these calculations is in fair agreement with experimental data, although a metallic surface is obtained contrary to the experimental findings, even for the asymmetric [11] dimers and the calculated occupied surface band is about 0.5eV above the experimental one [11]. It is remarkable that, in spite of the effort made, there is no general agreement on the geometrical structure of this surface.

In this work we present a study of the electronic structure of the Si(100)-2x1 surface taking into account possible spin arrangements other than uniform since, as it was pointed out, all previous calculations have neglected spin correlations within the dimers. However, cluster calculations [9] and a recent model calculation [6] reveal that spin correlations can be of great importance. The main aim of this work is not to establish the geometrical arrangement of the atoms at the surface but rather to understand the influence of spin in the calculations.

2. Si 2*p* core level shift

Core level shifts give valuable information about charge transfers that, in our case, would occur in the dimer if it is essentially asymmetric. In Fig. 1 (a) experimental results for the Si 2*p* core level emission are shown as obtained by Rich *et al.* [4]. They claim that the S labeled peak is emitted by both atoms in each dimer rather than just one of them. This already favors the symmetric dimer geometry against the asymmetric one. To confirm this result we have calculated the 2*p* energy levels for the Chadi symmetric dimer model [7] and for the Yin and Cohen asymmetric model [8]. We perform the calculations in a cluster with nine silicon atoms simulating a dimer and part of the first layers of the silicon crystal. The rest of it is simulated by appropriate saturators. We have performed an *ab initio* all electron Hartree-Fock calculation using a single-ζ basis with each basis orbital expanded in four gaussian functions. The calculated chemical shifts, convoluted with Doniach-Šunjić [13] functions, together with the XPS experimental data are shown in Fig. 1 for both dimer models. We observe an excellent agreement for the symmetric dimer model and a two peak structure, not observed experimentally, for the asymmetric one, favoring symmetric arrangement at the dimer.

3. Spin-polarized valence ground state

Contrary to the conclusions of the previous section, many theoretical works claim that dimer asymmetry stabilizes the surface with a charge transfer from one atom to the other. We propose that it is spin asym-

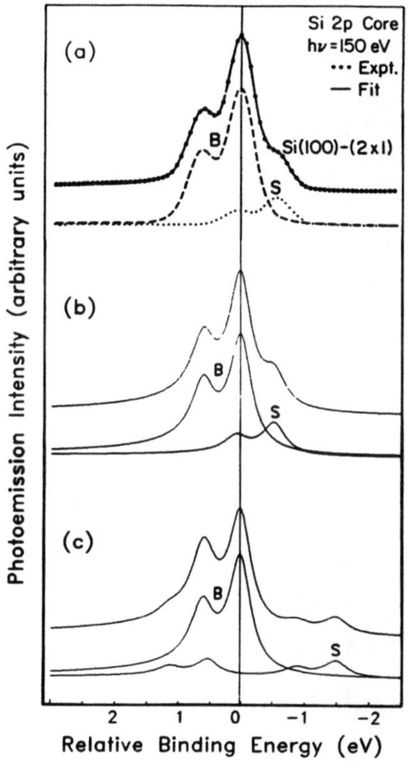

Figure 1. Silicon $2p$ core level at the Si(100)-2x1 surface. In (a) the experiments of Ref. 4 are presented. (b) and (c) are the theoretical results obtained for Chadi symmetric and for Yin and Cohen asymmetric dimer models, respectively. The binding energies obtained for the $2p$ levels of surface and bulk silicon atoms, including a spin-orbit splitting of 0.61 eV [4] are convoluted with properly weighted Doniach-Šunjić [13] functions. S stands for surface and B for bulk signal, with the sum of both shifted upwards for clarity.

metry (in the sense of spin correlation) rather than charge and geometric asymmetry what characterizes the electronic structure of the ground state of the surface. To pursue this idea we have performed spin-polarized electronic structure calculations [14] for symmetric and asymmetric dimers geometries in actual semi-infinite silicon crystals by means of a non-parametrized linear combination of atomic orbitals method [15].

To check whether the method is capable of describing the chemistry of a spin polarized system we have studied a single dangling bond in the Si(111) surface passivated by one monolayer of As by replacing one

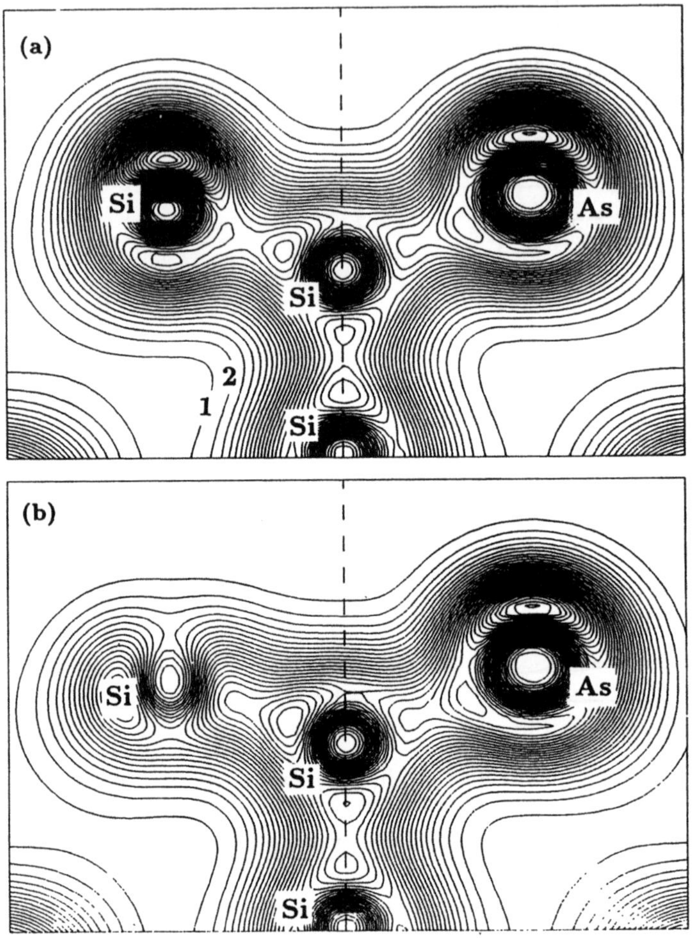

Figure 2. Spin-resolved valence electronic charge distribution (multiplied by two) for a single Si atom replacing an As atom at the Si(111)-1x1:As passivated surface, corresponding up (down) panel to spin up (down) electrons. In each panel the charge is plotted in two different planes in order to simultaneously see the impurity atom and a surface As atom, the dashed lines separating the plots. The left side presents a $(01\bar{1})$ plane passing through the impurity atom, the right side presenting a 120 degrees rotated plane passing through the same second and third layer atoms than the preceding and also through a surface As atom. The units are electrons per bulk unit cell and the contour spacing is 1.

surface arsenic atom with a Si atom. The results for the density of "up" and "down" electrons are shown in Fig. 2 (a) and (b), respectively. The sum and difference of them give the total charge and the spin densities respectively. The dangling bond on the surface silicon atom pointing outwards the surface is apparent when comparing both densities, the spin polarization disappearing rapidly away from this atom. The net spin population on the surface silicon atom is 1.01 e^- with a very limited spreading of 0.09 e^- on the neighboring atoms.

We now apply the method to the Si(100)-2x1 surface. In Fig. 3 we show the total charge and spin densities for the Chadi symmetric

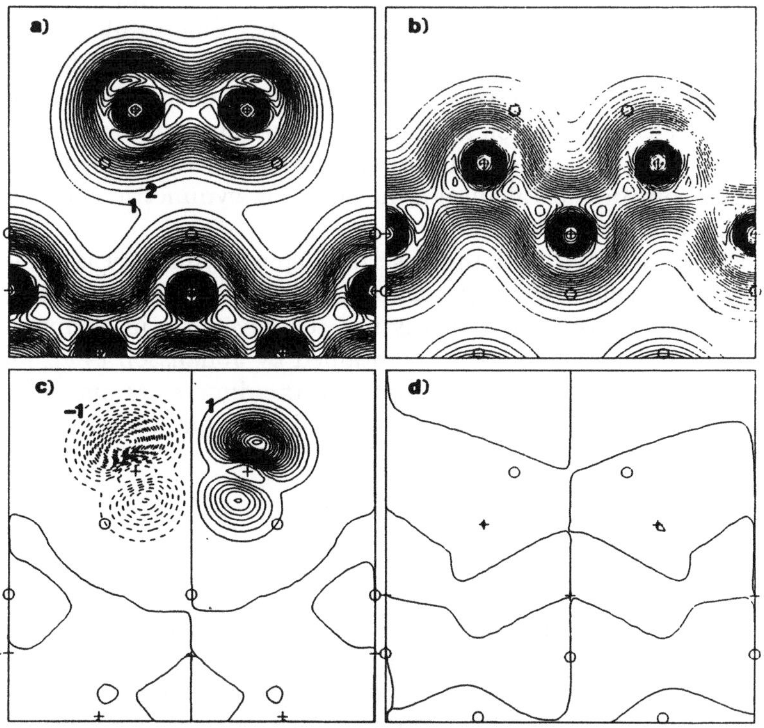

Figure 3. Charge and spin distributions near the surface for the symmetric Chadi model in the spin unrestricted calculation plotted in (011) planes cutting the surface at right angles. Crosses and open circles represent silicon atoms contained in the plane and out of the plane respectively. (a) Total charge distribution at the (011) planes containing surface dimers. (b) Total charge distribution at the (011) planes containing atoms of the second layer. (c) Spin density at the (011) planes containing surface dimers. (d) Spin density at the (011) planes containing atoms of the second layer. The units are electrons per bulk unit cell and the contour spacing is 1.

dimer model. An important spin polarization appears in the dimer, the spin populations being of opposite sign in each atom. For the Yin and Cohen asymmetric dimer these results are essentially the same (1.07 e^- and −1.07 e^- for the symmetric model and 1.01 e^- and −1.14 e^- for the asymmetric one). The total charge distribution is very similar to the restricted spin calculation the main difference being the spin asymmetry (correlation) mostly localized at the surface decaying rapidly towards the bulk.

Another interesting result of the calculations is that for both symmetric and asymmetric dimer models there is substantial gain of electronic energy in the cluster calculations by releasing the equal spin constraint. We obtain an energy gain of 2.08 and 1.05 eV per dimer for the symmetric and the asymmetric models respectively. This result makes the symmetric dimer being the more stable one, its energy being 0.92 eV lower than that of the asymmetric dimer. This large energy difference has to be considered with caution since no geometrical minimization including spin has been performed. This result indicates the importance of including the spin in the calculation in both the symmetric and the asymmetric dimer models.

It has to be stressed that the described spin asymmetry is compatible with experiments observing symmetric dimers: STM essentially measures the charge density, and this is not sensitive to spin asymmetry. The photoemission results recently obtained by Johansson *et al.* [5] concerning a mirror plane symmetry normal to the dimers are also fully compatible with our findings: what we have in the system are local spin correlations (see section 5).

4. Surface states bands

The effect of the inclusion of spin in the surface states bands is mainly to remove states from the gap. To show this, since the spin polarization is almost completly localized on the dangling bonds, and since the method used in the previous section does not describe properly unoccupied bands, we use a simple tight-binding two-dimensional model in which the spin correlations are included via a Hubbard term treated in the mean field approximation. In Fig. 4 the surface states bands associated to the dangling bonds are shown for a symmetric dimer array with and without equal spin constraint. The release of this constraint opens a gap at the Fermi energy making the surface semiconducting, as observed experimentally [16].

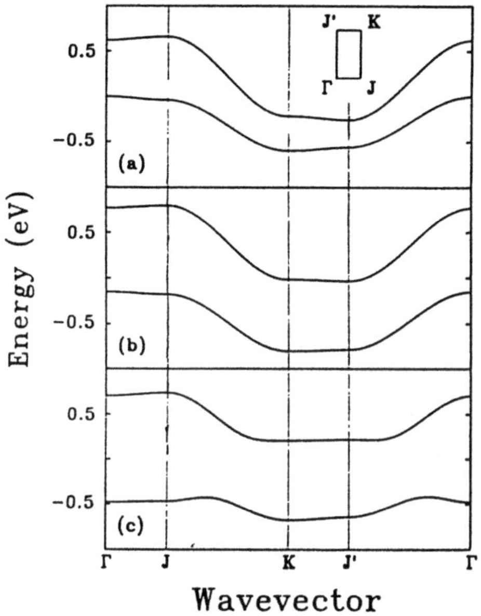

Figure 4. Surface states bands for the symmetric dimer configuration for three different spin arrangements: (a) no spin ordering, (b) layer antiferromagnetic and (c) antiferromagnetic.

5. Spin ordering

The main fact responsible of the effects described in the previous sections is the antiferromagnetic spin arrangement within the dimer. However, if the preceding ideas hold, a new question appears: what is the spin ordering among dimers? In the ground state calculations of section 3, for making the calculation practicable, a 2x1 unit cell is considered also for the spin polarization; this implies a layered antiferromagnetic surface. In Sect. 4, together with this 2x1 magnetic surface, a full antiferromagnetic structure, $p(2 \times 2)$, is also considered in order to show that the relevant fact for the gap opening is the spin correlation within the dimer. The latter structure has a total energy of 0.1 eV per atom lower than the former.

Nevertheless, all these calculations impose long range order which may not be the case. In fact, the ground state associated with the Hamiltonian used in the previous section is not expected to present long range spin ordering. The interdimer versus intradimer effective magnetic interaction ratio lies below the value for which a transition to Néel ordering occurs [17] and, in addition, the interaction between rows of dimers is rather weak. Therefore, the ground state will be an adiabatic continuation of the state associated to the isolated dimers limit.

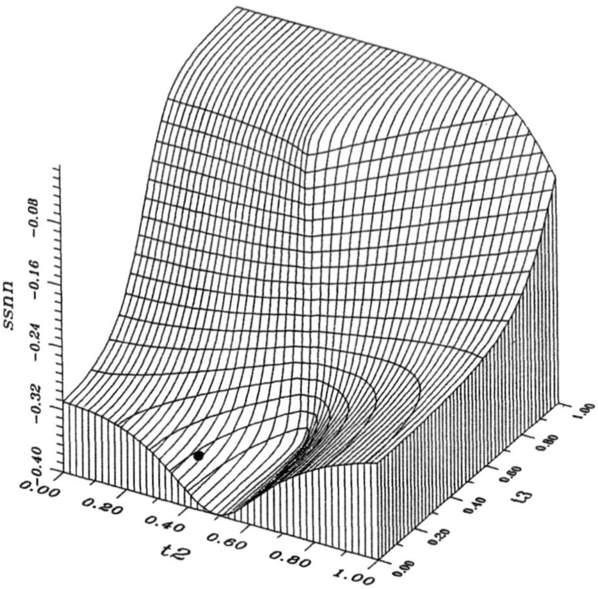

N.N. SPIN CORRELATION

Figure 5. Nearest neighbors spin-spin correlation function (*ssnn*) versus t_2 and t_3 for $U = 0.85$ and $t_1 = 0.5$. The black dot corresponds to the values of t_2 and t_3 used in section 4.

However, local spin correlation may exist among dimers at a finite distance. The question of how this correlations look like is quite interesting, specially noting that the dispersion of the bands shown in Fig. 4 requires that the two different hopping interactions between two nearest neighbor dimers have opposite sign; this gives rise to bond frustration effects . To study the local correlations between two dimers of the same row we have considered the Hamiltonian used in the previous section and solved it by exact diagonalization for just the four atoms forming these two dimers [18].

Considering as nearest neighbors the atom in the same dimer (interacting through the hopping term t_1) and also the closest atom in the other dimer (t_2), and second nearest neighbor the remainder (t_3), we compute nearest neighbors spin-spin correlation (*ssnn*) and second nearest neighbors spin-spin correlation (*sssn*). The former is shown in Fig. 5 for varying t_2 and t_3 and for $t_1 = 0.5$ and the Hubbard integral $U = 0.85$, values used in the calculations of section 4.

In the figure we see regions in parameter space where the system is locally antiferromagnetic. This local ordering is destroyed by the frustration introduced in the dynamics by the t_3 hopping term. The transition is smooth except in the case $t_1 = t_2$, where the states involved in the transition correspond to different irreducible representations of the symmetry group associated with the Hamiltonian. For the actual values of t_2 and t_3 local antiferromagnetic order is obtained.

6. Conclusions

From our calculations we can conclude the following:

i) Core level shift analysis indicates that the charge transfer associated with the buckled dimer models is not compatible with experimental findings.

ii) Inclusion of spin correlation when calculating the electronic structure and total energies for different atomic geometries at the Si(100)-2x1 surface seem essential to obtain reliable results.

iii) The spin arrangement within the dimers is antiferromagnetic for all current buckled and non-buckled dimer models. Therefore it is essential to include the spin when calculating atomic and molecular adsorption at the surface. There are also local antiferromagnetic spin correlations among dimers.

iv) The inclusion of spin correlation in the calculations makes the symmetric models more stable than the asymmetric ones.

v) Spin polarization makes the surface semiconducting in agreement with experimental findings, even for symmetric dimers.

I am indebted to Dr. Ynduráin and to Dr. Miláns del Bosch for their collaboration. Work supported in part by Comisión Interministerial de Ciencia y Tecnología.

References

1. R. E. Schlier and H. E. Farnsworth, J. Chem. Phys. **30**, 917 (1959).
2. R. M. Tromp, R. J. Hamers, and J. E. Demuth, Phys. Rev. Lett. **55**, 1303 (1985) and Phys. Rev. B **24**, 5343 (1986); R. J. Hamers and U. K. Kohler, J. Vac. Sci. Technol. **A7**, 2854 (1989).
3. R. Pinchaux, M. Sauvage-Simkin, J. Massies, N. Jedrecy, N. Greiser and V. H. Etgens, to be published.
4. D. H. Rich, T. Miller, and T.-C. Chiang, Pys. Rev. B **37**, 3124 (1988).
5. L. S. O. Johansson, R. I. G. Uhrberg, P. Mårtensson, and G. V. Hansson, Phys. Rev. B **42**, 1305 (1990).
6. E. Artacho and F. Ynduráin, Pys. Rev. Lett. **62**, 2491 (1989).
7. D. J. Chadi, Phys. Rev. Lett. **43**, 43 (1979) and J. Vac. Sci. Technol. **16**, 1290 (1979).
8. M. T. Yin and M. L. Cohen, Phys. Rev. B **24**, 2303 (1981).

9. A. Redondo and W. A. Goddard, J. Vac. Technol. **21**, 344 (1982).
10. K. C. Pandey, in *Proceedings of the 17th ICPS*, edited by D. J. Chadi and W. A. Harrison (Springer, Berlin, 1984), p.55.
11. Z. Zhu, N. Shima and M. Tsukada, Phys. Rev. B **40**, 11868 (1989).
12. I. P. Batra, Phys. Rev. B **41**, 5048 (1990).
13. S. Doniach and M. Šunjić, J. Phys. C **3**, 285 (1970).
14. E. Artacho and F. Ynduráin, accepted in Phys. Rev. B.
15. P. Ordejón, E. Martínez, and F. Ynduráin, Phys. Rev. B **40**, 12416 (1989); E.Artacho and F.Ynduráin, to be published.
16. F. J. Himpsel and Th. Fauster, J. Vac. Sci. Technol. A **2**, 815 (1983).
17. R. R. P. Singh, M. P. Gelfand, and D. A. Huse, Phys. Rev. Lett. **61**, 2484 (1988).
18. L. Miláns del Bosch, E. Artacho, and F. Ynduráin, to be published.

HRTEM of Decahedral Gold Particles

M. Avalos-Borja[1], *F.A. Ponce*[2], *and K. Heinemann*[3]

[1]Instituto de Física, Universidad Nacional Autónoma de México, Apdo. Postal 2681, 22800 Ensenada, B.C., Mexico
[2]Xerox Palo Alto Research Center, 3333 Coyote Hill Rd., Palo Alto, CA 94304, USA
[3]Eloret Institute, 1178 Maraschino Dr., Sunnyvale, CA 94087, USA

Abstract. Decahedral gold particles supported on carbon are analysed by high resolution TEM. Evidence is found that supports the existence of non–fcc structures in this type of particles. The existence of angles larger than 70.5 degrees for {111} planes supports a model based on a body centered orthorhombic structure. No gaps or dislocations were found between the building tetrahedra.

1. INTRODUCTION

The decahedral particle has been the subject of extensive studies in the past (see for example [1–15]), mainly, perhaps, due to the unique 5–fold symmetry axis, which is forbidden in classical crystallography. It is evident from TEM micrographs that it is composed of five building units. The simplest model would suggest that the building units are perfect fcc tetrahedra with {111} planes as outer faces (which are well known to have the lowest energy). The relationship between any two tetrahedra would be a twin relationship, and the five tetrahedra have a common <110> direction coincident with the 5–fold axis. The major problem with this model is that the angle between tetrahedral {111} planes is 70.5°, and not 72.0° as required for the perfect decahedral polygon. Several theories have been proposed to explain this disagreement; the most common can be summarized as follows: a) the gap is filled with extra planes bound by dislocations; b) the gap is filled by inhomogeneous strain; and c) the structure is not fcc but a slight distortion from it, i.e., body centered orthorhombic (bco), constructed in such a way as to produce an angle of 72.0° among the building tetrahedra, making gaps or strains

Springer Proceedings in Physics, Volume 62
Surface Science Eds.: F.A. Ponce and M. Cardona

TABLE I
Relationship between the Miller indices and interplanar spacings for fcc (left side) and bco. Lattice parameters are normalized to a = 1.0, b = 1.0, and c = √(2) for fcc and a = 1.0, b = 1.0515, and c = 1.3764, for the bco case.

(hkl)	d	$(khl)_O$	d_o
(111)	$\sqrt{(2/3)}=0.8165$	(011)	0.8355
$(11\bar{1})$		$(01\bar{1})$	
$(1\bar{1}1)$		(101)	0.8090
$(\bar{1}11)$		$(\bar{1}01)$	
(200)	$\sqrt{(1/2)}=0.7071$	(110)	0.7246
(020)		$(\bar{1}10)$	
(002)		(002)	
$(2\bar{2}0)$	$\sqrt{(1/4)}=0.5000$	(200)	0.5000
$(20\bar{2})$		$(11\bar{2})$	0.4990
(220)		(020)	0.5257

unnecessary. In this paper we present new data that may help to resolve this controversy.

A summary of model (c) will be appropriate for the interpretation of results and subsequent discussion. The fcc structure can also be described by means of a body centered orthorhombic cell [1,2], with lattice parameters $a=\sqrt{2}/2\ a_0$, $b=\sqrt{2}/2\ a_0$, and $c=a_0$ (where a_0 is the original fcc lattice parameter). For simplicity we can normalize the parameters dividing by $\sqrt{2}/2\ a_0$. Therefore, without any loss of generality we can consider the bco unit cell (equivalent to fcc) as having parameters $a=1.0$, $b=1.0$ and $c=\sqrt{2}=1.4142$. A slight distortion of the previous structure with parameters $a=1.0$, $b=1.0515$, and $c=1.3764$,

TABLE II

Interplanar angles for the {111} fcc-type planes in the bco structure.

	$(01\bar{1})_O$	$(101)_O$	$(\bar{1}01)_O$
$(011)_O$	74.76	69.09	69.09
$(01\bar{1})_O$	0.0	69.09	69.09
$(101)_O$	69.09	0.0	72.00

produces the angle of 72.0° between the {111}–type planes and that is the main reason why it is adopted. There is, of course, a new nomenclature of planes and different interplanar spacings and angles. To avoid any confusion, the bco planes will be denoted with an O subscript. Table I gives the equivalence between fcc and bco planes with the respective d–spacings. It is interesting to note that, since the new {111} planes are no longer equivalent, different angles between them are feasible. Table II shows the possibilities, illustrating that only one combination yields 72.0°. More surprisingly, another combination produces 74.76°.

2. EXPERIMENTAL

Gold particles were evaporated onto vacuum cleaved NaCl crystals under near UHV conditions. The substrate temperature was 200°C; with a deposition rate of 0.1 nm/min, and average film thickness of 1.0 nm. The particles were subsequently coated with carbon by electric–arc discharge (estimated thickness of about 20 nm). The NaCl substrate was dissolved in distilled water and the films were mounted on Cu grids for TEM observation.

Electron microscopy was performed at 400 kV in a JEOL 4000 EXV transmission electron microscope. Observations were performed under axial illumination and under conditions yielding better than 0.18 nm point to point resolution.

Local structure analysis was done by optical diffractometry, using an optical bench, directly from TEM negatives. Digital Fourier transforms were performed for large areas.

3. RESULTS AND DISCUSSION

Fig. 1a shows a decahedral particle with its zone axis closely parallel to the 5–fold axis. We can see that the planes are straight when viewed radially, i.e., from the center towards the edge; and no dislocations, distortions, or defects of any kind are evident, as shown in the enlargement in Fig. 1b. The structure of this boundary is basically the same as a twin along a {111} plane.

Figs. 2a–2c show the expected diffraction patterns for the bco and fcc structures considering the {111} and {002} planes only. In the fcc case (Fig. 2c) {111} double spots arise from the "gap" between the tetrahedra. The bco model predicts, on the other hand, a diffraction pattern with single spots (Figs. 2a and 2b). A digital diffractogram from the whole particle in Fig. 1a is shown in Fig. 2d. No evidence of elongated or double spots is observed, although any evaluation based on this figure is at risk due to the lack of sharpness of the spots in the diffractograms. Even enlargements from single spots, like the one in Fig. 2e, fail to provide definite answers.

In order to improve the resolution, we performed selected area optical diffractograms from the original negatives. Fig. 3a shows a diffractogram from a region inside one of the composing tetrahedra, Fig. 3b shows the usual fcc indexing, and Fig. 3c the corresponding bco indices (those indices are unique, since only one combination of planes yields 72°,

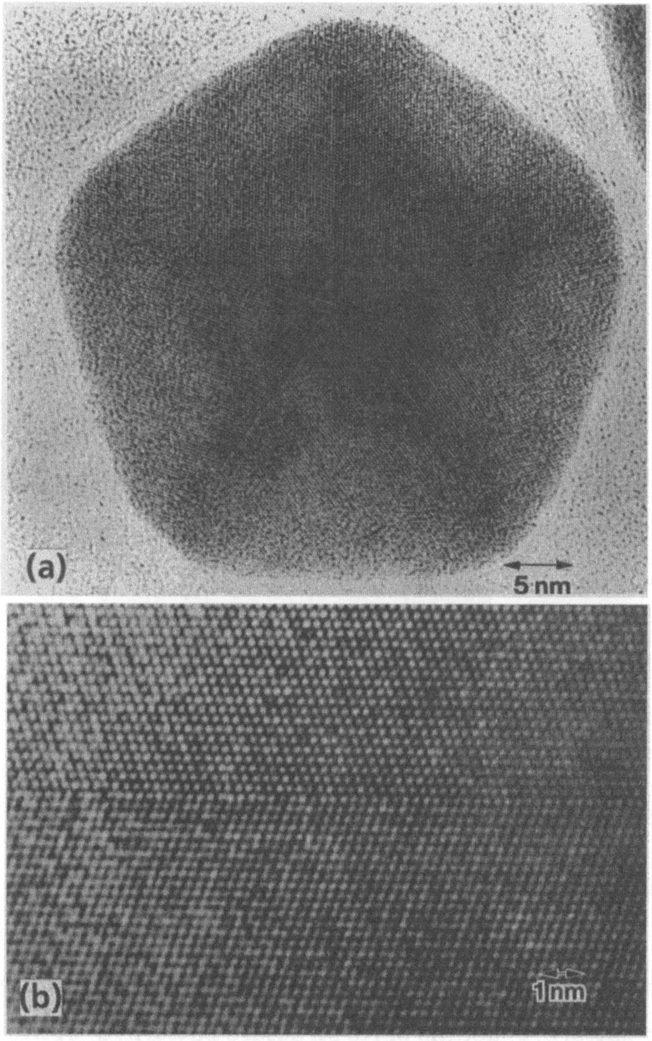

Fig. 1. High resolution lattice images: a) Decahedral particle seen along the 5–fold axis, b) enlargement of twin boundary

as explained above). The angles between the $(\bar{1}0\bar{1})_O$ and $(\bar{1}01)_O$ planes (obtained from figures like 3c) for the five tetrahedra in Fig. 1a were measured as 71.9°, 72.0°, 72.4°, 71.9°, and 71.9° (with a ±0.3° experimental error), clearly above the value of 70.5° that would be expected for the fcc case. (Similar values were measured from published

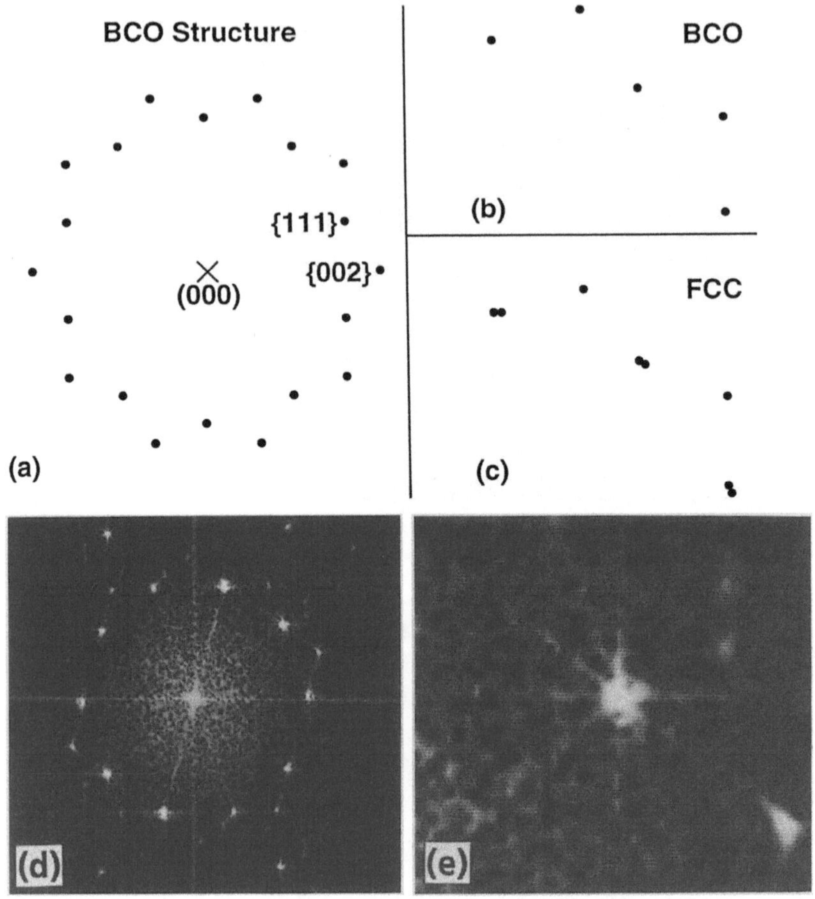

Fig. 2. Simulated diffraction patterns of particle in Fig. 1a: a) corresponding to bco structure; b) enlargement from (a) showing details of bco pattern; c) details of fcc structure pattern; d) optical diffractogram from the whole particle in Fig. 1; e) enlargement from (d) showing no evidence of elongated or double spots.

atomic resolution micrographs from Marks [6] and Renou [12], although they never addressed the fact that this measurement clearly deviates from the expected fcc value.) Additionally, the ratio of interplanar spacings between the $(002)_O$ and $(\bar{1}01)_O$ is measured as 1.17 from Fig. 3a. It would theoretically be 1.15 and 1.17 for fcc and bco, respectively.

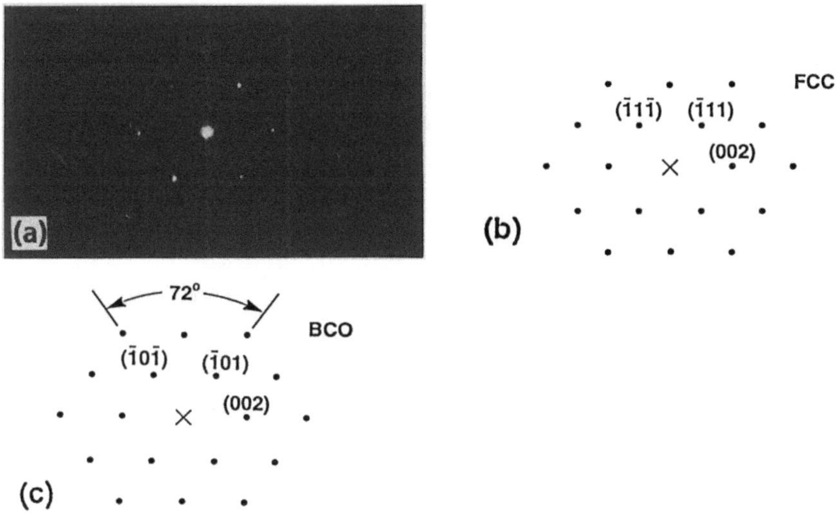

Fig. 3. a) Optical diffactogram from a single tetrahedron in Fig. 1, and corresponding b) fcc indexing, and c) bco indexing.

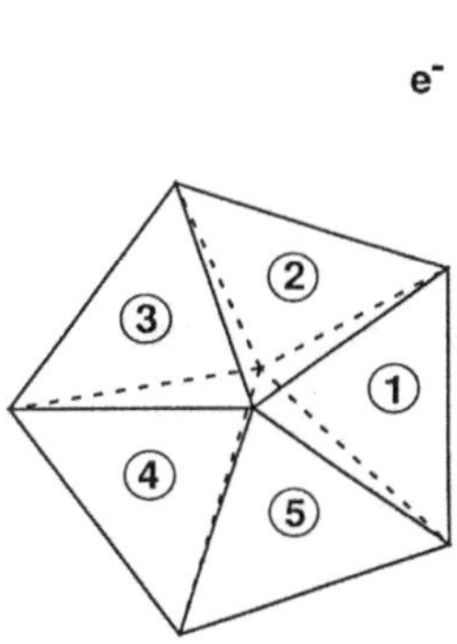

Fig. 4. Model of decahedral particle on edge, corresponding to the orientations seen in Figs. 5(a) and 5(b).

Although the evidence in favor of the bco model is substantial at this point, we consider that actual experimental demonstration of the 74.76° angle would even be stronger. This angle is formed by the external faces comprising the "wedge" in any tetrahedron in Fig. 1a. In other words, this angle would only be measured when the particle is seen on edge, as is illustrated with the arrow in Fig. 4. Fig. 5a shows a decahedral particle in such an orientation, with one of the composing tetrahedra observed in a direction parallel to the [110] external edge. The fact that Figs. 1a and 5a correspond to the same type of particle has previously been thoroughly

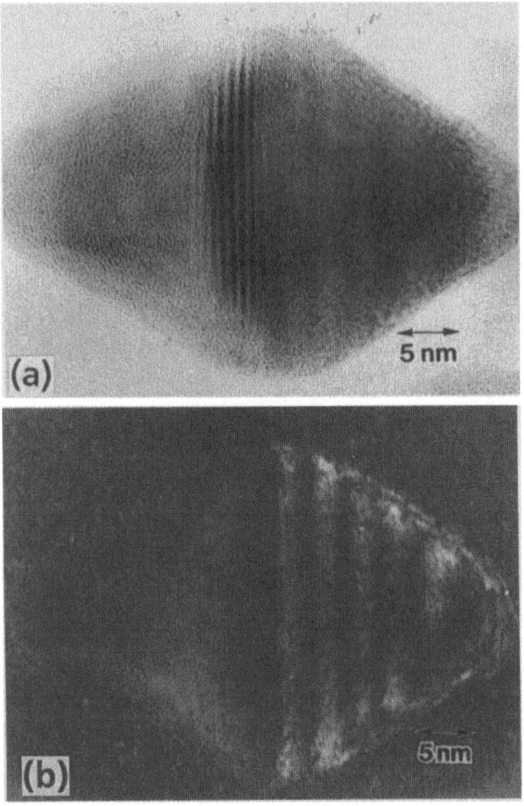

Fig. 5. a) Decahedral particle seen on edge, with tetrahedron 1 in Fig. 4 oriented with a <110> zone axis parallel to the electron beam, b) weak–beam–dark–field image of same particle as in (a).

documented in the literature [3, 4, 13, 15], and we feel that no additional proof is needed. Fig. 5b illustrates the expected weak–beam–dark–field contrast under these conditions; the fringes are periodically spaced, which indicates a constant change in thickness. Fig. 6a is the selected area optical diffractogram from a region in Fig. 5a, and Fig. 6b is a reproduction with bco indices. The angle measured between the $(011)_O$ and $(01\bar{1})_O$ is 74°, which is reasonably close to the expected bco value of 74.76°.

Thickness contours in Au occur at ~ 16 nm. The broad finges observed in the right hand side of Fig. 5a (with spacings of 4.3 nm) are moiré

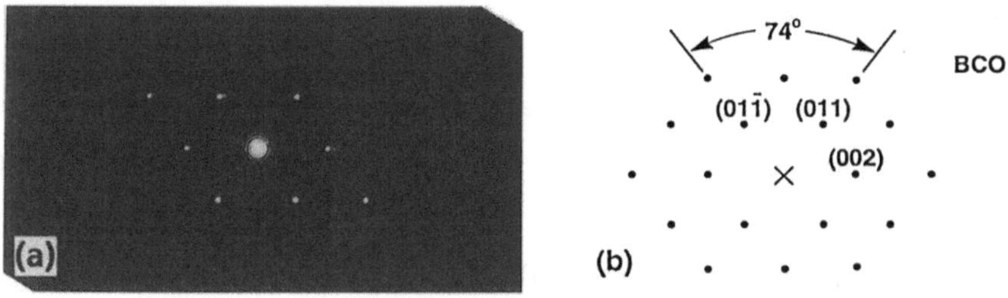

Fig. 6. a) Optical diffractogram from the <110> tetrahedron in Fig. 5(a); b) Bco indexing of (a).

fringes associated with the superposition of the crystallographic planes of two adjacent tetrahedra. The planes involved in these moires are the {008} in tetrahedron 1 and the {442} in tetrahedron 2, of Fig. 4. If the structure was fcc, these planes would produce a moiré pattern with fringes spaced by 0.202 nm, i.e., a distance that corresponds exactly to the {002} spacing, thus being indistinguishable from the ordinary lattice planes. The moiré pattern separation is $D = d_1 d_2/(d_1 - d_2)$. As discussed above, one set of planes is of the $\{002\}_{fcc}$ type ($d \sim 0.202$ nm). Using $D \sim 4.3$ nm, from Fig. 5a, we get $d_1/d_2 \sim 1.5$. This value is consistent with the orthorhombic model where $b/a = 1.0515$.

The moiré fringes on the left side of the particle in Fig. 5a have a spacing of 0.7 nm, which corresponds to moires produced by the interference between tetrahedron 2 and 5, and tetrahedra 3and 4 in Fig. 4. The planes involved are of the {442} and {224} fcc types or equivalent $\{042\}_0$ and $\{024\}_0$ bco types. In the fcc structure the moiré patterns between {442} and {224} (lowest allowed indices of these type) have spacings of 0.367 nm. In the bco structure, the equivalent planes of the types $\{042\}_0$ and $\{024\}_0$ (lowest allowed indices) produce moiré spacings of 0.675 nm. Our measurements on Fig. 5a of 0.7 nm come fairly close to the calculated value of the bco structure.

4. CONCLUSIONS

We have studied decahedral gold particles using HRTEM. Our results indicate that: a) the structure is homogeneous in each of the five composing tetrahedra,

b) the boundaries between tetrahedra are twin boundaries,

c) we have observed no defects such as dislocations, nor inhomogeneous strain fields near or at these twin boundaries,

d) the angle between the {111} planes is 72° (on the planar view) rather than 70.5° which would correspond to the cubic case,

e) images in lateral view allow us to make more accurate determinations of the lattice parameter, and underline the bco particle model.

ACKNOWLEDGMENTS

Two of the authors (M. A. B. and K. H.) gratefully acknowledge support of this work throughout NASA grant NCC2-283 to Eloret Institute.

REFERENCES

[1] B. G. Bagley, Nature, 208 (1965), 674.
[2] C. Y. Yang, J. Cryst. Growth, 47 (1979), 274.
[3] C. Y. Yang, M. J. Yacaman, and K. Heinemann, J. Cryst. Growth, 74 (1979), 283.
[4] K. Heinemann, M. Avalos-Borja, H. Poppa, and M. J. Yacaman, Inst. Phys. Conf. Ser. No. 52/6 (1980), 387.
[5] L. D. Marks and D. J. Smith, J. Cryst. Growth, 54 (1981), 425.
[6] D. J, Smith and L. D. Marks, J. Cryst. Growth, 54 (1981), 433.
[7] L. D. Marks and D. J. Smith, J. Micros., 130 (1983), 249.
[8] L. D. Marks, Phil. Mag., 49 (1984), 81.
[9] A. Howie and L. D. Marks, Phil. Mag., 49 (1984), 95.
[10] L. D. Marks, Surf. Sci., 150 (1985) ,302.
[11] P. L. Gai, M. J. Goringe, and J. C. Barry, J. Micros., 142 (1985), 9.
[12] S. A. Nepijko, V. I. Styopkin, H. Hofmeister, and R. Scholtz, J. Cryst. Growth, 76 (1981), 501.
[13] A. Renou and J. M. Penisson, J. Cryst. Growth, 78 (1986), 357.
[14] S. Iijima, Jpn. J. Appl. Phys. 36 (1987), 365.
[15] M. Avalos-Borja, Ph. D. Thesis, Stanford University, 1983.

The Structure of Gold Icosahedral Nanoclusters

M. José-Yacamán, R. Herrera, C. Zorrilla, S. Tehuacanero, and M. Avalos

Instituto de Física, Universidad Nacional Autónoma de México,
Apdo. Postal 20-364, 01000 México, D.F., Mexico

In the present work we report on the characterization of small icosahedral particles (~10 nm size) using HREM and image processing. Fourier power spectra of similarly oriented particles present different spectra. The results are interpreted in terms of a continuous oscillation of particle structure in agreement with the quasimelting concept of Marks and Ajayan (10).

1. INTRODUCTION AND EXPERIMENTAL METHODS

In the present work we report on the characterization of gold clusters in the size range up to ~100nm which is the upper size limit for what can be considered a nanostructured material. In particular we will focus our attention on the so called multiple twinned particles (MPT'S) (1) which appear as an important component of nanostructured materials in many preparation methods. In spite of the extensive effort in characterization of the MPT'S (2-6), many aspects of these clusters are still not clear. In particular, the issue of how the particle is strained to "close" the gap produced by the regular FFC structure is still unresolved since the two models that have been proposed, including homogeneous and inhomogeneous (2-3) strain, produce identical results as recently shown by Fluei (6). On the other hand, the structure of the clusters should be examined in view of the sudden changes in particle structure that can occur during TEM observation (7-8) which where termed quasimelting by Ajayan and Marks (9). This phenomenon is believed to happen also in a free cluster under non irradiation conditions as recently reported (10).

In the present work we used a high resolution JEOL 4000 TEM to produce images. We also applied image processing in order to improve the image contrast. Fast Fourier transforms of the high resolution images were obtained, after digitizing them with CCD camera using a Micro Vax III computer. The samples were produced by evaporation of the gold metal under a 2 Torr He gas pressure.

In order to obtain the best possible image of gold particles we have used defocus conditions to set either the first or the second maximum on the contrast transfer function.

In figure 1 we plot the contrast transfer function versus the defocus. As can be seen the second maximum appears more favorable to image simultaneously (111) and (200) planes. Therefore in many pictures a "white" atomic column condition was used. Experimental images were compared with calculations of the particle structure. We used the software package developed by Herrera and Gómez (11). This program, which includes an N-

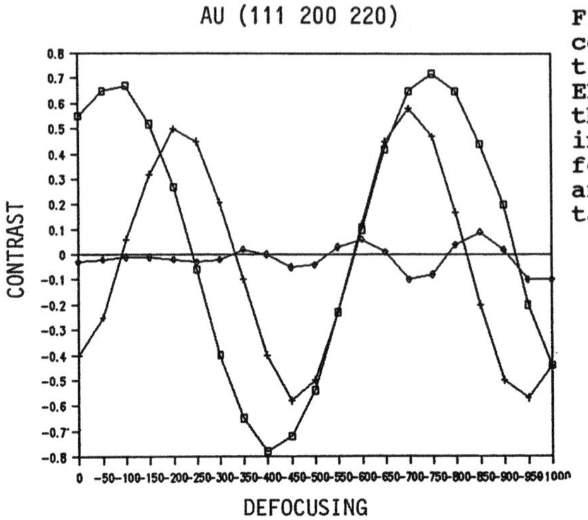

Figure 1. Plot of the contrast transfer function for the JEOL 4000 EX vs defocus. Note that in the second maximum optimum conditions for imaging of (111) and (200) planes is obtained.

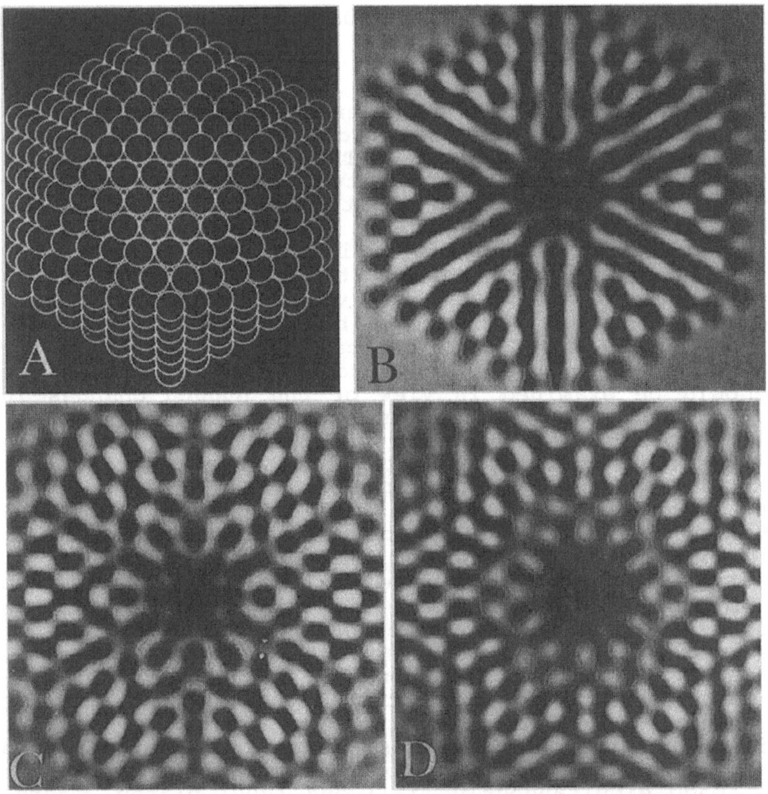

Figure 2. Calculated high resolution images for an icosahedron in six-fold orientation. a) Atomic arrangement. b) Image for defocuss o Δf=+700Å. c) Image for defocuss of = -400Å. d) Image for Δf = -700Å.

beam dynamical calculation is specially convenient for calculations of contrast in small particles with non-periodic features such as the MPT'S.

2. IMAGE CALCULATIONS

We have calculated the high resolution image for an icosahedron in the three-fold orientation at different defocus and aperture values. The results are indicated in figure 2 a)-e). The basic atomic structure used in the calculation is in figure 2a). As can be observed, the images have three-fold symmetry as expected and the central portion of the particle shows six-fold symmetry features.

3. EXPERIMENTAL RESULTS

We have studied extensively the high resolution images of the different kinds of particles. In this report we will mainly focus on the case of a six-fold oriented icosahedral nanocluster. A series of images are shown in figure 3.

The high resolution image of the particle is shown together with the Fourier transform power spectrum. Basically three different situation of spots can be observed. In figure 3a) we have a situation with the six spots predicted for a six-fold oriented icosahedral particle (2) in which the spots are arced. Figure 3b) shows a situation in which the spots are split and figure 3c) shows the situation in which spots are arced in all directions but one (in which are split). The pictures were obtained in neighboring areas of the same sample. We have measured the split on a number of spots using microdensitometric methods and the average value is 7.6° ± 0.6° which is in excellent agreement with the theoretically expected value for an fcc structure (1). On the other hand the arc of intensity covered by the non-split spots corresponds to a similar value.The experimental evidence indicates that a number of different states are presented on a six-fold oriented particle.

4. DISCUSSION

It is well known that in the icosahedral structure the composing tetrahedra will not close the structure if they remain fcc. This problem has been addressed by Yang et al. assuming that each unit is distorted homogeneously in such a way that each unit is distorted to a rombohedral structure. An alternative model has been presented by Howie and Marks(3) in which inhomogeneous strain is introduced in the form of disclinations. Six 7.5° wedge disclinations are enough to produce a space filling structure. It is hard to separate between these two models because they produce the same atomic positions (6) identical diffraction patterns and almost identical high resolution images (12). Therefore very little can be gained arguing in favor of any model. However in both cases the power spectra are expected to be composed by arced spots. On the other hand the fcc structure with "gaps" will generate well defined split spots in all directions.

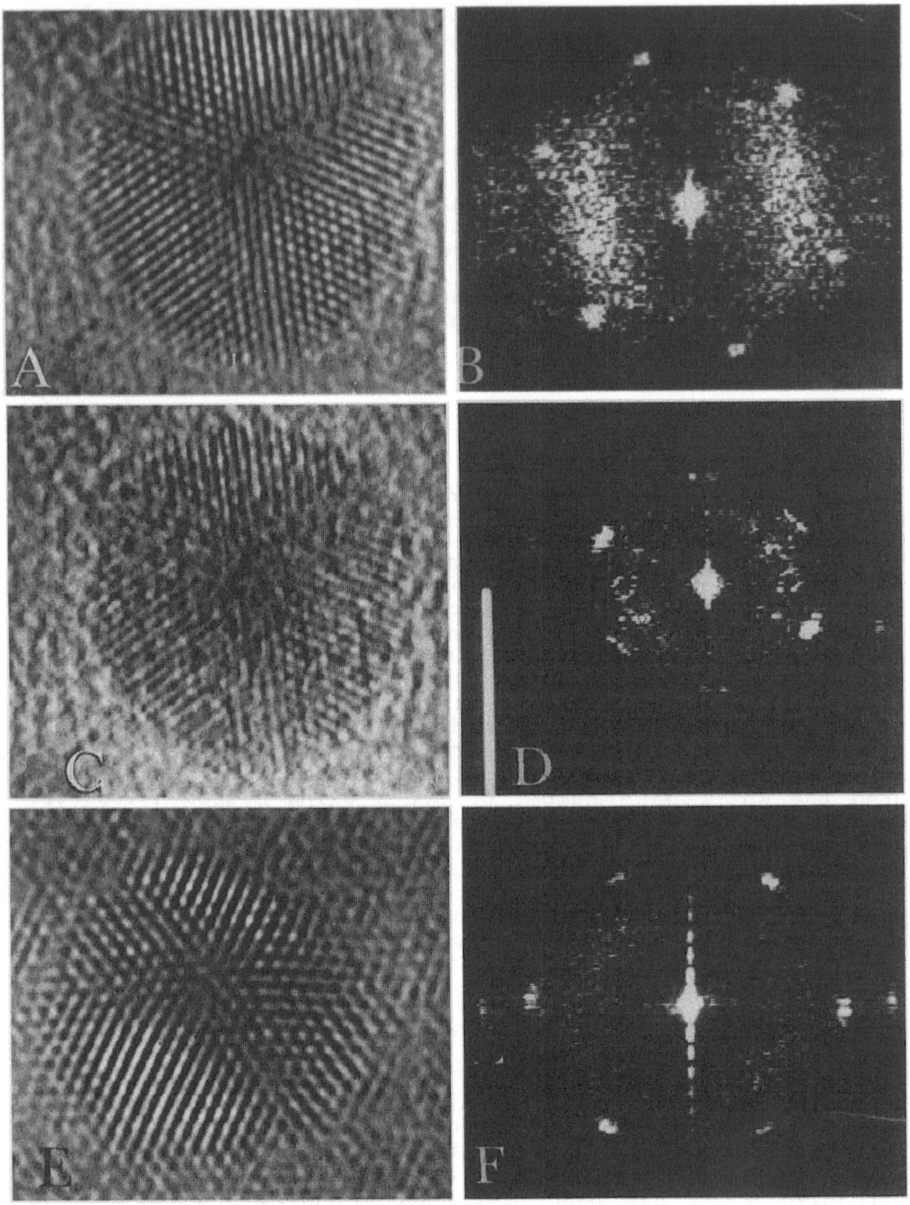

Figure 3. Examples of high resolution images of an icosahedral particle in six-fold orientation showing its corresponding FFT power spectrum. As can be noted the three spectrums are different.

The results of this work indicate that in some cases in the six-fold oriented particle six arced spots are observed. In other cases split spots are seen and in others splitting is observed in only one direction.

We must therefore conclude that there is no clear cut situation for the particle structure i.e; in similar evaporation conditions the particles have a number of different configurations.

The results can be understood in terms of a quasimelting concept (10) since the particle is oscillating between two extreme situations, fcc and icosahedral structure. In between a continuum of states seems to exist. Our results therefore seem to give further support to the idea of quasimelting occurring at room temperature which was recently suggested by Ajayan and Marks (10).

REFERENCES

[1] S. Ino. J. Phys, Soc. of Japan 27, 941(1967).

[2] C.Y. Yang, M. José-Yacamán, K. Heinemann and H. Poppa. J. Cryst. Growth 47, 283(1979).

[3] A. Howie and L.D. Marks, Phil. Mag. 49, 95(1984).

[4] E. Gillet and M. Gillet, Thin Sol. Films 4, 171(1969)

[5] M. Fluei, R. Spycher, P. Stadelman, P. Buffat and J.P. Borel, J. Micros. and Spec. Elect. 14, 35(1989).

[6] M. Fluie, Ph. D. Thesis EPL Lausanne (1989).

[7] S. Ijima and T. Ichihashi, Phys. Rev. Lett. 56, 616(1986).

[8] O. Bovin, R. Wallemberg and D. Smith, Nature 317, 47(1985).

[9] J. Dundurs, L.D. Marks and P.M. Ajayan, Phil. Mag. 57, 605(1988).

[10] L.D. Marks and Ajayan, Phys. Rev. Lett. 63, 279(1989).

[11] R. Herrera and A. Gómez. Acta Crystallographica in Press.

[12] D. Romeu, Private communication.

Reflection Electron Microscopy and Reflection Electron Diffraction in the Electron Microscope

J.A. Eades

Center for Microanalysis of Materials, Materials Research Laboratory, University of Illinois at Urbana-Champaign, 104 S. Goodwin, Urbana, IL 61801, USA

Abstract. If the transmission electron microscope is used, not in transmission but in reflection (that is with the electron beam incident at near-grazing incidence onto the surface), images and diffraction patterns of the surface can be obtained. The results are dramatic and informative. In the images, surface steps one atom high can be clearly seen. The diffraction patterns are essentially the same as RHEED patterns except that there is better control of the illuminating conditions and camera length. Diffraction patterns can be obtained from small areas of the specimen, correlated with specific areas in the image. Similar results can be obtained at lower resolution in scanning electron microscopes. This field has been slower to develop than normal transmission microscopy but, with the development of UHV microscopes, is now advancing rapidly.

1. Introduction

It has long been known that the presence of steps on surfaces is critical in determining surface properties. For example, catalytic function and the growth of epitaxial overlayers are both very strongly dependent on the disposition of steps on the surface. Most studies in these and related fields have characterized the surfaces used by LEED, X-ray diffraction or conventional RHEED. Each of these techniques provides only statistical information over a large area of surface (on the order of a mm).

There are now several techniques that permit local imaging of step configurations. This ability to see the detailed structure of surfaces, opens up the possibility of exploring mechanisms in surface science with a detail that was not previously possible. Among these techniques are:

a) Low energy electron microscopy. This technique, associated with the name of E. Bauer, uses low energy electrons incident at normal incidence to the surface [1]. The outgoing electrons (also near normal incidence) are used to form LEED patterns or images. The difference between this system and a normal LEED system is the presence of a complex lens system that permits the formation of high resolution (~ 15nm) images. The images reveal, with excellent contrast, such features as surface steps, regions of reconstructed surfaces, decorated steps, the structure of overgrowths etc. [1,2,3].

b) Scanning Tunneling Microscopy. The STM has shown spectacular images of surfaces in its few years of existence. It has been so widely reported that no further comments will be made here.

c) Reflection Electron Microscopy. This technique, which can be based on either the transmission electron microscope [4,5,6] or the scanning electron microscope [7,8] is the subject of this paper. The images and diffraction patterns are formed by the elastic scattering of electrons incident on the surface at near-grazing incidence. The diffraction configuration is akin to traditional RHEED but the patterns are obtained from much smaller areas. Furthermore the surface can be imaged at high resolution (1-20 nm) by scanning the beam or by the use of lenses.

2. Reflection Diffraction

Reflection electron diffraction patterns (RHEED) are complex and not fully understood. There are three regimes that can be identified. If the surface is rough, then the diffraction patterns correspond to transmission of the electrons through asperities on the surface. The patterns are characterized by the formation of a two dimensional net of reflections just as in transmission patterns. The diffraction is essentially three dimensional. The positions of the reflections can be explained on the basis of a three-dimensional reciprocal lattice and an Ewald sphere construction.

The diffraction pattern from an atomically smooth surface has a geometry that is quite different. Because the sample is two dimensional, the reciprocal lattice used with the Ewald sphere construction can be thought of as a set of rods. The rods cut the Ewald sphere on a circle and the reflections in the diffraction pattern lie on arcs of circles. In this case, a convergent incident beam can be used to produce reflection convergent-beam diffraction patterns (figure 1) that can be used to characterize the surface.

If the surface roughness lies between the two cases described above, the diffraction pattern consists of a set of streaks perpendicular to the surface. For the case of a rough surface and of a smooth surface the mechanism of diffraction is fairly well understood. However, the detailed mechanism that gives rise to streaking as the surface passes from rough to smooth is currently unknown.

The most likely explanation is that the streaking arises from the scattering at surface steps. This would be consistent with the observation of RHEED oscillations during crystal growth. The technique of RHEED oscillations is widely used for the *in situ* characterization of MBE growth. If the intensity at a particular position along the specular diffraction streak is monitored, that intensity can show an oscillation with a period corresponding to the addition of one atom layer to the specimen. This occurs only for certain growth regimes, for example when there is the formation of many additional steps on the surface as the overlayer nucleates and then a reduction in the number of steps as the layer fills up.

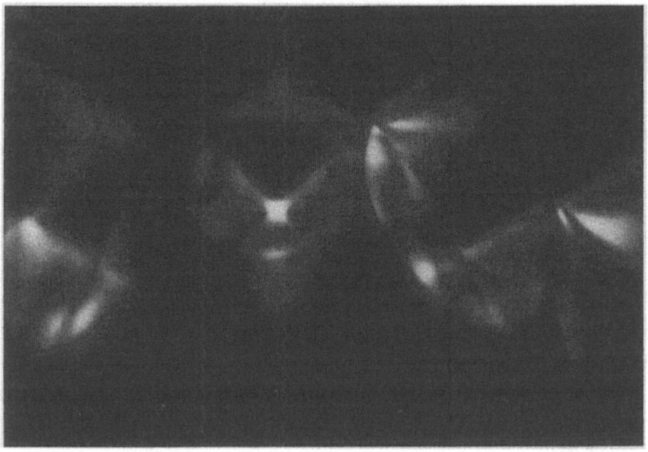

Figure 1. Reflection convergent-beam diffraction pattern from a (111) facet of an annealed gold sphere. 120 kV.

3. Reflection Microscopy

REM is concerned with forming images of surfaces that are close to atomically smooth. The images are formed using the electrons of the specular beam (or, more rarely, one of the other diffracted beams). Under these circumstances the sample consists of regions of the perfect surface, which show uniform intensity in the image, with occasional single-atomic-height steps which appear in the image as dark or bright lines. Thus the positions of the steps are clearly seen. The images reveal whether the steps are straight or curved, irregularly or equally spaced and so on. See, for example, figure 2.

Another valuable use of the technique is the imaging of reconstructed surfaces and overlayers. If the surface undergoes a reconstruction a new periodicity is developed and this gives rise to additional reflections in the diffraction pattern. An image formed in such a reflection reveals those parts of the surface that are reconstructed and those that are not.

Images have been formed using these techniques for many years in transmission electron microscopes and more recently, in scanning electron microscopes.

Reflection imaging, especially when linked with small area diffraction, is thus a very powerful technique for the characterization of surfaces. However, it has not yet had very much impact in surface science. This is because, until recently, the observations were all made in poor vacuum (by surface science standards) and with very little scope for *in situ* observation of the processes of interest. There is very little space in the specimen area of a transmission electron microscope. This situation is now changing. There are several UHV TEM's now in operation and there will soon be more. The scanning electron microscopes with field emission guns that have been

Figure 2. Reflection electron-microscope image of sapphire (α - Al_2O_3). The image shows atomic-height steps on the basal plane (0001), 100kV. Photograph courtesy of Yootaek Kim (University of Utah - see also Electron Microscopy 1990 Proceedings of the XII International Congress for Electron Microscopy Vol 1, p 328-329).

specifically designed for work of this kind [7,8] provide both a UHV environment and space for incorporating evaporation systems (for example) into the specimen chamber.

Acknowledgements

This work was supported through the University of Illinois Materials Research Laboratory by the Department of Energy contract DE-AC02-76ER01198.

Bibliography and References

There is an excellent compendium of articles on this subject in: "Reflection High-Energy Electron Diffraction and Reflection Electron Imaging of Surfaces" Edited by P.K. Larsen and P.J. Dobson (Plenum, New York 1988). For convenience this volume is referenced here as "Larsen and Dobson".

Papers of a more tutorial nature are given in: "Surface and Interface Characterization by Electron Optical Methods" Edited by A. Howie and U. Valdré (Plenum, New York 1988).

There are many papers in "Electron Microscopy 1990." The Proceedings of the XII International Congress for Electron Microscopy held in Seattle, USA (San Francisco Press, San Francisco 1990), especially in volumes 1, 2, and 4. These provide an up-to-the minute view of the field.

For a good review of the whole field of surface imaging see [6].

[1] W. Telieps and E. Bauer, Ultramicroscopy **17**, 57-66, 1985.
[2] E. Bauer and W. Telieps, Larsen and Dobson 381-384.
[3] E. Bauer, M. Mundschau, W. Swiech and W. Telieps, Ultramicroscopy **31**, 49-57 (1989).
[4] J.M. Cowley, Larsen and Dobson, 261-284.
[5] K. Yagi, S. Ogawa and Y. Tarishiro, Larsen and Dobson, 285-301.
[6] K. Yagi in "High-Resolution Transmission Electron Microscopy and Associated Techniques", Eds. P. Buseck, J. Cowley and L. Eyring (Oxford University Press, Oxford, 1988) 568-606.
[7] T. Ichinokawa, Larsen and Dobson, 385-394.
[8] M. Ichikawa and T. Doi, Larsen and Dobson, 343-369.

Studies of Chemisorption with the Scanning Tunneling Microscope

M. Salmeron

Center for Advanced Materials, Materials Science Division,
Lawrence Berkeley Laboratory, 1 Cyclotron Road,
Berkeley, CA 94720, USA

Abstract. In this paper we review recent advances in the study of the structure of chemisorbed layers on transition metal surfaces in real space using the Scanning Tunneling Microscope (STM). The examples include Carbon on Pt(111) and Sulfur on Mo(001) and on Re(0001). The new phenomena that were unraveled by the STM and other surface science techniques include the passivation of the surfaces by a single layer of adsorbates that prevented oxidation and contamination in air. In Ultra High Vacuum, the STM revealed the multiple structures formed by S on Re as a function of coverage. It revealed also that at local coverages above 0.25 monolayers, isolated S adatoms are not stable and coalesce into structures formed by trimers and other aggregates.

1. Introduction

The scanning tunneling microscope (STM) has made it possible to study in real space the atomic structure of surfaces of metals and semiconductors [1]. One area that is particularly relevant to epitaxy and catalysis is that of the crystallographic structure of surface layers. Until recently, this has been studied mostly with low energy electron diffraction (LEED) [2].

LEED, as all diffraction techniques, is sensitive to the ordered areas of the surface while the defect structures (steps, domain boundaries, point defects, etc.) are not easily detected. However, these defect structures are thought to play a major role both in crystal growth and in catalysis [3].

In addition to providing local atomic structure information, the STM is capable of operating in a wide range of environments, including ultra high vacuum, liquids, and atmospheric.

In this paper we will show how the STM has been applied in the author's laboratory to unravel the structure of carbon layers on platinum and sulphur layers on molybdenum and rhenium single crystals. One outcome of these studies is the observation of multiple defect structures that could not be observed by LEED before and also of new phenomena like the coalescence of adsorbate atoms into clusters in disordered and ordered arrangements. While the ordered structures are in principle resolvable by diffraction methods, the size of the new unit cell is often too large for present day computational programs.

2. Principle of STM Operation

The STM is based on the measurement and control of the tunneling current between a sharp metal tip (W and Pt-Rh alloys in these studies) and a conductive surface. The

Springer Proceedings in Physics, Volume 62
Surface Science Eds.: F.A. Ponce and M. Cardona

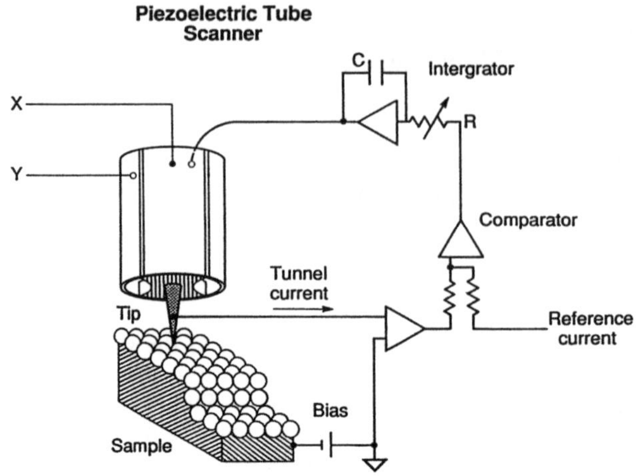

Fig. 1 Schematic diagram of the STM experiment. A piezoelectric tube supports a sharp tip and displaces it over the surface at a distance of a few Angstroms. The tunnel current is measured and maintained constant during scanning by means of an electronic feedback. The images are generated by plotting the xy and z displacements of the tip.

electrons "tunnel" through the potential barrier between tip and surface. The height of this barrier is derived from the work function of the two materials. The tunnel current therefore depends exponentially on both the tip-surface distance (z) and the barrier height Ø through:

$$I \propto e^{-\sqrt{\text{Ø}} . z}$$

The exponential dependence of the tunneling current I on the distance z is the basis for the high spatial resolution of the STM in the direction perpendicular to the surface z. This resolution is limited by noise to a typical value of 0.1 Å.

The sharpness of the tip determines the "in-plane" resolution. This is provided by natural asperities at the apex of tips that often terminate in one or a few atoms. Because of the exponential decay of I with z, the atomic size asperity that is closest to the surface will carry most of the tunnel current I. The fine displacements of the tip over the surface are controlled by a piezoelectric transducer that is capable of fine displacements of the order of 10 Å per applied volt. A schematic of the experimental set up is shown in figure 1.

Because of the nature of this paper, we will not describe many important experimental details, including isolation from mechanical and other electric perturbations, approach of tip to sample from macroscopic (1 cm typically) to microscopic distances (5 to 10 Å), and others. The interested reader might consult several reviews on those subjects [4,5].

The images are formed by a three-dimensional plot of x, y, and z coordinates obtained by scanning the tip over the surface while maintaining the tunnel current I constant.

3. Carbon on Pt(111)

The first system that we present is the monolayer formed by C on Pt(111). This system is important in hydrocarbon catalysis since it is the by-product from many reactions after high temperature (larger than 300° C) treatments. In the present experiments, the carbon overlayer was obtained by heating in ultra high vacuum (UHV) a hydrocarbon contaminated Pt(111) crystal. After heating, the sample was analyzed by Auger Electron Spectroscopy (AES) and LEED. Our AES calibration indicated that carbon coverage θ was 1 monolayer on the average and LEED indicated the presence of rings in the diffraction pattern characteristic of graphite [4]. Then the sample was placed in the STM located in the same UHV chamber and imaged. Large terraces and multi- atom height steps were observed over the whole surface. At the atomic scale, the images clearly revealed the graphitic nature of the carbon layers as shown in figure 2. The higher density of valence electronic states in the regions between carbon atoms is clearly observed as elevations in the current image. In this image we observe also that the distribution of tunneling current intensity is not six-fold symmetric, as expected from the honeycomb lattice of a single graphite layer, but rather three-fold symmetric. Some authors have proposed that such distribution, that is also observed in graphite crystals, is due to an electronic interaction between first and second carbon layers. Our observation of three-fold symmetry can then be due to a local coverage of carbon of two monolayers in spite of the AES observation of one monolayer average coverage over mm^2 areas.

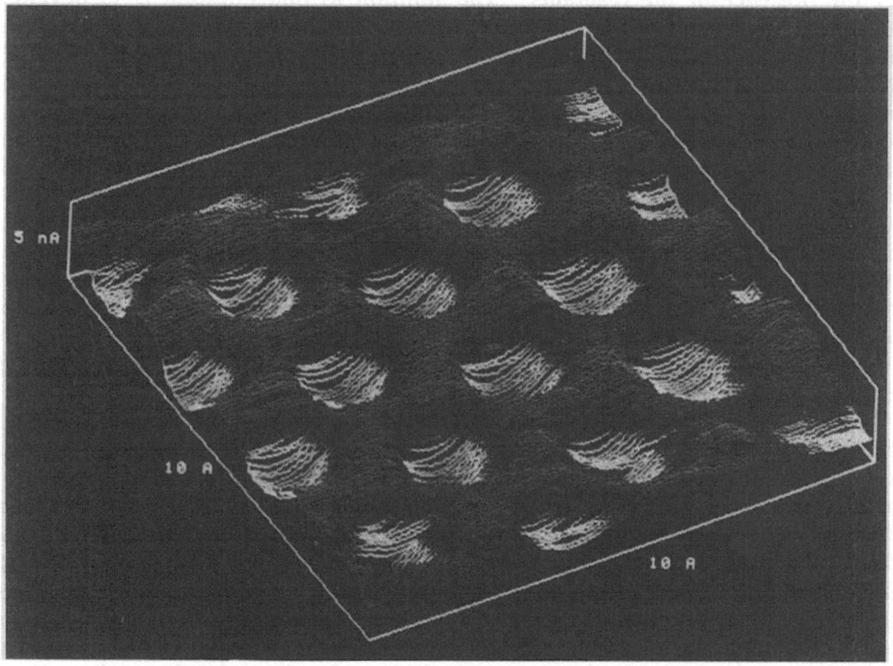

Fig.2 Current image of a 10x10 Å area of a Pt(111) surface that is covered by a monolayer of graphitized carbon. The higher tunnel current is found over the C-C bonds.

The results are revealing both of the power of STM as a local probe and also of its limitations. The very small area examined (a few tens of Angstroms with atomic resolution) might not be representative of the average surface structure. Further experiments on this system are needed to insure that the area imaged is one carbon layer thick. One way to ascertain this is by finding a region consisting of an island of graphite where the substrate outside it is also imaged. One additional observation on this system was the stability of the C structure to air exposure. The images obtained immediately after extracting the crystal from the UHV chamber revealed the graphitic structure still present in air.

4. Sulphur on Mo(100)

Sulphur overlayers on molybdenum single crystals have been studied in the past by LEED [6,7]. This system is interesting because it was shown that its chemical activity in hydrodesulphurization (HDS) reactions parallels that of the industrial catalyst based on promoted MoS_2. In many basic surface science studies of model catalysts, a host of surface science techniques are used to characterize the state of the catalyst surface before (in UHV) and after the reaction. Since this reaction takes place in an environmental cell, no characterization is done during the reaction at high pressure (≥ 1 torr) and high temperature ($> 200^{\circ}$ C). One of the main open questions in catalysis science is how the structure of the catalyst surface is modified by the presence of the reactants. The solution to that question is to develop techniques that can reveal the atomic structure of the catalyst in high pressure environments. The STM is one possible answer. With this aim in mind, a special version of this instrument is being developed in the author's laboratory that will operate at high reactant pressure and at high temperatures of up to 200° C. As a test to investigate whether the STM can resolve the atomic structure of chemisorbed layers at high pressure we have studied the structure of a passivating monolayer of sulphur on Mo(001) [8]. This layer was prepared in UHV and characterized by LEED and Auger. Then it was studied by STM after transfer of the crystal to atmospheric environments. We found that S protects the surface from oxidation in air as checked by LEED and AES after reinsertion of the crystal in the UHV chamber. This passivating layer corresponds to the saturation coverage (1 S atom per surface Mo atom). Its structure has a periodicity of (1x2) which means that the new surface lattice has 1 times and 2 times the dimensions of the clean Mo(100) surface along each unit cell vector.

The STM images obtained revealed the atomic structure of this overlayer as shown in figure 3. The unit cell is indicated by the rectangle. As we can see, the sulphur atomic arrangement is a compact pseudohexagonal lattice with sulphur-sulphur distances of 3.1 Å. This is the short side of the rectangular cell. This distance is comparable to the 3.16 Å found in the basal planes of molybdenum disulfide (MoS_2). The compactness of this pseudohexagonal layer and the strong binding of S to Mo are responsible for the passivating properties that S imparts to the surface towards contamination and oxidation. The oxide is thermodynamically more stable than the sulphide. However, to initiate the oxidation of the surface, a pair of exposed nearest Mo sites are required where the O_2 molecule can dissociate. These sites are not available on the perfect S overlayer structure. The oxide formation is therefore retarded by a kinetic barrier. After prolonged exposure to the atmosphere, eventually the oxide nucleates at defects and expands to the whole

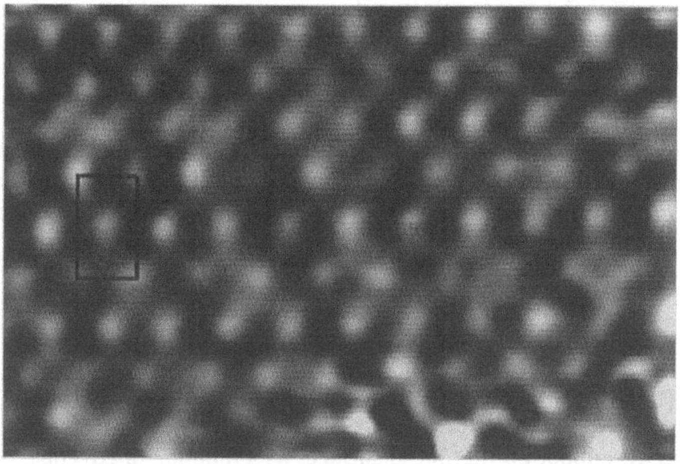

Fig.3 Image of a Mo(001) surface that is covered by a monolayer of Sulfur. The unit cell of the S structure is marked by the rectangle and contains two S atoms. The S-S distance is 3.1 Å. The image is taken in air and the Mo surface is protected from oxidation by the saturation monolayer of S.

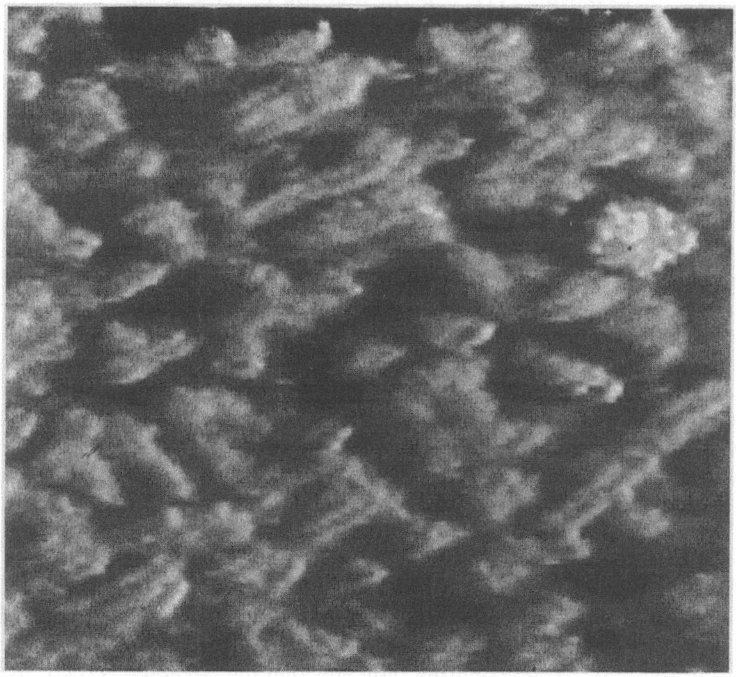

Fig.4 1000x1000 Å image of a surface of Mo(001) that has oxidized after exposure to air for three days. The S overlayer has been displaced. The oxide crystallites are oriented along the [001] and [010] directions. Corrugation is approximately 20 Å.

surface, displacing the S. The time scale for this process is variable since it is controlled by the number of defects. This depends in turn on the preparation conditions. One example of surface oxide formed after three days of exposure to air is shown in figure 4. The oxide crystallites grow as elongated rectangles along the two principal crystallographic directions and have an average size of 20 x 60 Å.

These results are important because they showed for the first time that high pressure environments in this case 1 atmosphere of O_2 and N_2 and other gases found in air, do not prevent the operation of the STM and chemisorbed layers can still be imaged with atomic resolution.

Further details on the S structure found with the STM are described elsewhere [8,9].

5. Sulfur on Re(0001)

The structures formed by sulphur on Re(0001) surfaces as a function of coverage θ, have been studied by LEED and AES [10]. Recently [10], STM was applied to study the structure of the saturation coverage layer at θ = 0.50. At saturation S imparts to the Re(0001) surface similar passivating properties as in the previous example of Mo(001). To study the structures formed at lower coverage, the experiments have to be performed in UHV. We have recently completed studies of the structures formed from around θ = 0.25 up to θ = 0.50 [12,13]. The outcome of these studies constitute a dramatic demonstration of how STM helps solve structures with large unit cells. The capability to study complicated structures is not only an incremental refinement of crystallographic techniques, but it allows to uncover new phenomena, like new phases formed by adsorbate aggregation. We describe now briefly these findings. Up to a coverage of 0.25, S-S repulsive pairwise interactions dominates the structure of the adsorbed layers. The two structures observed, the c(√3 x 5)Rect and the (2x2) conform to the rule of maximum spreading of the S adatoms. As the coverage increases, S-S distances would be maximized and energy minimized by forming structures like the (√3x√3)R30 when θ = 0.33, as observed on many other single crystal surfaces [14,15]. Instead, on Re(0001) a new phase forms in which S coalesces into trimers, first near domain boundaries and then at all regions as shown in figure 5. In these trimers, each sulphur atom sits on the same three-fold hollow site as in the low coverage monomer structures. The fact that dimers do not form except as occasional defects indicates that three-body forces are determinant at this stage. Three-body forces were previously introduced as necessary corrections to the dominant pairwise interactions in Monte Carlo simulations of gas lattice dynamics in order to fit the experimental data [16]. Here, these forces are dominant and cause the local coverage to change abruptly from 0.25 to near 0.45 (the value for the ordered (3√3x3√3)R30° structure described below). Initially, the trimers are identical and have their center on a three-fold hollow site. As the coverage continues to increase, trimers of a different structure form that are rotated 60° relative to the first ones and have their center on a top site. At maximum trimer density an ordered (3√3x3√3)R30° structure forms that is composed of trimers of the two types in a 3 to 1 ratio, as shown in figure 6. Atom counting indicates that the coverage is 0.45. By adding more sulphur to the surface, the trimers incorporate a fourth sulphur atom to form tetramers in a transformation requiring little surface rearrangement. The image of figure 7 shows the new structure formed at completion, a (3,1;1,3) in LEED matrix notation. Notice the asymmetry of intensity of tunnel current at the sulphur

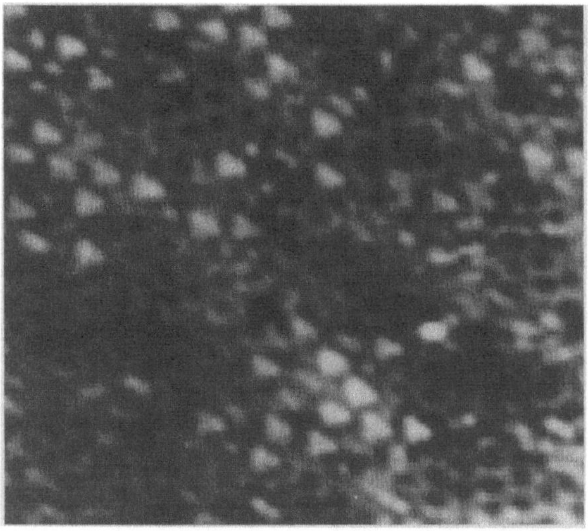

Fig.5 A 100x100 Å current image of a Re(0001) surface covered by S with a coverage between 0.25 and 0.3. The structure is mostly (2x2) sprinkled with numerous trimers. Trimers form preferentially near domain boundaries and appear brighter than the (2x2) region. Notice also that all trimers have the same orientation.

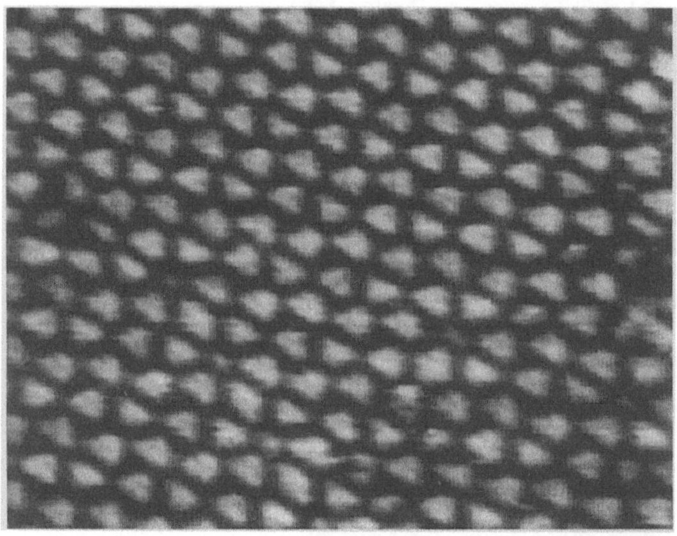

Fig.6 Topographic 100x100 Å image of a Re(0001) surface saturated with trimers. An ordered $(3\sqrt{3}\times3\sqrt{3})R30^{\circ}$ structure is formed. Notice that 1/4 of the trimers have a different orientation. Some point defects are also observed. Vertical gray scale = 2 Å

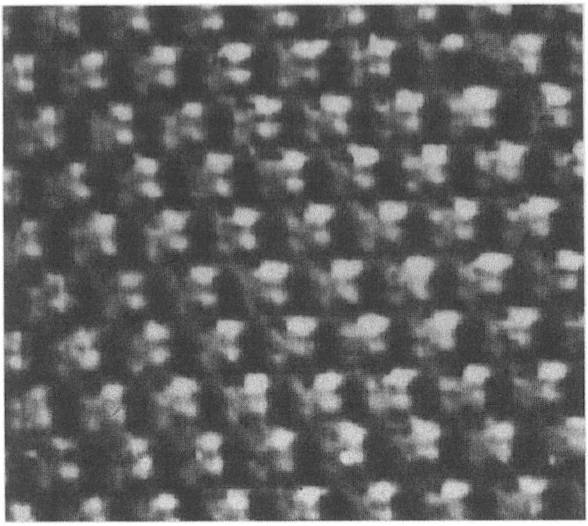

Fig.7 68x68 Å Current image showing the tetramers of S formed at coverage above 0.45. Notice the asymmetry of the intensity of the S atoms that reflect the symmetry of the unit cell with only one mirror plane across the long diagonal in the diamond shaped tetramers.

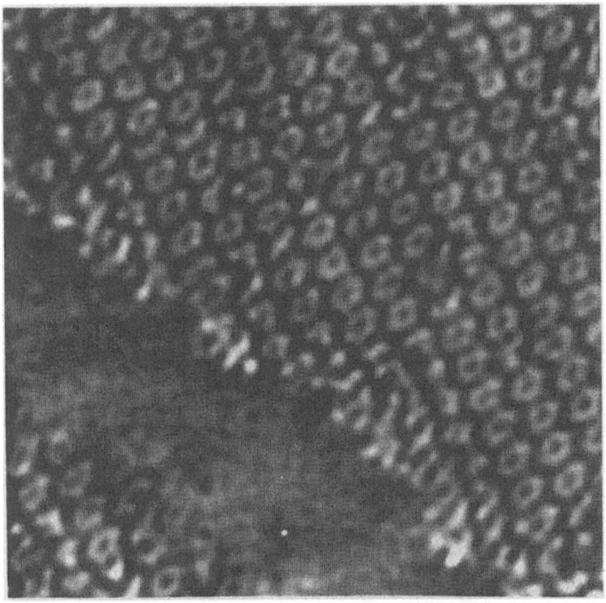

Fig.8 Topographic image of the $(2\sqrt{3}x2\sqrt{3})R30°$ structure showing the aggregation of S into hexagonal units. The distance between centers of adjacent hexagons is approximately 9 Å. A step separates two terraces and produces the blurred region of the image. Notice also numerous defects and domain boundaries.

atoms in the tetramers. It reflects the reduced substrate symmetry with only one mirror plane. The coverage, determined by counting atoms, is 0.5. This observation came also as a surprise. The highest coverage structure (and saturation) was thought to be the one obtained by further dosing, with a structure of$(2\sqrt{3}x2\sqrt{3})R30^{\circ}$, where S forms hexagonal units as shown in figure 8. The coverage here is again 0.5. To form this last structure, a higher exposure to H_2S or S_2 is required. By heating this saturation structure to increasingly higher temperatures, all the lower coverage ones can be generated, including the tetramers that coexist with the trimers and defects. The $(2\sqrt{3}x2\sqrt{3})R30^{\circ}$ structure might be energetically more stable than the (3,1;1,3), but it requires a substantial rearrangement of sulphur atoms in order to break the trimer unit. Thus the tetramers structure is kinetically stabilized against the more stable hexagonal ring structure.

6. Conclusions

The examples presented in the previous sections are representative of the qualitative and quantitative change in our understanding of surface structure and phenomena brought about by the application of the STM. We believe that the application of STM and probably also AFM, the Atomic Force Microscope not discussed in this paper, to studies of catalyst surfaces under realistic conditions of pressure and temperature, will bring similar or perhaps even more dramatic changes to our understanding of catalyst structure and operation.

Acknowledgments

This work was supported by the Director, Office of Energy Research, Office of Basic Energy Sciences, Material Sciences Division, U.S. Department of Energy under contract No. DE-AC03-76SF00098.

REFERENCES

[1] G. Binnig, H. Rohrer, Ch. Gerber and E. Weibel. Phys. Rev. Lett. 49, 57 (1982).

[2] M.A. Van Hove and S.Y. Tong, *Surface Crystallography by LEED.* Springer Series in Chemical Physics 2. Berlin. 1979.

[3] G.A. Somorjai. *Chemistry in Two Dimensions: Surfaces.* Cornell University Press. Ithaca. New York. 1981.

[4] M. Salmeron. *Emerging Techniques for Catalyst Characterization.* Chapter 5. Edited by Catalytica. Mountain View. CA. 1989.

[5] Y. Kuk and P.J. Silverman, Rev. of Sci. Instr. 60, 165 (1986).

[6] M. Salmeron, G.A. Somorjai and R.R. Chianelli, Surf. Sci. 127, 526 (1983).

[7] A. Gellman, W.T. Tysoe, F. Zaera and G.A. Somorjai, Surf. Sci. 191, 271 (1987).

[8] B. Marchon, P. Bernhardt, M.E. Bussell, G.A. Somorjai, M. Salmeron and W. Siekhaus, Phys. Rev. Lett. 60, 1166 (1988).

[9] B. Marchon, D.F. Ogletree, M.E. Bussell, G.A. Somorjai, M. Salmeron and W. Siekhaus, J. of Microsc. 152, 427 (1988).

[10] D.G. Kelly, A.J. Gellman, M. Salmeron, G.A. Somorjai, V. Maurice and J. Oudar, Surf. Sci. 204, 1 (1988).

[11] D.F. Ogletree, C. Ocal, B. Marchon, G.A. Somorjai and M. Salmeron, J. Vac. Sci. Technol. A8, 297 (1990).

[12] R.Q. Hwang, D.M. Zeglinski, A. López Vázquez-de-Parga, C. Ocal, D.F. Ogletree, G.A. Somorjai, M. Salmeron and D.R. Denley, Phys. Rev. Lett. (submitted).

[13] D.F. Ogletree, R.Q. Hwang, D.M. Zeglinski, A. López Vázquez-de-Parga, G.A. Somorjai and M. Salmeron. J. Vac. Sci. Technol. A, (in press).

[14] H. Ohtani, C.-T. Kao, M.A. Van Hove and G.A. Somorjai, Progress in Surface Science, 23, 155 (1986).

[15] M.A. Van Hove, S.W. Wang, D.F. Ogletree and G.A. Somorjai, Advances in Quantum Chemistry, 20, 1 (1989).

[16] T.L. Einstein, Surf. Sci., L497 (1979).

Automation and Control of a Commercial Scanning Tunneling Microscope

J. Valenzuela and J. Rodríguez

Instituto de Física, Universidad Nacional Autónoma de México, Laboratorio de Ensenada, Apdo. Postal 2681, 22800 Ensenada, Baja California, Mexico

We present processed images obtained with a commercial Scanning Tunneling Microscope (STM) using a personal computer and a data acquisition card. With the PC-acquisition card scheme, we are able to digitize and process images for display on a high resolution monitor as well as obtain and store scanning parameters. Modification of the microscope control unit was also necessary in order to incorporate the lock-in technique for imaging. Images of graphite and gold scanned in air are presented.

INTRODUCTION

The Scanning Tunneling Microscope (STM) has developed very quickly into an important tool for surface science since its invention by Binnig and Rohrer [1]. Topographical and spectroscopic information can be obtained by scanning a fine tip over a surface and monitoring a tunneling current or a z displacement while a feedback network maintains a constant average current (in a constant current mode), achieving atomic resolution on metal [2], semiconductor and superconductor materials in air, liquid or UHV enviroments [3], at room temperature or down to a few degrees Kelvin [4].

Given this versatility of the STM, computer automation for image and data acquisition in real time has become an important part in tunneling microscopy. Many of the reports in STM computer automation deal with home built microscopes [5,6], with D/A converters for generating scanning, z-piezo element, offset and bias voltages, and A/D converters for digitizing tunneling current and feedback voltage. This arrangement makes the microscope a computer dependent instrument in the imaging of samples, which has its advantages and disadvantages.

DESCRIPTION

We used a commercial STM, the Nanoscope I [7] which is operated in air, for our automation and control. This instrument is one of the first microscopes that came out on the market, with its operation based entirely on mannual control knobs and provided with some output and input signals for the user.

A general view of our arrangement is shown in Figure 1. The main task of the acquisition card is to take data from the STM electronics for the computer to store and manipulate, like scanning parameters and current or z-piezo voltage that make up a complete image, that can be sent to a larger computer for porcessing.

The acquisition of data and control of the instrument was made through a data acquisition card, Data Translation Mod. 2801-A [8]. This card has a 12 bit resolution A/D converter with 27,500 samples per second conversion rate and 16 multiplexed single-ended channels (or 8 differential input channels), two D/A converters with a $\pm$10V range and two digital I/O ports. All of the software for accessing the acquisition card was written in Pascal language.

Springer Proceedings in Physics, Volume 62
Surface Science Eds.: F.A. Ponce and M. Cardona

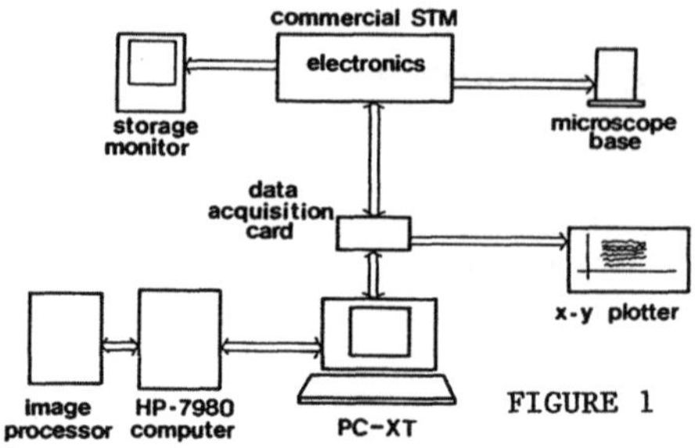

FIGURE 1

FIGURE 2

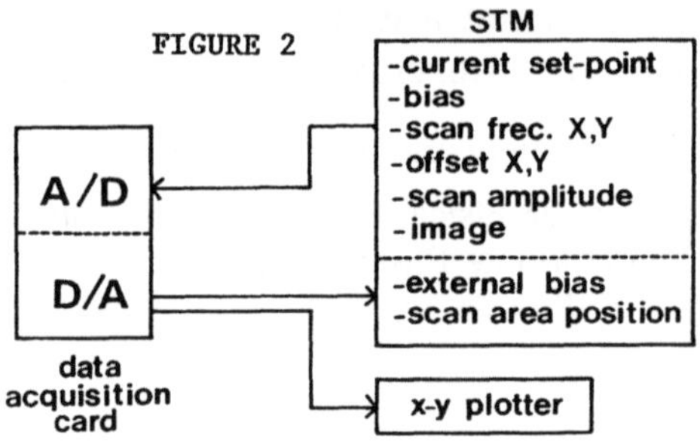

FIGURE 1. General view of the STM-PC system.

FIGURE 2. Distribution of signals between the STM, X-Y plotter and the data acquisition card.

The distribution of electronic signals from the STM to the A/D and D/A - converters is shown in Fig.2. An image is formed during scanning in a storage monitor which is a standard output for STM images, and can be digitized (128x128 points typical) using one of the channels of the A/D converter with Direct Memory Access (DMA) [9] mode and stored on disk for processing. A TTL pulse generated by the STM electronics indicates the beginning of a scan that triggers the A/D conversions.

Actually very little control of the microscope is done with the computer, namely the positioning of scanning area and external bias used for spectroscopic measurements. We believe that having manual control of all the scanning parameters instead of a complete computer control instrument can save time in getting a good image, particularly in noisy samples, where the operator has to make use of tricks known to improve an image, like rapidly increasing and decreasing the bias voltage and tunneling current, varying the scanning frequency in the X direction that is best suited for a good image and also changing the position of the scanning area. All of the changes can be relatively slow to perform in a computer controlled STM, even with a mouse driven menu.

A simple program was written for controlling an analog X-Y plotter through the D/A outputs (Figure 2) for drawing the images in a line scanned mode.

Serial communication is used for sending image files to a HP-7980 computer where they can be filtered using FFT and displayed on a high resolution monitor. We have found it necessary to adopt a standard format for image files: all (x,y) image points (i.e., the tunneling current or z-piezo voltage) are normalized to 127, this simplifies the software that uses these files, like the X-Y plotter and the image processor.

When an image is digitized a second file is generated that contains all the scanning parameters associated with that image, like tip speed, tip-sample bias voltage, size of scanning area and tunneling current set point (used in constant current mode).

For the lock-in technique we used the Stanford Research Systems Mod.SR510 lock-in amplifier. In this scheme, a small a.c. modulation is injected on the bias voltage or in the z-piezo element with a frecuency higher than the cutoff point of the feedback network, thus modulating the tunneling current with a better signal-to-noise ratio.

Figures 3 and 4 show filtered images of graphite and of Au(100) respectively obtained with our system.

Our future plans for upgrading our system include a more powerful PC computer for in-situ image processing and viewing.

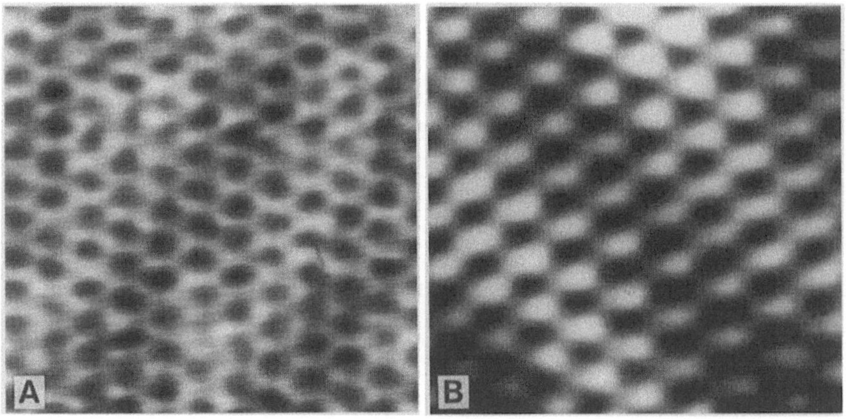

FIGURE 3. A digitized image of a graphite surface, a) raw data; b) filtered image.

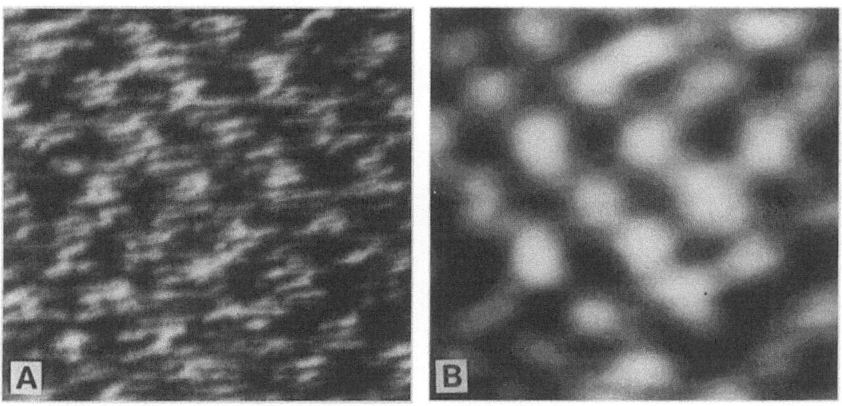

FIGURE 4. Image of a gold crystal showing the (100) face. a) raw data; b) filtered. Atomic spacing is about 4.0Å.

ACKNOWLEDGEMENTS

We would like to thank Mr. G. Vilchis for the photographic work and the support given by the Instituto de Física, Laboratorio de Ensenada, U.N.A.M. and the Centro de Investigación Científica y Educación Superior de Ensenada (C.I.C.E.S.E.). The help given by G. Soto is also appreciated.

REFERENCES AND NOTES

1.- G. Binnig, H. Rohrer. Helv.Phys.Acta 55, 726 (1982); G. Binnig, H.Rohrer, Ch. Gerber, and E. Weibel. Phys. Rev. Lett. 49, 57 (1982).

2.- V.M. Hallmark, S. Chiang, J.F. Rabolt, J.D. Swalen, and R.J. Wilson. Phys. Rev. Lett. 59, 25 (1987).

3.- P.K. Hansma and J. Tersoff. J. Appl. Phys. 61, 2 (1987).

4.- Ch. Renner, Ph. Niedermann, A.D. Kent, and Ø. Fischer. J. Vac. Sci. Technol. A. 8, 1, (1990).

5.- S. Grafström, J. Kowalski, R. Neumann, O. Probst, and M. Wörtge. J. Vac. Sci. Technol. A. 8, 1, (1990).

6.- P.H. Schroer, and J. Becker. IBM J. Res. Develop. 30, 5, (1986).

7.- Digital Instruments, Inc. Santa Barbara, Calif. 93117.

8.- Data Translation, Inc. Marlborough, Mass. 01752-1192.

9.- Operation in which data is transfer to or from a computer's memory without passing through the CPU, speeding up the computer's overall performance.

Part IV

Interaction Between Radiation and Surfaces

Electron and Photon Stark Ladders in Finite Solids

F. Claro

Facultad de Física, Pontificia Universidad Católica de Chile,
Casilla 6177, Santiago 22, Chile

Abstract. Stark ladders in finite model systems are discussed. Ladders are well known to occur in the energy spectrum of crystalline electrons in an electric field if use of the tight binding approximation is justified. We address a case where this approximation fails and discuss periodic structures that arise in the density of states, and localization of eigenstates. For the propagation of electromagnetic waves in a layered dielectric with the property $\epsilon(x+a) = \epsilon(x) + \tau$ we discuss periodic Stark-like structures that arise in the transmission coefficient.

I. Introduction

The concept of a Stark ladder in solids was introduced by G. H. Wannier some thirty years ago.[1] He proposed that such ladders are likely to exist as metastable states or as a periodic structure in the energy spectrum of an electron in the presence of both a periodic potential and a uniform external electric field.[2] An argument for the latter statement is the following. Consider an electron in the crystal potential $V(x) = V(x+a)$, where a is the lattice period, and an external uniform field E. If $\psi_\varepsilon(x)$ is an eigenstate of energy ε then it must obey

$$\left(\frac{p^2}{2m} + V(x) + eEx\right)\psi_\varepsilon(x) = \varepsilon\psi_\varepsilon(x). \tag{1}$$

A space translation in the lattice constant a leaves this equation in the form

$$\left(\frac{p^2}{2m} + V(x) + eEx\right)\psi_\varepsilon(x+a) = (\varepsilon - eEa)\psi_\varepsilon(x+a), \tag{2}$$

where the periodic property of the lattice potential has been used. This equation says that if ε belongs to the energy spectrum, then $\varepsilon - eEa$ also belongs to the spectrum, with the new eigenfunction $\psi_{\varepsilon - eEa}(x) = \psi_\varepsilon(x+a)$. Thus, any structure present in the density of states within an energy interval $\Delta = eEa$, the so called Stark period, is repeated periodically in the spectrum.

The above argument holds when the crystal is infinite. A long controversy, not yet resolved, has resulted over whether metastable states exist in such case.[2–12] A definite answer has been found if a one band model is adopted[13–15], in which case the electric field splits the band in a ladder of bounded eigenstates and the density of states is a sequence of equally spaced delta functions with the period Δ. This exact result holds even in the presence of a magnetic field if the electric field is sufficiently large.[16] What limits the validity of this model is the possibility of Zener tunneling between bands.

If the crystal is finite the argument in support of the existence of a ladder and based on Eqs. (1) and (2) is not rigorous since edge effects impair the discrete translational invariance of the lattice potential. Still, studies of this case show that under restricted

Springer Proceedings in Physics, Volume 62
Surface Science Eds.: F.A. Ponce and M. Cardona

conditions the ladder does exist.[5,11,17] Furthermore, recent experiments in supperlattices in an external electric field indeed show a ladder structure in the density of states.[18,19] In Section II we shall illustrate the finite case with a simple model and show explicitely the structure that develops in the energy spectrum for such model. As explained below, we choose a potential profile that permits addressing the question of the existence of the ladder when the tight binding approximation may not be used.

The question of the existence of a Stark ladder may be viewed more generally as characteristic of the spectrum of differential operators containing a linear term, of which Eq. (1) is just an example.[6] In particular, the wave equation for propagation of light in layered media may be discussed in the same context.[20] Section III treats such case, and we show that a structure with a characteristic Stark period may be obtained in the transmission coefficient as a function of angle of incidence of the light beam.

II. Electrons in a Quantum Pool

Consider an electron moving in the following potential

$$V(x) = \begin{cases} V_0, & x < -L \\ n\Delta, & na < x < (n+1)a < Na \\ V_0, & x > Na \end{cases} \tag{3}$$

where n=0,1,..., and $N = V_0/\Delta$. We call this model a quantum pool on account of the stepladder portion it contains. Its profile is shown in Fig. (1) for $N = 10$. The stepladder may be obtained as the sum of a periodic sawtooth potential of period a and a linear term $\Delta x/a$. The simple model (3) thus contains a finite region where the essential features of a periodic plus a linear term in the potential occurs, and provides confinement for energies $\varepsilon < V_0$. Also, since in each step the potential is flat, it is not suited for a tight binding treatment and therefore enhances many band effects which are of special interest here since the tight binding case has already been solved.[13-15] The model is also of special interest since it may be realized experimentally as an effective potential for electrons in the conduction band of an heterostructure with stepwise doping.[21].

A quantum mechanical treatment for the dynamics of an electron in potential (3) is straightforward since only plane waves are needed to construct an eigenstate in the classicaly allowed region $-L < x < Na$. We prefer to discuss eigenstates within the semiclassical approximation, however. Consider the classical orbits for energies in the interval $0 < \varepsilon < V_0$. Applying the Bohr-Sommersfeld quantization rule to such orbits one obtains for the allowed energies the equation

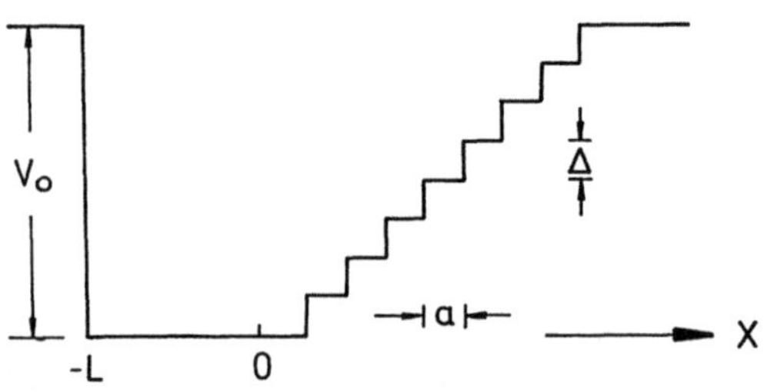

Fig.1 Potential profile in the quantum pool model. The energy spacing Δ plays the role of a Stark cell shift.

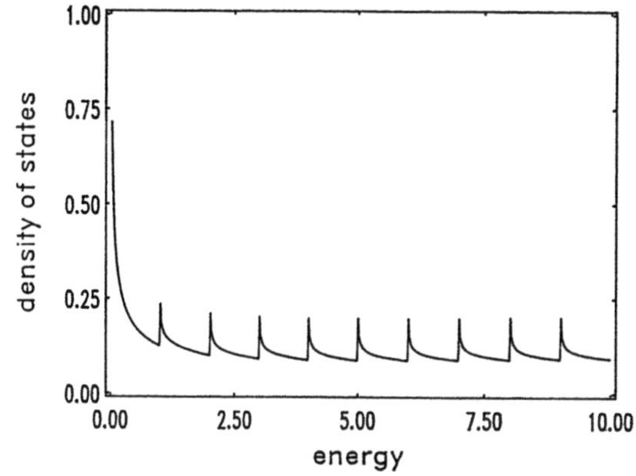

Fig.2 Density of states (Eq. 5) for the model of Fig. 1. The energy is in units of the Stark spacing Δ.

$$2a\sqrt{2m\varepsilon}\left(L/a+\sum_{s=0}^{M}\sqrt{1-\frac{s\Delta}{\varepsilon}}\right)=(l+\gamma)h\ . \tag{4}$$

Here l labels the state, γ is some constant, h is Planck's constant, and $M=[\varepsilon/\Delta]$, where $[u]$ denotes the integral part of u. The separation between two states can be made arbitrarily small by letting L approach a sufficiently large value. Calling $\delta\varepsilon=\varepsilon_{l+1}-\varepsilon_l$, a density of states may then be defined as the inverse separation between neighboring levels $\rho=\delta\varepsilon^{-1}$. We obtain from (4)

$$\rho(\varepsilon)=\frac{1}{2\pi}\sqrt{\frac{1}{\beta\varepsilon}}\left(\frac{L}{a}+\sum_{s=0}^{M}\frac{1}{\sqrt{1-\frac{s\Delta}{\epsilon}}}\right) \tag{5}$$

where $\beta=\hbar^2/2ma^2$ is a unit of energy. Figure (2) shows the density of states obtained from Eq. (5) for $L=10a$ and $\Delta=200\beta$. Inspection of the figure shows that there is a periodic term of period Δ, which includes a square root singularity whenever the energy is a multiple of the Stark period. The singular terms are explicitely exhibited in the sum of Eq. (5). This periodic term is the signature of a Stark ladder for our model. In addition there is an envelope composed of a free-electron like square root singularity at low energies and a high energy component that grows as $\varepsilon^{1/2}$.[17] It represents the density of states of (3) with the steps replaced by a smooth linear potential Δx. Thus, the presence of a periodic term in the potential introduces an also periodic component in the spectrum, which is the manifestation of a Stark ladder for the model (3).

The semiclassical expression (5) has been verified by solving accurately the full quantum mechanical problem, always for the potential (3) and for energies $\varepsilon < V_0$.Details of this solution will be published elsewhere.[21] The eigenfunctions are free electron like in the interval $-L < x < a$ and resemble an Airy function for $x > a$, except possibly as x approaches $([\varepsilon/\Delta]+1)a$, where the electron may become somewhat localized at certain values of the energy. These correspond to states where constructive interference develops in the rightmost step within the classically allowed region.

To illustrate this behaviour we show in Fig. (3) two (not normalized) solutions at different energies $\varepsilon_1=9.05\Delta$ (Fig.3a), and $\varepsilon_2=9.37\Delta$ (Fig. 3b), both within the same Stark period. Notice that they exhibit a different degree of localization. The values of parameters for the figure are as in Fig.2. We note that although we have chosen to exhibit

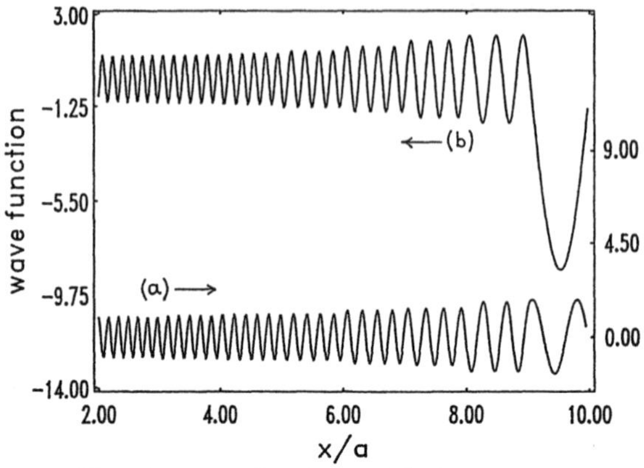

Fig.3 Wave functions for the model of Fig. 1 at energies $\varepsilon_1 = 9.05\Delta$ (a), and $\varepsilon_2 = 9.37\Delta$ (b). Beyond the figure to the right they decay exponentially.

the most localized solution in the period the degree of localization is still relatively small. Numerical solutions for a tilted Kronig-Penney potential yield strongly localized solutions[22] which may be interpreted as follows: If the potential is a sequence of wells having a low lying bound state when isolated, the effect of the electric field is to upset the resonant tunneling condition that makes the crystalline bands develope in an energy interval around such bound state. Wave functions are then no longer extended. Localization is enhanced for states in the neighborhood of such energy as the electric field increases.

III. Photons in a Rising Rampart

Stark ladders in the propagation of light was first studied by Monsivais *et al.*[20] Consider an electromagnetic wave of frequency ω incident on a layered structure with a dielectric function that obeys the relation $\epsilon(x + a) = \epsilon(x) + \tau$, and is a constant in the y-z plane. The electromagnetic field is a plane wave in such plane, with wave numbers k_y, k_z, and the dependence on x of any of its vector components F(x) is determined by the equation

$$[d^2/dx^2 + \epsilon(x)k^2]F(x) = Q^2 F(x). \tag{6}$$

Here $k = \omega/c$ with c the speed of light, and $Q^2 = k_y^2 + k_z^2$. Equation (6) has the same form as Eq. (1) and we may repeat the argument given in the introduction to show that the spectrum of the operator in the left hand side has the period τk^2. Physical solutions for fixed k are restricted however to the interval $0 < Q^2 < k^2$ in this case. As we did in Sec. II for electrons, we shall here discuss only a finite model.

Consider the propagation of electromagnetic waves in the dielectric structure shown in Fig. (4).[20] We call the rising profile a tilted rampart. A linearly polarized plane wave enters from the left with the magnetic field normal to the plane of the figure. The transmission coefficient $T = \sqrt{\epsilon_t} \mid E^t \mid^2 / \sqrt{\epsilon_i} \mid E^i \mid^2$ may be obtained by solving Eq. (6) for the transmitted electric field E^t in terms of the incident field E^i.

Figure (5) shows T as a function of $q^2 = (Q/k)^2$ for several values of the tilting parameter τ, and fixed 'barrier' height $h = 10$, and width $w = 0.1a$. The results are for a structure of 201 layers of width $a = 10^{-4}cm$, $\lambda = 2\pi \cdot 10^{-5}cm$ and $\epsilon_i = 1$. At zero tilting (curve (a)) the dielectric function is periodic and there are well defined transmission windows, looking just like the allowed energy bands of electrons in a crystal. As the tilting

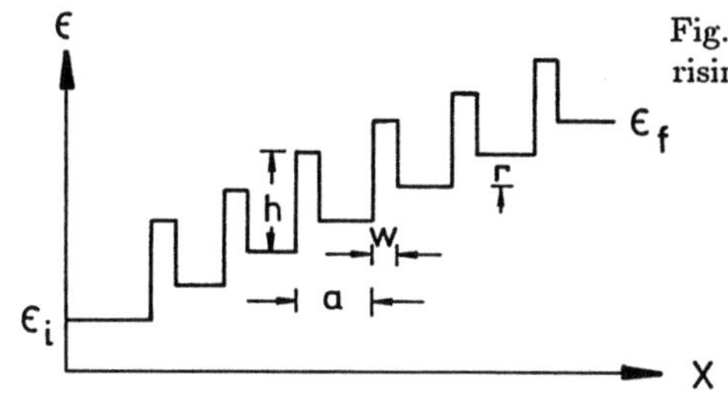

Fig.4 Profile of the dielectric constant for the rising rampart model.

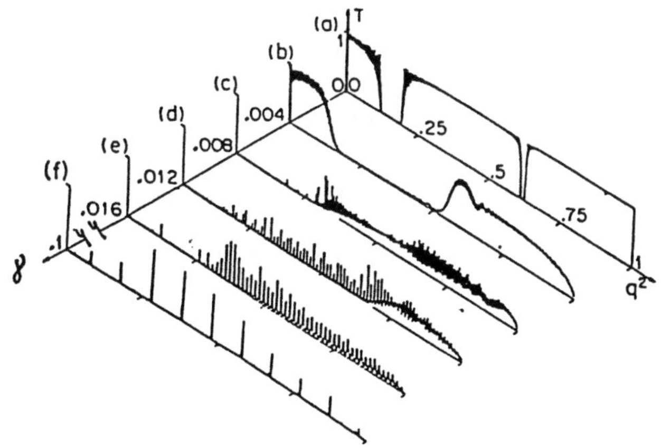

Fig.5 Optical transmission coefficient of a multilayered system with a profile as in Fig. 4. For values of the parameters see text.

is increased the transmission profile changes and one observes the development of equally spaced resonances in the transmission curve, their separation being just τk^2. These resonances become well established at sufficiently large tilting. They are the optical analog of the electronic Stark ladders. We note in passing that the ladder becomes less noticeable as the barrier height h decreases.[20] Just as in the case of electrons resonances correspond to a condition where constructive interference occurs within the structure, and this is enhanced by the abruptness of the steps in the dielectric profile.

An interesting feature of such optical Stark-like resonances in layered structures is the high resolution that may be achieved in the transmission of electromagnetic waves. Experiments with electrons must cope with line broadening that arises from phonon and impurity scattering, as well as from imperfections in the widths of the layers. It is hoped that the reported results for photons will stimulate experimental work in this area.

Acknowledgments. The author is indebted to Fundacion Andes and CONICYT for support. This work was supported in part by Fondo Nacional de Ciencias, Grant 90/0375.

References

[1] G.H.Wannier, *Phys. Rev.* **117**, 432 (1960)
[2] G.H.Wannier, *Phys. Rev.* **181**, 1364 (1969)
[3] J.Zak, *Phys. Rev. Lett.* **20**, 1477 (1968)

[4] C.A.Moyer, *Phys. Rev. B* **7**, 5025 (1973)
[5] W.Shockley, *Phys. Rev. Lett.* **28**, 349 (1972)
[6] J.E.Avron, *Phys. Rev. Lett.* **37**, 1568 (1976)
[7] I.W.Herbst and J.S.Howland, *Commun. Math. Phys.* **80**, 23 (1981)
[8] D.Emin and C.F.Hart, *Phys. Rev. B* **36**, 7353 (1987)
[9] P.W.Argyres, *Phys. Lett.* (1990)
[10] J.E.Avron, *Ann. of Phys.* **143**, 33 (1982)
[11] J.N.Churchill and F.E.Holmstrom, *Physica* **123**, 1 (1983)
[12] J.B.Krieger and G.J.Iafrate, *Phys. Rev. B* **33**, 5494 (1986)
[13] K.Hacker and G.Obermeir, *Z. Phys.* **234**, 1 (1970)
[14] H.Fukuyama, R.A.Bari, and H.C.Fogedby, *Phys. Rev. B* **8**, (1973)
[15] M.Luban and J.H.Luscombe, *Phys. Rev. B* **34**, 3674 (1986)
[16] Z.Barticevic and F.Claro, *Phys. Rev. B* **38**, 361 (1988)
[17] M.Pacheco and F.Claro, *J. Phys. C:Solid State Phys.* **21**, 739 (1988)
[18] E.E.Mendez, F.Agullo-Rueda and J.M.Hong, *Phys. Rev. Lett.* **60**, 2426 (1988)
[19] P.Voisin, J.Bleuse, C.Bouche, S.Gaillard, C.Alibert and A.Regreny, *Phys. Rev. Lett.* **61**, 1639 (1988)
[20] G.Monsivais, M.Castillo-Mussot and F.Claro, *Phys. Rev. Lett.* **64**, 1433 (1990)
[21] F.Claro and M.Pacheco, unpublished
[22] J.Bleuse, G.Bastard and P.Voisin, *Phys. Rev. Lett.* **60**, 220 (1988)

Electron Energy Loss in STEM Spectra

P.M. Echenique[1], *A. Rivacoba*[1], *N. Zabala*[2], *and R.H. Ritchie*[2]

[1]Dpto. de Física de Materiales, Universidad del País Vasco, Facultad de Química, Apdo. Postal 1072, San Sebastián 20080, Euskadi, Spain

[2]Dpto. Electricidad y Electrónica, Universidad del País Vasco, Facultad de Ciencias, Apdo. Postal 644, Bilbao 48080, Euskadi, Spain

[3]Oak Ridge National Laboratory, P.O. Box 2008, Oak Ridge, TN 37831, USA and Department of Physics, University of Tennessee, Knoxville, TN 39996, USA

Abstract. The interaction of fast electrons incident at a fixed impact parameter on targets of different geometries is studied. The connection between classical and quantum descriptions of the probe electrons is discussed. The relative contribution of surface and bulk modes is studied as a function of the impact parameter.

1. Introduction

Recent developments in the scanning transmission electron microscope (STEM) have made it possible to study electronic excitations of inhomogeneous systems in highly localized regions. This is achieved by recording changes in energy loss distributions measured when a well-focused ~0.5 nm probe of swift (~100 keV) electrons is scanned slowly across a specimen. The interpretation of these data raises a number of questions about the description and localization of excitations in inhomogeneous systems [1,3] and involves a wave-particle duality in an interesting context. Published treatments of excitations in periodic crystalline media have used broad-beam wave mechanical descriptions of the fast electron states [4] whereas studies of excitations in other systems bombarded by swift charged particles [5,11] have usually employed a classical description of the fast electron together with dielectric response theory for the solid.

In this paper we begin by establishing the connection between classical and wave mechanical description in STEM [12] and later discuss the application of classical dielectric theory to materials of different geometries.

2. Broad coherent irradiation

Using time dependent perturbation theory to evaluate the cross section for transitions from the ground state (energy ω_0) to the nth excited state (ω_n) of the target caused by an incident electron plane wave basis set one finds[1]

$$\sigma_{n0} = \frac{4}{v} \int \frac{d\mathbf{q}}{q^4} \,|\rho_{n0}(\mathbf{q})|^2 \,\delta(\,\mathbf{v}.\mathbf{q} - \frac{q^2}{2} - (\omega_n - \omega_0)) \tag{1}$$

where

$$\rho_{n0}(\mathbf{q}) = \langle n|\sum_{j=1}^{N} e^{i\mathbf{q}.\mathbf{r}}\, |0\rangle \tag{2}$$

is the matrix element of the density operator, and we use atomic units throughout:

$\hbar=e^2=m=1$. Note that $\mathbf{q} = \mathbf{k}_0 - \mathbf{k}_f$: where $\mathbf{k}_0$ and $\mathbf{k}_f$ are the initial and final electron momenta and

$$\frac{1}{2}(k_0^2 - k_f^2) = \mathbf{v}.\mathbf{q} - \frac{q^2}{2} \quad . \tag{3}$$

To introduce an impact parameter conjugate to momentum transfer, we first neglect recoil of the incident electron; that is, we drop, for the moment, the term proportional to q^2 in the argument of the delta function. The larger v becomes, the less important will be the neglected term. We take v in the direction of the z-coordinate axis and rewrite equation (1) in the equivalent form

$$\sigma_{n0} = \frac{4}{v}\int \frac{d\mathbf{q}}{q^2}\ \rho_{n0}(\mathbf{q}) \int \frac{d\mathbf{q}'}{q'^2}\rho^*_{n0}(\mathbf{q}')\ \delta(\omega_{n0} - v\ q_z)\ \delta(\mathbf{q} - \mathbf{q}') \ . \tag{4}$$

We now make use of the identity

$$\delta\ (\mathbf{Q}-\mathbf{Q}') = \frac{1}{(2\pi)^2}\int d\mathbf{b}\ e^{-i\,\mathbf{b}.(\mathbf{Q}-\mathbf{Q}')} \tag{5}$$

where $\mathbf{q} = (\mathbf{Q}, q_z)$, $\mathbf{Q}$ is the variable conjugate to $\mathbf{b}$ [13,14]. Then equation (4) may be written

$$\sigma_{n0} = \int d\mathbf{b}\ |\ a_{n0}(\mathbf{b})\ |^2 \tag{6}$$

where

$$|a_{n0}(\mathbf{b})|^2 = \frac{1}{\pi^2}\left|\int \frac{d\mathbf{q}}{q^2}\, e^{-i\,\mathbf{b}.\mathbf{q}}\ \rho_{n0}(\mathbf{q})\ \delta(\omega_{n0} - \mathbf{v}.\mathbf{q}_z)\ \right|^2 . \tag{7}$$

Note that $a_{n0}(b)$ is precisely the probability amplitude that the many-electron system experiences a transition under the influence of the Coulomb field of a classical point electron travelling with constant velocity **v** along a path specified by the impact parameter **b** beginning at $z=-\infty$ and ending at $z=+\infty$.

In the early days of quantum mechanics Frame [15] and Mott [16] using first order time-independent perturbation theory showed that for a swift incident proton identical results are obtained when an infinite-plane-wave representation is chosen and when a classical trajectory is assumed in computing excitation of a target.

So we have shown that in the case of broad beam coherent irradiation of the target, energy analysis of swift imaged electrons that have generated localized electronic excitations in the target gives nearly the same result as if the electrons were classical projectiles moving on rectilinear trajectories and with a uniform distribution in impact parameter. This equivalence is more accurately satisfied the larger the electrons speed. In reference 1 quantal corrections to this were evaluated for various velocities and for different targets.

3. Excitation of electronic transitions by a microprobe electron

Now we consider excitation of the target by an electron prepared in the form of a narrow beam centered at the impact parameter **b** relative to the target. The electron may be represented by the wave packet

$$\psi_0(\mathbf{r}) = \phi(\rho-\mathbf{b})\,\frac{e^{i k_0 z}}{\sqrt{L}} \quad . \tag{8}$$

By analogy with the above development Ritchie and Howie [12] showed that when all inelasticaly scattered electrons are included, the transition probability is given by

$$P_n^q(\mathbf{b}) = \int d\rho \, | \phi(\rho-\mathbf{b}) |^2 P_n^c(\rho) \tag{9}$$

where P_n^c is the classical probability for excitation at impact parameter ρ.

This shows that when all inelastically scattered electrons are collected the measured probability of exciting a given transition may be computed theoretically as if the microprobe consisted of an incoherent superposition of classical trajectories distributed laterally to the beam direction according to the probability density $|\phi(\rho-\mathbf{b})|^2$.

Saying this in another way, provided that the spectrometer aperture in the STEM is large enough, classical excitation functions are correct when averaged over a range of impact parameters corresponding to the current distribution in the probe. The classical approximation could therefore fail when the spectrometer semi-angle of aceptance θ_m is very small. For typical values in a STEM experiment $\theta_m >> 1$ mrad. In practice, the tendency is to maximize detection efficiency by taking $\theta_m \geq \theta_a \cong 8$ mrad, where θ_a is the maximum probe convergence half-angle.Thus this condition is well satisfied in practice.

In view of the above it is of interest to review work on the local dielectric treatment of the interaction with various targets of the electrons that are assumed to move on classical trajectories.

4. A classical electron moving parallel to a plane

One may compute the retarding field, and hence the rate of energy loss experienced by a fast electron travelling parallel to and at distance b from a surface using ordinary local dielectric theory of excitation by a moving charge. We take the charge density due to the swift electron to be

$$\rho(\mathbf{r},t) = -\delta(z-b)\,\delta(x)\,\delta(y-vt)\ . \tag{10}$$

From the force acting on the particle we define a probability of energy loss $P(\omega)$ in terms of the probability of the momentum transfer k_x and energy transfer ω as [8,17-20]

$$\frac{dW}{dy} = \int_0^\infty P(\omega)\,\omega\,d\omega \quad , \quad P(\omega) = \int_0^\infty dk_x\, P(k_x,\omega) \tag{11}$$

where

$$P(k_x,\omega) = \frac{2}{\pi v^2}\,\frac{e^{-2Qb}}{Q}\,\mathrm{Im}\left[\frac{\varepsilon-1}{\varepsilon+1}\right] = \frac{2}{\pi v^2}\,\frac{e^{-2Qb}}{Q}\,\mathrm{Im}\left[\frac{-2}{\varepsilon+1}\right] \tag{12}$$

where $Q^2 = k_x^2 + \omega^2 v^{-2}$. $P(\omega)$ is given as

$$P(\omega) = \frac{d\,W(b)}{dz\,d\omega} = \frac{2}{\pi v^2}\,\mathrm{Im}\left[\frac{-2}{\varepsilon+1}\right] K_0\!\left(\frac{2\omega b}{v}\right)\ . \tag{13}$$

The properties of the K_0 Bessel function give rise to interesting behaviour in the spectra. For large values of its argument $x = 2\,\omega b/v$, it falls off like $\exp(-x)/\sqrt{x}$ so that the probability of interaction will be appreciable out to distances of the order v/ω and which are typically 50 Å in the valence loss region. On the other hand, at small values of x, the function has a more rapid logarithmic variation so the losses at larger ω, in particular, can decrease rapidly with impact parameter. It is important to realize that v/ω is an upper estimate of the impact parameter corresponding to zero scattering angle and thus zero momentum transfer perpendicular to the initial velocity. Larger values of momentum transfer are associated with smaller interaction distances. The function $K_0(2\omega b/v)$ varies quite rapidly for small arguments thus allowing a high spatial resolution [21] that may explain recent experimental findings[22,23]. The error incurred by using a local $\varepsilon(\omega)$ has been evaluated [24,27]. It is found that for distances greater than 5, 10 Å no appreciable errors may be expected at typical STEM conditions. Generalization of Eqs. (12,13) to take account of relativistic effects have been made [28]. The relativistic corrections are expected to be small unless $Re(\varepsilon(\omega))$ becomes large enough that the criteron for the emission of Cherenkov photons ($\varepsilon v^2/c^2 > 1$) is satisfied for a substantial range of frequencies.

Equation (13) can be adapted to deal with glancing angle trajectories by putting $dy=\theta^{-1}dz$ and integrating over z. The probability of exciting a surface mode in such a trajectory is given by

$$Q(\theta) = \frac{2}{v\theta} \int_0^\infty \frac{1}{\omega} \, \mathrm{Im}\left[\frac{-2}{\varepsilon+1}\right] d\omega$$

$$= \frac{\pi}{v\theta} \quad \text{in the free electron case.} \tag{14}$$

When $z < 0$ and the electron travels in the dielectric

$$P(\omega) = \frac{2}{\pi v^2} \left\{ \mathrm{Im}\left[\frac{-1}{\varepsilon(\omega)}\right] \ln\left(\frac{q_c v}{\omega}\right) + \left(\mathrm{Im}\left[\frac{-2}{\varepsilon+1}\right] - \mathrm{Im}\left[\frac{-1}{\varepsilon}\right]\right) K_0\left(\frac{2\omega b}{v}\right) \right\}. \tag{15}$$

The logarithmic term yields the ordinary loss rate to volume excitations, q_c is a cut-off wave number, and the terms containing the K_0 function describe boundary corrections to these losses.

The effect of the boundary as first pointed out by Ritchie [29,30] is twofold: a surface mode appears via the term $-2/(1+\varepsilon(\omega))$ while on the other hand, a reduction in loss due to excitation of bulk modes is introduced via the term $-1/\varepsilon(\omega)$. Howie and Milne [9] from their experience in applying the dielectric theory to the silicon-silica and other interfaces have concluded that the experimental results can be a critical check of the dielectric function used since they must reproduce both features of the individual media and the interface features.

For the case of a plane of thickness "a" having a thin coating of a different dielectric material it can easily be shown (see Eq. 12) that for small values of a, when the thin coating might correspond to one or two anomalous atomic layers near the interface the result only is sensitive to this surface layer for $Qa > 0.1$, i.e. $k_x a > 0.1$ [31]. The conventional axial spectroscopy technique is rather insensitive to thin surface or interfacial layers because typically $\omega/v \cong 0.01$. Larger values of ω/v and hence Q are important in low energy electron spectroscopy, which is, for example, able to detect changes in the spectra due to surface reconstruction [32] when these cannot be detected by axial experiments with 100 keV electrons [33].

The spatial resolution of axial high energy spectroscopy improves at higher loss values. In the case of the silicon-silica interface, Howie and Milne [9] have been able to shown that the Si L edge, which occurs at 100eV in Si and is shifted to ~107eV in SiO_2, has a value of ~105eV at the interface.

5. Spherical particles

The probability function for this case is given by [34]

$$P(\omega) = \frac{4a}{\pi v^2} \sum_{L=0}^{\infty} \sum_{m=0}^{L} \frac{2-\delta_{m0}}{(L+m)!\,(L-m)!} \left(\frac{\omega a}{v}\right)^{2L} K_m^2\left(\frac{\omega b}{v}\right) \mathrm{Im}\left[\frac{L(\varepsilon(\omega)-1)}{L\varepsilon(\omega)+L+1}\right] . \qquad (16)$$

Here δ_{0m} is the Kronecker delta, and K_m is the modified Bessel function of the second order of order m.

Equation (16) allows us to understand the relative importance of a given mode in the loss process. For small values $K_0 \sim -\ln(x)$ and $K_m \sim 0.5\Gamma(m)(x/2)^{-m}$, while for large values $K_m \sim \sqrt{(\pi/2x)} \cdot e^{-x}$. When $\omega a/v \ll 1$, the L=1 term dominates,and we regain the dipole approximation [35]. When $\omega a/v \ll 1$ but $\omega b/v \gg 1$, no m term prevails and $P_\omega(b)$ goes like $(\omega a/v)$[36]. If $\omega a/v \gg 1$ many L's are necessary. Even if $\omega a/v \gg 1$ when $\omega b/v \gg 1$ only small $\omega \ll v/b$ are appreciably excited. The essential dipole condition $\omega a/v \ll 1$ therefore still applies provided that the sphere has sufficient dielectric loss at these frequencies.
In many experimental situations b ~ a and $\omega a/v \sim 1$ and one therefore needs to include many L values in Eq.(16) [36,40]. The most important characteristic frequencies involved in an energy loss process will vary both with L and with the dielectric function $\varepsilon(\omega)$. Since the excitation of the various modes also depends on $\omega a/v$ for a given electron energy, spheres of different materials can also show substantial differences in energy loss spectra as well as in the number and nature of the modes excited.

To illustrate these points we show in Fig. 1 the probability of losing energy ω for a 50 keV electron at grazing incidence on a sphere of radius a=10nm. We display the contribution of the dipole mode, the first two L's, the ten L's, and the total probability. For small radii and low energy losses, the dipole contribution dominates, while even at small radius, it is inadequate to describe the high energy losses. Experimental optical data [41] has been used in the calculations of the results shown in Fig 1.

When the electron penetrates the spheres, the expression for the probability becomes more complicated. Explicit expression for the energy loss probability valid for any $\varepsilon(\omega)$ can be found in the literature[42]. For a free electron gas

$$P(\omega) = \frac{2a}{v^2} \sum_{L,m} (2-\delta_{m0}) \frac{(L-m)!}{(L+m)!} \left\{ \omega_L \delta(\omega-\omega_L) \left[A^o_{Lm}+A^i_{Lm}\right]^2 - \right.$$
$$\left. -\omega_p \delta(\omega-\omega_p) \left[A^o_{Lm}+A^i_{Lm}\right] A^i_{Lm} \right\} \qquad (17)$$

where $\omega_L = \omega_p \sqrt{L/(2L+1)}$ and

$$A^o_{Lm} = a^L \int_{\sqrt{a^2-b^2}}^{\infty} dz \frac{1}{r^{L+1}} P^m_L\left(\frac{z}{r}\right) g_{Lm}\left(\frac{\omega z}{v}\right); \qquad A^i_{Lm} = \frac{1}{a^{L+1}} \int_0^{\sqrt{a^2-b^2}} dz\, r^L P^m_L\left(\frac{z}{r}\right) g_{Lm}\left(\frac{\omega z}{v}\right) \qquad (18)$$

where P_L^m stands for the Legendre functions, $r=\sqrt{z^2+b^2}$ and the functions $g_{Lm}(x)$ are sin(x) if (L+m) is odd, or cos(x) otherwise.

Figure 2 shows for the case of a free electron gas model the probability of surface and bulk loss for axial trajectory as a function of the radius of the sphere. The planar limits of π/v and $\pi/(2v)$ respectively first derived by Ritchie are clearly obtained for big radii.

Excitation functions for dielectric bodies bounded by other coordinate systems have been found. These include some allowance for the effect of the support of a

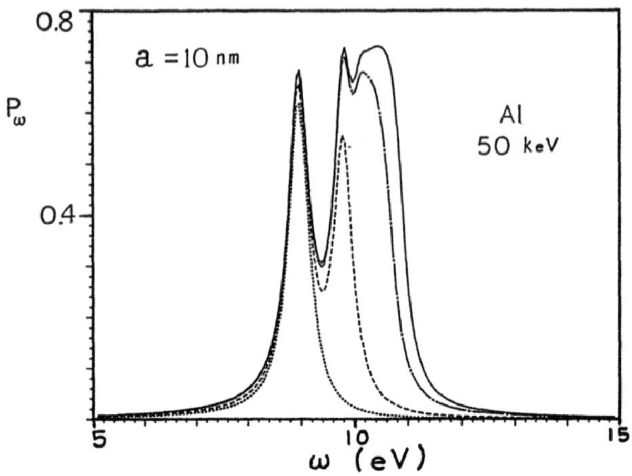

Fig. 1 Energy loss spectra of 50 eV electrons moving at grazing incidence on an aluminum sphere (a=10 nm).

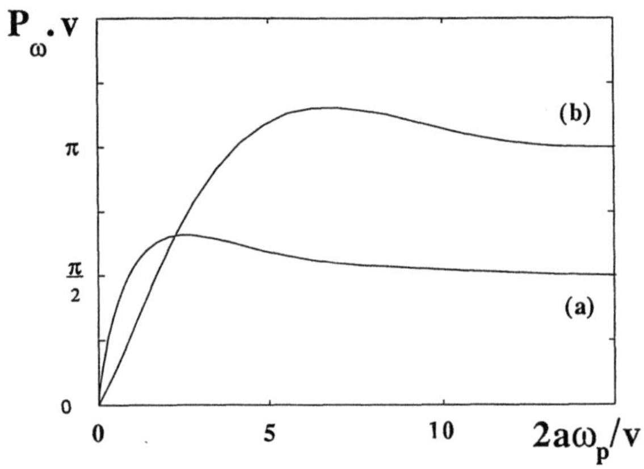

Fig. 2 Dependence of the bulk and surface losses on the radius of the sphere for axial electron trajectories. Curve (a) corresponds to $-P(\omega_p)$ while curve (b) is the $\Sigma\ P(\omega_L)$. A free electron response function $\varepsilon(\omega)$ has been used.

spherical particle [43]. The interactions between closely-spaced pairs of spherical particles has been analyzed as well [38]. A spheroidal dielectric has been studied [44]. Solutions found for a cylindrical wedge have relevance to the case of a fast electron passing near a corner of a cube [45,46]. Results for excitation of a dielectric by a fast electron passing through a cylindrical cavity in the medium have also been obtained [47-49]. It has been possible to interpret in considerable detail the energy loss spectra obtained experimentally in this geometry [50].

6. Deflection effect

STEM observations have indicated a deflection or flaring effect when a 100 keV, ~1nm diameter electron beam travels parallel to, but at a distance x of up to 5 nm outside of a surface face of a ~100nm MgO cube[51] or an Au [52] particle. For a 100 keV electron moving parallel to and outside the surface of MgO or Au the potential well generating the image force has a depth of about 0.1eV and extends about 10nm into the vacuum. The deflection angle for an electron travelling outside an Al cube of edge 150nm can be easily estimated as

$$\Delta\theta = \frac{v_x}{v} = \frac{F_{\perp} t}{v} \approx \frac{\omega_s^2 . \text{length}}{v^4} \approx 10^{-6} \text{rad} , \qquad (19)$$

which is about three orders of magnitude less than the experimental result. More elaborate calculations by Echenique and Howie [53], corraborate this result. Calculations by other authors in different geometries also confirm this result [54,55]. For instance Echenique and Howie [53] conclude that although the classical dielectric model is able to account for the energy losses observed in the STEM experiments, the image force it yields is much too small to explain the beam deflection effects observed by Cowley [52] and that the explanation must be sought elsewhere. In the case of MgO particle charging should perhaps be considered but this may be somewhat less likely in the case of Au particles.

Acknowledgements

The authors gratefully acknowledge Iberduero S.A., Gipuzkoako Foru Aldundia and the Education Department of the Basque Goverment for help and support. The Office of Health and Environmental Research of the U.S. Departament of Energy has given partial support for this research.

References

[1] R.H. Ritchie, Phil. Mag. **A44**, 931 (1981)
[2] P.E. Batson, Phys. Rev. Lett. **49**, 936 (1982)
[3] C. Colliex, Ultramicroscopy **18**, 131 (1985)
[4] J. Taffø, O.L. Krivanek, J.C.H. Spence, and J.M. Honig, Phys. Rev. Lett. **48**, 560, (1982)
[5] P.E. Batson, Solid State Comm. **34**, 477 (1980)
[6] P.E. Batson, Ultramicroscopy **9**, 277 (1982)
[7] J.M. Cowley, Phys. Rev. **B25**,140 (1982)
[8] P.M. Echenique and J.B. Pendry, J. Phys. **C8**, 2936, (1975)'
[9] A. Howie and R.H. Milne, Ultramicroscopy **18**,427 (1985)
[10] M. Schmeits, J. Phys. C14, 1203 (1981)
[11] P.C. Das and J.I. Gersten, Phys. Rev. **27**, 54 (1983)
[12] R.H. Ritchie and A. Howie, Phil. Mag. **58**, 753 (1988)
[13] N.P. Chang and K. Raman,Phys. Rev. **181**, 2048 (1969)
[14] U. Fano, *Chargeď Particle Tracks in Solidis and Liquids*,(The Institute of Physics, London, 1970)
[15] J.W. Frame, Proc. Camb. Phil. Soc. **27**, 551 (1931)
[16] N.F. Mott, Proc. Camb. Phil. Soc. **27**, 553 (1931)
[17] A. Howie, Ultramicroscopy **11**,141 (1983)
[18] R. Nuñez, P.M. Echenique and R.H. Ritchie, J.Phys **C13**,4229 (1980)
[19] J.P. Muscat and D.M. Newns, Surf. Sci. **64**, 641,(1977)
[20] P.M. Echenique, R.H. Ritchie, N. Barberan and J. Inkson, Phys. Rev. **23**, 6486 (1981)
[21] R.H. Ritchie,A. Howie,P.M. Echenique, and G.J. Basbas. Scann. Microsc.
[22] SC. Cheng, Ultramicroscopy **21**, 291 (1987)
[23] M. Scheinfein, A Murray and M. Isaacson, Ultramicroscopy **16**, 233 (1985)
[24] P.M. Echenique, Phil. Mag. **B52**, 9 (1985)
[25] N. Zabala and P.M. Echenique,Ultramicroscopy **32**, 327 (1990)
[26] R. Fuchs and F. Claro, Phys. Rev. **B35**, 3722 (1987)

[27]R. Rojas, F. Claro and R. Fuchs, Phys. Rev. **B37**, 6799 (1988)
[28]R. Garcia Molina, A. Gras-Marti, A. Howie and R.H. Ritchie, J. Phys. **C18**, 5335 (1985)
[29]R.H. Ritchie, Phys. Rev **106**,874 (1957)
[30]H. Boersch, J. Geiger and W. Stickel, Z. Phys. **212**, 130 (1968)
[31]A. Howie and R.H. Milne, J. Micros. **136**,279 (1984)
[32]T. Ichinokawa, Y.Ishikawa, N. Awaya and A. Onoguchi, *Scanning Electron Microscopy*, part 1, pp 271(Ed O. Johari, SEM, AMF O'Hare, IL, 1981)
[33]O.L. Krivanek, Y. Tanishiro, K. Takayanagi and K. Yagi, Ultramicroscopy **11**, 215 (1983)
[34]T.L. Ferrell and P.M. Echenique, Phys. Rev. Lett. **55**, 1526 (1985)
[35]J.D. Jackson, *Classical Electrodinamics* 2nd ed. Chap.13(Wiley, New York, 1975)
[36]P.M. Echenique, A. Howie and D.J. Wheatley, Phil. Mag. B56, 335 (1987)
[37]A. Acheche, C. Colliex, H. Kohl, A. Nometier and P. Trebia, Ultramicroscopy **20**, 99 (1986)
[38] P.E. Batson, Surf. Sci. **156**, 720 (1985)
[39]D.B. Tran Thoai, Phys. Stat. Sol. (b)**133**, 329 (1986)
[40]D.B. Tran Thoai and E. Zeitler, Phys. stat. Sol. (a)**107**, 791 (1988)
[41]H.J. Hagemann, W.Gudat and C. Kunz, Deutsches Electronen Synchrotron Report Nº DESY-SR-74/7 (unpublished)
[42]A. Rivacoba and P.M. Echenique, Scann. Microsc. **4**, 73 (1990)
[43]Z.L. Wang and J.M. Cowley, Ultramicroscopy **21**,77 (1987); ibid. **21**, 335 (1987)
[44]B.L. Illman, V.E. Anderson,R.J. Warmack and T.L.Ferrell, Phys. Rev. **B38**,3045 (1988)
[45]A.O. Boardman, R. Garcia-Molina, A. Gras-Marti and E. Louis, Phys. Rev. **B32**,162 (1985)
[46]R. Garcia-Molina, A. Gras-Marti and R.H. Ritchie, Phys. Rev. B31, 121 (1985)
[47]Y.T. Chu, R.J. Warmack, R.H. Ritchie, J.W. Little, R.S. Becker and T.L. Ferrell, Particle acelerators **16**, 13 (1984)
[48]D. De Zutler and D. De Vleeschauwer, J. appl. Phys. **59**, 4146 (1986)
[49]N. Zabala, A. Rivacoba and P. M. Echenique, Surf. Sci. **209**, 465 (1988)
[50]M.G. Walls, *Electron Energy Loss Spectroscopy of surfaces and interfaces,* Ph.D. Thesis, University of Cambridge (unpublished) (1988)
[51]J.M. Cowley, Ultramicroscopy **9**, 231 (1982)
[52]C.S. Tan and J.M. Cowley, Ultramicroscopy **12**, 333 (1983)
[53]P.M. Echenique and A. Howie, Ultramicroscopy **16**, 269 (1985)
[54]A. Rivacoba and P.M. Echenique, Ultramicroscopy **26**, 389 (1988)
[55]A. Martinez-Torregrosa, R. Garcia-Molina and A. Gras Martí, to be published.

Chemical Information from Auger Electron Spectroscopy

J. Ferrón and R. Vidal

INTEC, Universidad Nacional del Litoral and
Consejo Nacional de Investigaciones Científicas y Técnicas,
Güemes 3450, C.C. 91, 3000 Santa Fe, Argentina

Abstract. We have analyzed refinements in the use of Auger electron spectroscopy to obtain chemical information. We introduce a novel form, the sequential way, of applying the principal component analysis (PCA) and target transformation (TT) methods. In conjunction with a novel form to analyze the experimental error, also based on PCA, we enhance the capability of PCA and TT for extracting the Auger line shape of unknown components. In order to show the capability of this method, we apply it to the study of the palladium crystalline and amorphous silicon interfaces.

1. Introduction

Auger electron spectroscopy (AES) and X-ray photoelectron spectroscopy (XPS) are the two most widely used spectroscopies for surface characterization. The element identification is produced in both techniques through the analysis of the kinetic energy of the emitted electrons. In addition, the chemical state of the elements can be determined through the shifts (chemical shifts) in the binding energies of the atomic levels induced by the changes in the surrounding of the atom. Thus, at first sight the information obtained from AES and XPS, limited to chemical analysis, is very similar. However, the fact that XPS involves only one atomic level, against the three involved in AES, produces that the photoemission lines are sharper than the Auger ones. If we add that chemical shifts amount only up to 1–2 eV, we have that XPS is a more suitable technique for studying chemical compounds. On the other hand, the use of electrons as excitation source gives to AES an enormously better spatial resolution [1]. There are however some cases, for instance in depth profiling of reactive interfaces, where chemical information is required together with good spatial resolution. In the following, we will describe a data analysis method, the Factor Analysis, and how a simple application of this , the sequential way, can largely improve the obtention of chemical information when AES is used in depth profiling analysis.

2. Data treatment method

The method of principal component analysis (PCA) [2] has been successfully used on different analysis techniques including AES [3–10]. In AES depth profiling, the aim of PCA together with the target transformation approach (TT) [4] is to decompose the data matrix D in two matrices

$$D = R\,C, \tag{1}$$

where D is a $m \times n$ matrix whose n columns are formed by the Auger spectra acquired sequentially during sputter etching, each spectrum being formed by n channels (rows in the data matrix); R is a $m \times c$ matrix whose columns are formed by the Auger spectra of the c pure components, and C is a $c \times n$ matrix whose rows are formed by the weights of each of the c pure components.

The first step in PCA treatment is the construction of the covariance matrix

$$A = D^T D. \tag{2}$$

If the number of independent chemical components of D is c, the rank of A will be also c. Therefore, for a perfectly noise-free spectra set, the determination of the number of independent chemical components is reduced to the determination of the number of non-zero eigenvalues of A. In the real experimental case, noise is always present and all eigenvalues are different from zero. The key point in PCA is then to determine those eigenvalues which have physical meaning, i. e. those eigenvalues which are statistically different from zero.

The second step in PCA analysis is then to diagonalize A:

$$\begin{aligned} Q^T A Q &= [\lambda_{ij}\delta_{ij}], \\ A[Q_j] &= \lambda_j [Q_j], \end{aligned} \tag{3}$$

where the columns of Q (Q_j) are the eigenvectors of A.

One of the possible decompositions of D can be achieved by associating the diagonalizing matrix Q with the component matrix C in eq. 3. Of course, this decomposition is only one of the multiple abstract possible decomposition of D in two matrices. The physical meaning is achieved in the third and last step, known as target transformation [4]. Using the complete rotation matrix Q we would reproduce the data matrix, even with the statistical errors. The goal of PCA is just to lower the dimension of Q to the physical meaningful values, i.e. the use of the minimum number of eigenvectors (eigenvalues) to reproduce D within the experimental error. In addition, the very simple idea of the sequential method is to fix the number of the eigenvalues and to study the evolution of the error, defined as the error performed in reproducing the data matrix, as we add points of the depth profile.

The final step of Factor Analysis consists in performing a suitable rotation of the matrices R and C in order to obtain the physically meaningful composition matrix C' and pure components matrix R',

$$D = RC = R'T^T TC'. \tag{4}$$

In doing that, the so called target transformation (TT) [4], extra information is required. In the conventional TT we need to know the concentration of $c-1$ components in at least one point of the profile. Usually, in interface depth profiling one knows the spectra corresponding to the substrate and overlayer. Then, TT can be performed when PCA reveals up to three independent compounds along the profile [4]. On the other hand, the sequential PCA can be applied if either the substrate or the overlayer spectrum is known, and there are interface regions where only two components simultaneously coexist.

3. Application

In order to show the capability of this method to extract chemical information from the variation of the Auger line shapes along an interface depth profiling, we have applied it to study the interface between a near noble metal (palladium) and silicon. The formation of silicide at the interface is of critical importance, since it determines the macroscopic electrical properties of the interface. We study two different cases: first we analyze the different reactivity of the interface formed by Pd on crystalline silicon (c-Si) and amorphous silicon (a-Si), and second we study the evolution of the interface Pd/c-Si under a mild annealing treatment.

3.1 Experimental approach

The interfaces were prepared by evaporating a thin (70 nm) Pd layer onto the silicon substrate. The c-Si was cleaned by a standard method (etched in HF dilute solution and rinsed in deionized water), and the a-Si film was prepared by DC sputtering. The temperature evolution of the interface was followed in a single sample, heated to 200 °C for fixed time periods, and the Auger depth profilings were performed immediately after the annealing was ended. The Auger spectra were acquired in the differentiated mode, using a single pass cylindrical mirror analyzer with an energy resolution of 0.6 %. Due to charging effects in the a-Si sample, the conditions for analyzing both substrates were different. For c-Si we use a 4.5 μA, 3 keV electron beam current, and for a-Si a 2.5 μA, 1 keV one. The modulation amplitude was 2 V. The sputter etching was produced by using a 2 keV Ar ion beam. All measurements were performed under UHV conditions, i.e. in the low 10^{-10} Torr range.

3.2 Results and discussion

In Fig. 1a and 1b we show the depth profiles corresponding to the Pd/c-Si and Pd/a-Si interfaces, respectively. These correspond to the evolution of the peak–to–peak heights in the differentiated spectrum for the Si_{LVV} and Pd_{MNN} Auger transitions. We observe that both profiles are very similar. The yield detected at the surface in the Si_{LVV} energy range corresponds to a Pd low energy structure. The decay of this signal while approaching the interface overlaps with the appearance of the Si_{LVV} feature.

From the depth profiles depicted in the Figs. 1a and 1b we could not infer any difference between the chemical reactivity of both interfaces. However, a simple visual inspection of the evolution of the Auger line shapes along the depth profiles, Figs. 2a and 2b, shows differences for both

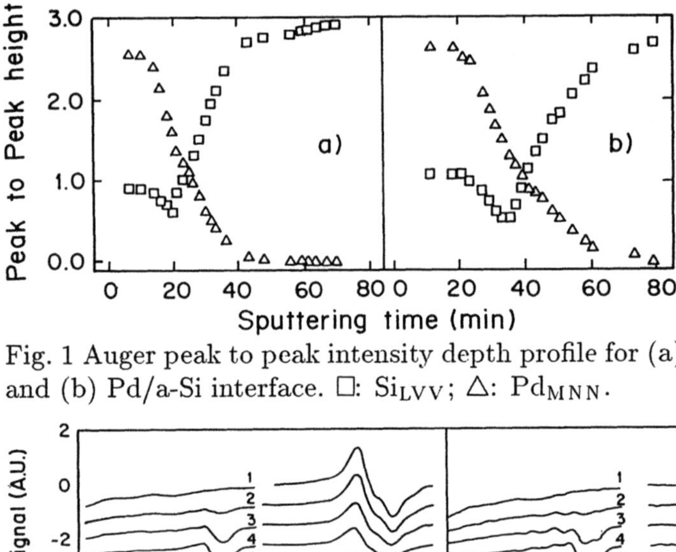

Fig. 1 Auger peak to peak intensity depth profile for (a) Pd/c-Si interface and (b) Pd/a-Si interface. □: Si_{LVV}; △: Pd_{MNN}.

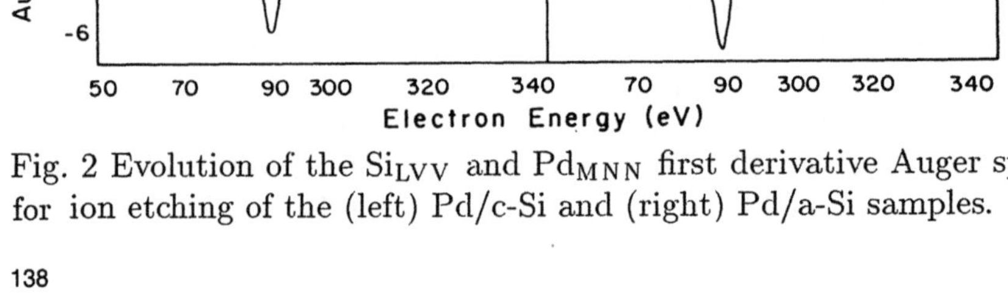

Fig. 2 Evolution of the Si_{LVV} and Pd_{MNN} first derivative Auger spectra for ion etching of the (left) Pd/c-Si and (right) Pd/a-Si samples.

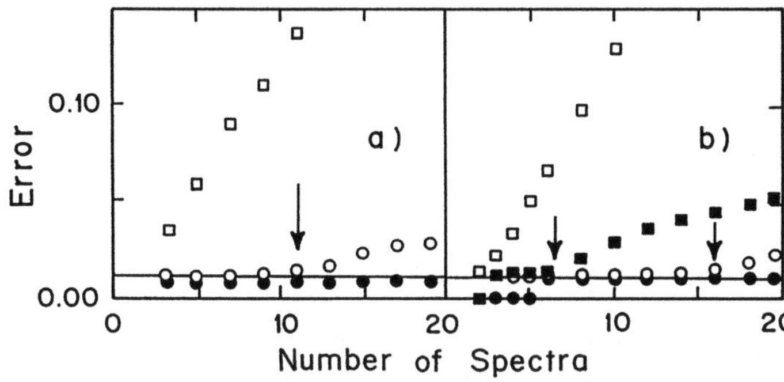

Fig. 3 Evolution of the error in reproducing the data matrix as a function of the number of spectra included (see text) for (a) Pd/c-Si and (b) Pd/a-Si samples. The parameter is the number of factors. a) □: 1 factor; ○: 2 factors; ●: 3 factors; b) □: 1 factor; ■: 2 factors; ○: 3 factors; ●: 4 factors. The thin horizontal line represents the experimental error.

these substrates. In fact, the characteristic features of Pd_2Si, appearing as a structure in the low energy side of the Si_{LVV} peak, seem to be present only when Pd is evaporated onto amorphous silicon.

The differences between both these profiles clearly appear when we analyze the evolution of the error in PCA following the sequential method. In Fig. 3a and 3b we show the evolution of the error, defined as the error produced in reproducing the data matrix D by using only one, (two, three...) factors, as we are progressing in the sputter etching, i.e. as we add columns to the D matrix.

Following the analysis of PCA, developed in the preceding paragraphs, the number of independent components will be equal to the minimum number of factors which allows to reproduce the data matrix D within the experimental error. The experimental error associated with PCA is determined by taking a series of 10 spectra at a fixed point in the profile. Since in this case we know the real existence of only one factor, the error determined by PCA with one factor is equal to the experimental error. This value is represented in Fig. 3 as a thin horizontal line.

From the set of Fig. 3, we see that the complete depth profile of the interface Pd/c-Si can be reproduced, within the experimental error, by using 3 factors, while in order to reproduce the interface of Pd/a-Si we need at least 4 factors. The points of appearance of each different factor are also clear from this analysis, and are pointed by arrows in the graphs. The simplicity of this method rests on our capability of identifying the number of different compounds in the profile, and their in–depth distribution by performing a very simple algebra.

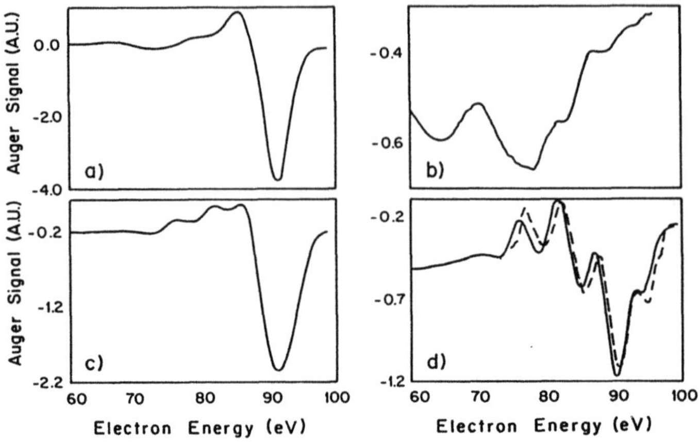

Fig. 4 Auger spectra after target transformation (see text). a: silicon; b: palladium; c: Si rich silicide; d: continuous line: palladium silicide obtained through TT, dashed line: experimental spectrum of silicide.

At this point we know that the interfaces of c-Si and a-Si are different. The presence of up to 3 different factors in c-Si, together with the knowledge of the presence of pure Pd (the overlayer) and pure Si (the substrate) would allow the application of the target transformation in the normal way. On the other hand, the presence of 4 factors in the a-Si case prevents us on the application of TT in the normal form [6].
However, by inspection of the profiles in Fig. 3 we observe that the factors appear in sequential way. Then the TT method can be applied on zones of the profile where only two different factors coexist. Once the unknown compound is identified, the zone is enlarged in order to allow the inclusion of the next factor, and so on. In this way, the Auger line shapes of the different pure components can be obtained.

The application of this method to both interfaces gives only 4 different line shapes, i.e. the 3 components obtained in the Pd/c-Si interface are also present in the amorphous one. In Fig. 4, the AES line shapes corresponding to the 4 different components are depicted. Three Auger line shapes are easily identified, they are Si(5a), Pd(5b) and $SiPd_2$(5d). For comparison, the Auger line shape obtained from a thick silicide substrate is shown [6]. A very good agreement is observed between both line shapes: the experimental spectrum and the one extracted from the data treatment. The remaining spectrum corresponds to a Si Pd compound with a stoichiometry different from the silicide ($SiPd_x$ with $x < 2$) [8].

In Fig. 5, the distribution of these different compounds along the profile is shown. Some interesting points can be extracted from these improved depth profiles. We found that Pd silicide appears only at the

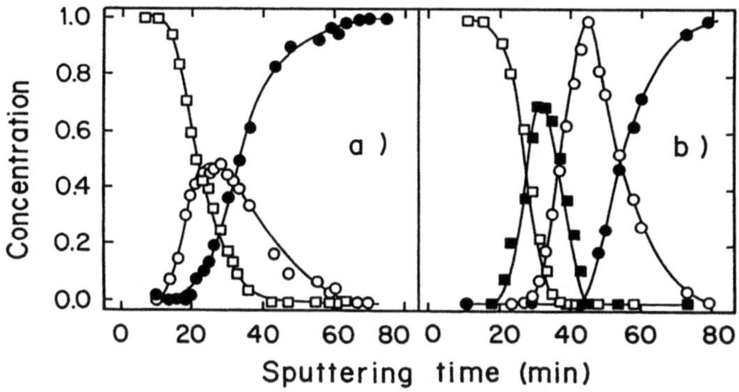

Fig. 5 Composition depth profile of the (a) Pd/c-Si and (b) Pd/a-Si interfaces obtained after PCA. □: unreacted Palladium; ○: Si rich silicide; ■: Palladium silicide; ●: Silicon.

Pd/a-Si interface. This result is in agreement with the high reactivity observed in the case of amorphous Si when it is compared with the crystalline one [11]. The Si-rich silicide, which appears at both interfaces has been predicted and found in several theoretical and experimental works [6,12–14]. However, due to the lack of standard spectra, the complete interface depth profile and the Auger line shape of such a compound could not be obtained [6]. It is clear however that the appearance of this Si rich SiPd compound at both interfaces could be produced by the ion bombardment through the displacement of Pd atoms into the Si matrix, instead of a normal chemical reaction as predicted theoretically [12].

The second application of this method consists in the study of the evolution under thermal treatment of a Pd/c-Si interface. In the set of Fig. 6 (a, b and c) we show the depth profiles corresponding to the thermal treatment of the Pd/c-Si sample; (a) without annealing, (b) after 30 minutes, and (c) 120 minutes of heating at 200 °C. The profiles, similarly to those of Fig. 1, were taken by measuring the peak–to–peak height of the Pd and Si differentiated Auger spectra. The chemical information contained in these depth profiles, although limited, is more important than in the previous case. In fact, the reaction between Pd and Si can be inferred from the variation in the peak-to-peak ratio between Pd and Si Auger yields. The lowering of the Pd yield and the appearance of Si at the surface, the broadening of the interface and the change in the Auger line shape (not shown) points out the occurrence of a chemical reaction. In Fig. 6, (a', b' and c') the depth profiles obtained with PCA in the sequential way are also shown. The differences are important, and the information can be directly extracted. The non-annealed interface (6a')

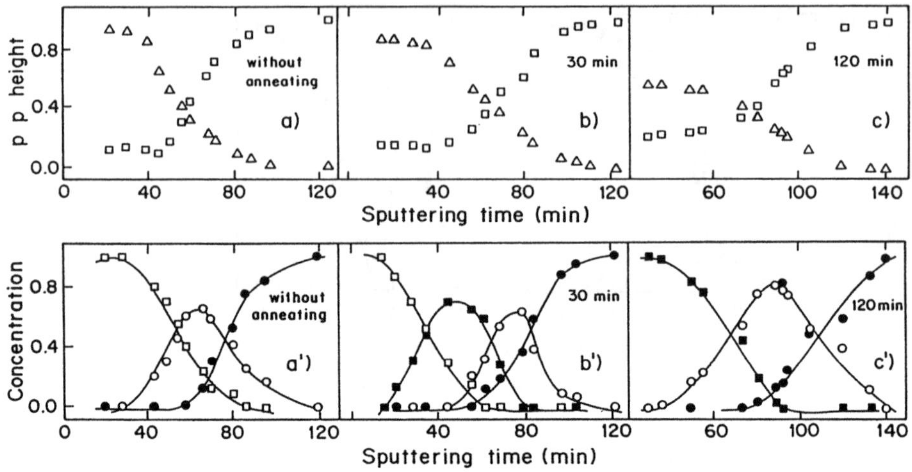

Fig. 6 Upper panel: Auger peak to peak intensity depth profiles for Pd/c-Si interface annealed at 200 °C. a) without annealing; b) 30 min., and c) 120 min. of annealing; □: Si_{LVV}; △: Pd_{MNN}. Lower panel; after PCA treatment; □: unreacted Palladium; ○: Si rich silicide; ■: Palladium silicide; ●: Silicon.

shows the presence of three compounds, in agreement with the result obtained for this interface above (Fig. 5a). The three components are in the same way identified with Si, Pd and $SiPd_X$. The less annealed sample (6b') presents four compounds, revealing the appearance of palladium silicide at the interface $SiPd_x$-Pd, as in the case of amorphous silicon. And finally, the reaction is complete in the most annealed sample, showing that the metallic Pd has reacted completely to form silicide (6c'). The Si rich compound is, however, always present in these depth profiles, as in all the previously analyzed interfaces.

4. Conclusions

We have developed a different way of applying the principal component analysis and the target transformation method. We have applied this method to the Pd/c-Si and Pd/a-Si interfaces and found that: i) there is formation of a $SiPd_x$ ($x < 2$) compound at all the Pd/Si studied interfaces; ii) palladium silicide is spontaneously formed only over the amorphous silicon substrate, confirming its higher reactivity compared with crystalline silicon; iii) Under a mild annealing (200 °C), the Pd film reacts completely with c-Si forming palladium silicide, but leaving always $SiPd_x$ compound at the interface; iv) we can not rule out the possibility of this compound forming under the influence of the ion bombardment.

Acknowledgments. We are deeply indebted to Dr. J.L. del Barco for his continuous help. This work has been partially supported by CONICET through grant PID 75300.

5. References.

1. For a review on XPS and AES, see for instance: Practical Surface Analysis (D.Briggs and M.P.Seah eds., Wiley, Norwich, 1983)
2. E.Malinowski and D.Howery, Factor Analysis in Chemistry (Wiley, New York, 1980).
3. S.W.Gaarenstroom, Appl.Surface Sci. 7, 7 (1981).
4. S.W.Gaarenstroom, J. Vacuum Sci. Technol. 20, 458 (1982).
5. V.Atzrodt, T.Wirth and H.Lange, Phys. Status Solidi 62, 531 (1980).
6. V.Atzrodt and H.Lange, Phys. Status Solidi a 79, 489 (1983).
7. V.Atzrodt and H.Lange, Phys. Status Solidi a 79, 373 (1984).
8. R.Vidal and J.Ferron, Appl. Surface Sci. 31, 263 (1988).
9. L.Steren, R.Vidal and J.Ferron Appl. Surface Sci. 29, 418(1987).
10. J.Steffen and S.Hofmann, Surface Sci. 202, L607 (1988)
11. L.Hung, E.Kennedy, C.Palmstron, J.Olowolafe, J.Mayer, and H.Rhodes, Appl. Phys. Letters 47, 236 (1985).
12. G.W.Rubloff, P.S.Ho, J.F.Freeouf and J.E.Lewis, Phys.Rev. B23, 4183 (1981).
13. J.Roth and C.Crowell. J.Vacuum Sci. Technol., 15, 1317 (1978).
14. U.del Pennino, P.Sassarolli and S.Valeri, Surface Sci. 122, 307 (1982).

Electron Energy Loss Studies of Surface Phonons on Crystal Surfaces

B.M. Hall and D.L. Mills

Department of Pysics, University of California, Irvine, CA 92717, USA

Abstract. A new generation of off specular inelastic electron scattering experiments allows one to measure surface phonon dispersion curves, throughout the entire two dimensional surface Brillouin zone. A multiple scattering description of the surface phonon excitation process accounts quantitatively for the observed loss intensities, and their (substantial) variation with beam energy and scattering angle. The theory allows one to identify features in the loss cross section found at various energies, and can guide the choice of experimental conditions. We review this body of work briefly, then present new calculations which explore the sensitivity of the loss spectra to the nature of the phonon eigenvectors at the surface. This is done for two models discussed recently for the Cu(111) surface: (i) one within which surface force constants differ only slightly from their values in bulk Cu, and (ii) one in which very large (60%) force constant softenings are proposed between atoms within the surface, and very large (75%) stiffenings of the force constant between the first and second layer. Electron energy loss intensities are in excellent accord with (i), but there is poor agreement for (ii).

1. Introduction

Since the appearance of high resolution electron energy loss spectroscopy two decades ago, [1,2], the technique has become a primary tool in the study of the vibrational motion of atoms or molecules on crystal surfaces [3]. The experiment, at least in principle, is a surface analogue of the well known neutron scattering studies of phonons in bulk crystals. The electron strikes the crystal with energy $E^{(I)}$, and exits with energy $E^{(S)}$ after creating a vibrational quantum such as a surface phonon, whose frequency is $\omega_s(\vec{q}_\parallel)$, where $\vec{q}_\parallel$ is a wave vector parallel to the surface. Energy is conserved in the scattering process, so

$$E^{(S)} = E^{(I)} - \hbar\,\omega_s\left(\vec{q}_\parallel\right). \tag{1}$$

In such an experiment, where the electron penetrates only two or three atomic layers into the crystal before backscattering from it, wave vector components normal to the surface are not conserved. However, parallel components are conserved to within a (surface) reciprocal lattice vector, so we have also

$$\vec{k}_\parallel^{(S)} = \vec{k}_\parallel^{(I)} - \vec{q}_\parallel + \vec{G}_\parallel \tag{2}$$

Springer Proceedings in Physics, Volume 62
Surface Science Eds.: F.A. Ponce and M. Cardona

with $\vec{k}_{\parallel}^{(S)}$ and $\vec{k}_{\parallel}^{(I)}$ the projection of the wave vector of the scattered and incident electron onto the surface plane; $\vec{G}_{\parallel}$ is a surface reciprocal lattice vector.

For many years, such inelastic electron scattering experiments were carried out in a very special scattering geometry, the near specular geometry. Electrons which suffer inelastic scatterings, then emerge very near (within one or two degrees) of the specular direction are collected. The reason is that very close to the specular direction, the intensity of the inelastically scattered electrons is very high, perhaps as much as three or four orders of magnitude higher than those which suffer large angle deflections.

The origin of the very intense, near specular loss peak is the following [4]. When an atom in or on a crystal surface vibrates, it has an electric dipole moment that oscillates in time with the frequency $\omega_S(\vec{q}_{\parallel})$. The oscillating dipoles produce an electric field in the vacuum above the crystal; if z is normal to the surface, the oscillating electric field has the spatial variation $\exp(-q_{\parallel} z)$. We thus realize very long ranged fields for small values of $q_{\parallel}$. The long ranged fields give a contribution to the matrix element for inelastic scattering that peaks sharply around the specular direction.

The theory of the near specular dipole scattering [4] shows that the typical angular deflection suffered by the electron is the order of $\Delta\theta = \hbar\omega_S/2E^{(I)}$, which is typically a few tenths of a degree. In a manner characteristic of coulomb scattering, the excitation cross section increases with decreasing beam energies. In a typical experiment, rather low energies in the range of 1-10eV are utilized. The angular deflection $\Delta\theta$ is so small that the wave vector $\vec{q}_{\parallel}$ is not resolved in such near specular studies. One estimates that typical values of $q_{\parallel}$ probed in these studies lie in the range of 10^6 cm^{-1}.

It follows that in the near specular studies, one explores only the near vicinity of the center of the surface Brillouin zone. It is not possible to measure the dispersion curves of surface phonons in such experiments. While this first generation of electron energy loss studies has proved most useful, particularly for the study of vibrations of adsorbed atoms and molecules, the information obtained on the surface phonons of clean crystals was limited.

A few years ago, the first in a new generation of off specular electron energy loss studies of the complete dispersion curve of a surface phonon was reported by Ibach and collaborators [5]. At the time of this writing, off specular studies of surface phonons, and the influence of bulk phonons on the vibrational motions of surface atoms, have been reported for a considerable number of clean and adsorbate covered surfaces.

There are fundamental differences between the new experiments, and the first generation of electron energy loss studies, carried out in the near specular geometry outlined above. One is the use of much higher beam energies, in the range 50-300eV employed also in low energy electron diffraction (LEED) studies of surfaces. Under these conditions, the contribution to the excitation matrix element from the long ranged Coulomb fields that are controlling in the near specular geometry are quite small; these fall off in magnitude with increasing beam energy, as remarked above. Also, as indicated earlier, they fall off rapidly with increasing deflection angle (increasing $q_\parallel$).

The interpretation of the off specular experiments requires a fully microscopic theory of the surface phonon excitation process, which recognizes that the electron penetrates into the substrate, and engages in multiple scattering from the ion cores, as it excites a surface (or possibly a bulk) phonon before backscattering from the crystal. Such a theory has been formulated, [6] and implemented. It has proved remarkable in its ability to reproduce the experimentally measured loss spectra on diverse surfaces, to the point where the theory has predicted in advance of experiment energy regimes and scattering geometries within which the excitation cross section of selected modes is particularly large [7]. Early calculations [8] demonstrated that in the LEED range of energies, the off specular excitation cross sections are substantially larger than at the low energies employed in the near specular dipole scattering dominated regime. Also, these studies showed that the energy and angle variation of the off specular loss cross sections are very sensitive to surface structure, a conclusion reinforced by recent comparisons between theory and experiment, for the c(2x2)S overlayer on Ni(100) [9].

The ingredients of the theory are as follows, in schematic form. The adiabatic approximation is invoked by supposing the electron encounters the crystal in a disordered state, with the disorder a consequence of the thermal motions of the atoms. The atoms thus sit at positions $\{\vec{R}(\vec{\ell})\}$, which differ from the equilibrium lattice positions $\{\vec{R}_o(\vec{\ell})\}$. We write $\vec{R}(\vec{\ell}) = \vec{R}_o(\vec{\ell}) + \vec{u}(\vec{\ell})$, with $\vec{u}(\vec{\ell})$ the displacement produced by the thermal motions. The electron approaches the surface in a state with wave vector $\vec{k}^{(I)}$, and is scattered to the final state $\vec{k}^{(S)}$ which, by virtue of the disorder, need not be the specular or a Bragg direction.

Let $f\left(\vec{k}^{(I)}, \vec{k}^{(S)}; \{\vec{R}(\vec{\ell})\}\right)$ be the scattering amplitude for this process. One can write down the formal expression for f, through use of multiple scattering theory. Our procedure is to expand f in powers of $\vec{u}(\vec{\ell})$, recognizing that the single phonon losses are described by the leading term in the perturbation expansion:

$$f\left(\vec{k}^{(I)}, \vec{k}^{(S)}; \{\vec{R}(\vec{\ell})\}\right) = f\left(\vec{k}^{(I)}, \vec{k}^{(S)}; \{\vec{R}_o(\vec{\ell})\}\right) + \sum_{\vec{\ell}} \sum_{\alpha} \left(\frac{\partial f}{\partial R_\alpha(\vec{\ell})}\right) u_\alpha(\vec{\ell}) + \cdots . \tag{3}$$

The first term in Eq. (3) describes the elastic scattering studied in LEED; it vanishes unless $\vec{k}^{(S)}$ describes a Bragg beam. The sum on α in the second term ranges over the Cartesian directions x, y and z.

The procedure is to use multiple scattering theory to compute the scattering amplitude derivatives $(\partial f/\partial R_\alpha(\vec{\ell}))$. When the consequences of translational invariance in the two directions parallel to the surface are taken into account,(6) for each atomic layer we require in general three derivatives, corresponding to the choices x, y and z for α, for simple crystals with one atom per surface unit cell. We find that calculating $(\partial f/\partial R_\alpha(\vec{\ell}))$ for five layers provides fully converged results.

We also need the eigenvectors and frequencies of the various surface and bulk phonons of the crystal of interest, to generate a representation of $u_\alpha(\vec{\ell})$. This we do by a lattice dynamical "slab" calculation. In nearly all our work to date, we have found that relatively simple force constant models prove adequate.

In the calculations of $(\partial f/\partial R_\alpha(\vec{\ell}))$, we use the "muffin tin" picture, within which potentials generated by electronic structure calculations are imbeded. For many of the simple surfaces, the surface geometry has been well established through LEED studies, photoemission and various atom beam probes. Thus, there are no adjustable parameters in this phase of the analysis, for such systems. We then concentrate our attention on only the lattice dynamics. The surface phonons at the Brillouin zone boundary often involve rather simple atomic motions, and thus are influenced sensitively by relatively few force constants. For instance, on Ni(100) at the $\bar{X}$ point, the Rayleigh surface phonon involves perpendicular motions of the atoms in the surface layer. The frequency of this mode is uninfluenced by couplings between atoms within the surface layer as a consequence. There is a high frequency surface phonon at $\bar{X}$, the S_6 mode, which resides in a gap in the projected bulk phonon density of states. For this mode, the atomic motions in the outer layer are parallel to the surface, and the frequency of the mode is thus influenced importantly by the force constants within the surface.

The outline of the remainder of this paper is as follows. Section II is devoted to a review of one example that we have studied in detail, the nature of surface vibrations on the Ni(110) surface. Here we show the role excitation cross section calculations play in the analysis. We present new calculations for the Cu(111) surface in section III. Two lattice dynamical models have been proposed to describe this surface; they each give rather similar frequency spectra, but very different eigenvectors for the various modes. A consequence is that the theoretical excitation cross sections differ very dramatically.

2. An Example: Surface Vibrations on Ni(110)

As an example of the points made above, we shall discuss some results presented earlier [11] for the Ni(110) surface. In ref. (11), the reader will find the complete set of results, including dispersion curves of various surface phonons, and a considerable number of comparisons between calculated and measured loss cross sections.

The Ni(110) surface is unreconstructed, and consists of rows of closely spaced atoms. In this section, we limit our attention to off specular energy loss data taken with the wave vector $\vec{q}_\parallel$ set to the $\bar{X}$ point of the surface Brillouin zone. Thus, $\vec{q}_\parallel$ is directed along the closely packed rows. A selection rule [8] limits contributions to the loss cross section to features produced by atomic motions either normal to the surface or parallel to $\vec{q}_\parallel$, when $\vec{k}^{(I)}$ and $\vec{k}^{(S)}$ both lie in the plane that contains $\vec{q}_\parallel$ and the normal to the surface. From the examples below, all for this value of $\vec{q}_\parallel$, one can appreciate how sensitive the loss cross sections are to variations of both beam energy and angle. Also, it will be evident that the theory provides an excellent account of the data.

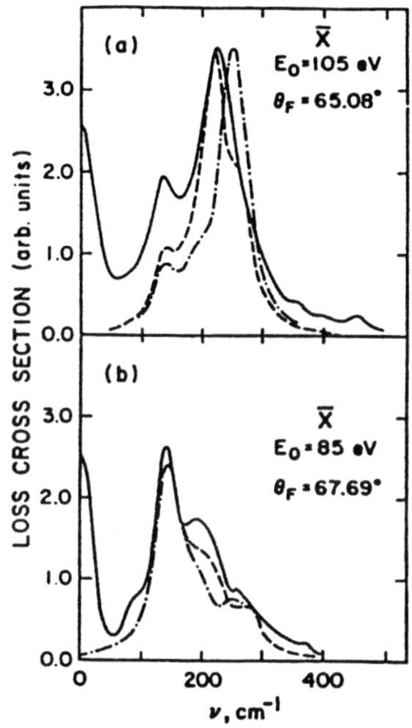

Fig. (1): At the two energies indicated, and for $\vec{q}_{\parallel}$ at the $\bar{X}$ point of the surface Brillouin zone, of the Ni(110) surface, we show two experimental loss spectra (solid lines), one at (a) a beam energy of 105eV and one at (b) a beam energy of 85eV. In each figure, the dot-dash line is calculated with no softening of surface force constant, and the dashed line for a 30% softening of force constant between atoms within the rows on the surface. The figure is reproduced from Ref. (11).

In Fig. (1a), we show off specular loss data (solid line) taken at a beam energy of 105eV, with the scattering geometry such that the scattered electron wave vector makes and angle θ_F with respect to the surface, where $\theta_F = 65.08^{o}$. We see a large, prominent peak just above 200 cm^{-1}, and a weaker feature just above 100 cm^{-1}. The former lies close in frequency to a surface phonon which lies in a gap in the projected bulk phonon density of states. This mode has atomic motions parallel to the surface, and to $\vec{q}_{\parallel}$, in the outermost layer. The dot-dash line is our theoretical loss cross section, calculated by assuming the force constant between atoms in the rows equals the bulk Ni nearest neighbor force constant in value. The overall shape of the theoretical spectrum is quite good, but clearly the gap surface phonon is too high in frequency. This mode has frequency quite sensitive to the force constant between nearest neighbors within an atomic row, as one would suspect from its character. Reduction of this force constant by 30% produces the dashed curve, which agrees very well with the data. The low frequency feature is a zone boundary Rayleigh wave with surface atom motion normal to the surface. As one sees by comparing the two theoretical curves, its frequency is insensitive to this force constant.

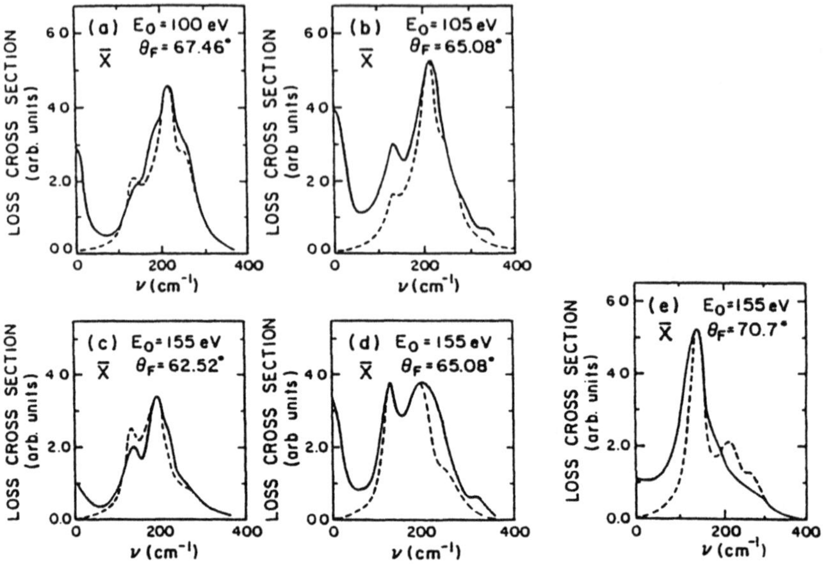

Fig. (2): For various beam energies and exit angles, and for the wave vector transfer $\vec{q}_{\parallel}$ set to the $\bar{X}$ point of the Brillouin zone of the Ni(110) surface, we show comparison between theory (dashed line) and experiment (solid line), for Ni(110). The theory is for the preferred model discussed in the text, in which the force constant between atoms within the surface rows is reduced by 30%.The figure is reproduced from ref. (11).

One can see from this example that the loss cross section calculations allow one to identify the origin of the various features in the data. The gap mode appears as the dominant feature in the data only near 105eV, for example. Then from the character of the atomic motions associated with the mode, one can isolate particular force constants that affect its frequency sensitively, and construct a picture of the surface lattice dynamics. When this is done for several modes, at more than one inequivalent point on the surface Brillouin zone boundary (where the pattern of atomic motions is simple), a complete picture of the various surface force constants can be inferred. One can test the result by requiring the model to reproduce the dispersion curves throughout the surface Brillouin zone.

In Fig. (1b), we show the loss cross section at the beam energy of 85eV, again with the wave vector transfer $\vec{q}_{\parallel}$ at $\bar{X}$. The loss data in Fig. (1b) differs dramatically from that in Fig. (1a), even though precisely the same point in the Brillouin zone is being probed. Now the Rayleigh wave stands out prominently. The theory with the preferred model (dashed line) again works nicely. The structure near 200 cm^{-1}, a bit more prominent in the data than the theory, has its origin in the scattering from bulk phonons reflected off the surface, near the edge of the bulk phonon spectrum that lies below the gap surface phonon seen in Fig. (1a). This structure is absent completely from the spectrum calculated with no softening of the force constant. This reinforces the view that there is indeed softening of the force constant between atoms within the surface rows.

In Fig. (2), we show a comparison between theory and experiment for various energies, and also for various incident angles at fixed energy (Fig. (2c) - (2e)). From Fig. (2a) and Fig. (2b), we see that even a 5eV change in beam energy can lead to appreciable changes in the measured and calculated loss spectra (of course, the angles then must change if $\vec{q}_{\parallel}$ is fixed); the last three panels exhibit the strong dependence on scattering angle.

When one realizes that all the above spectra are probing the surface lattice dynamics at one selected value of the wave vector $\vec{q}_{\parallel}$, one can appreciate that a truly substantial amount of information on the surface response is contained in the off specular electron energy loss studies of surface vibrations. Furthermore, the multiple scattering description of the excitation cross section tracks the features in the data very well indeed. We have found this to be the case in every surface we have examined so far.

The two lattice dynamical models compared in Fig. (1) provided different frequencies and dispersion curves for the various modes that control the surface lattice dynamics, but the character of the eigenvectors are roughly similar for the two. We now turn to a case where the nature of the eigenvectors associated with the surface modes has been the topic of discussion.

3. Cu(111): Sensitivity of Off Specular Electron Energy Loss Spectra to Surface Phonon Polarization

There has been considerable interest in the lattice dynamics of the (111) surfaces of the noble metals Cu, Ag and Au, in response to very interesting data presented by the Toennies group. These authors explored the dispersion relation of Rayleigh surface phonons on the above surfaces, by the method of inelastic helium atom scattering, a technique pioneered by this group, and which stands as a powerful form of surface vibrational spectroscopy complementary to the electron energy loss studies described here. In addition to the Rayleigh wave, at higher frequencies in the case of Cu and Ag, the authors found a well defined resonance feature within the substrate phonon bands. It displayed dispersion characteristic of an acoustical phonon. The spectrum of Au(111) is more complex, but displays related anomalies.

An explanation for the anomaly on Ag(111) was put forward by Bortolani et al. [13]. These authors suggest that the force constants between Ag atoms within the surface layer are softened with respect to their bulk value by roughly 50%. The softening leads to a resonance, in which the motion of the outermost Ag atoms is primarily longitudinal (parallel to both $\vec{q}_{\parallel}$ and the surface). The resonance has the correct dispersion to account for the data, and the authors present calculated loss spectra with relative intensities in good accord with the data.

These authors then analyzed the case of Cu(111), and argued that a similar very large reduction in force is required in this case, to obtain an account of the measured dispersion and intensities [14].

This proposal for such a substantial force constant softening within a simple, unreconstructed surface with virtually no interlayer relaxation stimulated the present authors, in collaboration with Kesmodel and Mohamed, [15] to explore this surface with off specular electron scattering. A primary aim of the work was to search for a high frequency surface mode at the $\bar{M}$ point of the surface Brillouin zone. This mode, predicted by simple models of the surface lattice dynamics, is highly localized on the surface, and involves motion of surface atoms in the outermost layer that is primarily longitudinal. The frequency of the mode is thus a sensitive measure of the force constant between atoms within the surface layer, in a manner similar to the high frequency gap phonon at $\bar{X}$ on Ni(110), discussed in section II. Such high frequency modes are difficult to see with He scattering.

The longitudinal surface phonon was found in the electron scattering work, at a frequency of 210 cm^{-1}. Also, the "longitudinal" resonance and Rayleigh wave were studied, and their frequencies are in excellent accord with the He data where the two data sets overlap. The off specular electron loss data could follow the resonance out to the Brillouin zone boundary, while the original He data followed it to about 2/3 of the way out.

A simple nearest neighbor force constant model provides an excellent fit to the overall phonon spectrum of bulk Cu [15]. This model, applied to the (111) surface with all force constants equal to their bulk value, gives 220 cm^{-1} for the high frequency surface phonon. A small (15%) softening reduces this to the observed 210 cm^{-1}. This picture gives the resonance near the zone boundary to be mainly longitudinal, and the Rayleigh wave at $\bar{M}$ to be mainly perpendicular in polarization. An embedded atom analysis of the Cu(111) surface provides a very similar description in the end [16].

Bortolani et al. have refined their original model of Cu(111), to argue that in addition to the large force constant softening (62%) within the surface, there is a large stiffening (75%) of the force constant between the outermost two atomic layers. This new feature, the stiffening of the force constant between the first and second layer, is necessary to generate the correct frequency for the gap surface phonon, once the very large softening of the intra planar force constant is introduced. With this model, very different than the simple picture set forth in ref. (15), they obtain an account of the frequency of the gap mode, along with the dispersion relation of the Rayleigh wave and the surface resonance.

It is, of course, well known from bulk lattice dynamics that two different force constant models can give very similar frequency spectra. However, the eigenvectors of any two such models must differ, and in this case they differ very substantially. In the model of Bortolani et al., at the $\bar{M}$ point, <u>both</u> the Rayleigh wave <u>and</u> the longitudinal resonance have mixed character, with parallel and perpendicular atomic displacements of comparable magnitude. As discussed earlier, the former is primarily a perpendicular mode and the latter is primarily a longitudinal mode in the outermost layer, in the model that invokes only the modest change in surface force constant.

A consequence is that the off specular electron cross sections are dramatically different for the two pictures of Cu(111), according to theoretical calculations we have carried out [17]. We summarize some of the results here.

In Fig. (3), we show experimental data on the off specular loss cross section for Cu(111), at three different beam energies. In each case, the momentum transfer $\vec{q}_{\parallel}$ is arranged

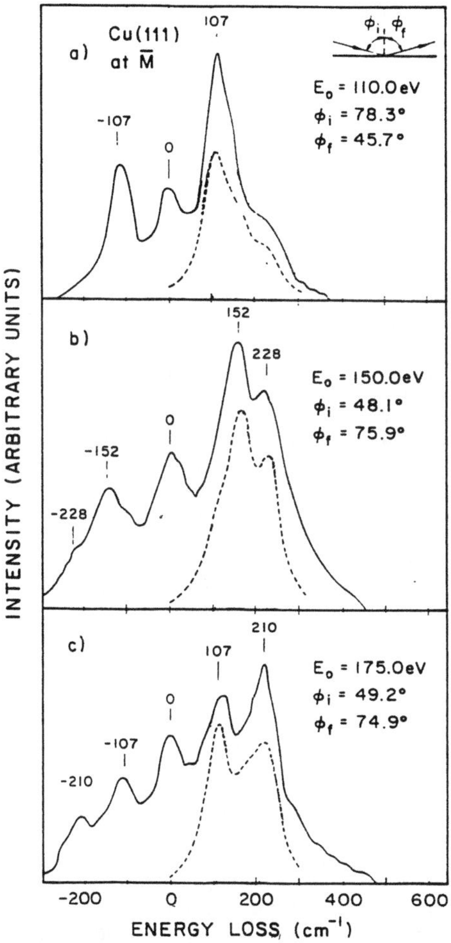

Fig. (3): The electron energy loss spectra measured for Cu(111) by Mohamed and Kesmodel (solid lines). In each figure, the phonons at the $\bar{M}$ point of the surface Brillouin zone are excited. We show spectra at three beam energies, (a) 110eV, (b) 150eV and (c) 175eV. The dashed curves are generated theoretically, from a model where all force constants assume their bulk values, save those within the surface layer, which is reduced by 15%. The figure is adapted from reference (15).

so the phonons excited are at the $\bar{M}$ point of the surface Brillouin zone. The solid line is the data, with both the Stokes (energy loss) and anti Stokes (energy gain) side of the spectrum shown. In Fig. (3a), the 107 cm^{-1} loss is the Rayleigh wave, and the high frequency shoulder comes from bulk phonons that reflect off the surface, and excite vibrational motion of the surface atoms. At 150eV beam energy, the 152 cm^{-1} feature is the zone boundary frequency of the resonance seen earlier at smaller wave vectors by the Toennies group, and the high frequency 228 cm^{-1} peak has its origin in

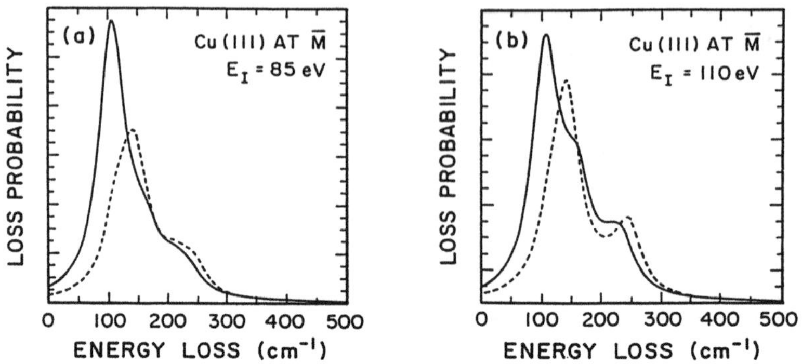

Fig. (4): For excitation of phonons at the $\bar{M}$ point in the surface Brillouin zone of Cu(111), we show theoretical loss cross sections for beam energies (a) E_I = 85eV and (b) E_I = 110eV. The solid lines are calculated for the model proposed in ref. (15), where all force constants assume bulk values save for those within the surface, that are reduced 15%. The dashed lines are for the model of the Modena group, where the force constant between atoms in the surface is reduced 62%, that between first and second layer atoms increased by 75%. In the language of ref. (15), the calculations are for geometry 1.

high frequency bulk phonons, above the gap within which the longitudinal surface phonon resides. At 175eV, the Rayleigh wave shows clearly, along with the gap mode at 210 cm^{-1}.

The dashed lines are theoretical loss spectra, for the simple model wherein the force constants are equal to their bulk value everywhere, save for a slight 15% reduction of that between atoms within the surface layer. The agreement between theory and experiment is very good indeed, as in the case of Ni (110) and the several other surfaces we have examined. The energy loss spectra from Cu(111) share a common feature with all other low index metal surfaces examined to date: at nearly all beam energies, the (predominantly perpendicular) Rayleigh wave is seen clearly and strongly. Longitudinal modes, such as in this case the resonance (to use the description provided by the simple model) and the high frequency 210 cm^{-1} surface phonon, appear only in narrow "energy windows" within which interference effects render the excitation cross section of the Rayleigh wave to be rather small. For Ni (100), the reader may appreciate this from the comparison between theory and experiment given in Fig. (3) of reference (7), and for Cu(111) from Fig. (4) of reference (15); see also the rather detailed commentary in reference (18).

When we calculate loss spectra for the Modena model for Cu(111), the results disagree strongly with the data. At $\bar{M}$, the 152 cm^{-1} resonance appears in the spectrum at many beam energies. In this picture, it contains a very strong admixture of perpendicular motion. One either finds a spectrum dominated by the resonance, or a structure which contains contributions from both the Rayleigh wave and the resonance. We have only occasional spectra where the Rayleigh wave dominates alone while, as remarked above, the experiments (and the theory provided by the simple model) provide spectra where the Rayleigh wave is seen over very broad ranges of energy.

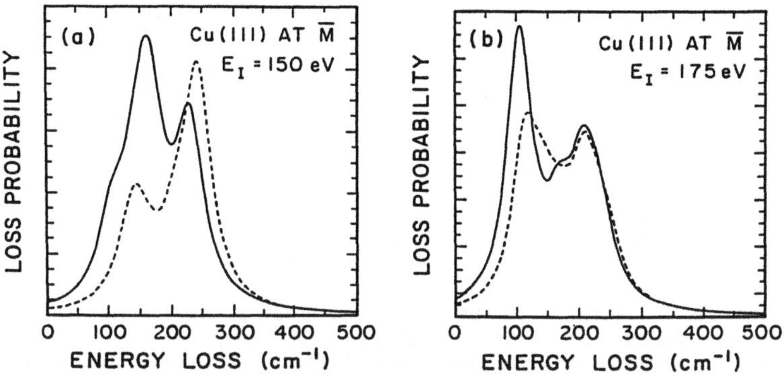

Fig. (5): The same as Figure (4), except now the calculations are performed at the beam energies of (a) 150eV, and (b) 175eV. Also, these calculations are for geometry 2, in reference (15). The theoretical curves are to be compared to the data in Fig. (3b) and Fig. (3c).

We illustrate these points in Fig. (4), where we show loss spectra calculated with the Modena model for $\bar{M}$, at the beam energies of 85eV, and 110eV; the latter is the same energy as illustrated in Fig. (3a). We see that the Modena model (dashed line) leads to the prediction of a spectrum in which the resonance is the dominant feature, rather than the Rayleigh wave as found in the experiments. (In this version of the Modena model, the frequency of the resonance is a few wave numbers lower than that in the model used in reference (15)).

In Fig. (3b), there is also a loss structure near 250 cm^{-1}, at the top of the Cu phonon bands, in the spectra calculated for the Modena model. This has its origin in a resonance associated with the perpendicular motions of surface atoms which is located at the high frequency edge of the bulk phonon bands; this is produced, we believe, by the large stiffening of the force constant between the first and second layer. Such a structure appears in a number of our calculations, but so far as we know no such feature has been observed in the experiments. Our analyses have been carried out for a simplified version [17] of the full many parameter Modena model of the surface lattice dynamics of Cu(111). We understand that in the full model, there is in fact a surface phonon of perpendicular character pushed out of the top of the Cu phonon bands by the force constant stiffening [19]. Thus we would suppose that in a loss spectrum calculated with the full model, this high frequency loss structure would be moved farther up in frequency.

In Fig. (5), we compare the predictions of the two models, at the beam energies where data is displayed in Fig. (3b) and Fig. (3c), 150eV and 175eV respectively. At 150eV, both models provide a structure from the resonance, but in the Modena model, the most prominent structure is associated with the perpendicular resonance near the top of the phonon band at 250 cm^{-1}. The data and the "simple model" produce a less dramatic feature near 228 cm^{-1}, in the middle of the band of bulk phonons which lie above the gap at $\bar{M}$. As remarked in the previous paragraph, in the full multi-parameter Modena model, the prominent high frequency loss would be shifted upward in frequency, much farther than the 228 cm^{-1} feature in the data.

At 175eV, we have an energy where there is no qualitative difference between the two models. In each, we see the high frequency gap phonon, and a loss peak associated with the Rayleigh wave.

We can see from this example that one can extract information not only on the frequencies or dispersion curves of surface phonons from off specular electron energy loss spectra, but in addition the energy variation of the loss intensities associated with various modes provides information on their polarization, a point emphasized earlier on in our theoretical work on this topic [8]. Here, of course, we have real data to compare with theory.

Our conclusion is that for Cu(111), a very simple model of the lattice dynamics of the surface, in which the force constant between atoms in the surface layer is softened only very slightly (~ 15%), provides a very good account not only of the measured dispersion relations for the various modes, but in addition it gives a remarkably quantitative account of the energy variation of the mode intensities. It predicts accurately the "energy windows" within which the resonance is observed, and in which the high frequency longitudinal phonon is observed.

The loss spectra calculated for the Modena model, with its very large surface force constant softening, and very large stiffening of the force constant between first and second layer, are in qualitative disagreement with the electron energy loss data. It is our understanding that the features contained in this model are essential to the understanding of the He scattering data. It follows that important questions are yet to be resolved; the reader must keep in mind that the frequencies and dispersion relations produced by the two methods agree well, where ever the data sets overlap.

Acknowledgments

We have enjoyed stimulating discussion of the lattice dynamics of Cu(111) surfaces with Prof. V. V. Bortolani. We are grateful to him and his colleagues for supplying us with details of their model, in advance of publication.

This research was supported by the U. S. Department of Energy, through Grant No. DE-F603-84ER45083.

References

1. H. Ibach, Phys. Rev. Letters 24, 1416 (1970).

2. H. Ibach, Phys. Rev. Letters 27, 253 (1971).

3. See Chapter 3 of Electron Energy Loss Spectroscopy and Surface Vibrations, H. Ibach and D.L. Mills, (Academic Press, New York, 1982).

4. E. Evans and D.L. Mills, Phys. Rev. B5, 4126 (1972).

5. S. Lehwald, J. Szeftel, H. Ibach, Talat S. Rahman and D.L. Mills, Phys. Rev. Letters 50, 518 (1983).

6. C.H. Li, S.Y. Tong and D.L. Mills, Phys. Rev. B21, 3057 (1980).

7. This was first done for a high frequency surface phonon (the S_6 mode) on the Ni (100) surface. This mode has surface atom displacements parallel to the surface, and a small excitation cross section save in selected "energy windows". See M.L. Xu, B.M. Hall, S.Y. Tong, M. Rocca, H. Ibach, S. Lehwald, and J.E. Black, Phys. Rev. Letters 54, 1171 (1986).

8. C.H. Li, S.Y. Tong and D.L. Mills, Phys. Rev. Letters 44, 407 (1980), and S.Y. Tong, C.H. Li and D.L. Mills, Phys. Rev. B24, 806 (1981).

9. S.Y. Tong et al., Phys. Rev. B39, 3116 (1989).

10. Burl M. Hall and D.L. Mills, Phys. Rev. B34, 8318 (1986).

11. S. Lehwald, F. Wolf, H. Ibach, Burl M. Hall and D.L. Mills, Surface Science 192, 131 (1987).

12. U. Harten, J.P. Toennies and Ch. Wöll, Faraday Discuss. Chem. Soc. 80, 137 (1985).

13. V. Bortolani, A. Franchini, F. Nizzoli and G. Santoro, Phys. Rev. Letters 52, 429 (1984).

14. V. Bortolani, A. Franchini and G. Santoro, private communication.

15. Burl M. Hall, D.L. Mills, Mohamed H. Mohamed and L.L. Kesmodel, Phys. Rev. B38, 5856 (1988).

16. J.S. Nelson, M.S. Daw and Erik C. Sowa, Phys. Rev. B40, 12618 (1989).

17. We are grateful to Prof. V.V. Bortolani for providing us with details of a nearest neighbor model which reproduces the essential features of the full multi-parameter model of Cu (111) developed by the Modena group.

18. S.Y. Tong and D.L. Mills, Phil. Trans. Roy. Soc. London A318, 179 (1986).

19. V.V. Bortolani, private communication.

Auger Electron Spectroscopy Measurements on Na β''-Alumina Crystals

C.A. Achete[1] *and F.L. Freire, Jr.*[2]

[1]LEMI/PEMM/COPPE, UFRJ, Cx. Postal 68505, 21910 Rio de Janeiro, RJ, Brazil

[2]Departamento de Física, Pontifícia Universidade Católica do Rio de Janeiro, Cx. Postal 38071, 22453 Rio de Janeiro, RJ, Brazil

Abstract: Auger analysis made along the conduction planes of sodium β"-alumina crystals shows an increase with time of the Na-signal intensity. Scanning microscopy analysis reveals the growth of sodium metallic clusters on the surface at the impact point of the primary electron beam. Results obtained bombarding the surface parallel to the crystallographic c-axis direction show that the sodium concentration either remains stable or decreases, depending on the electron fluxes.

1. Introduction

Electron bombardment of surfaces during Auger Electron Spectroscopy (AES) causes electronic excitations of both surface and bulk atoms and molecules. The electronic excitations may result in damage to the surface monolayer by a number of processes, including dissociation of surface molecules and desorption of ionic and neutral fragments. It should be noted that such excitations influence not only surface monolayers, but can also result in changes to thicker surface layers.

The examination of surfaces of dielectric materials with AES is complicated by the appearance of a surface charge [1]. Up to now, few attempts to characterize the surface of solid-state ionic conductors have been made, despite the importance of these materials in the development of electrochemical devices where surface and interface characterizations are mandatory [2].

2. Experimental Procedure

Single crystals of sodium β"-alumina were obtained by melting $NaCO_3$, MgO and Al_2O_3 with standard procedure. Their composition was $Na_{1.67}Mg_{0.67}Al_{10.33}O_{17}$ and typical dimensions were 10 x 8 x 0.2mm^3.

A Physical Electronics Model 590 Scanning Auger Electron spectrometer was used. The electron beam employed for bombardment and subsequent AES measurements made an angle of 60° with the sample

Springer Proceedings in Physics, Volume 62
Surface Science Eds.: F.A. Ponce and M. Cardona

surface. The beam diameter was determined to be of the order of 25μm and its profile was nearly homogeneous.

3. Results and Discussions

Figure 1 shows Auger-electron spectra from the edge surface of the sodium β"-alumina crystal. The spectrum presented in figure 1a is from the crystal as loaded in the UHV analysis chamber. No trace of the aluminum signal is seen in this spectrum, which probably means that the real surface of the crystal was covered by sodium compounds. After one minute of 1keV argon ion sputtering, these surface contaminations were almost completely removed (figure 1c). Meanwhile, the previous state of the surface is of no importance if it is subjected to a heavy electron bombardment. This is clear from figures 1b and 1d, which show Auger spectra obtained after two minutes of 3keV, $100mA/cm^2$ electron-irradiation. The former corresponds to the edge surface of the crystal, as inserted in the analysis chamber, and then, irradiated. The last one corresponds to a spectrum obtained in a zone of the edge surface previously subjected to sputtering. Clearly, within the detection limits of AES, sodium became the only element present at the surface.

It is very important to note that no Auger-line energy shifts were observed in the spectra shown in figure 1. But this is not a general rule. We also observed an energy shift of the order of 150ev, accompanied by a sodium increase of only twenty per cent with respect to the initial sodium concentration.

In figure 2 we present two images from the edge face of the crystal made by Scanning Electron Microscopy. Note that the dimensions of the sodium cluster (30μm of diameter) were of the order of the beam size.

The electron bombardment of the cleavage face, could provoke surface charging, as in the case of the edge irradiation. However, in the range

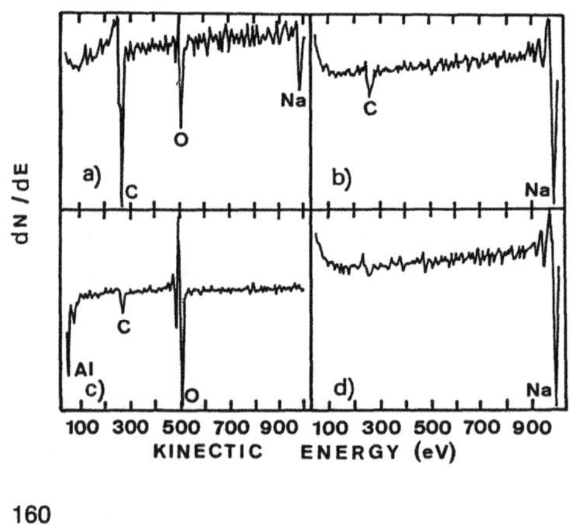

FIGURE 1:
Auger spectra from the Na β"-alumina edge face.

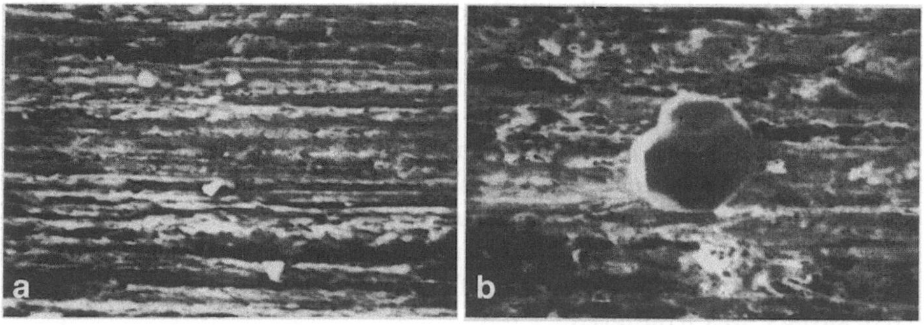

FIGURE 2: SEM images from the edge face of the Na β"-alumina crystals, (a) as inserted in the analysis chamber, (magnification: 400x), and (b) after two minutes of electron-bombardment with 3keV, $100mA/cm^2$ (magnification: 800x)

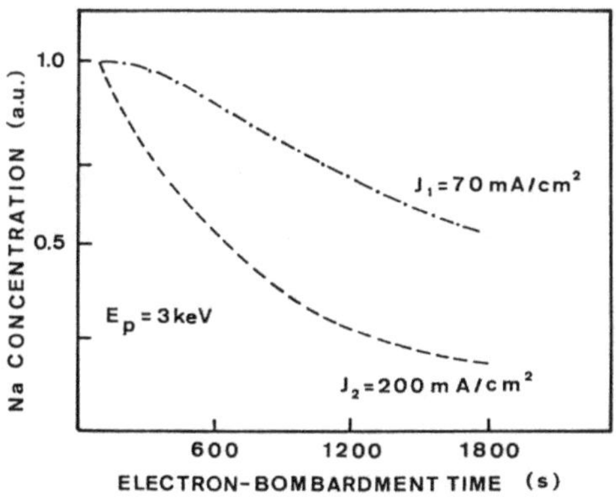

FIGURE 3: Experimental Na_{KLL}-Auger peak-to-peak height as a function of electron-bombardment for different current densities.

of electron fluxes employed in this work no increase of sodium intensity was observed (figure 3).

As the differences between the crystal structure of β– and β"-alumina are of minor importance, we can expect that the sodium transport processes in these materials, when submitted to electron bombardment, present some similarities. Livshits and Polak [3] determined a correlation between the appearance of a net surface charge and the segregation of sodium at the irradiated surface of β-alumina crystals.

Although we worked in a range of current densities lower than that used by Livshits and Polak, we did not observe the correlation between the formation of sodium clusters at the irradiated surface and the energy shift in the Auger spectrum. The charging of the irradiated surface was observed to be dependent upon many beam parameters, and the status of the surface.

Mariotto and Miotello [4] described these results with a continuity equation which includes both ordinary and electric-field assisted migration, taking account of the Electron Stimulated Desorption (ESD) through the boundary conditions. So, we can think of the behaviour of the time evolution of the sodium concentration at cleavage surface as resulting from the Na^+ electric-field assisted transport process through the spinel blocks of β"-alumina structure and the ESD. This transport process was strongly enhanced by electron irradiation. However, this model fails in describing the formation of metallic sodium clusters when the edge surface was irradiated.

More work is needed to explain the formation of metallic sodium clusters when the edge face was irradiated with an electron-beam.

References

1. C.G. Pantano and T.E. Madey, Appl. Surface Sci. **7** (1981) 115.
2. C. Julien, Mat. Scienc. Eng. **B6** (1990) 9.
3. A. Livshits and M. Polak, Surface Sci. **119** (1982) 314.
   M. Polak and A. Livshits, Appl. Surface Sci. **10** (1982) 446.
4. G. Mariotto and A. Miotello, in: Solid State Ionics, eds. G. Nazri, R.A. Huggins, D.S. Schriver and P.T. Hu (Materials Research Society, Pittsburgh, 1989) vol. 135, p. 505.

Forward Focusing Effect in the Elastic Scattering of Electrons from Cu(001)

H. Ascolani, M.M. Guraya, and G. Zampieri*

Centro Atómico Bariloche, C.N.E.A.,
8400 S.C. de Bariloche, Río Negro, Argentina

We have measured the intensity of the elastic reflection of electrons impinging on Cu(001) as a function of the polar angle of emergence along the [010] azimuth and of the energy of the electrons in the range of 200 to 1500 eV. We have found that the forward focusing effect, which occurs in Auger electron diffraction (AED) and x-ray photoelectron diffraction (XPD) /1-3/, is the dominant effect at electron energies greater than 500 eV. The importance of this finding is that both the reciprocal and the real lattices can be explored in the same experiment by simply varying the energy of the electrons.

A Polar Intensity Plot (PIP) of the LMM Auger electron emission (E = 917 eV) was included in figure 1 for comparison with our results of the elastic reflection. The structures observed in this PIP are due to interference effects involving relatively few atoms. The most prominent of these structure occur at 0 and 45° (emission along [001] and [011] axes) and have a simple explanation /2/ : The outgoing Auger electron is focused along the bond directions by the attractive potentials of the near neighbors of the emitting atom. This forward focusing effect gives rise to maxima in the PIPs of the electron emission at angles connected with the main crystallographic axes. The structure at $\Theta \simeq 20°$ has two sources; besides an enhancement due to forward focusing along the [013] axis, there is a first order interference effect which is expected to be important at these polar angles /4/.

The PIP of the elastic peak at 200 eV presents very sharp structures due to LEED effects. The arrows indicate the polar angles at which LEED directions are expected to enter into the acceptance cone of the energy analyzer.

When the energy of the electrons is increased to 400 eV, the PIP is still dominated by LEED effects. An important loss of intensity is observed; this is due to the loss of coherence in the scattering caused by the thermal vibrations of the atoms; an effect which is more important the smaller the wavelenght of the electrons.

At the energy of 900 eV the PIP has taken the form characteristic of the AED and XPD experiments. Note that the anisotropies are a factor of two greater than in the emission of LMM Auger electrons. The LEED structures have disappeared as expected on the basis of the loss of coherence discussed above. Increasing the energy to 1300 eV does not affect the angular position of the two major peaks, only their intensities. This strongly suggests that these anisotropies are due to forward focusing along the [001] and [011] directions. The structure at $\Theta \simeq 20°$ does change and therefore seems to be connected more to an interference effect and less to enhanced emission along the [013] axis.

To know if a picture similar to that of AED and XPD is applicable to the case of elastic reflection of electrons at intermediate energies, we have performed a simple calculation and compared its results with those of fig. 1. We assume that the LEED spots have disappeared completely because of the loss of coherence discussed above, and that the surface atoms behave as independent scatterers. Next, we have adapted the single scattering cluster

* Fellow of CONICET

Springer Proceedings in Physics, Volume 62
Surface Science Eds.: F.A. Ponce and M. Cardona

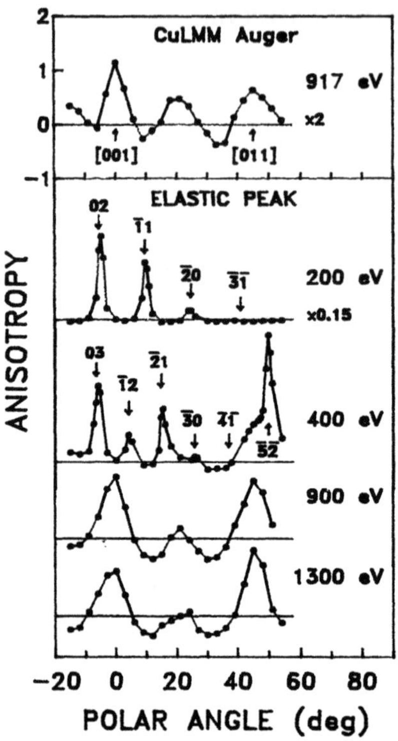

Figure 1: PIPs of the LMM Auger electrons and the elastic peak at the energies indicated on the right. The anisotropies are defined with reference to the emission from a polycrystalline sample: A = (Ic - Ip)/Ip. The vertical scale is valid for all the PIPs.

calculations /1,4/ to our experiment replacing the outgoing Auger/photoelectron wave by the scattering wave $f(\Theta)\exp(ikr)/r$, where $f(\Theta)$ is the plane wave scattering amplitude. The results of these calculations, using very simple clusters, are in excellent agreement with the PIPs of figure 1.

In conclusion, we have found that when the LEED effects disappear at high energies, the angular distribution of the elastically reflected electrons is not smooth but full of anisotropies due to simple interference effects. Interestingly enough, the most prominent of these anisotropies are due to forward focusing and thus simply related to the main crystallographic axes, making the phenomena an excellent complement to LEED for the determination of surface structures.

References

1. S. Kono, S.M. Goldberg, N.F.T. Hall, and C.S. Fadley, Phys. Rev. B 22, 6085 (1980).
2. W.F. Egelhoff, Phys. Rev. B 30, 1052 (1984).
3. S.A. Chambers, S.B. Anderson, and J.A. Weaver, Phys. Rev. B 32, 4872 (1985).
4. E.L. Bullock and C.S. Fadley, Phys. Rev. B 31, 1212 (1985); R.A. Armstrong and W.F. Egelhoff, Surf. Sci. 154, L225 (1985).

XPS Characterization of Nitrogen-Implanted Titanium with Pulsed Ion Beams

C.O. de González[1], G. Scordia[2], and J. Feugeas[2]

[1]Comisión Nacional de Energía Atómica, Dpto. Ciencia de Materiales, Av. Libertador 8250, 1429 Buenos Aires, Argentina and C.I.C., Provincia de Buenos Aires

[2]Université de Technologie de Compiègne, France

[3]Universidad Nacional de Rosario and CONICET, Argentina

Abstract. Samples of pure titanium were nitrogen-implanted with a pulsed ion beam accelerator BD-1 [1]. The nitrogen fluencies were varied between 2 and 30×10^{16} ions/cm^2. The atomic concentrations of O,N and Ti as function of depth were uniform until about 3000 Å in all cases and independent of implantation fluencies. The binding energy position of the N1s peak (397 eV) corresponds to nitrides TiN_x. A direct relation between the binding energies difference (Ti 2p3/2 - N1s) and the value of x was established.

1. Introduction

Nitrogen implantation improves surface properties of metals such as hardness, corrosion and wear resistance. Nitrogenated Ti has a wide range of technical applications. Besides, this transition metal can be easily nitrogenated by ion implantation. In this work we summarize the results obtained after nitriding pure titanium with a pulsed accelerator. This system allows the acceleration of ions up to high energy values (above 1 MeV) and the energy profile of the beams is continuous. These features contribute to the creation of uniform and deep implantation profiles.

2. Experimental

Commercial Ti of T60 grade cut in samples of 7x14x1 mm was used. The samples were mechanically and electrolytically polished before the implantations. The implantations were carried out in the Plasma Focus (PF) device detailed in [1]. The specimens were located at 40 mm from the end of the gun and exposed to different flrencies: 2, 4, 20 and 30×10^{16} ions/cm^2 (samples A) and placed at 100 mm of the "pinch" and implanted with 4×10^{16} ions/cm^2 (samples B).

The XPS measurements were performed with an ESCA3 Mk II spectrometer (V.G. Scientific Ltd.). The spectra were excited with Mg K_α X-rays (1253.6eV). The Au $4f_{7/2}$ line at 83.9eV with respect to the Fermi level was used for binding energy calibration of the spectrometer. The depth profiles were obtained by the sequential application of sputtering with A^+ ions (5kV, 10μA/cm^2) and XPS analysis. The erosion rate was estimated in 10 Å/min.

3. Results and discussion

To determine the atomic percentage concentrations of elements, XPS high resolution narrow spectra of N1s, Ti2p, C1s and O1s were taken and the calculation method described in [2] was applied. The O1s signal, rather important at the surface, was formed by two contributions: adsorbed O (E_b = 532±0.1eV) and oxide (E_b = 530.4±0.1eV). The former was eliminated with ion sputtering while the later remained along the analyzed depth. In A type samples the N and Ti concentrations, 43%at. and 48%at. respectively, as well as the concentrations ratio N/Ti~0.9 did not change along the analyzed depth (~3000Å) and were the same for the different fluencies. In B type samples the N atomic concentration (39%) was lower and constant until 2500 Å and then it decreased to 28% at the end of the analysis (5400 Å). The concentrations ratio N/Ti varied between 0.75 and 0.5.

The chemical shift of the XPS peaks indicated the formation of Ti nitrides. The binding energy (E_b) of the N1s peak was constant, E_b = 397.0 eV for all the samples (A and B). Ti2p3/2 peak kept the energy E_b = 454.7 eV for all implantation fluencies in A type samples. In B type samples, the Ti energy was lower and constant E_b = 454.2 eV until 2500 Å and then it shifted to lower values.

The different E_b values observed are related to the formation of different Ti-N compounds. Nitrides, carbides and oxides of transition metals can exist in a wide range of stoichiometries. Particularly, TiN_x can be obtained with $0.5 \leq x \leq 1.1$. To relate our results with those of Porte et al [3], the difference of binding energies between $Ti2p_{3/2}$ and N1s peaks, ΔE_b, was calculated. Porte [3] observed an increase in ΔE_b with x, reflecting an increase in charge transfer from Ti to N as x increases. A value of ΔE_b = 58.3 eV corresponds to the stoichiometric compound TiN. In A samples ΔE_b = 57.7 eV was constant. We can assume that only one substoichiometric nitride is formed, independent of the implantation fluence. This conclusion agrees with the fact that the ratio N/Ti is ~0.9 and always constant. For a B sample, the difference E_b is constant (57.5 eV) until 250 min of erosion and then it decreases to 57.2 eV. This behaviour is similar to that of the ratio N/Ti showing again the relation between ΔE_b and the stoichiometry. The nitrides formed were TiN_x with $0.5 \leq x \leq 0.7$.

4. Conclusions

Samples of pure titanium were nitrogen-implanted with different fluencies using a pulsed ion implanter. High concentrations of N were observed along the analyzed depth in all samples. Nitrogen was combined forming substoichiometric nitrides. The relation between chemical shifts, charge transfer and compound stoichiometry was confirmed.

Acknowledgements

We are grateful to S. Hild for his help in XPS measurements.

References

[1] J. Feugeas, E. Llonch, C.O. de González and G. Galambos, J. Appl. Phys. 64, 2648 (1988).
[2] C.O. de González and E.A. García. Anais do Coloquio Latinoamericano de Fisica de Superficies, Niteroi, Brasil, 203 (1980).
[3] L. Porte, L. Roux and J. Hanus, Phys. Rev. 28, 3214 (1983).

Part V

Processes at Surfaces: Interface Formation

Microscopic Phenomena in Epitaxy

A.A. Chernov

Institute of Crystallography, USSR Academy of Sciences,
SU-117333 Moscow, Leninsky Prospect 59, USSR

Abstract. Selected surface phenomena controlling epitaxial growth and the structure of epitaxial thin films are briefly reviewed and discussed with emphasis on recent achievements. Some basic statements on epitaxy are presented.

1. INTRODUCTION

Various techniques within the general frames of the vapour, liquid, solid and, especially, molecular beam epitaxy (VPE, LPE, SPE, MBE, respectively) were invented and drastically developed during the last decade. Among those are normal and low pressure chemical vapor deposition (CVD, LPCVD) in chloride-hydrogen and metalorganic-hydrogen systems (CVD, LPMOCVD), atomic and molecular layer epitaxy (ALE, MLE) both in vapour and solutions, gas source or chemical beam epitaxy (CBE), and migration enhanced molecular beam epitaxy (MEMBE).[1-6] Photo, electron and ion-assisted CVD, MBE, and MLE are becoming a popular challenge. Electric current assisted LPE is receiving some attention too. All these technologies continue to support an interest in the fundamentals of epitaxy and pose many new problems in surface physics and chemistry.[9,10]

Epitaxy includes two interrelated groups of phenomena. First, the phenomena that are thermodynamically or kinetically influenced by the substrate directly. These are adsorption, nucleation of a new epitaxial phase (epiphase), expansion of these nuclei or just continuation of the substrate lattice. The surface defect states and impurities are of primary importance in these initial stages.[11]

After complete coverage, the epiphase (epifilm) continues growing following the general mechanisms of crystal growth. However, as a result of elastic stress due to the substrate-epifilm lattice misfit, dislocations and stacking faults born at the substrate-epifilm boundary and substrate-epifilm thermal expansion difference continue to propagate and sometimes even multiply.

Both aspects of epitaxy, including the crystal growth mechanism, will be very briefly touched in this overview, emphasizing selected subjects

Springer Proceedings in Physics, Volume 62
Surface Science Eds.: F.A. Ponce and M. Cardona

which are most attractive from my own scientific and technological points of view.

2. MODES OF EPITAXY

Traditionally, epitaxy is classified into three modes, depending on the film-to-substrate "affinity," as listed in Table I. The "affinity" is actually the driving force for wetting of the substrate by the solid film

$$\Delta\alpha = \alpha + \alpha_{fs} - \alpha_{sm} \tag{1}$$

where α is the interfacial energy between the epifilm (f) and the medium (m) ($\equiv \alpha_{fm}$), α_{fs} is the film-substrate energy, and α_{sm} is the substrate-medium energy. The wetting free energy $\Delta\alpha$ is the work needed to replace the substrate-medium interface by the film-substrate and the film-medium interfaces. The wetting energy may be expressed via the adhesion energy $\alpha_{adh} = \alpha + \alpha_{sm} - \alpha_{sf}$ which is needed to separate the film from the substrate

$$\Delta\alpha = 2\alpha - d_{adh} = (\psi_1 - \psi_s)/a^2. \tag{2}$$

Here the adhesion free energy is expressed by the work ψ_s needed to separate one film particle from the substrate, their contact area being a^2. The doubled surface energy per particle, $\psi_1 = 2\alpha a^2$, may be estimated from the sublimation heat of the crystallizing substance.

All the cases of epitaxy known at present occur on atomically smooth substrate surfaces and include the formation of crystals also covered with atomically smooth surfaces. Thus the concepts of 2D and 3D nucleation and layer growth are most fundamental for the known epitaxies.

Epitaxy in systems with atomically rough interfaces are interesting topics. Fig. 1[13] gives an example of poor wetting (weak "affinity"), known as the Volmer-Weber (VW) mode. It is known to operate, for instance, in noble metals (Au, Ag, Pt, etc.) on alkali halides. In reference [13], after the KBr crystal substrate was cleaned in UHV (10^{-10} torr) and after deposition of Au, a subtle secondary decoration by Au-Pd molecular beam was made up to 0.3 nm effective layer thickness. The primary Au large particles are seen in Fig. 1, to be located at the upper of the two terraces divided by the evaporation spiral step decorated by fine Au-Pd particles. The "bond saturation" model suggests the opposite: the Au deposition on the lower terrace in the step re-entrant corner. A possible but not yet finally proven explanation is that the Au particle induces a stress in the KBr lattice and thus the upper location minimizes the particle–substrate elastic interaction. An additional contribution comes from the step-induced strain in the KBr lattice.

Contrary to the poor wetting typical of the VW mode, the lattice mismatch in the Frank-van-der Merve (FvdM) and Stransky-Krastanov

Table 1. Epitaxy mechanisms

Mechanism and its successive stages	Energy condition	Minimum supersaturation $\Delta\mu$ for beginning of condensation	Examples of condensate–substrate pairs	Investigation technique[a]
Frank-van-der-Merve (strong adhesion, complete wetting) 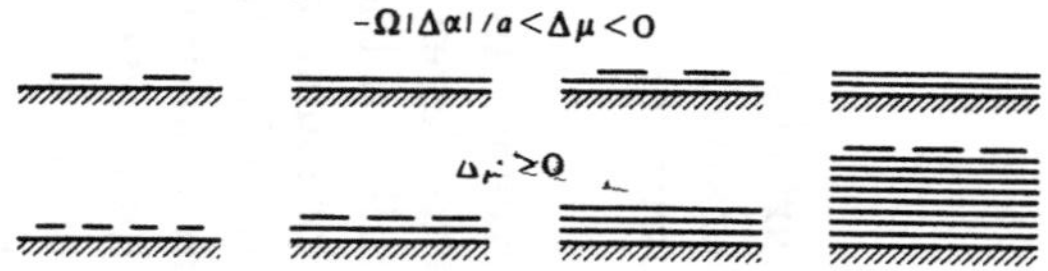	$\Delta\alpha<0$ or $2\alpha<a_S$	$\varepsilon_1-\varepsilon_s<\Delta\mu<0$ undersaturation relative to macrophase	Au/Ag, Ag/Au, Fe/Au, Au/Pd, Xe, Kr/graphite	LEED, RHEED, AES, TEM, ellipsometry, measuring of adsorbate amount
Stransky-Krastanov 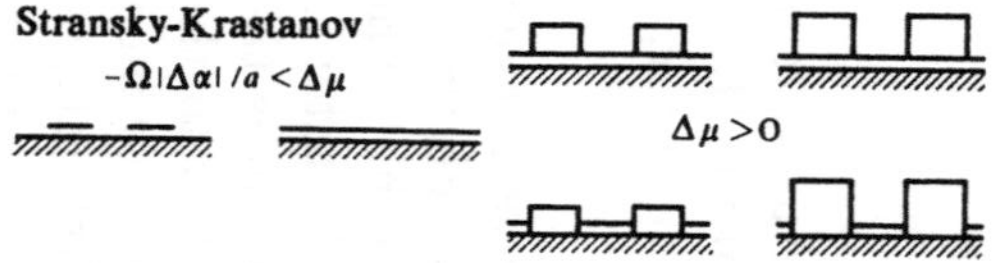	First layers: $\Delta\alpha<0$ Three-dimensional crystals: $\Delta\alpha>0$	First layers: $\Delta\mu<0$ Three-dimensional crystals: $\Delta\mu>0$	Ag/Si, Au/Si	LEED, RHEED, AES, TEM, TDS, isothermal TDS
Volmer-Weber (weak adhesion) 	$\Delta\alpha>0$ or $2\alpha>a_S$	$\Delta\mu>0$	Au/NaCl Ag/KCl	RHEED, AES, TEM

[a] LEED: low-energy electron diffraction. RHEED: reflection high-energy electron diffraction. AES: Auger electron spectroscopy. TEM: transmission electron microscopy. TDS: thermodesorption

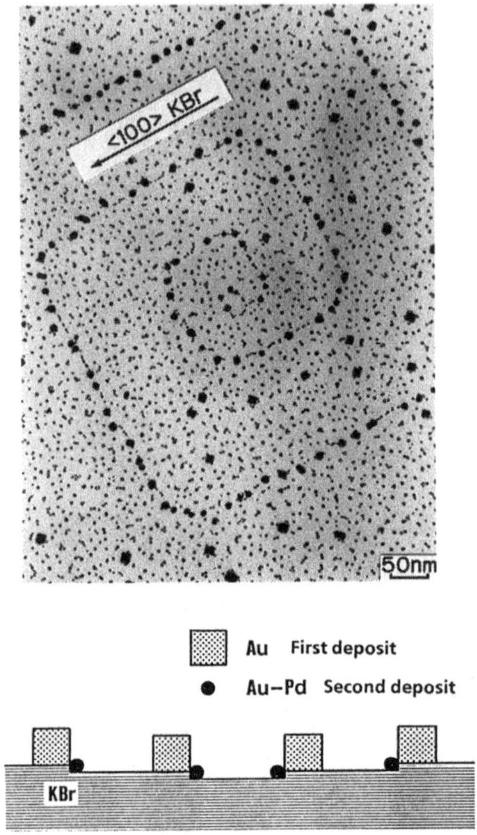

Fig. 1 Au (big particles) and Au-Pd (small particles) deposited on the KBr (100) face containing the evaporation spiral. The Au particles are sitting on the upper, not lower, terrace at the step.[13]

(SK) modes usually does not exceed several percent. In addition, if the types of binding within both the substrate and the layer are similar, the two epilayer lattice constants in the plane parallel to the interface may coincide with the ones of the substrate; i.e., a pseudomorphous epilayer grows. The lattice mismatch induces the strain and stress in the film with energy which is proportional to the film thickness. When this thickness exceeds a critical value (typically, hundreds or thousands of Angstroms), a grid of misfit dislocation arises at the substrate–film interface, thus providing the necessary stress relaxation. Nucleation of the misfit dislocations may be difficult in very perfect and clean films on dislocation–free clean substrates, thus obtaining strained superlattices even for large film thicknesses.

Creation of the misfit dislocations at the interface means an increase in the substrate-epifilm interfacial energy, α_{fs}, in (1); thus wetting becomes

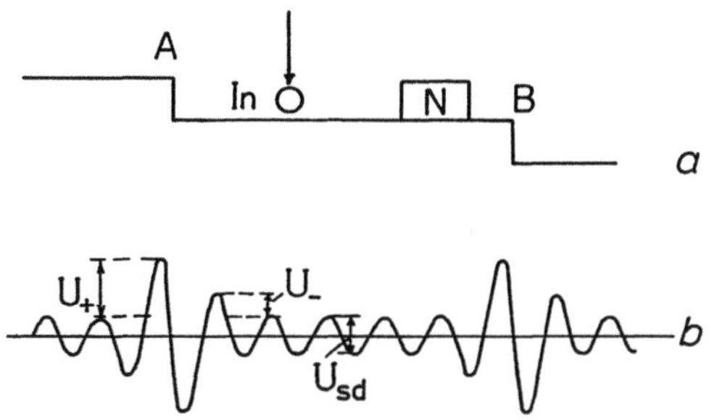

Fig. 2 Asymmetry in the barriers (b) for adatoms to join crystal lattice at steps (a).

worse. If strong, the latter effect may provide coarsening of the continuous film into islands.

An example of this mode, classified by authors[14] as the SK type, is given by the epitaxy of $In_xGa_{1-x}As$/(100)GaAs studied by RHEED. The GaAs and InAs lattice constants are 5.65 and 6.05 Å, respectively. The linear Vegard's law allows to find the constants for any intermediate atomic fraction x. At x = 1; 0.78; 0.66; and 0.23, the calculation gives a critical thickness for lattice relaxation of 4.5; 6.0; 7.0; and 35 Å, correspondingly. The initial RHEED pattern from the GaAs(100) – 2×4 disappears but the surface remains flat (streaky RHEED pattern). As the film thickness reaches 6-7 Å, the diffraction intensity increases; later at 35 Å, the streaky pattern becomes spotty; the electrons are now diffracted both by reflection from the flat surface and by transmission through the $In_{0.66}Ga_{0.34}As$ crystallites into which the formerly grown epifilm is modified or which appear at the later growth stage. Thus, pseudomorphic growth of the first one of two epitaxial layers is replaced by the island growth on top of these layers, which is a definition of the SK mode.

From the quantitative standpoint, however, this interpretation is not unambiguous since the SK mode definition (like the FvdM and VW ones) considers only the thermodynamic driving force and does not pay attention to kinetics. The latter is especially important for MBE controlled by high supersaturations. More specifically, the picture may be reconstructed as follows. An In atom impinged on a terrace between Steps A and B (Fig. 2a) has a very low probability of reevaporation but is sufficiently mobile on the surface at the epitaxial temperature of ~480°C. Step A is a good trap for the In atom, where it may be incorporated pseudomorphically together with Ga and As atoms. Step B is less attractive for the atoms on the AB terrace, since joining the lattice at this step includes passing a relatively high barrier U_+ needed to overcome the convex "edge" of Step B, as seen in Fig. 2b. The asymmetry for joining the step from the lower (–) to the upper (+) terrace (Fig. 2a) (called the Schwöbel effect) is seen, e.g., in a field emission microscope when surface diffusion of single atoms of Re, W and some other metals is observed on various steps surrounding W facets.[15] Other evidence that the barriers U_+ and U_- in Fig. 2b are unequal comes from the asymmetric distribution of Au nuclei on the Ag (111) terraces. The density maximum is shifted to the step B, the barrier asymmetry $U_+ - U_- = 0.67$ eV.[16] The In atom under consideration may also participate in building the pseudomorphous nuclei N (Fig. 2a). However, due to the asymmetry between U_+ and U_- (Fig. 2b), it seems less probable to build two, three or more stories high 3D nuclei rather than to expand the 2D ones. If such a picture is correct, annealing of a rather thick homogeneous pseudomorphic layer would provide its coarsening into 3D islands. Thus the kinetic effects like the ones described above have a tendency to force the systems belonging to the SK type to behave as though they were the FvdM type.

3. EPITAXIAL ORIENTATIONS

Supersaturation, impurities, temperature and deposition rate influence the film to substrate orientation, especially in the VW mode. Atomic defects on the substrate may also influence the orientation of nuclei. The nuclei may not be oriented at all at the initial stages. The regular film to substrate orientation appears only in the coalescence stage.[11] These and many other effects in the "real" systems are not yet well understood.

In the FvdM and SK modes, the lattice fitting provides strict mutual orientation. However, X-ray double crystal diffractometry shows that the epifilm lattice planes deviate from the ones of the substrate by the angles of the order of arc minutes or less and, in addition, lose its bulk symmetry. For example, both magnetic $(YGaTm)_3\,(GaFe)_5O_{12}$ and nonmagnetic $(EuGd)_3\,Ga_5O_{12}$ garnet films grown by LPE on GGG (Gd_3Ga_5–O_{12}) (111) substrates are rhombohedral rather than cubic.[17] The LPE grown $Ga_{1-x}Al_xAs$ films on the GaAs(100), (110), (111) and (113) substrates are also deformed.[18] The relative distance Δd/d between the film lattice planes increases from zero for the planes normal to the growth surface to the values of $\Delta d/d \simeq 5\cdot 10^{-4}$ for the planes inclined from this surface. The lattice plane orientations in the film and the substrate coincide if these planes are normal to the epitaxial interface and are tilted by up to 1.5 arc min from each other. Similar effects have been found in the $Ga_{1-x}Al_x$ As films grown by MBE on the GaAs vicinal surfaces inclined from (100) by several degrees.[19,20]

In all these cases, misfit dislocations were absent. This important fact allows one to strictly define boundary conditions for elasticity theory and to determine the film strain and distortions which evidently are homogeneous. Qualitatively, these distortions and deformations may be understood from Fig. 3. Suppose we have a substrate with the lattice

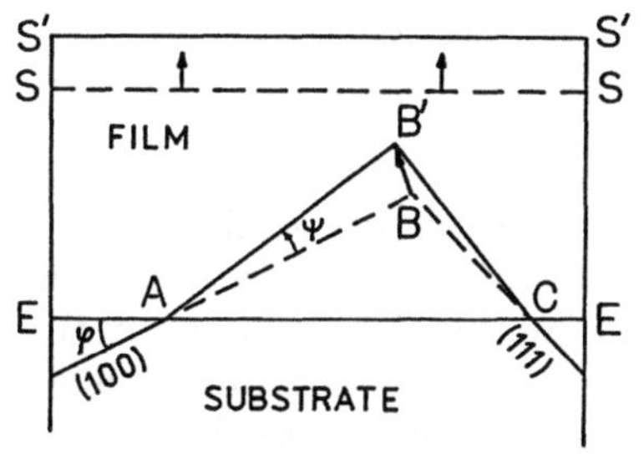

Fig. 3 Lattice misfit induced strain in the epifilm may cause various tilts of various film lattice planes with respect to the substrate ones, because of higher stiffness along (111) as compared with the (100) plane. Epitaxy on a vicinal face.

constant a_0, less than the one of the film, a ($a > a_0$). Let the epitaxy occur along the infinite plane EE inclined from the singular one by an angle ϕ. To fit the film with the substrate along the plane EE, the film may be first compressed homogeneously in all directions by $\varepsilon = (a-a_0)$ and then glued to the substrate. The external film surface is SS. At that stage, the substrate and film lattice planes should strictly coincide. At the second stage, the film may be allowed to relax, its external surface shifting from SS to S'S', as shown by the vertical arrows. During this expansion, each film point moves not necessarily along the normal to EE. Since the stiffness of the (111) planes exceeds the one of the (100), the point B in Fig. 3 should move up and left, as shown by the arrow BB'. Such asymmetrical shift (absent if EE is a symmetry plane) provides film lattice tilt by an angle ψ. Making use of this procedure, lattice deformations for the epitaxy planes (100), (110), (111) and (113) were calculated,[21] providing explanation for results obtained in Ref. 16 for the epitaxies at simple index planes. The same picture is applicable to the epitaxy at vicinal planes, as seen in Fig. 3.

The calculations[22] of deformation (not distortion!) tensor in the epifilm of cubic material grown on a vicinal inclined from the (100) plane by the angle ϕ give for the shear deformation

$$\varepsilon_{xy} = ((C_{11}-C_{12})(C_{11}+2C_{12}))/4\, C_{11}C_{44}\, \varepsilon\phi \qquad (3)$$

$C_{11} = 11^8$, $C_{12} = 53.5$, $C_{44} = 59.4$ dyn/cm^2 being the elastic constants for GaAs. At $\varepsilon \simeq 2 \cdot 10^{-3}$, $\phi = 3° \simeq 5 \cdot 10^{-2}$, one has $\varepsilon_{xy} \simeq 10^{-4}$; i.e., the cubic unit cell becomes triclinic with the angle deviating from 90° by about arcsec, which agrees with [19] where a similar expression was obtained purely geometrically.

4. LAYER GROWTH KINETICS

4.1 *General*

Fig. 4 presents schematically the growth rate dependence on the driving force for crystallization, $\Delta\mu/\kappa_T$, $\Delta\mu$ being the difference in chemical potentials of the crystallizing substance in the mother medium and in the crystal. Curve 1 in Fig. 4 is related to the screw dislocation assisted growth, Curve 2 to the 2D nucleation mechanism. For LPE, the supersaturation even in the bulk solution typically does not exceed $\simeq$ 50% ($\Delta\mu/\kappa_T \simeq 0.5$) and is much less at the growing interface due to diffusion resistance in the liquid.[23,24] Some higher values are employed in the bulk in CVD techniques. Therefore, during LPE and CVD, the steps are generated mainly by dislocations (though 2D nucleation is possible for some substrate orientations) and supersaturations.

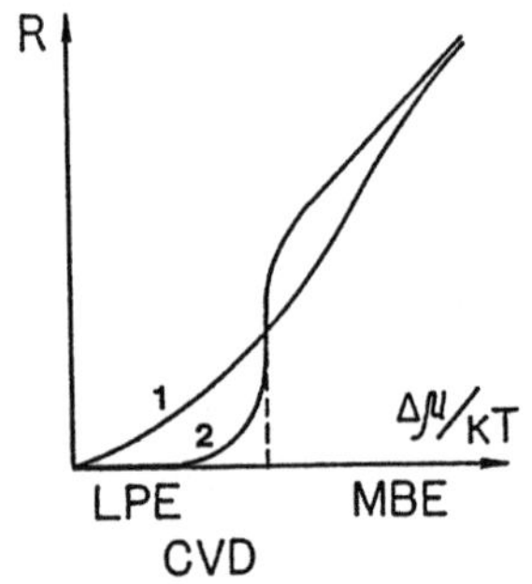

Fig. 4 Growth rate vs. supersaturation for the dislocation and 2D nucleation assisted growth (curves 1 and 2, respectively).

On the other hand, for MBE, $\Delta\mu/\kappa_T \gg 1$ (for Si, at a substrate temperature of 1100 K and beam source temperature of 1700 K, $\Delta\mu/\kappa_T \simeq$ 18). Therefore MBE is controlled by 2D nucleation unless artificially made steps are too dense to consume all the adatoms before they become organized into the 2D nuclei.

4.2 *Periodic nucleation and diffusivity*

The 2D nucleation starting on a surface free of steps occurs periodically for the generation of each new layer. This is the reason for oscillations in the molecular beams evaporated or reevaporated from the crystal surface. The phenomenon discovered first in 1972 for KCl[25] was reopened 10 years later in the form of RHEED spots intensity oscillations during the growth of III-V compounds by MBE and then became an important tool to investigate atomic surface processes.[26,1,2]

When the substrate temperature increases (up to ~600°C for a 1° off-axis (100) GaAs), the surface diffusion of Ga becomes so fast that the adatoms reach the steps (existing due to the surface vicinal misorientation) before they get together to form a new 2D nuclei. It was assumed[27,28] that the oscillation disappears when the diffusion path becomes equal to half the interstep distance, the diffusion time τ being the ratio of Ga surface sites density to the Ga flux density. With these assumptions, the Einstein formula gives the surface diffusivity D_s. However, deeper analysis must take into account nucleation probability. With this probability in mind, 10^4 larger diffusivity (at 580°C) from the same type of experiments was obtained.[29,30] This is because the time needed for nucleation actually includes the period when the supersaturation reaches some critical value after previous nucleation; this is much less than τ defined above.[27,28] According to Ref. 30, $D_s^{Ga} = 1.6 \cdot 10^{-2} \exp(-1.1 \text{ eV/kT}) \text{ cm}^2/\text{s}$, and the surface diffusion path is $\lambda_s^{Ga} = 4 \exp(3.48 \cdot 10^3/T)$, $\lambda_s^{As} = 1.7 \exp(8.12 \cdot 10^2/T)$.[29] Still the data are not definite since the step kinetic coefficient in Refs. 29 and 30 was assumed to be infinite.

4.3 *Reconstruction and kinetics*

It is evident that, for instance, (2×4) surface reconstruction on the GaAs(100) or Si(100)-1×2 may cause anisotropy in surface diffusivity and influence the diffusivity behavior. Indeed, temperature dependence of the D_s^{Ga} on the GaAs(100) shows a steep change at a temperature (~900K) at which the (2×4) → (1×1) transition occurs at As^4 flux of 0.49 ML/S and Ga flux of 5.10-2 ML/S.[28]

A diffusivity of Ga in the <110> direction four times higher than the <110> direction was found on the GaAs(100) within the temperature range 550-650°C, with 2.8 eV activation energy for both azimuths.[31] Contrary to Ref. 28, no correlation with the GaAs(100) reconstructions (3×1)-(2×4)-(1×1) sequence was found. In both studies, neither the nucleation probability nor the step kinetic coefficients was taken into account, thus leaving the question on anisotropy of surface diffusion open.

Some additional data come from the Si(100) surface. Si atoms deposited on the Si(100) at room temperature exhibit a strongly (~20:1) elongated STM pattern,[32,33] as shown in Fig. 5. The chains are the dimer rows on the reconstructed Si(100)-2×1. Annealing at ~300°C for 5 min[33] causes the deposit to coarsen and the growth of more isotropic (~2:1) islands. No monomers between the islands and no anisotropic depletion in island population near the two non-equivalent sides of big islands (along and perpendicular to their dimer chains) were reported.[33] Thus the possibility of strongly anisotropic surface diffusion should be excluded.[31] Since surface energy and tension should not change seriously between room and annealing temperatures, the strong anisotropy of surface energy is also not probable, in my opinion. What seems to be more important is the anisotropy in the step kinetic coefficient and therefore the growth rate. Namely, the steps whose motion elongates the new dimer rows should possess a higher kinetic coefficient than the ones which direction is

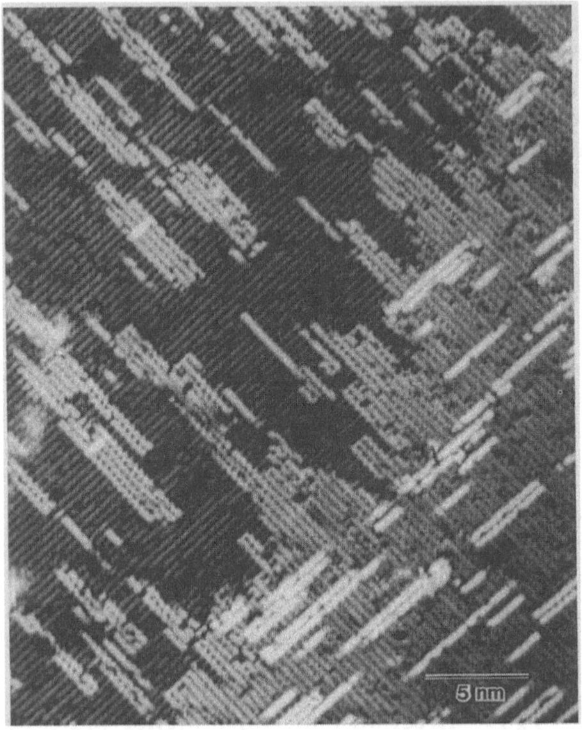

Fig. 5 STM picture of the Si(100)-1×2 surface on which Si was deposited. The dimer chains are clearly seen.[32]

parallel to the newly creating rows, thus demanding 1D nucleation of these rows. This effect, together with surface diffusion anisotropy,[34] was predicted to cause creation of double steps at sufficiently low temperatures.[35]

4.4 Reconstruction and lattice superstructure

A striking example of how surface growth processes may influence the internal crystal structure comes from $Ga_{0.5}In_{0.5}P$ films grown by MBE on the (100) related vicinal faces.[36] It turns out that the {111} planes are alternatively predominantly occupied by either Ga or In atoms. This follows from different superspot intensities in the transmission electron diffraction pattern taken in the electron microscope. If two (111) planes cross the growing vicinal surface symmetrically with respect to the growth steps, i.e., perpendicularly to the steps, then the two corresponding (111) superspot intensities belonging to these two planes are unequal. This result supports the view that the statistical selection of atoms building the lattice occurs at steps and is a non-equilibrium one.[11]

The described superstructures are typical also for the $GaAs_{0.5}Sb_{0.5}$ MBE-grown films[37] and other III-V's also grown by MOCVD.

5. TRAPPING OF IMPURITIES

During the growth from gaseous phase or molecular beams, the impurities behave essentially like the eigenspecies: they adsorb on the terraces between steps, diffuse on these terraces and may be either trapped by the steps or reevaporate.[38] Thus the impurity amount trapped is controlled by the interstep distances, impurity surface diffusivity, coefficient of trapping by steps and adsorption energy. Two modes are possible. In the first, the overall trapping is controlled by the step trapping coefficient; in the second, the final impurity content in the crystal is determined by the relation between the impurity surface diffusion length and the interstep distance. In the latter case, the growing surface orientation is important if the nucleation does not occur on the interstep terraces. If the nucleation does occur, the interstep distance and thus the impurity content in the film is determined by the nucleation of the subsequent lattice planes, i.e., on the growth rate.

In the step bunches appearing in some regimes of CVD, MOCVD and LPE growth,[37] the distances between elementary steps are much less than the ones on the growing vicinal face, on average. Therefore, the impurity concentration trapped by the step bunches may differ substantially from the concentration trapped by the parts of the rowing surface between the bunches where the step density is low.

The step bunching in solution growth does depend on relative flow directions of the steps and of the mother liquid.[40] The same phenomenon may be expected during CVD also.

6. CONCLUSION

Several examples mentioned in this overview show in particular that the epitaxial crystal structure is controlled not only by thermodynamic structural and phase relations but also by growth kinetics and surface superstructures. Further understanding of these phenomena remains a good challenge for surface and crystal growth sciences.

REFERENCES

1. "Molecular Beam Epitaxy 1988," J. Cryst. Growth 95 (1989).
2. M. A. Hermann and H. Sitter, "Molecular beam epitaxy," Springer Ser. Mat. Sci. (Springer, Berlin, 1989).

3. "Metallorganic Vapor Phase Epitaxy 1988," J. Cryst. Growth 93 (1988).
4. II-VI Compounds 1989, J. Cryst. Growth 101 (1990).
5. "Atomic Layer Epitaxy", ed. T. Suntola (Blackslee, 1989).
6. J-i Nishizawa, K. Aoki, S. Suzuki, and K. Kikuchi, J. Cryst. Growth 99 (1990), 502.
7. J. E. Greene, T. Motooka, J.-E. Sundgren, A. Rockett, S. Gorbatkin, D. Lubben, and S. A. Barnett, J. Cryst. Growth 79 (1986), 19.
8. *Laser Diagnostics and Photochemical Processing for Semiconductor Devices*, eds. R. M. Osgood, S. R. J. Bueck, H. R. Schlossberg (Oxford, Amsterdam, 1983).
9. *Crystal Growth of Electronic Materials*, ed. E. Kaldis (North Holland, Amsterdam, 1985).
10. Surfaces and surface reactions, Ultramicroscopy 31 (1989), N 1.
11. A.A.Chernov, *Modern Crystallography III. Crystal Growth*, Springer Ser. Solid St. Sci. 36 (Springer, Berlin, 1984).
12. E. Bauer, Kristallogr. 110 (1958), 372.
13. K. Yamamoto, T. Iijima, T. Kunishi, K. Fuwa, and T. Osaka, J. Crystal Growth 94 (1989), 629.
14. H. Nakao, T. Yao, Jpn. J. Appl. Phys. 28 (1989), N3, L352-355.
15. G. L. Kellog, T. T. Tsong,m P. Cowan, Surf. Sci. 70 (1978), 485.
16. M. Kloma, in: *Growth of Crystals Vol. 11*, ed. A. A. Chernov (Consultant Bureau, New York, 1979).
17. S. Isomae, S. Kishino, and M. Takahashi, J. Cryst. Growth 23 (1974), 253.
18. W. J. Bartels and W. Nijman, J. Cryst. Growth 44 (1978), 518.
19. P. Auvray, M. Baudet, and A. Regreny, J. Cryst. Growth 95 (1989), 288.
20. A. Leiberich and J. Levkoff, J. Cryst. Growth 100 (1990), 330.
21. J. Hornstra and W. J. Bartels, J. Cryst. Growth 44 (1978), 513.
22. A. A. Chernov, unpublished.
23. Liquid Phase Epitaxy, J. Cryst. Growth 27 (1974).
24. W. van Erk, H. J. G. J. van Hock-Martens, and G. Bartels, J. Cryst. Growth 48 (1980), 621.
25. H. Dabringhaus, H. J. Meyer, J. Cryst. Growth 16 (1972), 31; J. Cryst. Growth 61 (1983), 91.
26. Molecular Beam Epitaxy 1986, J. Cryst. Growth 81 (1987).
27. J. H. Neave, P. J. Dobson, B. A. Joyce, and Jing Zhang, Appl. Phys. Lett. 47 (1985), 100.
28. J. M. van Hove and P. I. Cohen, J. Cryst. Growth 81 (1987), 13.
29. T. Nishinaga, T. Shirata, K. Mochizuki, and K. I. Cho, J. Cryst. Growth 99 (1990), 482.
30. T. Nishinaga and K. I. Cho, Jpn. J. Appl. Phys. 27 (1988), L12.
31. K. Ohta, T. Kojima, and T. Nakagawa, J. Cryst. Growth 95 (1989), 71.

32. R. J. Hamers, V. K. Köhler, and J. E. Demuth, Ultramicroscopy 31 (1989), 10.
33. M. G. Lagally, R. Kariotis, B. S. Swartzentruber, and Y.-W. Mo, Ultramicroscopy 31 (1989), 87.
34. S. Stoyanov, J. Cryst. Growth 94 (1989), 751.
35. S. Stoyanov, Europhysics Lett. 11 (1990), 361.
36. T. Suzuki, A. Gomyo, and S. Iijima, J. Cryst. Growth 99 (1990), 60.
37. N. Otsuka, Y. E. Ihim, Y. Hirotsu, J. Klem, and H. Morkoc, J. Cryst. Growth 95 (1989), 43.
38. A. A. Chernov and S. S. Stoyanov, Kristallografiya 22 (1977), 248.
39. A. A. Chernov and T. Nishinaga, in: *Morphology of Crystals*, ed. I. Sunagawa (Terra Science, Tokyo, 1987), p. 207.
40. A. A. Chernov, Contemporary Physics 30 (1989) No. 4, 251.

Competition Between Nucleation and Two-Dimensional Step Growth in Molecular Beam Epitaxy

V. Fuenzalida[1] *and I. Eisele*[2]

[1]Departamento de Física FCFM, Universidad de Chile, Casilla 487-3, Santiago de Chile, Chile
[2]Institut für Physik, Universität der Bundeswehr, Werner Heisenberg Weg 39, W-8014 Neubiberg, Fed. Rep. of Germany

We consider ultra high vacuum growth from the vapor phase (MBE), under conditions far away from equilibrium. We show that at very high growth rates and low temperatures, nucleation plays an important role. Nucleation competes with growth at the surface steps, as described by the Burton,Cabrera and Frank (BCF) Theory. In addition, the surface misorientation towards a low indexed plane strongly influences nucleation.

1. Introduction

The purpose of this work is to find the physical limits for which 2-dim epitaxial growth is possible in MBE experiments. We introduce a model that includes adatom diffusion to surface steps as well as adatom nucleation, leading to a modified BCF growth mode [1]. We derive the continuity equation for the adatom current, including adatom "creation" (impingement) and "annihilation" (desorption and collisions).

We consider the steady state conditions in which the step flow is only slightly disturbed. In this situation we calculate the rate at which nuclei formed on the terraces are incorporated in the advancing steps.

2. Model

The surface is described by a periodic arrangement of steps as in ref. [1], see Fig.1. Let ϕ be the misorientation angle towards a smooth surface. Let z be a reduced non-dimensional adatom density, Z a reduced total cluster density and y a reduced distance. Adatoms on the terraces can migrate to step edges leading to a diffusion current proportional to $-\nabla z$. Considering the continuity equation for this current, there is a creation contribution from the impingement flux from the vapor, 1 in reduced units. Annihilation is due to collisions between adatoms, leading to a contribution proportional to z^2, or to collisions of adatoms with preformed clusters leading to a term proportional to zZ. New clusters are generated only by adatom-adatom collisions. Then, it turns out that the kinetics of crystal growth for monoatomic steps is described by the

Springer Proceedings in Physics, Volume 62
Surface Science Eds.: F.A. Ponce and M. Cardona

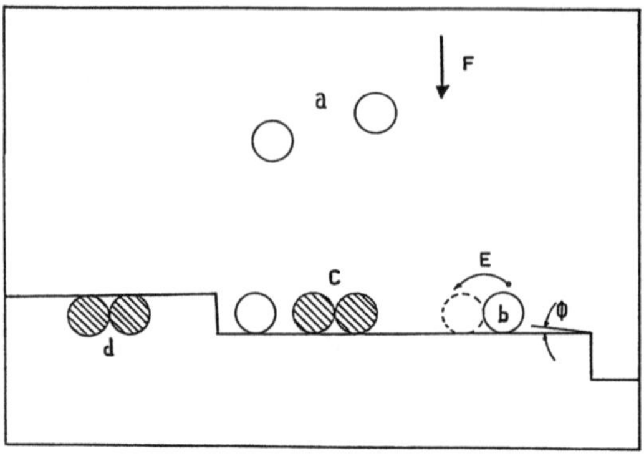

Fig.1: Advancing step and terrace showing impinging atoms (a), diffusing adatoms (b), clustered adatoms on the terrace (c) and incorporated nuclei (d).

following equations:

$$d^2 z/dy^2 = -1 + 2gz^2 + \cot^2 \phi z Z \quad (1)$$

$$dZ/dy = -gz^2 \quad (2)$$

$$g = Fn^{-1}\nu^{-1} \cot^4 \phi \exp(E/kT). \quad (3)$$

g plays the role of a nucleation factor, which determines if two or three dimensional growth takes place. It increases by decreasing the temperature T, decreasing the surface misorientation ϕ, or increasing the impingement rate F. ν is a vibration frequency of the order of $10^{13} s^{-1}$, n the density of surface sites (about $10^{19} m^{-2}$) and k the Boltzmann constant. E is the activation energy for surface diffusion.

As it is inferred from the equations, parameter g dominates the nucleation of new clusters, while $\cot\phi$ describes the further growth of already existing clusters and their mean size.

3. Results

The deviation from ideal two-dimensional step growth can be described by the fraction γ of the impinging beam which nucleates and is incorporated as preformed clusters rather than single atoms into the steps, and by their mean cluster size at the time they are captured by the step.

The numerical solutions [2] of equations (1,2) allow a rich variety of growth conditions ranging from perfect epitaxy (low g and b) to different kinds of nucleation. The growth of large amounts of small clusters may promote many uncorrelated nucleation sites and low ordered deposits.

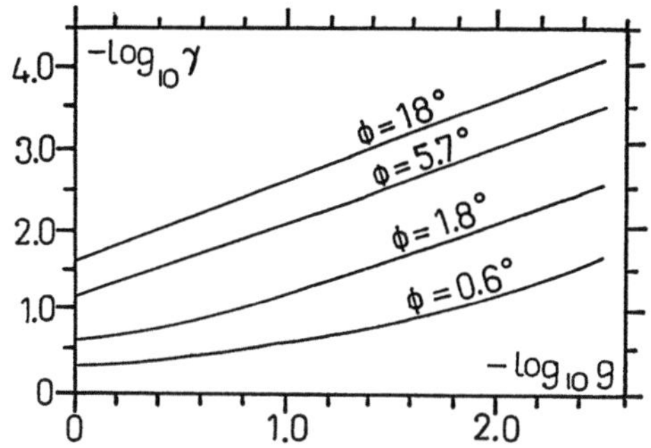

Fig.2: Fraction γ of the impingement beam that nucleates rather than incorporates into the steps as a function of the nucleation parameter g. Each curve corresponds to a different misorientation.

The fraction γ is depicted in Fig.2 as a function of g for several misorientations.

As an example let us consider the Si(111) surface at a rather low temperature T=900K, assuming $E = 1eV$. A typical impingement flux is $F = 2 \cdot 10^{19} m^{-2} s^{-1}$. At a misorientation of $0.5°$ we obtain $g = 18$ and $\gamma = 36\%$. An important part of the beam nucleates and layer growth is not possible leading to nucleation controlled growth (the formalism will break down before such a g value is reached).

Under the same conditions let us consider a strongly misoriented substrate with $\phi = 4°$. Then g=0.004 and $\gamma = 0.01\%$, which means that a only very small fraction of the beam is not incorporated into the steps, leading to smooth epitaxy.

4. Conclusions

The effect of temperature, growth rate and misorientation angle in MBE experiments is obtained quantitatively. The fraction of the impingement beam which does not contribute to step growth, but nucleates at random, is calculated. This provides a prediction of the quality of homoepitaxial overgrowths with the only knowledge of the diffusion energy. Misorientation strongly enhances step growth epitaxy.

5. References

[1] W.Burton, N.Cabrera and F.Frank, *Philosophical Transactions of the Royal Society (London)*, **A43**(1951)299-358.

[2] V.Fuenzalida and I.Eisele, submitted to *The Journal of Crystal Growth*

Fermi Level Readjustments on Adsorption and Interface Formation

C. Pinto de Melo

Departamento de Física, Universidade Federal de Pernambuco, 50739 Recife, PE, Brazil

Whenever a localized perturbation acts upon an extended system, charge readjustments should occur and a new Fermi level has to be defined. In this work we call attention to the fact that even when the relaxation of the Fermi level is infinitesimally small, significant changes of extensive properties could result. As a consequence, neglect of these relaxation effects could lead both to serious errors in the calculation of the interaction energies and to the nonconservation of total electronic charge. Through simple models we exemplify this point for the cases of surface formation and of atomic chemisorption on unidimensional chains.

1. Introduction

The equilibrium electronic distribution of an isolated N-electron system can be characterized by the Fermi level, E_F, an intensive property of the system. In this way, any perturbation which involves a small ($\ll N$) number n_s of electrons (such as the chemisorption of an atom or molecule on a infinite surface) can be considered as localized. In the general case, the perturbation causes a change on the total number of atoms and electrons, and a new Fermi level E_F' has to be defined for the system.

Common sense would tell us that $E_F = E_F' + O(n_s/N)$, and that therefore Fermi level readjustment corrections could be neglected. However, while this can be true for the calculation of local quantities, when determining extensive properties one should not overlook the fact that such infinitesimal Fermi level correction terms can add up to give significant contributions. Although the absolute values of extensive properties are seldom of interest for large systems, frequently one is required to estimate small *variations* of them. Interaction energies between two coupled subsystems, for example, have to be determined as the difference between the total energies before and after the coupling has occurred.

In the present work we illustrate the importance of properly accounting for these Fermi level readjustment terms by examining two simple examples. In the first, the cleavage of an infinite chain is used as a model of "surface formation". In the second, the interaction of a semi-infinite chain with an approaching atom mimics the chemisorption problem. In both cases our results confirm that the neglect of the Fermi level relaxation terms can lead to unphysical results, such as non-conservation of the total number of electrons and serious errors in the estimate of the interaction energy. By repeating the calculation for several occupancies of the substrate band we examine how the Fermi level position affects these conclusions. It is shown that for certain values of E_F the correction term can be comparable to the value of the interaction energy obtained by a "frozen" Fermi level approximation.

2. Specific Examples

2.1 Cleavage of an Infinite Chain

A very convenient way to describe the electronic structure of one-dimensional chains is through the use of transfer functions [1]. In this way analytic expressions for the Green's function matrix elements in the site representation can usually be found [2].

Springer Proceedings in Physics, Volume 62
Surface Science Eds.: F.A. Ponce and M. Cardona

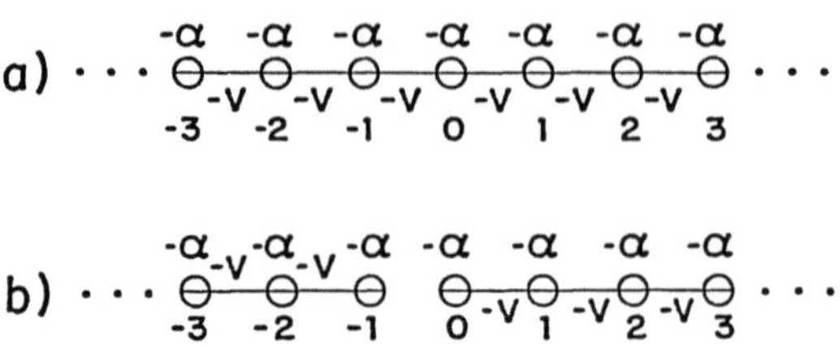

Figure 1: Cleavage of the infinite chain (a) produces two semi-infinite chains (b)

Consider the infinite one-dimensional chain of Fig. 1a, described by a first-neighbors tight-binding hamiltonian. A perturbation $\mathcal{V}(0,-1) = +V$ which decouples sites 0 and -1 would cleave the original system to produce the two semi-infinite chains of Fig. 1b. The translational symmetry allows the transfer matrix of the problem to be defined as

$$T \;=\; -\,e^{\mp i\theta} \tag{1}$$

where

$$\theta = \arccos\frac{E+\alpha}{2V} \quad . \tag{2}$$

For the infinite chain a generic expression for the diagonal elements of the Green's function at any site can be easily obtained in terms of the element for site zero as

$$G^I_{n,n} \;=\; 2\,\frac{1-e^{\mp i2n\theta}}{1-e^{\mp i2\theta}}\,e^{\mp i\theta} \;+\; e^{\mp i2n\theta}\,G^I_{0,0} \tag{3}$$

where

$$G^I_{0,0} \;=\; \mp i\,\frac{1}{\sin\theta} \quad . \tag{4}$$

For each semi-infinite chain the structure of the equations connecting the Green's function elements for all sites beyond the "surface" atom is also periodic, and the same transfer function as in (1) can be defined. For the *rhs* truncated chain, for instance, one can find in a manner similar to (3) a general expression for the diagonal element at any site as a function of the element for site zero, $G^S_{0,0}$. Here, though,

$$G^S_{0,0} \;=\; 2\,e^{\mp i\theta} \quad . \tag{5}$$

Since the total number of electrons of the perfect chain is the sum of the integrals over the density of states at each site, it can be represented as

$$\begin{aligned}\mathcal{N}^I \;&=\; -\frac{1}{\pi}\int_{E_i}^{E_F} ImG^I_{0,0}(E)\,dE - \frac{2}{\pi}\sum_{n=1}^{\infty}\int_{E_i}^{E_F} ImG^I_{n,n}(E)\,dE \\ &=\; \pm\frac{1}{\pi}(2\mathcal{N}^0_m-1)[\pi-\theta_F]\end{aligned} \tag{6}$$

where $\mathcal{N}^0_m \longrightarrow \infty$ is the number of sites on the *r.h.s.* semi-infinite chain.

No localized states result from the cleavage of the chain. Hence, the total number of electrons for each semi-infinite chain can be represented by

$$\mathcal{N}^S = -\frac{1}{\pi}Im\int_{E_i}^{E'_F}\sum_{n=0}^{\infty}G^S_{n,n}(E)\,dE \tag{7}$$

where changes on the Fermi level position after cleavage were allowed for.

After some algebra the infinite sum in (7) can be calculated [3] to produce

$$\mathcal{N}^S = \pm\frac{1}{\pi}(\mathcal{N}^0_m+\frac{1}{2})(\pi-\theta'_F)+\mathcal{J}(\theta'_F) \tag{8}$$

Table 1: Error in the total number of electrons ($\Delta\mathcal{N}$), cleavage energy for frozen Fermi level approximation ($\Delta\mathcal{E}(E_F = E_F')$) and correction term for the cleavage energy due to Fermi level relaxation ($-E_F\Delta\mathcal{N}$).

η	0_+	1/4	1/2	3/4	1
$\Delta\mathcal{N}$	∓ 0.5	∓ 0.25	0	± 0.25	0
$\Delta\mathcal{E}(E_F = E_F')$ (eV)	± 3.0	± 1.65	± 1.09	± 1.65	0
$-E_F\Delta\mathcal{N}$ (eV)	∓ 3.0	∓ 0.75	0	∓ 0.75	0

where

$$\mathcal{J}(\theta_F') = \mp\frac{1}{2}\int_{\theta_F'}^{\pi}(1 - 2\sin^2\theta)\sum_{n=-\infty}^{\infty}\delta(\theta - n\pi)d\theta \quad . \tag{9}$$

Taking into account the fact that the number of sites is preserved during the cleavage process, one should expect the change of the total number of electrons, $\Delta\mathcal{N}(E_F, E_F')$, to be identically equal to zero. However, from (6) and (8) one obtains

$$\Delta\mathcal{N}(E_F, E_F') = \frac{2\mathcal{N}_m^0}{\pi}(\theta_F - \theta_F') \pm (1 - \frac{\theta_F}{\pi}) + 2\mathcal{J}(\theta_F') \quad . \tag{10}$$

It is evident from the above expression that the neglect of Fermi level relaxation will lead to the absurd result $\Delta\mathcal{N}(E_F = E_F') \neq 0$. In fact, to assure charge conservation the first term on the *r.h.s.* of (10) must exactly cancel the remaining two others. Table 1 shows, for different values of E_F, the error on the computation of the total number of electrons which results from the neglect of relaxation effects.

To estimate the effect of Fermi level relaxation on the surface formation energy, $\Delta\mathcal{E}$, consider first the case of a frozen Fermi level approximation. By a treatment similar to the above one obtains

$$\begin{aligned}\Delta\mathcal{E}(E_F = E_F') &= -\frac{2}{\pi}\sum_{n=0}^{\infty}\int_{E_i}^{E_F} E\, Im[G_{n,n}^S(E) - G_{n,n}^I(E)]\, dE \qquad (11)\\ &= -\frac{4V}{\pi}[\pm\frac{1}{2}\sin\theta_F + \mathcal{I}_1(\theta_F)]\\ &\quad +\frac{2\alpha}{\pi}[\mp\frac{1}{2}(\pi - \theta_F) + \mathcal{I}_2(\theta_F)] \qquad (12)\end{aligned}$$

where

$$\mathcal{I}_1 = \pm\frac{\pi}{2}\int_{\theta_F}^{\pi} d\theta \sum_{n=-\infty}^{\infty}\delta(\theta - n\pi)\cos 2\theta\, \cos\theta \tag{13}$$

$$\mathcal{I}_2 = \pm\frac{\pi}{2}\int_{\theta_F}^{\pi} d\theta \sum_{n=-\infty}^{\infty}\delta(\theta - n\pi)\cos 2\theta \quad . \tag{14}$$

If Fermi level relaxation was allowed for, one would have obtained instead

$$\begin{aligned}\Delta\mathcal{E}(E_F, E_F') &= \Delta\mathcal{E}(E_F = E_F')\\ &\quad +\frac{2}{\pi}\sum_{n=1}^{\infty}\int_{E_F'}^{E_F} E\ ImG_{n,n}^S(E)\, dE\\ &\quad -\frac{1}{\pi}\int_{E_F'}^{E_F} E\ ImG_{0,0}^I(E)\, dE \quad .\end{aligned} \tag{15}$$

Since the Fermi level relaxation is small, the third term on the *r.h.s.* should vanish as $\mathcal{O}(1/\mathcal{N}_m^0)$. The middle term on the *r.h.s.*, however, gives a significant contribution to the surface energy since it represents an infinite sum of infinitesimal terms. This is the error which results from a frozen Fermi level approximation. After a bit of algebraic manipulation it can be shown that this term is related to the error in the computed number of electrons (10) through

$$\Delta\mathcal{E}(E_F', E_F) = \Delta\mathcal{E}(E_F' = E_F) - E_F\,\Delta\mathcal{N}(E_F', E_F) \ . \tag{16}$$

To illustrate these points, in Table 1 the surface creation energy calculated in the frozen Fermi approximation for different fractional occupancies of the band (η) is compared to the correction term $-E_F\Delta\mathcal{N}$, for $\alpha = 0$ and the typical value $V = 3.0eV$.

2.2 Atomic Chemisorption on a Semi-infinite Chain

Consider now the interaction of an approaching atom with a semi-infinite chain, as depicted in Fig. 2. We can work in units of $2V$, and define the parameters

$$\epsilon = \frac{E + \alpha}{2V} \tag{17}$$

$$\delta = \frac{\alpha_a - \alpha}{2V} \tag{18}$$

and

$$\lambda_a = \frac{V_a}{2V} \ . \tag{19}$$

Since the periodic structure of the chain remains the same, the transfer matrix (1) can be used and the diagonal element of the Green's function for the adsorbate written as

$$G^A_{a,a} = \frac{1}{\epsilon + \delta + 2\,\lambda_a^2\,T} \ . \tag{20}$$

Now, a generic expression for the diagonal element of the Green's function at any site along the chain can be found as

$$G^A_{n,n} = G^S_{n,n} + 4\,\lambda_a^2\,T^{2n+2}\,G^A_{a,a} \ . \tag{21}$$

The interaction with the foreign atom will induce charge readjustments on the substrate. As a result, the local density of states for different sites will vary along the chain and a state localized below the band region will appear at the position

$$\epsilon_p = \frac{-\delta\,(1 - 2\lambda_a^2) \pm \sqrt{4\,\lambda_a^4\,(4\lambda_a^2 + \delta^2 - 1)}}{1 - 4\lambda_a^2} \ . \tag{22}$$

The total number of electrons before and after the interaction is switched-on can be written as

$$\mathcal{N}_i^A = n_a - \frac{1}{\pi} Im \int_{\epsilon_i}^{\epsilon_F} \sum_{n=0}^{\infty} G^S_{n,n}(\epsilon)\,d\epsilon \tag{23}$$

and

$$\begin{aligned} \mathcal{N}_f &= -\frac{1}{\pi} Im \int_{\epsilon_i}^{\epsilon_F'} G^A_{a,a}(\epsilon)\,d\epsilon + Z^A_{a,a}(\epsilon_p) \\ &\quad -\frac{1}{\pi} \sum_{n=0}^{\infty} Im \int_{-\infty}^{\epsilon_F'} G^A_{n,n}(\epsilon)\,d\epsilon + \sum_{n=0}^{\infty} Z^A_{n,n}(\epsilon_p) \ , \end{aligned} \tag{24}$$

respectively, where n_a is the initial electronic charge on the adsorbate and $Z^A_{i,i}(\epsilon_p)$ is the residue of the diagonal element of the Green's function at site i at the energy ϵ_p. Then, one can write

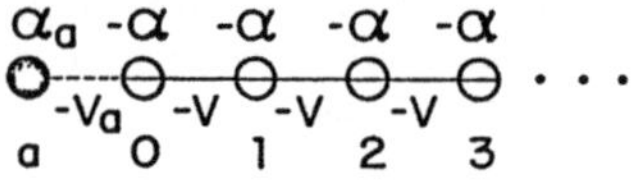

Figure 2: Chemisorption of a single atom on a semi-infinite chain

$$\begin{aligned}\Delta\mathcal{N} &= \mathcal{N}_f - \mathcal{N}_i \\ &= Z^A_{a,a}(\epsilon_p) + \sum_{n=0}^{\infty} Z^A_{n,n}(\epsilon_p) \; - \; n_a \\ &\quad + \rho^A_{a,a} \; + \; \Delta q_n \; + \; \Delta\mathcal{N}(\epsilon_F - \epsilon'_F) \quad , \end{aligned} \tag{25}$$

where

$$\rho^A_{a,a} \;=\; -\frac{1}{\pi}\int_{\epsilon_i}^{\epsilon'_F} G^A_{a,a}(\epsilon)\,d\epsilon \quad , \tag{26}$$

$$\Delta q_n \;=\; -\frac{1}{\pi}\sum_{n=0}^{\infty} Im \int_{\epsilon_i}^{\epsilon_F} [\,G^A_{n,n}(\epsilon) - G^S_{n,n}(\epsilon)]\,d\epsilon \tag{27}$$

and

$$\Delta\mathcal{N}(\epsilon_F - \epsilon'_F) \;=\; \frac{1}{\pi}\sum_{n=0}^{\infty} Im \int_{\epsilon'_F}^{\epsilon_F} G^A_{n,n}(\epsilon)\,d\epsilon \quad . \tag{28}$$

The first two terms on the *r.h.s.* of (25) correspond to the total electronic charge on the localized state and therefore should add up to 1. Then, since the total number of electrons in the system should be preserved we can write

$$\Delta\mathcal{N}(\epsilon_F - \epsilon'_F) \;=\; n_a - 1 - \rho_{a,a}(\epsilon'_F) - \Delta q_n \quad . \tag{29}$$

This is the term neglected by a frozen Fermi level approximation; one can see from (28), that since it involves an infinite sum of infinitesimal contributions, it can indeed have a nonvanishing value. In practice, its value is computed through the determination of the individual terms in (29). Equation (26) expresses the integral in the band region of the adsorbate wavefunction, while the change on the total electronic charge of the substrate (27) can, after a bit of algebra, be written as

$$\Delta q_n \;=\; \pm\,\frac{2\lambda_a^2}{\pi}\int_{\epsilon_i}^{\epsilon_F} d\epsilon\,\frac{(\epsilon+\delta)\cos\theta \;-\; 2\lambda_a^2}{\sin\theta\,[(\epsilon+\delta)^2 \;+\; 4\lambda_a^4 \;-\; 4\lambda_a^2(\epsilon\;+\;\delta)\cos\theta]} \quad . \tag{30}$$

In Fig. 3, the behavior of $\Delta\mathcal{N}$ for different values of λ_a is depicted, where the typical values $\alpha_a = 13.6$eV, $\alpha = 4.6$eV, and $V = 2.5$eV were adopted to model the hydrogen chemisorption on a transition metal.

For a frozen Fermi level approximation, the chemisorption energy will be given by

$$\begin{aligned}\Delta\mathcal{E}(E_F = E'_F) \;&=\; E_p \;-\; E_A - \frac{1}{\pi}\, Im \int_{E_I}^{E_F} E\,G^A_{a,a}(E)\,dE \\ &\quad -\frac{1}{\pi}\sum_{n=0}^{\infty} Im \int_{E_I}^{E_F} E\,[G^A_{n,n}(E) - G^S_{n,n}(E)]\,dE \quad , \end{aligned} \tag{31}$$

while according to (16) relaxation effects will introduce corrections related to $\Delta\mathcal{N}(E_F - E'_F)$.

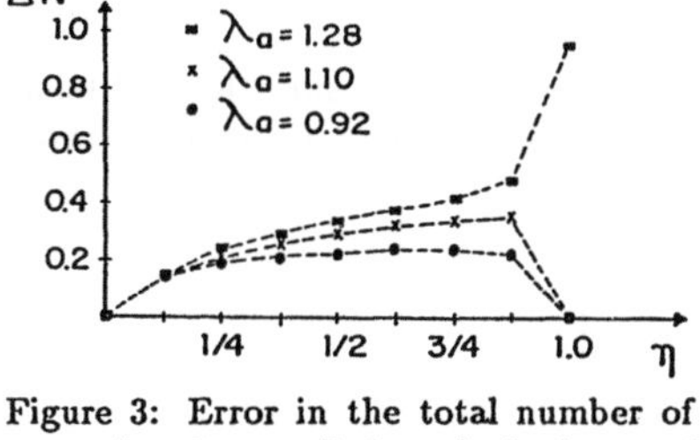

Figure 3: Error in the total number of electrons which results from the frozen Fermi level approximation applied to the hydrogen chemisorption on a semi-infinite chain

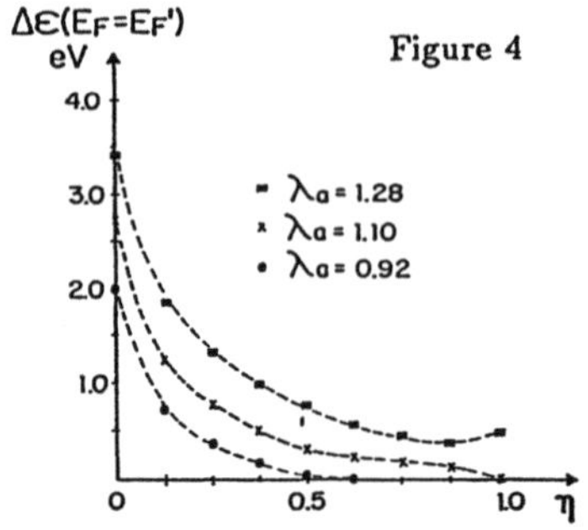

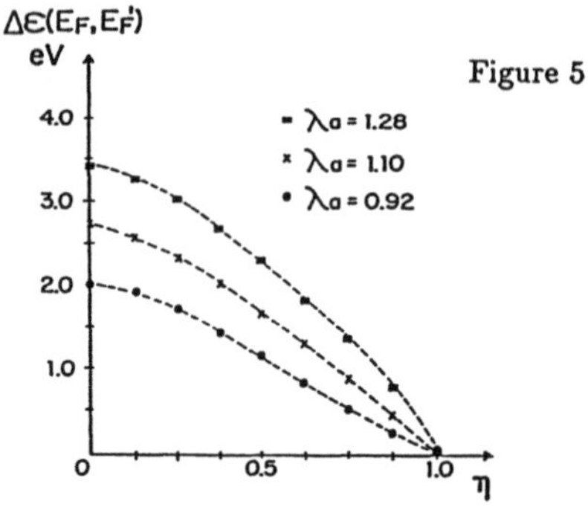

Figure 4: Binding energy for hydrogen chemisorbed on a semi-infinite chain within the frozen Fermi level approximation

Figure 5: Binding energy for hydrogen chemisorption on a semi-infinite chain after inclusion of Fermi level relaxation corrections

Figure 4 shows the absolute values of the chemisorption energy of a hydrogen atom on a transition metal, for the same set of parameters as before, computed within the frozen Fermi level approximation. Comparison with Fig. 5, where relaxation effects are included, reveals that the correction terms for the binding energy can dominate the behavior of the chemisorption energy with the change on the fractional band occupancy η.

3. Discussion

The above results are absolutely general, whenever localized perturbations act on extended systems. Since the charge and spin readjustments on individual sites are of long range [4], even infinitesimal contributions to the charge and spin distributions over all affected sites have to be taken into account if extensive properties are of interest. Interface formation, atomic and molecular chemisorption and the indirect interaction between adsorbates [5] are examples of problems where such relaxation effects can become important.

Introduction of self-consistency or the different dimensionality of more realistic models could affect the relative contribution of the Fermi level relaxation corrections. However, since they are of a fundamental nature, care must be exercized to estimate them properly, before assuming they could be neglected. For these more realistic models an explicit calculation such as the one performed here could be difficult to implement; however, in these cases techniques as the Local Space Approximation [4,6] have an specific advantage for providing the automatic Fermi level readjustment.

4. Acknowledgements

This work was partially supported by the Brazilian Agencies FINEP and CNPq. I thank Dr. M. Matos for some helpful discussions at the early stages of this work about the behavior of atomic chemisorption.

5. References

1. C.T. Papatriantafillou: Phys. Rev. B7, 5386 (1973).
2. C.P. de Melo: In *Electronic Structure of Atoms, Molecules and Solids*, ed. by F. Paixão, J.A. Castro and S. Canuto, (World Publishing, Singapore,1990), p. 137.
3. G.P. Tolstov: *Fourier Series*, (Dover, New York,1967).
4. C.P. de Melo, M.C. dos Santos, M. Matos, and B. Kirtman: Phys. Rev. 35, 7847 (1987).
5. S.R. de Freitas and C.P. de Melo: in preparation.
6. B. Kirtman and C.P. de Melo: J. Chem. Phys. 75, 4592 (1981).

Effective Dielectric Response of a Composite with Aligned Ellipsoidal Inclusions

J. Giraldo[1], *R.G. Barrera*[2], *and W.L. Mochán*[2]

[1]Departamento de Física, Universidad Nacional de Colombia, Bogotá, Colombia
[2]Instituto de Física, Universidad Nacional Autónoma de México, Apdo. Postal 20-364, 01000 México, D.F., Mexico
[3]Laboratorio de Cuernavaca, Instituto de Física, Universidad Nacional Autónoma de México, Apdo. Postal 139-B, 62191 Cuernavaca, Mor., Mexico

Abstract. A previously developed theory for the calculation of the effective dielectric response ϵ_M^γ of composites is further extended to the case of a composite with ellipsoidal inclusions in an otherwise homogeneous, isotropic matrix. An alternative derivation of the mean-field approximation is also presented and it is shown that depending on the choice of the two-particle distribution function different expressions for ϵ_M^γ are obtained. Results are presented for a system of metallic (Drude) inclusions in dispersionless gelatin.

I. Introduction

Shape effects in the dielectric response of composites has been a problem which has attracted the interest of many researchers over the years [1-5]. Here we will deal with composites prepared as small inclusions embedded in an otherwise homogeneous, isotropic matrix. From the theoretical point of view the simplest model has been to regard the inclusions as identical spheres [6], but obviously, in actual samples, there is a distribution of sizes and shapes. Although the results so far obtained for a system with identical spherical inclusions cover a wide range of methods and approximations [7], extensions to non-spherical inclusions has been restricted, almost entirely, to the mean-field approximation (MFA); and even in this case the problem has not been thoroughly examined. It seems that the main difficulty lies in the mathematical complexity. Here we will consider a composite consisting of a homogeneous isotropic matrix with aligned identical ellipsoidal inclusions. Since the electromagnetic field produced by a polarized ellipsoid is constant within the ellipsoid, our choice of an ellipsoidal shape lies in its mathematical handiness; besides, by varying the eccentricity of the ellipsoid one can span an ample variety of forms, from flat dishes to needles. Furthermore, we only consider the simplest angular correlation thus we take the axes of the ellipsoids all aligned.

Our objective is to calculate the effective (macroscopic) dielectric response of the composite, beyond the mean-field approximation, in terms of the components of the polarizability tensor of the inclusions and on the properties of the statistical distribution of its centers. We do this within the long wavelength limit, which means that the size of the inclusions and their typical separation are much smaller than the wavelength of the electromagnetic radiation. Now, since the axes of all the ellipsoids are aligned, the effective physical properties of the system are anisotropic. Therefore the effective dielectric response will be described by a second rank tensor. Our calculation here follows the procedure given by the renormalized polarizability theory (RPT) [8] developed, first, for the case of identical spherical inclusions and extended later to the case of spherical inclusions with a given distribution of radii [9]. The merits and shortcomings of RPT have been already discussed [7,8,9]; here we may only remark that RPT yields

Springer Proceedings in Physics, Volume 62
Surface Science Eds.: F.A. Ponce and M. Cardona

an extremely simple and adequate way of dealing with the fluctuations of the field (beyond MFA) which makes it possible to actually extend the theory to more complicated geometries.

The paper is structured in the following manner: in section II we derive an expression for the principal components of the effective dielectric tensor of the composite in the mean-field approximation. In the case of a composite with spherical inclusions this approximation is known as Maxwell Garnett theory (MGT) [10] which is also equivalent to the celebrated Clausius-Mossotti relation (CMR) [11]. There is a popular way, first introduced by Lorentz, [12] of deriving MFA through the use of a fictitious spherical cavity, known as the Lorentz sphere (LS), which is centered at a reference spherical inclusion. The MFA is obtained when the contribution to the field, which polarizes the central inclusion, coming from all the inclusions contained within LS is ignored and the contribution from the ones outside LS is taken in the continuous limit. When one tries to extend this method to the case of aligned ellipsoidal inclusions, it is not clear whether the fictitious cavity to be chosen should be an ellipsoid or a sphere. First, Galeener [1] derived an expression for the effective dielectric tensor by taking a spherical cavity, but it was shown later [2], that the correct limits of dishes and needles was obtained only if an ellipsoidal cavity was chosen. Here we derive MFA in a different way and demonstrate that the effective dielectric tensor depends not only on the volume fraction of the inclusions, as in the case of spheres, but also in their two-particle distribution function. The results obtained by taking an ellipsoidal or a spherical cavity, in the Lorentz' method, correspond then to two different choices of the two-particle distribution function.

In section III we extend the formalism developed in Ref. [8] to the case of a composite with aligned ellipsoidal inclusions. We show that the relationship between the effective dielectric tensor and the components of the polarizability tensor of the inclusions, is the same as in MFA but with renormalized polarizability components which obey a set of coupled second-order algebraic equations. Then the theory is applied to the case of metallic inclusions in dispersionless gelatin and the results are compared with MFA. Finally, we present our conclusions.

II. Mean field approximation

We consider a system of $N \gg 1$ aligned identical polarizable ellipsoids with semiaxes a and b, polarizability components α^γ (in principal γ-axis) with centers located at random positions $\{\mathbf{R}_i\}$ within an homogeneous, isotropic matrix characterized by a dielectric function ϵ_h. The system is excited by an external electric field $\mathbf{E}_{ex}$ with wave-vector $q \gg 1/a, 1/b$ and frequency ω. Setting our coordinate system along the principal axis of the polarizability tensor of the aligned ellipsoids, with the z-axis along a, we have that the induced dipole moment of the i-th ellipsoid obeys the equation

$$p_i^\gamma = \alpha^\gamma [E_i^{0,\gamma} + \sum_{j,\delta} t_{ij}^{\gamma\delta} p_j^\delta], \qquad (1a)$$

where the superscripts $\gamma, \delta = x, y, z$ indicate cartesian components, $\mathbf{E}_i^0$ is the electric field at $\mathbf{R}_i$ in the absence of the ellipsoids and $t_{ij}^{\gamma\delta}$ is the dipole-dipole interaction tensor which relates the electric field at $\mathbf{R}_i$ produced by a polarized ellipsoid at $\mathbf{R}_j$ with dipole moment $\mathbf{p}_j$. It is given by

$$t_{ij}^{\gamma\delta} = \partial_i^\gamma \partial_j^\delta (1/R_{ij}), \qquad (1b)$$

where $R_{ij} = |\mathbf{R}_i - \mathbf{R}_j|$. The polarizability components are

$$\alpha^\gamma = \frac{1}{3} ab^2 \frac{\epsilon_m - \epsilon_h}{L_\gamma \epsilon_m + (1 - L_\gamma)\epsilon_h}, \qquad (1c)$$

where ϵ_m and ϵ_h are the dielectric functions of inclusions and host, respectively, and L_γ are the depolarization factors [13] of an ellipsoid which fulfill $\sum_\gamma L_\gamma = 1$. For a prolate ellipsoid (the only one treated here) $a > b$ and

$$L_z = \frac{1-e^2}{e^2}\left(-1+\frac{1}{2e}\log\frac{1+e}{1-e}\right), \tag{3d}$$

where $e^2 = 1 - a^2/b^2$ is the eccentricity.

The actual electric field produced by a polarized ellipsoid is dipolar only at large distances thus the use of this approximation will limit the validity of our results to low filling fractions and/or low eccentricities of the inclusions. A more detailed study using the expressions for the exact field produced by the ellipsoids will be reported elsewhere.

Now we choose a longitudinal external field of the form

$$\mathbf{E}_{ex} = E_{ex}\hat{q}e^{i(\mathbf{q}\cdot\mathbf{r}-\omega t)}, \tag{2}$$

where $\hat{q} \equiv \mathbf{q}/q$, and thus when $\mathbf{E}^o = \mathbf{E}^{ex}/\epsilon_h$ is substituted in Eq. (1) yields

$$P_i^\gamma = \alpha^\gamma[E_{ex}\hat{q}_\gamma/\epsilon_h + \sum_{j,\delta} T_{ij}^{\gamma\delta}P_j^\delta], \tag{3a}$$

where we have defined

$$P_i^\gamma = p_i^\gamma e^{-i\mathbf{q}\cdot\mathbf{R}_i} \qquad \text{and} \qquad T_{ij}^{\gamma\delta} = t_{ij}^{\gamma\delta}e^{-i\mathbf{q}\cdot(\mathbf{R}_i-\mathbf{R}_j)}, \tag{3b}$$

in order to get rid of trivial exponential factors. The principal components of the effective macroscopic dielectric response ϵ_M^γ are now obtained through an immediate extension of a relation derived in Ref. (8), that is

$$\frac{\epsilon_h}{\epsilon_M^\gamma} = 1 - 4\pi\epsilon_h\chi_{ex}^{\ell,\gamma}(q\to 0,\omega), \tag{4a}$$

where $\chi_{ex}^{\gamma\delta}$ is the Fourier transform of the longitudinal projection of the external susceptibility tensor defined through

$$n < P >^\gamma (q,\omega) = \sum_\delta \chi_{ex}^{\gamma\delta}(q,\omega)\cdot E_{ex}^\delta(q,\omega). \tag{4b}$$

Here $< \ldots >$ means ensemble average, the superscript ℓ means longitudinal projection and n is the number density of ellipsoids.

In the mean field approximation (MFA) it is assumed that all the ellipsoids acquire the same average dipole moment $< \mathbf{P} >$; that is, one neglects the fluctuations of the dipolar moments around its average (dipolar fluctuations). Since the average of the off-diagonal components of $T_{ij}^{\gamma\delta}$ vanish, the average dipole moment in MFA obeys

$$< P >^\gamma = \alpha^\gamma[E_{ex}\hat{q}_\gamma/\epsilon_h + < \sum_j T_{ij}^{\gamma\gamma} >< P >^\gamma], \tag{5}$$

which can be trivially solved for $< P >^\gamma$. Therefore combining Eqs. (4) and (5) one obtains readily the following expression for the effective dielectric tensor

$$\frac{\epsilon_M^\gamma - \epsilon_h}{(1+ < T >^\gamma)\epsilon_M^\gamma - < T >^\gamma \epsilon_h} = 3f\tilde{\alpha}^\gamma, \tag{6}$$

where $4\pi n < T >^\gamma \equiv lim_{q\to 0} < \sum_j T_{ij}^{\gamma\gamma} >$ and is independent of i due to homogeneity. Here $f = 4\pi nab^2/3$ is the volume fraction of ellipsoids and $\tilde{\alpha}^\gamma \equiv \alpha^\gamma/ab^2$.

The average of the dipole–dipole interaction tensor $4\pi n < T >^\gamma$ is now calculated as

$$< T >^\gamma = \frac{1}{4\pi}\int t^{\gamma\gamma}(\mathbf{R})e^{-i\mathbf{q}\cdot\mathbf{R}}\rho^{(2)}(\mathbf{R})d^3R, \tag{7}$$

which contains the two-particle distribution function $\rho^{(2)}(\mathbf{R})$ of the ellipsoids as defined in Ref. [8]. It can be shown [8] that in the case of spherical inclusions $< T >^{\gamma}$ turns out to be independent of $\rho^{(2)}(R)$ thus ϵ_M depends only on the volume fraction of the spheres; this is not the case for ellipsoidal inclusions and therefore, even in MFA, we will obtain different results depending on our choice of $\rho^{(2)}(\mathbf{R})$.

First we will assume that $\rho^{(2)}(\mathbf{R})$ is given by the low-density limit of the two-particle distribution function of hard ellipsoids, that is

$$\rho^{(2)}_{HE}(r,\mu) = \Theta(R - 2r_0(\mu)), \tag{8}$$

where Θ is the unit step function, $\mu = \cos\theta$ in spherical coordinates, $r_0(\mu) = b/(1 - e^2\mu^2)^{\frac{1}{2}}$. In this case it can be shown, by direct integration of Eq. (7), that $< T >^{\gamma}$ turns out to be related to the depolarization factors L_γ by

$$< T >^{\gamma} = -1 + L_\gamma. \tag{9}$$

When we substitute this expression in Eq. (6) we obtain, for the principal components of effective dielectric response, exactly the same expression as the one derived in Ref. (2) using the Lorentz method with an ellipsoidal fictitious cavity. On the other hand if we use in the averaging procedure $\rho^{(2)}(\mathbf{R}) = \Theta(R - 2a)$ we obtain Galeener's expression [1], as derived by the Lorentz method with a spherical fictitious cavity.

Obviously, the two-particle distribution function which should be used is the one which actually appears in the samples being examined.

III. Beyond the mean field approximation

Following the procedure of Ref. (8) we include the dipolar fluctuations through the scalar parameters α^*_r, called renormalized polarizability components, given by

$$P_i^{\gamma} = \alpha^*_\gamma[E_{ex}\hat{q}_\gamma/\epsilon_h + \sum_j T_{ij}^{\gamma\gamma} < P >^{\gamma}]. \tag{10}$$

When one substitutes this expression in the right hand side of Eq. (3a) one obtains a relationship between α^*_γ and α^γ given by

$$\begin{aligned} \frac{\alpha^*_\gamma}{\alpha^\gamma} &= 1 - (\tilde{\alpha}^*_\gamma)^2 < \sum_j T_{ij}^{\gamma\gamma}) >^2 \\ &+ \alpha^*_\gamma \sum_\delta \alpha^*_\delta \langle \sum_{jk} T_{ij}^{\gamma\delta} T_{jk}^{\delta\gamma} \rangle. \end{aligned} \tag{11}$$

It can seen that the renormalized polarizability components are given in terms of the fluctuations of the dipole-dipole interaction tensor. If we restrict ourselves to the low-density regime, we can further assume that the three-particle distribution function $\rho^{(3)}(\mathbf{R}_1, \mathbf{R}_2, \mathbf{R}_3)$, which is required in the calculation of $< \sum_{jk} T_{ij}^{\gamma\delta} T_{jh}^{\delta\gamma} >$, can be approximated by

$$\rho^{(3)}(\mathbf{R}_1, \mathbf{R}_2, \mathbf{R}_3) \cong \rho^{(2)}(\mathbf{R}_{12})\rho^{(2)}(\mathbf{R}_{23}). \tag{12}$$

In this case Eq. (11) yields

$$\frac{\tilde{\alpha}^*_\gamma}{\tilde{\alpha}^\gamma} = 1 + \frac{1}{4}\tilde{\alpha}^*_\gamma \sum_\delta \tilde{\alpha}^*_\delta f^{\gamma\delta}, \tag{13}$$

where $\tilde{\alpha}^*_\gamma \equiv \alpha^*_\gamma/ab^2$ and the coefficients $f^{\gamma\delta}$ are integrals which are displayed in the appendix.

Assuming again that $\rho^{(2)}(\mathbf{R})$ is given by $\rho^{(2)}_{HE}(R,\mu)$ [Eq. (8)], we calculate the coefficients $f^{\gamma\delta}$ and we solve (numerically) the system of coupled quadratic equations given by Eq. (13).

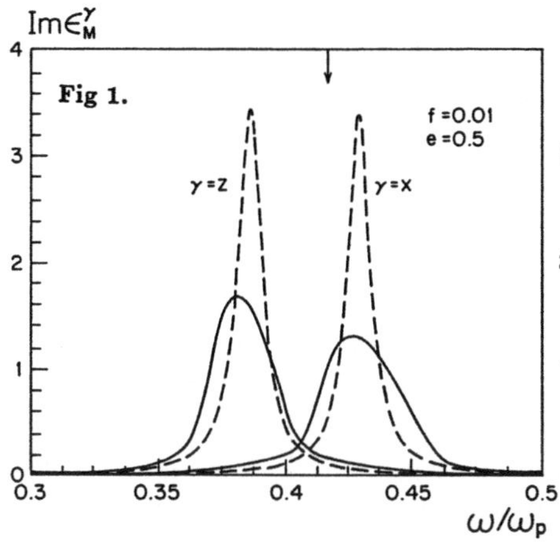

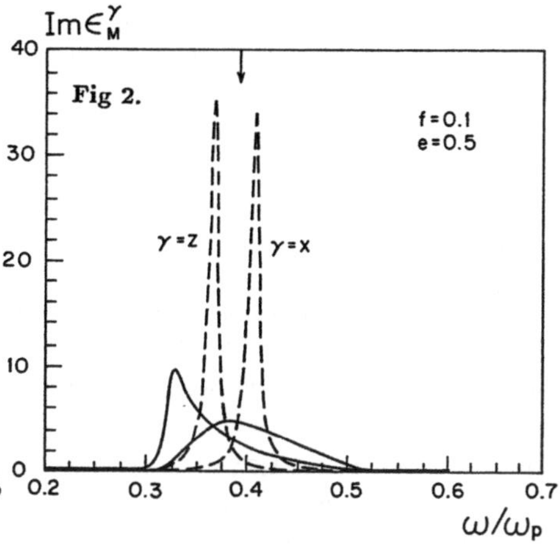

Fig 1. $Im\,\epsilon_M^\gamma$, for $\gamma = x$ and z, as a function of ω/ω_p for a composite with aligned ellipsoidal inclusions. The eccentricity of the ellipsoids is $e = 0.5$ and their volume fraction is $f = 0.01$. The continuous (dashed) line corresponds to RPT (MFA). The arrow shows the position of the corresponding peak in the case of spherical inclusions in MFA.

Fig 2. $Im\,\epsilon_M^\gamma$, for $\gamma = x$ and z, as a function of ω/ω_p for a composite with aligned ellipsoidal inclusions. The eccentricity of the ellipsoids is $e = 0.5$ and their volume fraction is $f = 0.1$. The continuous (dashed) line corresponds to RPT (MFA). The arrow shows the position of the corresponding peak in the case of spherical inclusions in MFA.

The theory is now applied to a system of metallic ellipsoidal inclusions in dispersionless gelatin. We choose for the inclusions a simple Drude dielectric function

$$\epsilon_m = 1 - \omega p^2/\omega(\omega + i/\tau), \tag{14}$$

where ω_p is the plasma frequency and τ the electronic relaxation time. For gelatin we take $\epsilon_h = 2.37$ independent of frequency.

In Fig. 1 and 2 we show the imaginary part of the effective dielectric function $Im\epsilon_M^\gamma$ as a function of ω/ω_p for prolate ellipsoids with eccentricity $e = 0.5, \omega_p\tau = 92$ and filling fractions f of 0.01 and 0.1, respectively. The dashed line corresponds to MFA and the solid line corresponds to the solution of Eqs. (13), which we will call renormalized polarizability theory (RPT). As it can be seen, there is a main peak for $\gamma = x$ and $\gamma = z$ and both cases are displayed together in the figures; the lower-frequency peak corresponds to $\gamma = z$. In MFA for the case of spherical inclusions ($e = 0$) there is only one component of $\epsilon_M^\gamma(\epsilon_M^x =, \epsilon_M^y = \epsilon_M^z \equiv \epsilon_M)$ whose peak in $Im\epsilon_M(\omega)$ is located $\omega/\omega_p = 0.415$ for $f = 0.01$ and $\omega/\omega_p = 0.391$ for $f = 0.1$. These values are shown with arrows in the figures. As the eccentricity increases ($e > 0$) for a given volume fraction, the frequencies at which $Im\epsilon_M^z(Im\epsilon_M^x)$ peaks shift to the red (blue) with respect to its corresponding value at $e = 0$. On the other hand, for a given eccentricity both peaks shift to red as the volume fraction increases.

Although the corresponding peaks in RPT follow the same general behavior as the ones in MFA, as a function of volume fraction and eccentricity, it can be clearly seen that the main effect of the field (dipolar) fluctuations is to reduce the height of the peaks, to increase its width in an asymmetric way and to shift both of them ($\gamma = x, z$) to the red with respect to their locations in MFA.

It can be easily shown that in MFA and in the long wavelength limit there is only one electromagnetic mode which is opticaly active along each principal direction. The main reason for this is that in MFA all the dipoles acquire exactly the same dipole moment; in this sense, this approximation regards the system more like a crystal than like a disordered system. Therefore the physical origin of the broadening of the peaks comes from the excitation of a larger number of modes which become now available due to the allowance of the dipolar fluctuations. In a normal-mode representation what actually happens is that instead of having a single-isolated mode (in MFA) we now have (in RPT) a continuous branch cut of modes.

Unfortunately, up to now, there are no experiments on this type of systems, due, essentially, to the difficulty in the preparation of the samples. The samples usually have non-spherical inclusions with random orientations. Therefore, our predictions, athough sensible and with a clear physical basis, cannot be quantitatively tested against experiment. Nevertheless, we are presenting, what to our knowledge is the first calculation of the dielectric response of a composite with ellipsoidal inclusions beyond MFA, thus we believe that our predictions could stimulate further progress in this fascinating problem.

In Summary:

(i) We showed that in MFA the components of the effective dielectric tensor ϵ_M^γ of a composite with ellipsoidal inclusions depends, besides their volume fraction, on their two-particle distribution function. Furthermore, we also demonstrated that two different expressions for ϵ_M^γ, which have been derived in the literature using the Lorentz' method, correspond to two specific choices of the two-particle distribution function.

(ii) We presented a procedure for calculating ϵ_M^γ, beyond MFA, using a simple theory (RPT) and results were displayed for the case of metallic (Drude) ellipsoidal inclusions in gelatin.

Appendix

The coefficients $f^{\gamma\delta}$ which appear in Eq. (13) are given by:

$$f^{xx} = 3ab^2 f \int_0^\infty \frac{dR}{R^4} \int_{-1}^1 d\mu 9\mu^2(1-\mu^2)\rho^{(2)}(R,\mu) \tag{A.1}$$

$$f^{zz} = 6ab^2 f \int_0^\infty \frac{dR}{R^4} \int_{-1}^1 d\mu (3\mu^2-1)^2\rho^{(2)}(R,\mu) \tag{A.2}$$

$$f^{xx} = 6ab^2 f \int_0^\infty \frac{dR}{R^4} \int_{-1}^1 d\mu [\frac{27}{8}(1-\mu^2)^2 - 3(1-\mu^2)+1]\rho^{(2)}(R,\mu) \tag{A.3}$$

$$f^{xy} = \frac{3ab^2}{4} \int_0^\infty \frac{dR}{R^4} \int_{-1}^1 d\mu 9(1-\mu^2)^2\rho^{(2)}(R,\mu). \tag{A.4}$$

Acknowledgements

We would like to acknowledge the partial support of Dirección General de Asuntos del Personal Académico of the Universidad Nacional Autónoma de México through grant IN-01-4689-UNAM.

References

1. F.L. Galeener, *Phys. Rev. Letters* **27**, 421 (1971).

2. R.W. Cohen, G.D. Cody, M.D. Couts and B. Abeles, *Phys. Rev. B* **8**, 3689 (1973).

3. D. Stroud, G. W. Milton and B.R. De, *Phys. Rev. B* **34**, 5145 (1986); X.C. Zeng, P.M. Hui, D.J. Bergman and D. Stroud, *Phys. Rev. B* **39**, 13224 (1989); S. Torquato, *Phys. Rev. B* **35**, 5385 (1987); G.A. Niklasson and C.G. Granqvist, *J. Appl. Phys.* **55**, 3382 (1984); *Physica A* **157**, 364 (1989); Z. Chen and P. Sheng, *Phys. Rev. B* **39**, 9816 (1989).

4. U. Kreibig and L. Genzel, *Surface Sci.* **156**, 678 (1985); M. Quinten and U. Kreibig, *Surface Sci.* **177**, 557 (1986); U. Kreibig, M. Quinten and D. Schoenaver, *Physica A* **157**, 244 (1989).

5. J. Giraldo, in *Thin Films and Small Particles*, ed. by M. Cardona and J. Giraldo (World Scientific, Singapore, 1989) p. 138; S.P. Apell, J. Giraldo and S. Lundquist, *Phase Transitions* **24/ 26**, 577 (1990).

6. See for example: *Electrical Transport and Optical Properties of Inhomogeneous Media*, ed. by J. C. Garland and D.B. Tanner, AJP Conference Proceedings, Number 40 (American Institute of Physics, New York, 1978); *ETOPIM 2*, ed. by J. Lafait and D.B. Tanner, Proceedings of the Second International Conference on Electrical Transport and Optical Properties of Inhomogeneous Media (North Holland, Amsterdam, 1989); *Electrodynamics of Interfaces and Composite Systems*, ed. by R.G. Barrera and W.L. Mochán, Advanced Series in Surface Science Vol. 4 (World Scientific, Singapore, 1988).

7. See for example R.G. Barrera, G. Monsivais, W.L. Mochán and E. Anda, *Phys. Rev. B* **39**, 9998 (1989) and references therein.

8. R.G. Barrera, G. Monsivais and W.L. Mochán, *Phys. Rev. B* **38**, 537 (1988).

9. R.G. Barrera, P. Villaseñor-González, W.L. Mochán, M. del Castillo-Mussot and G. Monsivais, *Phys. Rev. B* **39**, 3522 (1989); **41**, 7370 (1990).

10. J.C. Maxwell Garnett, *Philos. Trans. R. Soc. Lond.* **203**, 385 (1904).

11. See for example: J.D. Jackson, *Classical Electrodynamics* second edition (J. Wiley, New York, 1975) p. 155.

12. H.A. Lorentz, *The Theory of Electrons* (B.G. Teubner, Leipzig, 1909; Reprint: Dover, New York, 1952).

13. See for example: C.F. Bohren and D.R. Huffman, *Absorption and Scattering of Light by Small Particles* (J. Wiley, New York, 1981) pp. 141-152.

Heat Capacity Measurements of p-H_2 and o-D_2 Adsorbed on Graphite at Low Temperatures

M.E. Bassols and *F.A.B. Chaves*

Instituto de Física, Universidade Federal do Rio de Janeiro,
Cidade Universitária, Bloco A, 21945 Rio de Janeiro, RJ, Brazil

This work is a comparative study of heat capacity measurements performed on adsorbed films of p-H_2 and o-D_2 on graphite. A standard cryostat was used to measure the specific heat at low temperatures by the adiabatic method, for coverages below the monolayer in the temperature range of 3 to 30 K.

From the maximum obtained in the specific heat curves, as a function of temperature, we were able to determine the transition temperatures and consequently to construct the corresponding phase diagrams for the two isotopes, as shown in Fig. 1. It can be seen, in the figure, that both isotopes show the same qualitative behavior characterized by three distinct regions, which we propose to be: an ordered phase in registry with the substrate (a), a fluid phase with the characteristics of an imperfect 2D gas (c) and a coexistence region of these two phases (b).

A commensurate solid phase (a) was identified in the two systems as a lattice gas in a $\sqrt{3}\times\sqrt{3}R30$ superstructure, from the order-disorder transition. This transition, which occurs between the ordered phase and the fluid one, is characterized by a sharp peak which has its maximum when 1/3 of the

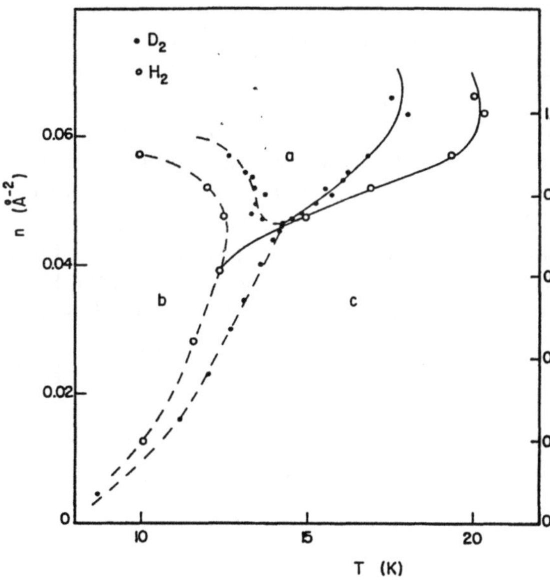

Fig. 1 Phase diagrams for p-H_2 and o-D_2

Springer Proceedings in Physics, Volume 62
Surface Science Eds.: F.A. Ponce and M. Cardona

graphite sites are occupied by the adatoms, corresponding to a critical density of 0.0636 Å^{-2}.

From these sharp peaks we were able to calculate the critical exponents $\alpha(H_2) = 0.37$ and $\alpha(D_2) = 0.32$, for the corresponding Potts temperatures of $T_c = 20.4$ K and $T_c = 18.1$ K, respectively. These values are in good agreement with the theoretical critical exponent of 1/3 obtained for the three-state Potts transition model, which occurs between an ordered phase with triple degeneracy and a disordered phase in which the average occupation is equal for the three sublattices.

H_2, being a quantum solid at very low temperatures, and, because it is lighter than D_2, having a greater zero point motion, stays in the $\sqrt{3}$ phase for a larger range of density and temperatures than its heavier isotope. This fact is shown in the phase diagrams, where the Potts temperature is higher for the lighter isotope, meaning that the heavier isotope disorders first, providing strong evidence for the quantum nature of the $\sqrt{3}$ structure for certain molecules [1].

From our experimental data we calculated the Einstein excitation energies for the two systems, considering the harmonic model for Einstein oscillators. We obtained 40 K and 51 K for D_2 and H_2, respectively, which are in reasonably good agreement with the values obtained from neutron scattering experiments [2]. Our results confirm the theoretical calculations by *Ni* and *Bruch* [3] that the correlation between the oscillator displacements is stronger for the H_2 isotope.

A fluid phase indicated in the phase diagrams (c) extends from the low density region to the more dense region around the monolayer in registry. The specific heat data for temperatures above the transition indicate an asymptotic behavior towards the NK value of the specific heat, characteristic of a 2D-gas-type fluid [4]. This behavior is present in both systems, although a difference between them is introduced by a certain coverage dependence. From these results and based on the theoretical discussion developed by *Siddon* and *Schick* [5] for the experimental results of helium adsorbed on graphite [6], we can consider that both p-H_2 and o-D_2, for low densities and high temperatures, behave as a two-dimensional imperfect quantum gas, as does helium.

In the proposed phase diagrams the full line, which limits the order-disorder transition, ends on the dashed line, which represents a first-order transition. This intersection is between the coexistence region and the ordered phase, above a critical density, and between the coexistence and fluid regions below this critical density. These points, according to our results, have coordinates $T = 12.4$ K, $n_c = 0.0390$ Å^{-2} and $T = 14.3$ K, $n_c = 0.0465$ Å^{-2} for p-H_2 and o-D_2, respectively. If in fact they correspond, as indicated in the figure, to the intersection of the two lines, these transitions represent, in both cases, the Potts tricritical point [7].

The preliminary data of this work were presented in [8], and our results are in good agreement with the work of *Mottler* and *Dash* [9] and *Freimuth* and *Wiechert* [10].

References

1. M. Schick, R.L. Siddon: Phys. Rev. A **8**, 339 (1973)
2. M. Nielsen, J.P. McTague, W. Ellenson: J. Phys. Colloq. **38**, 6Y-10 (1977)
3. X.Z. Ni, L.W. Bruch: Phys. Rev. B **33**, 4584 (1986)
4. J.G. Dash: *Films on Solid Surfaces* (Academic, New York 1975)
5. R.L. Siddon, M. Schick: Phys. Rev. A **9**, 907 (1974)
6. M. Bretz, J.G. Dash, D.C. Hickernell, E.O. McLean, O.E. Vilches: Phys. Rev. A **8**, 1589 (1973)
7. S. Alexander: Phys. Lett. **54A**, 352 (1975)
8. F.A.B. Chaves, M.E.B.P. Cortez, R.E. Rapp, E. Lerner: Surf. Sci. **150**, 80 (1985)
9. F.C. Mottler, J.G. Dash: Phys. Rev. B **31**, 346 (1985)
10. H. Freimuth, H. Wiechert: Surf. Sci. **162**, 432 (1985); ibid. **178**, 716 (1986)

The Growth of Cobalt on Cu(100): An Angle Resolved Auger Electron Spectroscopy Study

J.M. Heras, M.C. Asensio, G. Andreasen, and L. Viscido

Instituto de Investigaciones Fisicoquímicas Teóricas y Aplicadas (INIFTA),
Fisicoquímica de Superficies, Facultad de Ciencias Exactas,
Universidad Nacional de La Plata, C.C. 16, Suc. 4,
1900 La Plata, Argentina

1. Introduction

The growth mode of metal overlayers is of great scientific and technological importance because it determines the final layer morphology upon which, in turn, some physical and chemical properties depend, among them electronic, magnetic and chemisorptive ones. The handling of such properties is of fundamental importance in microelectronics and catalysis.

Very recently, many papers have been published in which the effect of forward scattering of energetic photoelectrons or Auger electrons is proposed as a very useful tool to assess the growth mode of metal overlayers and other adsorbates as well [1-5]. A more quantitative analysis of the forward scattering phenomena applying a simple single-scattering cluster model, has also been published [7, 8]. In the light of these papers, it is evident that the method provides a simple way to distinguish between different growth modes at the very early stages of deposition.

In this paper we shall deal with the growth of Co overlayers on Cu(100) at temperatures between 100 K and 373 K, studied by means of the forward scattering of the Auger electrons of the $L_3M_{45}M_{45}$ Cu peak at 918 eV kinetic energy, and the $L_3M_{23}M_{45}$ Co peak at 656 eV.

2. Experimental

The study was performed in a home-made stainless steel UHV-apparatus evacuated to about 10^{-8} Pa with turbomolecular and cryopumps, and equipped with a single pass cylindrical mirror analyzer (CMA) with coaxial electron gun (Physical Electronics), an ion sputtering gun Penning type (VG AGS2), an xyzθ manipulator, a two-filament metal vapor source provided with a water cooled shroud, and automated data acquisition capabilities. Cobalt films with thicknesses ranging from 0.3 monolayers (ML) to about 20 ML were evaporated onto the Cu(100) substrate at temperatures ranging between 100 to 373 K, from a Co coil (99.999% pure from Johnson & Matthey) resistively heated. Substrate temperature was monitored with a thermocouple type "E" attached to the reverse of the single crystal. Film thickness was monitored with a high sensitive quartz microbalance also attached to the sample holder such as to have the same temperature as the substrate. Its temperature was separately and accurately controlled by another thermocouple type "E". In this way,

Springer Proceedings in Physics, Volume 62
Surface Science Eds.: F.A. Ponce and M. Cardona

thicknesses of 0.3 ML could be routinely measured with a reproducibility better than 0.1 ML, according to the Auger signal.

In order to have a polar angle resolution, a diaphragm with an acceptance limiting aperture (8° in the polar direction, 70° in the azimuthal direction), is accurately displaced in front of the CMA, while the sample is rotated about an axis normal to both the electron beam and the horizontal. This rotation axis coincides with the [010] direction of the Cu(100) single crystal. Thus, the take-off angle of the CMA was reduced in the polar direction from 38° ± 7° to 38° ± 4° and in the azimuth from 360° to 70°. This high azimuthal aperture contributes only to the broadening of the diffraction peaks, but the signal-to -noise ratio is still favorable making a more sensitive detector unnecessary. The geometry actually allows the collection of only those Auger electrons which are emitted between 10° and 134° off the surface.

3. Results

Figure 1 shows the polar scans of two cobalt layers about 1.3 ML thick grown on Cu(100) at 128 K and 373 K, respectively. The Co peak at 656 eV was monitored in both cases. It is observable that in the case of the film grown at 128 K no marked features are present but in the film grown at 373 K there are features at 62°, 85° and 112°. At such low Co overlayer thickness, forward scattering of Co Auger electrons is not expected unless 3D islands are formed because of surface diffusion. This behavior would indicate that at 373 K the Co overlayer grows following a Volmer-Weber mechanism.

Figure 2 compares the polar scans of the Cu substrate signal at 918 eV, measured after each Co deposition with the Co Auger signal at 656 eV of layers with thicknesses of 1 and 6 ML Co grown at 365 K. At the beginning, the Co signal follows the features of the Cu substrate, but in the 6 ML overlayer (now the Co signal is more intense than the Cu signal) the peaks at 72° 82° and 90° become more intense. In the case of an epitaxial growth, all the substrate signals attenuate uniformly while

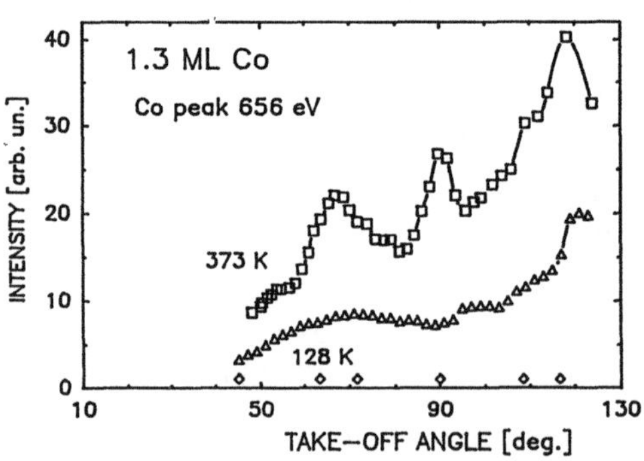

Figure 1. Deposition temperature effect on the polar scan spectra of two 1.3 ML thick Co films. The Co signal at 656 eV was monitored.

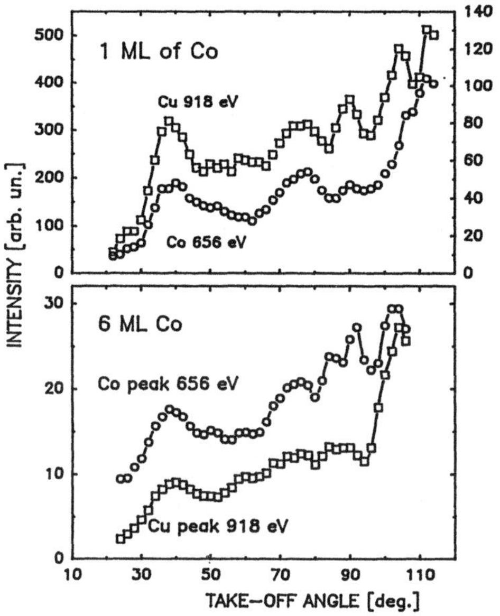

Figure 2. Comparison between the polar scans of Cu (Auger transition at 918 eV) and Co (Auger transition at 656 eV) for three different films grown at 665 K on Cu(100).

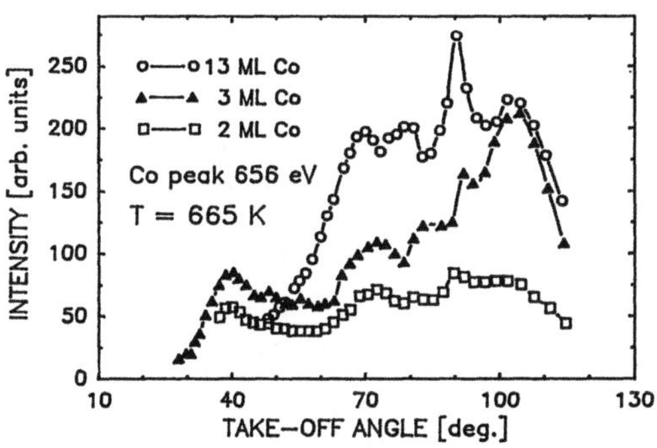

Figure 3. Polar angle scans of the Co $L_3M_{23}M_{45}$ at 656 eV in various film thicknesses.

overlayer signals increase in the same form, i.e. it is expected that the diffraction features remain stationary. Hence, from fig. 2 it follows that there is no epitaxial growth of Co in our experiments carried on Cu(100) at 365 K.

Figure 3 compares the Co Auger signal at 656 eV in films with thicknesses up to 13 ML. In this last overlayer, the peak at 90° clearly developed more intensely than the other two at 72° and 82°.

4. Discussion

The fcc form of Co has a lattice parameter of 0.3552 nm, while that of fcc Cu is 0.3615 nm, giving rise to a misfit of 1.77% between the Cu and Co lattice points. Consequently, the first Co layer must expand in order to fit the Cu lattice points. Hence, in this pseudomorphic growth mode, a relaxation of the second layer is expected to take place by diminishing its distance to the first, and so on in the following layers. This relaxation implies that the polar angles at which forward scattering is expected, do not coincide with the original of the Cu(100) substrate (45°, 63.4°, 71.6°, 90.0°, 108.4°, 116.6°), except that corresponding to 90°. Moreover, the stable form of Co at room temperature is the hcp and not the fcc. Hence, at high overlayer thicknesses, the forward diffraction features are expected to develop at other polar angles than those reported in the present paper.

Acknowledgements

The authors acknowledge the financial support of the CONICET (Argentina) as well as the donation of equipment by the A.v. Humboldt- and the Volkswagenwerk-Foundations (Fed. Rep. Germany). The authors are indebted to the Department of Condensed Matter of the University of Madrid for the loan of the Cu(100) single crystal.

References

1. W. F. Egelhoff, Phys. Rev. B, **30**, 1052 (1984)
2. W. F. Egelhoff, J. Vac. Sci. Technol., A **2**, 350 (1984)
3. S. A. Chambers, T. R. Greenlee, C. P. Smith and J. H. Weaver, Phys. Rev. B, **32**, 4245 (1985).
4. S. A. Chamber, S. B. Anderson and J. H. Weaver, Phys. Rev. B, **32**, 4872 (1985).
5. S. A. Chambers, H. W. Chen, I. M. Vitomirov, S. B. Anderson and J. H. Weaver, Phys. Rev. B, **33**, 8810 (1986).
6. S. A. Chambers and L. W. Swanson, Surface Sci. **131**, 385 (1983)
7. C. S. Fadley, Progress Surf. Sci., **16**, 275 (1984).
8. E. L. Bullock and C. S. Fadley, Phys. Rev. B, **31**, 1212 (1985).

Water Adsorption on Copper: Artifacts Emerging During AES

J.M. Heras, G. Andreasen, and L. Viscido

Instituto de Investigaciones Fisicoquímicas Teóricas y Aplicadas (INIFTA), Fisicoquímica de Superficies, Facultad de Ciencias Exactas, Universidad Nacional de La Plata, C.C. 16, Suc. 4, 1900 La Plata, Argentina

1. Introduction

It is well known that energetic electron beams strongly interact with adsorbates. As a consequence, adsorbates can dissociate and a variety of charged and uncharged species can either be desorbed or stay at the surface [1]. Metal surfaces in particular are known to stabilize electronically excited species [2]. Dissociative decays involving Auger transitions are very fast and may lead to stabilization of the adsorbates rather than to desorption.

The effect of beam irradiation on water adsorbed on polycrystalline Fe [3] indicates that the incident electron beam enhances the rate of oxidation and the passive layer which normally forms breaks down. In Al(100) it has been reported [4] that the Auger electron beam induces dissociation of the OH groups on the surface, with simultaneous desorption of H. The remaining O-atoms react with the surface and an oxide grows in an island mode.

The electron beam induced sample current between sample and ground depends on the integrated secondary electron emission, that is to say, it depends on surface morphology and chemical composition [5-7].

In the present paper, some results will be presented which show that water dissociation on Cu is induced by the electron beam of the Auger system, and that the target current constitutes a simple way to detect surface reactions.

2. Experimental

The experiments were performed in a self-assembled Auger system in which pressures down to $3x10^{-8}$ Pa can be routinely achieved by means of turbomolecular and cryopumps. The energy analyzer used was a single pass CMA with coaxial electron gun (Physical Electronics) and a resolution $\Delta E/E = 0.6$ %. The Cu sample was a polycrystalline sheet rated 99.999 % pure (Johnson, Matthey & Co.) with a marked recrystallization texture with the (100) planes parallel to the surface. The sample holder, rotatable by 250° allows direct Joule effect heating of the sample sheet to 1200 K and cooling down to 176 K by means of Cu braids attached to a liquid nitrogen reservoir.

The Auger spectra were recorded in the derivative form with the following settings: 2 keV primary electron beam energy with a beam current of 3 mA and normal incidence to the sample surface; 1 Vp-p modulation amplitude; and 775 V in the secondary electron multiplier. The maximum scanning speed compatible with the best resolution of the $M_{23}VV$ doublet of Cu at 59-61 eV

Springer Proceedings in Physics, Volume 62
Surface Science Eds.: F.A. Ponce and M. Cardona

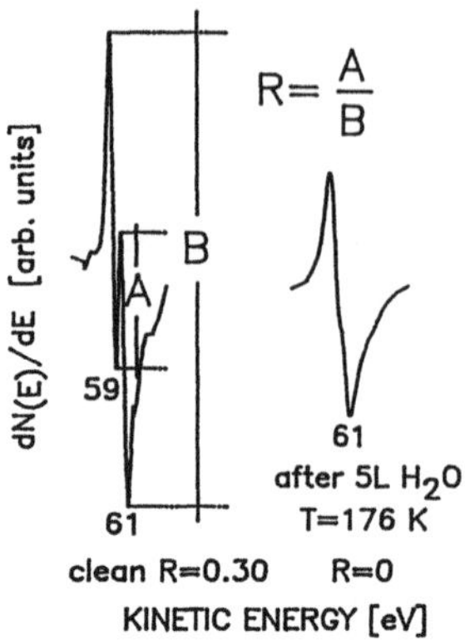

Figure 1. Definition of the ratio R taken in the Auger Cu doublet between the short excursion from the M_3VV to the M_2VV transitions to the overall peak-to-peak height. After 5 L H_2O exposure with the sample at 176 K, the doublet is unresolvable.

kinetic energy was 0.5 eV/s with a time constant of 0.3 s in the lock-in amplifier (EG&G). As the resolution of this doublet is taken as a measure of the surface reaction progress, a factor R is defined according to Vook et al. [8], taking the ratio of the short excursion from the M_3VV to the M_2VV transition to the overall peak-to-peak height, as fig. 1 shows. The best R-value for our textured polycrystalline Cu sample was 0.30. With a Cu(100) single crystal we found that the R value can be as high as 0.34. The resolution of these Cu Auger transitions is strongly dependent on surface roughness and cleanliness. For our purposes, we took a value R≈0.25 as a starting point for our experiments.

The electron beam induced sample current is a function of the incident electron beam energy. In the case of Cu(111) it is positive between 300-2500 eV, which indicates a net flux of electrons into vacuum [7]. In our clean and annealed polycrystalline Cu sample, with 2 keV electron beam energy, the picoammeter (Keithley 417) intercalated between sample and ground indicated ≈-2 mA.

3. Results

At water exposures routinely used, (≈150 L, L = Langmuir) slight effects were observed with the beam off (≈8% reduction in the value of R). However, water exposures with constant electron irradiation cause noticeable changes in R and sample current. At 176 K, with 5 L the Cu doublet at 59-61 eV can no longer be resolved (R = 0) and only one peak is found at 61 eV (see fig. 1). The peak-to-peak height of the Cu signal at 61 eV decreases with increasing exposure and remains constant above 5 L. The simultaneously monitored sample current and oxygen KLL Auger signal also reach the maximum value at this exposure. After 35 min annealing at 573 K, the starting conditions are practically restored.

When the experiment is performed at room temperature, beam effects are not so dramatic and saturation values are reached at about 60 L H_2O. The maximum peak-to-peak value of the oxygen KLL signal is 5 times greater in the experiments at 176 K than at 300 K (19 to 3.6 arbitrary units). Due to an unfavorable geometry, it was not possible to detect any desorbing species with the mass spectrometer, but it can be assumed that at both temperatures the OH species formed by electron impact dissociation are immediately stabilized by the Auger excitation process, giving rise to the

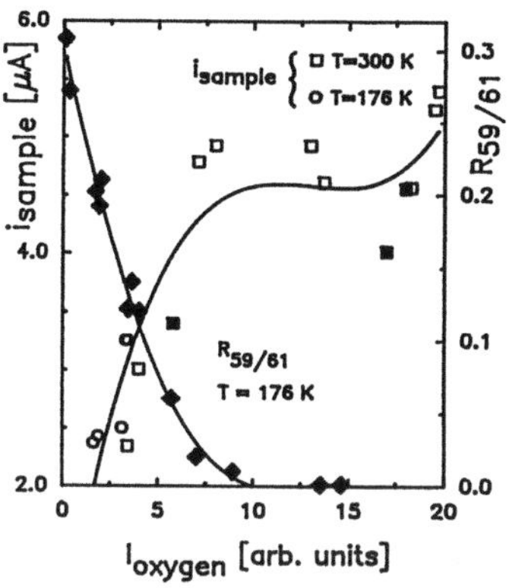

Figure 2. Relationship between the intensity of the oxygen KLL signal, I_{oxygen}, and the sample current, i_{sample}, as well as the ratio $R_{59/61}$, during water exposure under beam irradiation.

KLL oxygen signal [1b]. Nevertheless, it is known that at about 300 K water desorbs from many metal surfaces as a result of a disproportionating reaction between two previously formed OH groups [9], leaving O-atoms on the surface. Consequently, at 176 K the oxygen Auger signal should be more intense, according to the observed facts.

The correlation between sample current or R factor and peak-to-peak intensity I of the oxygen KLL signal is shown in fig. 2. There is a good correlation, also for the experimental points obtained at room temperature (circles) and in an experiment with high water exposures (≐300 L, filled squares).

It is known that Cu_2O does not show the doublet at 59-61 eV [10]. In our case a few minutes heating at 573 K causes the doublet to reappear. This rules out the possibility that an oxide is formed but suggests that the R factor becomes zero because an electronegative species is formed on the surface (O^-, OH^-). This species should also be responsible for the negative charge that the sample acquires, which saturates at the same time as the oxygen KLL signal. This is the expected effect of suppression of secondary electron emission by surface species that increase the work function. It is known that the principal effect of oxygen adsorption on metals is to smear out the fine structure of the N(E) distribution [6].

In order to determine the perturbed surface area, after beam irradiation of the Cu sample exposed to 40 L H_2O at 176 K followed by a 5 min annealing at 573 K, a lateral composition profile was taken scanning about 1 mm at both sides of the beam incidence point and monitoring the signal intensities of Cu at 61 eV, S at 156 eV, Cl at 180, and O at 513 eV, besides the R factor. Figure 3 shows the results. The signals are normalized to the respective maximum value. Clearly, while the Cu signal is practically constant, the R factor has a minimum where the oxygen signal shows a peak. Noteworthy, no Cl was found, but S, which has segregated from the bulk during annealing, shows a minimum where the oxygen signal has a maximum. Probably, S atoms reaching the surface from the bulk react with active O species and desorb as SO_2 (and/or SH_2). No S containing species could be detected in the gas phase with the mass spectrometer, due to its low sensitivity (10^{-6} Pa partial pressure) and unfavorable collection geometry.

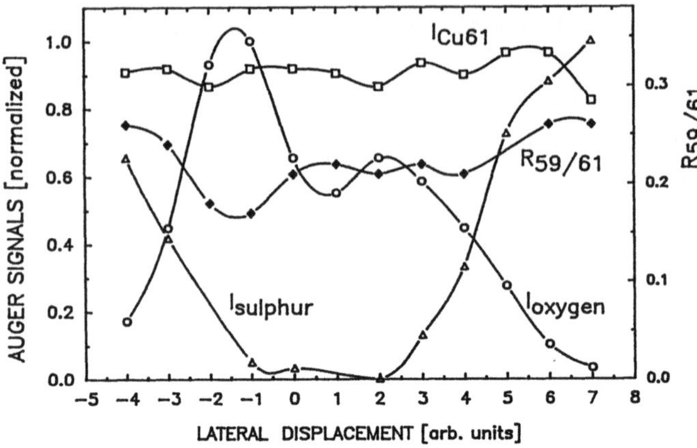

Figure 3. Surface composition profile lateral to the beam incidence point. The electron beam impinged at the position marked "0". 10 arbitrary units ≈1 mm.

4. Conclusions

Electron beam irradiation induces fragmentation of adsorbed water molecules on polycrystalline Cu. This is shown by: i) broadening of the Cu peaks at 59-61 eV kinetic energy (doublet $M_{23}M_{45}M_{45}$) which finally merge into only one peak; ii) increase of the negative charge on the sample, indicating a lower emission of secondary electrons: the work function of the sample is increased; iii) increase of the oxygen concentration at the beam incidence point, decreasing at 0.4 mm all around.

Annealing at 573 K causes the Cu feature at 59-61 eV to reappear and reduces surface oxygen concentration dramatically, suggesting that it is a labile species. Sulphur also appears on the surface because of diffusion from the bulk, though its surface concentration is maximum where oxygen is not detected. This suggests the formation of SO_2 as a possible eliminating mechanism.

Annealing at 673 K for about 1 h fully restores the Cu feature at 59-61 eV even if S and /or Cl are present on the surface. Moreover, the exposure of clean Cu at 300 K to 600 L O_2 also reduces the R factor by about 20% at the beam incidence point as 80 L H_2O does at this temperature. Hence, the suppression of the Cu signal splitting is only due to the presence of electronegative oxygen containing species which are responsible for the sample charging.

Acknowledgements

The authors acknowledge the financial support of the Argentine Research Council (CONICET) and the Argentine Science and Technology Secretariat (SECYT), as well as the donation of equipment by the A. von Humboldt and the Volkswagenwerk Foundations (Fed. Rep. Germany).

References

1. Workshops on the Desorption Induced by Electronic Transitions. a) DIET I, ed. by N.H. Tolk, M.M. Traum, J.C. Tully and T.E. Madey. Springer, Berlin, (1983). b) DIET II, ed. by W. Brenig and D. Menzel. Springer, Berlin, (1985). c) Diet III, ed. by R. H. Stulen and M. L. Knotek. Springer, Berlin, (1988).
2. H.D. Hagstrum, Phs. Rev., **96**, 336 (1954).
3. D.R. Baer and M.T Thomas, Appl. Surf. Sci., **26**, 150 (1986).
4. M.Q. Ding and E.M. Williams, Surface Sci., **160**, 189 (1985).
5. D. Chadwick, M.A. Karolewski and K. Senkiw, Surface Sci., **175**, L801 (1986).
6. M. A. Karolewski and D. Chadwick, Surface Sci., **175**. L806 (1986).
7. M.G. Barthès-Labrouse and G.E. Rhead, Surface Sci. **116**, 217 (1982).
8. S.S. Chao, E.A. Knabbe and R.W. Vook, Surface Sci., **100**, 581 (1980).
9. J. M. Heras and L. Viscido, Catal. Rev. Sci. Eng., **30**, 281 (1988).
10. S. W. Gaarenstroom, Appl. Surf. Sci. **7**, 7 (1981).

Model Calculations of the Indirect Interaction Between Chemisorbed Atoms

S.R. de Freitas and C. Pinto de Melo

Departamento de Física, Universidade Federal de Pernambuco,
50739 Recife, PE, Brazil

1. Introduction

When an atom or molecule approaches a clean metallic surface it experiments a translationally invariant potential. After it chemisorbs, however, the charge and spin distributions on the surface have to adjust to the transfer of charge between the two subsystems. These charge and spin disturbances, which are reminiscent of the Friedel oscillations, can be of very long range [1] such that if a second atom or molecule now approaches the system it will experiment a non-isotropic potential. It is an experimental fact that a first adsorbate determines a mesh of preferential sites for the chemisorption of a second atom [2]. If the final simultaneous adsorption sites are close enough to each other, the two adsorbates can interact and the local charge and spin distributions for the substrate may be substantially different from twice of those corresponding to the single atom chemisorption case.

A convenient parameter to use as criterium to investigate the range of the indirect interaction between the two adsorbates is the interaction energy defined as the difference between the binding energy of two adsorbed atoms and twice the chemisorption energy of an isolated adsorbate

$$\mathcal{E}^{int} = \Delta\mathcal{E}^{2adsorb.} - 2\,\Delta\mathcal{E}^{1adsorb.} \tag{1}$$

2. Model Calculation

In the present work we are interested in the indirect interaction between two hydrogen atoms on a one-dimensional chain (Fig. 1).

If the system is described by a first-neighbors tight-binding hamiltonian, transfer matrices techniques can be used to determine the Green's function elements in the site representation [3,4]. Changes on the electronic structure of the system can be investigated by determining the elements of the charge- and bond-order matrix connecting any two individual sites.

If we use renormalization ideas [4] the diagonal element on the adsorbate can be written as the continuous fraction [5]

$$G_{a,a} = \frac{1}{E+\alpha_A-}\,\frac{V_A^2}{E+\overline{\alpha}+VT-}\,\frac{\overline{V}^2}{E+\overline{\alpha}+VT-}\,\frac{V_B^2}{E+\alpha_B} \quad . \tag{2}$$

It is convenient to divide the problem into two "external" regions, where translational symmetry does exist and a single transfer matrix T can be defined, and an "internal" region, corresponding to the portion of the chain between the adsorbates. In this internal region site-dependent $T^{\mathcal{R}(\mathcal{L})}_{\bar{k}(-\bar{k})}$ matrices have to be determined by an iterative procedure after imposing the boundary condition $T^{\mathcal{R}(\mathcal{L})}_{n(-n)} = T$ for $n > 1$, where $\mathcal{R}(\mathcal{L})$ indicates the right (left) direction along the chain. For sites in the external regions one can find the generic diagonal element of G as

$$G_{\pm n,\pm n} = \frac{1 - VTG_{\pm(n-1),\pm(n-1)}}{E+\alpha+VT}, \quad n > 1 \quad . \tag{3}$$

Springer Proceedings in Physics, Volume 62
Surface Science Eds.: F.A. Ponce and M. Cardona

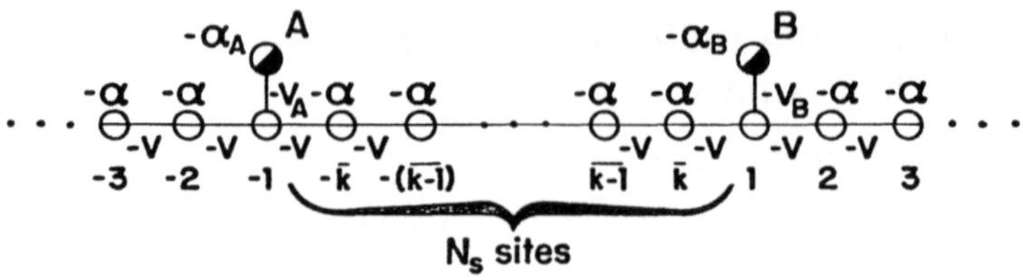

Figure 1: Simultaneous lateral chemisorption of two hydrogen atoms

The diagonal elements for sites in the internal region can be determined through the use of the site-dependent transfer functions. For elements to the right of the central site, for example, the generic element can be written as

$$G_{\bar{i},\bar{i}} = \frac{1 - VT^{\mathcal{R}}_{\bar{i}} G_{\overline{i-1},\overline{i-1}}}{E + \alpha + VT^{\mathcal{R}}_{\overline{i+1}}} \quad . \tag{4}$$

In the present model charge transfer effects are included through an Anderson-Newns one-center repulsion term [6] in the hydrogen atoms. The chain is assumed to be initially in a non-magnetic state with Fermi level position E_F. After chemisorption, charge and spin wave distributions centered at the substrate atoms immediately below the adsorbates are induced in the substrate; these densities reinforce each other at some points, while at other sites there occurs destructive interference [7]. This indirect interaction is examined as a function of the distance between the adsorbates. As a general rule, the range of interaction is shorter for parallel than for anti-parallel spin chemisorption. In Tables 1 and 2 we present,respectively, the adsorbate net charge and magnetization for single and double chemisorption for adsorption sites separated by $N_s = 11$ substrate atoms; the parameters used were $V_a = V_b = 4.6eV, \alpha_a = \alpha_b = 13.6eV, V = 2.5eV, \alpha = 4.6eV$, and $J = 12.9eV$, where J is the electron-electron repulsion term for the hydrogen atom.

Table 1: Relative charge on the adsorbate for $N_s = 11$

band fractional occupancy	1/4	1/2	3/4	1
1 adsorbate	0.0342	0.1878	0.3616	0.4575
2 adsorbates ↑↑	0.0320	0.1683	0.3698	0.4575
2 adsorbates ↑↓	0.0307	0.1926	0.3698	0.4575

Table 2: Adsorbate magnetization for $N_s = 11$

band fractional occupancy	1/4	1/2	3/4	1
1 adsorbate	0.4045	0.3413	0.0	0.0
2 adsorbates ↑↑	0.4677	0.4355	0.0	0.0
2 adsorbates ↑↓	0.4647	0.3817	0.0	0.0

According to Eq. 1, the interaction energy involves small differences between extensive quantitities. Hence, it is important to take into account even infinitesimal charge and spin rearrangements at each individual site [8]. It can be shown that in a frozen Fermi level approximation the error in the binding energy is related to the error in the computed total number of electrons, for each occupancy of the substrate band [7,8].

The total number of electrons in the initial (with no coupling among either atoms or adsorbates) and final (after simultaneous chemisorption) systems can be determined after summation over the charge densities and localized state contribution of each individual site [7]. Of course, the total number of electrons should be preserved in the chemisorption process. However, for a frozen Fermi level approximation this does not occur. Table 3 shows the error in the number of electrons for different values of E_F, for the same set of parameters as before.

Table 3: Error in the calculated number of electrons for $N_s = 11$

band fractional occupancy	0	1/4	1/2	3/4	1
1 adsorbate	0.0	-0.0173	0.1578	0.3231	-1.0000
2 adsorbates ↑↑	0.0	-0.0093	0.0987	1.1695	-2.0000
2 adsorbates ↑↓	0.0	-0.1017	0.7615	1.1695	-2.0000

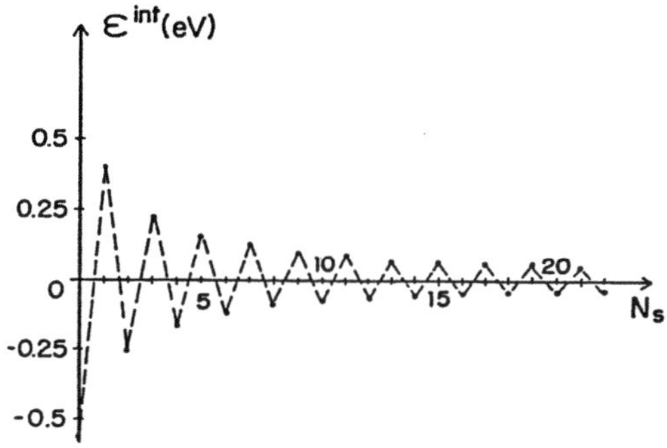

Figure 2: Interaction energy between two anti-parallel spin hydrogen atoms chemisorbed N_s sites apart

The interaction energy given by Eq. 1 has been computed as a function of the distance N_s between adsorbates, for different values of E_F. Results for anti-parallel spin chemisorption a half-occupied substrate band are presented in Fig. 2.

It can be observed from the oscillatory behavior of the curve that simultaneous chemisorption can be favored or not according to the distance between adsorption sites.

3. Discussion

In this work we have presented preliminary results for model calculations of the indirect interaction between adsorbed atoms in an infinite chain. The charge rearrangement induced on the substrate is analyzed as a function of the fractional occupancy of the band and, for a given value of E_F, the interaction energy is shown to be of oscillatory nature.

Although the choice of a universal criterium for examining the range of the indirect interaction is difficult, as a general trend one could say that if all other variables are the same the interaction is stronger for band occupancies close to one-half: for an almost empty or an almost totally full substrate band the induced spin and charge rearrangements and the interaction energies are of a much shorter range.

The importance of properly accounting for the small charge and spin rearrangements caused by Fermi level relaxation is reflected in the wrong values for the total number of electrons and for the interaction energy obtained in a frozen Fermi level approximation. Complete results for different values of E_F and extensive variation of N_s will be presented elsewhere [7].

4. Acknowledgements

This work was partially supported by the Brazilian Agencies FINEP and CNPq.

5. References

1. C.P. de Melo, M.C. dos Santos, M. Matos, and B. Kirtman, Phys.Rev.B35, 847(1987).
2. T.T. Tsong, Phys.Rev.B6,417(1972).
3. C.T. Papatriantafillou, Phys.Rev.B7,5386(1973).
4. C.P. de Melo, in *Electronic Structure of Atoms, Molecules and Solids*, ed. by F. Paixão, J. Castro and S. Canuto, (World Publishing, Singapore,1990).
5. We have used the notation suggested by G.H. Hardy and E.M. Wright, *An Introduction to the Theory of Numbers*, (Oxford University Press, London, 1954).
6. D.M. Newns, J.Chem.Phys. 50,4572(1969).
7. S.R. de Freitas and C.P. de Melo, in preparation.
8. C.P. de Melo, *Fermi Level Readjustments on Adsorption and Interface Formation*, previous communication.

Manifestation of Non-equilibrium Behavior in Thermal Desorption Dynamics

R. Almeida[1] *and E.S. Hood*[2]

[1]Departamento de Química, Facultad de Ciencias, Universidad de los Andes, Mérida, Edo. Mérida, 5101, Venezuela
[2]Department of Chemistry, Montana State University, Bozeman, MT 59717, USA

Abstract. The model proposed by Efrima et al./1-a/ is used to follow the time evolution of the bound states population of the Ar/W system and the vibrational relaxation is studied. The question of whether or not the adatom can be desorbed by a phonon pulse is addressed.

1. Introduction

In this work we have used the one-dimensional model introduced by Efrima et al. /1/ , to examine numerically the dynamics of the desorption of a weakly adsorbed system, where all the quantum state-to-state transition frequencies are smaller than the characteristic surface Debye frequency. We have also applied that model to study whether or not the desorption rate may be enhanced by using an acoustic pulse, which perturbes the equilibrium phonon population of the surface. The work is organized as follows. In section 2 we review the essential features of the model and the theory employed. In section 3.1 we study the time evolution of the population distribution of the bound states. Section 3.2 deals with systems whose initial population distribution differ from equilibrium and examines the vibrational relaxation. In section 4 the numerical results of the acoustic pulse enhanced desorption are presented.

2. The Model and the Theory

The theoretical model of the desorption system consists of an adsorbed particle, located at a distance z above the surface plane, interacting with a phonon heat bath through a Morse potential /2/

$$V(Q,z)= D\left\{\exp[-2\alpha((Q-z)-r)]-2\exp[-\alpha((Q-z)-r)]\right\} . \quad (1)$$

Here Q is the position of the nearest neighbor lattice atom and D, α and r are the usual Morse parameters defining the depth, anharmonicity and equilibrium distance of the interaction potential. Only the vibration of the adatom perpendicular to the surface is considered, all the lateral degrees of freedom are neglected.

The zero order description of the adsorbate is provided by the bound $|n\rangle$ and continuum $|\varepsilon\rangle$ states of a renormalized Morse potential $\langle V(Q,z)\rangle$, obtained by taking the thermal average of Eq. (1) over the lattice atomic positions. The thermal fluctuation in the interaction potential,

Springer Proceedings in Physics, Volume 62
Surface Science Eds.: F.A. Ponce and M. Cardona

$$U = V(Q,z) - \langle V(Q,z) \rangle , \tag{2}$$

is treated as a perturbation. Since U varies randomly, the oscillator evolves under the influence of a time dependent force that induces transitions of the adsorbed atom from an initial bound state $|n\rangle$ to a final bound state $|m\rangle$ or a continuum state $|\varepsilon\rangle$ .These adatom transitions are accompanied by energy conserving lattice transitions from an initial phonon state $|ph\rangle$ to a final phonon state $|ph'\rangle$. If U is considered as a small disturbance with respect to the adatom and lattice energies, the rate of transition $W_{n\to m}$ can be computed by using first order perturbation theory. Since only the $|n\rangle \to |m\rangle$ transitions are of specific interest, a thermal average over initial phonon states and a sum over final phonon states are performed. The expression for the transition rate can be shown /1-a/ to be of the following form:

$$W_{n\to m} = (2/\hbar^2)\mathrm{Re}\int_0^\infty dt\,(\langle U(t)U\rangle)_{nm}\exp(i\omega_{nm}t) \quad . \tag{3}$$

The transition rates W_n from bound to continuum state promotions is given by

$$W_n = \int_0^\infty W_{n\to\varepsilon}\,d\varepsilon \quad , \tag{4}$$

where $\varepsilon=(2\mu E)/(\hbar\,\alpha^2)$. The dynamical behavior of the system is described by the time evolution of the probability $P_n(t)$, that the particle occupies the vibrational state n at time t. The time rate of change of this occupation probability is described by a master equation of the Pauli type,

$$dP_n(t)/dt = \sum_m (W_{m\to n}P_m(t) - W_{n\to m}P_n(t)) - W_nP_n(t) \quad . \tag{5}$$

This equation permits promotion of bound particles to the continuum but forbids reverse transitions, thus all particles reaching the continuum are considered desorbed. The mean time required for desorption is

$$\langle t\rangle = \int dt\ t\ \eta(t) \quad , \tag{6}$$

where $\eta(t)$ is the probability that the system reaches a desortive state in a given time t /1-a/. The desorption rate constant k equals $1/\langle t\rangle$.

3. Results and Discussion

The calculations presented here were carried out with parameters mimicking the Ar/W system, thus for the interaction potential, D=1.9 Kcal/mol, $\alpha=1.44$ A^{-1}, the number of bound states is 25 and the energy corresponding to the 0→1 transition is 51.1 cm^{-1}. The surface Debye frequency is 127.7 cm^{-1}.

3.1 Bound state population during the desorption process

We compute $X_n(t)=P_n(t)/\sum_m P_m(t)$. If we consider an ensemble of oscillators distribu ted over the surface, X_n gives the fraction of those oscillators in the bound state n. Initially we have taken $P_n(0)$ to be the Boltzmann distribution at the temperature of the phonon bath. The dynamic behavior of the system is illustrated more clearly if we calculate the deviation of the instantaneous population of the level n from the Boltzmann distribution, P_n^{Bolt}, that is $(X_n(t)-P_n^{Bolt})/P_n^{Bolt}$. For a surface temperature of 100 K , we found that for the low energy levels ($n \leq 12$), X_n does not depart appreciably from its equilibrium value, however we notice substantial deviation for the levels $14 \leq n \leq 23$ as early as 0.1 picoseconds. It is also seen that the last bound level is drawn toward the steady state slower than the ones inmediately below it. The system reaches the steady state at about 10 picoseconds, a time short compared to the mean residence time of about one nanosecond (see table 1). Similar results were obtained for temperatures of the bath of 50 and 150 K.

In order to explain these observations we recall that a totally absorbing condition is applied at the boundary between the discrete states and the continuum. This boundary condition tends to drive the system out of equilibrium, while the energy exchange with the bath (surface phonons) leads it to equilibrium. If the rate of absorption is small compared to the rate of energy exchange with the surface, the system will be at equilibrium all the time. If the rate of absorption is comparable or larger than the equilibrium rate, the competition between these processes should establish a steady state distribution which may deviate substantially from the Boltzmann distribution. In addition, the rates of promotion from the bound state n to the continuum were found to be peaked at about n=18, with a region of maximum desorption for $14 \leq n \leq 23$. In this active desorption zone, the rate of promotion into the continuum is greater or equal than the maximum rate of transition between bound states. Therefore the rate of energy exchange with the surface will not be sufficient to compensate for the rate of passage into the continuum and the population of the levels in the active desorption zone will be depleted with respect to the equilibrium distribution characteristic of the surface temperature.

3.2 Vibrational Relaxation

We consider cases where the initial population of the adsorbed species is different than the equilibrium distribution at the surface temperature. We have followed the time evolution of a system where the adsorbed species initially have a Boltzmann distribution corresponding to a temperature half of the surface T_s and another with the initial distribution corresponding to $2T_s$, with T_s = 100k. This can be the case if by some mechanism the temperature of the bath is changed so fast that the adatom does not have time to equilibrate. For the first of the previous cases the results show that at very short time (less than 1 psec.), the transition from low energy levels to high energy levels are the dominant processes , however when the active desorption zone starts being populated the promotions to the continuum become more important and the system relaxes to a steady state indistinguishable from the one obtained in section 3.1 in a time close to 10 psecs. On the other hand, in the second case the upper states are overpopulated with respect to the equilibrium distribution and the system relaxes to the steady state by promoting transitions to the continuum as well as to lower energy levels. As in the other case it is reached in a time scale close to 10 psecs.

All these results suggest that after a very short time compared with the mean residence time, the desorption process becomes independent of the initial population. In order to further prove this point we have repeated the numerical tests described before using T_s=50,150 K and followed the time evolution of the population of a system that initially has $P_n(0) = \delta_{n1}$ and obtained similar results to the ones explained above This seems to confirm that the system loses memory of the initial state on a picosecond time scale, while the desorption occurs in a nanosecond time scale.

4. Phonon Pulse Induced Desorption

It has been observed /3/, room temperature desorption of residually adsorbed gas by application of low duty acoustic waves to solid samples. The observed desorption exhibits a strong dependence on the acoustic power. It has also been reported /4/ optical generation of ballistic acoustic phonon pulses of well defined frequencies which propagate macroscopic distances. Also, experiments have been carried out /5,6/ where desorption of helium films from sapphire crystals is induced by using nonequilibrium phonon pulses.

Here we assume that a steady acoustic pumping can be experimentally achieved, which creates a perturbation in the phonon population $n_p(\omega)$. As a consequence of this, the total phonon population, $n(\omega)$ will have the form $n(\omega)=n_T(\omega)+n_p(\omega)$, with $n_T(\omega)$ representing the equilibrium phonon distribution. The $n_p(\omega)$ will be taken as having gaussian form,

$$n_p(\omega) = A \exp(-(\frac{\omega - \omega_o}{\sigma})^2) \quad . \qquad (7)$$

where ω_o and σ are, respectively, the center and width of the acoustic phonon distribution and A is an intensity parameter that we assume is related to the strength of the perturbation. The new phonon population will go into Eq. (3), however due to the rather complicated expression that we obtain, the way it affects the transition rate is not easily predicted and numerical calculations are necessary.

The influence of altering ω_o is examined, the results using A=1 and a width arbitrarily set at one tenth of the surface Debye frequency, ω_D^{surf}, show that the inclusion of the acoustic signal induces an increase in the rate constant and the influence is more noticeable as ω_o exceeds the frequency corresponding to the energy difference between the ground and first excited state. Calculations were performed by changing the intensity prefactor while keeping ω_o at a fixed value (Table 1). The desorption rate constant increases proportionally to the pulse intensity. Under the assumption that the intensity of the acoustic signal can be directly related to the intensity of the phonon pulse, our results agree with the observation that the rate constant depends on the intensity of the applied signal. However, we must emphasize that we do not yet understand fully the details of the mechanism of phonon-pulse assisted desorption. It is probable that the

Table 1 : Desorption rate constant as a function of the intensity prefactor, A ($\omega_o=0.9\ \omega_D^{surf}$, $\sigma=0.1\ \omega_o^{surf}$)

T(K)	A= 0	1	6	10	20	30
50	$6.2\ 10^5$	$9.4\ 10^5$	$1.9\ 10^6$	$2.6\ 10^6$	$4.7\ 10^6$	$7.1\ 10^6$
100	$8.9\ 10^8$	$1.4\ 10^9$	$2.8\ 10^9$	$3.9\ 10^9$	$6.9\ 10^9$	$1.1\ 10^{10}$

anharmonicity dissipates the energy of the acoustic signal exciting phonons of high frequency, which are more effective in causing desorption.

References

1. (a)S. Efrima, C Jedrzejek, K. E. Freed, E. Hood and H. Metiu, J. Chem. Phys. 79, 2436 (1983) and references thereafter; (b)E. Hood, C. Jedrzejek, K. E. Freed and H. Metiu, J. Chem. Phys. 81, 3277 (1984).
2. P. M. Morse, Phys. Rev. 34, 57 (1927).
3. C. Krischer and D. Lichtman, Japan J. Appl. Phys., Supp. 2, Pt. 2 (1974); D. Lichtman, CRC Critics Review in Solid State Science, 395 (May 1974).
4. R. Ulbruch, V. Narayanamutry and M. Chin, J. Phys. Soc. of Japan A 49, 707 (1980); P. Hu, V. Narayanamutry and M. Chin Phys. Rev. Lett. 46, 19 (1981).
5. M. Sinvani, P. Taborek and D. Goodstein, Phys. Rev. Lett. 48, 1259 (1982); D. Goodstein, R. Maboudian, F. Scanamuzzi, M. Sinvani and G. Vivaldi, Phys. Rev. Lett. 54, 2034 (1985)
6. P. Taborek, Phys. Rev. Lett. 48, 1737 (1982).
7. A. Maradudin, E. Montroll, G. Weiss, J. Ipatova, Theory of lattice dynamics in the harmonic approximation (Academic , New York, 1975).

Table 1 : Desorption rate constant as a function of the [illegible]

[illegible]

The First Stages of Oxidation of Polycrystalline Cobalt Studied with Electron Spectroscopies

J.L. del Barco, R. Vidal, and J. Ferrón

INTEC, Universidad Nacional del Litoral and
Consejo Nacional de Investigaciones Científicas y Técnicas,
Güemes 3450, C.C. 91, 3000 Santa Fe, Argentina

Many studies of oxygen adsorption on cobalt have been performed [1–3]. They include adsorption on crystalline and polycrystalline surfaces. In general, these studies show the same sequence of events: oxide formation is preceded by the chemisorption of oxygen.

The present work extends our previous studies [4] on the oxidation of polycrystalline cobalt at room temperature as a function of oxygen dose. The aim of this work is to identify the different compounds formed during the oxidation process and to check the capability of Auger Electron Appearance Potential Spectroscopy (AEAPS) [5] to give reliable information when it is applied to the first stages of oxidation. To study the oxidation process we have applied the Principal Component Analysis (PCA) and Target Transformation (TT) methods [6] to both kinds of measurements, Auger Electron Spectroscopy (AES) [5] and AEAPS. These methods allow us to identify the number of different compounds appearing along the process and also to obtain the line shape of such compounds.

We have studied the first stages of low pressure and room temperature oxidation of polycrystalline cobalt using AES and AEAPS for exposures up to 40 L (1 L=10^{-6} torr seg) under an oxygen pressure of 10^{-9} torr.

The experiments have been carried out in an ultra high vacuum system equipped with AES, and AEAPS facilities. The sample was a high purity polycrystalline cobalt sample and the surface was cleaned by argon ion bombardment until contaminants could not be detected by AES.

The AES spectra were acquired in the first derivative mode, using a single pass cylindrical mirror analyzer with an energy resolution of 0.6 % and 4.5 μA, 3 keV electron beam current with a modulation amplitude of 2 V peak-to-peak. For the AEAPS spectra an electron beam with current density of 5 $\mu A/mm^2$, modulation amplitude of 1.0 V peak-to-peak and the second derivative mode was used.

The application of the PCA to the CVV-Co (AES) and $2p_{3/2}$-Co (AEAPS) spectra using the criteria of the real error [7] gives two independent components for both AES and AEAPS. Through the

Springer Proceedings in Physics, Volume 62
Surface Science Eds.: F.A. Ponce and M. Cardona

application of the TT method we can identify both these compounds Co and CoO, appearing the CoO component only for exposures greater than 5 L for both techniques. These results are in agreement with the model [8] about the formation of the oxide layer as a nucleation of oxide islands formed in a sea of chemisorbed oxygen and with our results [4,9]. The presence of the CoO and the appearance of islands should be detected at longer exposures (> 20 L) in order to be in accordance with other results [1–3], but in our case the CoO formation and the island growth could be promoted by the characteristics of the surface which is strongly bombarded during the the cleaning procedure and during the measurements.

However, our results show that AEAPS is able to give chemical information of a transition metal oxidation process of same quality as AES.

References.

[1] T. Matsuyama and A. Ignatiev, Surf. Sci., 102, 18(1981).

[2] A. Bogen and J. Küppers, Surf. Sci., 134, 223(1983).

[3] Nai-Li Wang, U. Kaiser, O. Ganschow, L. Wiedmann and A. Benninghoven, Surf. Sci., 124, 51(1983).

[4] J.L. del Barco and R.H. Buitrago, Anales Asoc. Física Argentina, 1989, in press.

[5] G. Ertl and J. Küppers, *Low Energy Electrons and Surface Chemistry*, (VCH Publishers, Weinheim, 1985).

[6] E.R. Malinowski and D.G. Howery, *Factor Analysis in Chemistry*, (Wiley, New York, 1980).

[7] R. Vidal, R. Koropecki, R. Arce and J. Ferrón, J. Appl. Phys., 62, 3, 1054(1987).

[8] P.H. Holloway, J. Vac. Sci. Technol., 18, 2, 653(1981).

[9] J.L. del Barco and R.H. Buitrago, in M. Cardona and J. Giraldo, *Thin Films and Small Particles*, (World Scientific, Singapore, 1989), p. 311.

Cluster Model for the Interaction of K with Si(100)

D.E. Rodríguez, E.C. Goldberg, and J. Ferrón

INTEC, Universidad Nacional del Litoral and
Consejo Nacional de Investigaciones Científicas y Técnicas,
Güemes 3450, C.C. 91, 3000 Santa Fe, Argentina

The adsorption of K on a Si(001)(2x1) surface is considered as a prototype of alkali metal over semiconductor surfaces. Two conflicting models have been proposed to explain experimental evidence. The first proposes an ionic bond between K and Si, and a complete donation of the K–4s electron that metallizes the surface [1] at coverages lower than one monolayer. The second one proposes a covalent bond K–Si, and the work function behaviour is explained by a coverage dependendent charge transfer from the adsorbate to the substrate [2]. With the aim of clarifying some aspects of this controversial matter, we have performed extensive calculations of the electronic structure of the K–Si system. These are based on an ab–initio unrestricted all-electron Hartree–Fock method, and are applied to K-Si clusters emulating various possible adsorption sites for the alkali over the Si(001)–(2x1) surface. We used three kinds of clusters: the sixfold hole between two parallel dimers (pedestal site), the cave between two adjacent rows of dimers (cave site), and the one over a dimer Si atom (on–top site). The spurious dangling bond of the clusters were saturated with H atoms at a distance of 1.48 Å from Si. We assume a symmetric reconstruction of the Si surface. We have used additional basis orbitals in the total energy calculations of the K and free Si surface to avoid the basis set superposition error in the binding energy.

In table I we summarize our results for the three studied adsorption sites of K. We show the binding energy, the K–Si equilibrium distance and the net K charge. We observe that the highest binding energy corresponds to the cave site with a value of 2.65 eV. Hereinafter we will refer to this configuration. The equilibrium distance compares very well with the experimental value of 3.14 Å obtained with SEXAFS [3]. The charge transfer from K to the Si surfaces is very small in all cases, as deduced from the Mulliken population analysis. However, there is an important change in the dipolar moment, which amounts to 2.03 Debyes. This value accounts for a change in the work function of 2.6 eV, against the experimental value of 3.4 eV measured for a half monolayer of K [4]. This dipole change together with a negligible charge transfer may be understood on the basis of a polarized bond model. This picture is

Springer Proceedings in Physics, Volume 62
Surface Science Eds.: F.A. Ponce and M. Cardona

Table 1: Binding energy, atomic distance and charge transfer for different adsorption sites

Site	E(eV)	d_{K-Si}(Å)	Charge (elec.)
Cave	2.65	3.28	0.01
On Top	1.21	3.02	0.11
Pedestal	0.25	3.34	0.02

supported by an analysis of the occupied molecular orbitals, which show a weakened participation of the 3s orbitals of Si. This feature indicates a lower directionality of the Si orbitals outward from the surface, that translates into a polarization of the electronic charges.

Acknowledgments. This work has been partially supported by CONICET through grant PID 75300.

References.

1. S.Ciraci and I.P.Batra, Phys.Rev.Lett. 56, 867 (1986).
2. R.Ramirez, Phys.Rev. B40, 3962 (1989).
3. T.Kendelewicz, P.Soukiassian, R.S.List, J.O.Woicik, P.Pianetta, I.Lindau and W.E.Spicer, Phys.Rev. B37, 7115 (1988).
4. E.M.Oellig and R.Miranda, Surface Sci. 177, L947 (1986).

A Model to Consider Clustering Effects for Composites

W.E. Vargas[1], *L.F. Fonseca*[1], *and M. Gómez*[2]

[1]Escuela de Física, Universidad de Costa Rica, San José, Costa Rica
[2]Physics Department, University of Puerto Rico, Rio Piedras, Puerto Rico 00931

A multiple scattering model to take into account multipolar interactions between adjacent particles is developed to describe the optical properties of granular materials when aggregation of the particles occurs. This is done by considering the material as made up of scattering units containing pairs of spherical particles. The T-matrix approach is used to describe the scattering properties of the units and the effective medium properties of the material are obtained from the multiple scattering equation of the medium after a statistical average is performed.

1. Introduction

Many of the theories proposed to describe the optical properties of media containing metal particles that are small compared with the radiation wavelength, assume that the metal particles are sufficiently far apart so that multipolar interaction between them can be ignored. More complex systems contain aggregate structures where particles tend to form clusters. The standard Maxwell-Garnett (MG) and Bruggeman (B) formulations [1] fail to describe these systems since they do not take into account the short-range higher-order multipolar interactions between the particles in the aggregate. These type of systems may exhibit two resonant peaks [2] that cannot be predicted by MG and B theories. Recently, some efforts have been made to describe these systems [3].

The purpose of this work is to extend a multiple scattering theory developed by V.Varadan et al.[4] to calculate the effective index of refraction of systems composed of metallic islands where, due to proximity of the particles, short-range high-order multipolar interactions must be considered. In reference [4] they limit their calculations to single spheres and spheroids and showed that the formalism tends to MG theory for very small spherical units. In our calculation, to obtain the effective response of the medium, we consider the system as formed by scattering units composed of two near spherical particles rather than isolated ones. The value of this theory is that it can take into account multipolar interactions within scattering units that are made-up of clusters of particles.

2. The formalism and results

In the T-matrix formalism developed by P.C.Waterman [5], the incident and the scattered fields are expanded in a convenient base of functions that are solution of the vectorial Helmholtz

Springer Proceedings in Physics, Volume 62
Surface Science Eds.: F.A. Ponce and M. Cardona

equation. The unknown expansion coefficients of the scattered field are obtained by the application of the T-matrix to the vector formed by the known expansion coefficients of the incident field. Peterson and Ström [6] extended the formalism to clusters of two particles and it has been successfully used to describe the multipolar interaction between two small metallic particles of different shapes [7]. Following Varadan et al. [4], the electric field $\vec{E}$ at any place in the medium is the sum of the incident field $\vec{E}_0$ plus the fields scattered by these units $\vec{E}^s$,

$$\vec{E}(\vec{r}) = \vec{E}_0(\vec{r}) + \sum_i \vec{E}^s_i(\vec{r}-\vec{r}_i) ,$$

where $\vec{r}_i$ is the position of the scattering unit "i". The field exciting unit "i" is:

$$\vec{E}^e_i(\vec{r}) = \vec{E}_0(\vec{r}) + \sum_{j\neq i} \vec{E}^s_j(\vec{r}-\vec{r}_j) , \qquad d \leq |\vec{r}-\vec{r}_j| \leq 2d ,$$

where "d" is the radius of the smaller sphere inscribing the unit to avoid superposition of the units.

Expanding $\vec{E}^s_j$ and $\vec{E}^e_j$ in terms of base functions, and relating the exciting field and the scattered field coefficients by the T-matrix of the units, a relation between the exciting field expansion coefficients B_n and the incident field expansion coefficients A_n is obtained,

$$B^i_n = \sum_{n''} T^i_{n'',n} \left\{ \exp(i\vec{K}_0\cdot\vec{r}_i) A_{n''} + \sum_{j\neq i} \sum_{n'} B^j_{n'} S_{n'n''}(\vec{r}_i-\vec{r}_j) \right\} ,$$

where $S_{n'n''}$ are the matrix elements describing the translation properties of the base functions. Finally, configurational averages are performed and the one and two fixed sites averages are assumed to be equal, $\langle B_n\rangle_{ij} \cong \langle B_n\rangle_j$. Considering no correlation between the scattering units other than the condition of impenetrability we obtain,

$$\langle B^i_n\rangle_i = \sum_{n''} T^i_{n'',n} \left\{ \exp(i\vec{K}_0\cdot\vec{r}_i) A_{n''} + \right.$$
$$\left. + \frac{1}{V} \int_{V'} \sum_{j\neq i} \sum_{n'} \langle B^j_{n'}\rangle_j \; S_{n'n''}(\vec{r}_i-\vec{r}_j) \, d\vec{r}_j \right\} ,$$

where V is the volume of the sample and V' is V minus a spherical volume of radius 2d to avoid penetrability. Using an effective medium approach,

$$\langle B^i_n\rangle_i = X_n \exp(i\vec{K}_{eff}\cdot\vec{r}_i) ,$$

where K_{eff} is the effective propagation wave vector, a final system of coupled equations is obtained for the unknowns X_n. From this system of equations, the dispersion relation is obtained by finding an adequate root of the determinant whose elements are,

$$\sum_{n''} T_{nn''} I_{nn''} - (K^2_{eff} - K^2_0)\, \delta_{nn'}/\nu$$

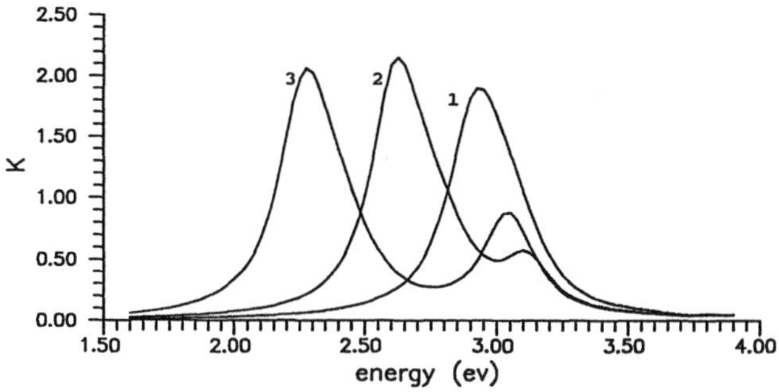

Fig.1. The imaginary part of the index of refraction of Al_2O_3/Ag granular material with 10%Ag as a function of energy. Curve 1 is the MG prediction, curves 2 and 3 are results from the proposed model with different interparticle separation.

where $T_{nn'}$ are the T-matrix elements of a cluster of two spherical particles, ν is the number of scattering units per volume, and

$$I_{nn''}=\int_{|\vec{r}|=2d}\left\{S_{nn'}(K_0r)\,\frac{\partial}{\partial r}\,e^{i\vec{K}_{eff}\cdot\vec{r}}-e^{i\vec{K}_{eff}\cdot\vec{r}}\,\frac{\partial}{\partial r}\,S_{nn''}(K_0r)\right\}\,ds\ .$$

Figure 1 shows the imaginary parts, K, of the index of refraction of a system composed of a Al_2O_3 matrix containing spherical particles of silver with a radius of 5nm and 0.1 metal volume fraction. Curve 1 is the MG prediction, curve 2 is obtained by modeling the system as consisting of scattering units constituted by two spherical particles with a separation of 11.4nm between their centers, while curve 3 is the same configuration but with a separation of 10.05nm. These results show that the effect of clustering is to shift the MG resonance toward the red and generate a second resonance. Curves 2 and 3 were obtained assuming all the scattering units aligned perpendicular to the incident wave vector. Calculations are in progress that will relax this restriction by considering random orientation of the scattering units and performing an orientational average.

This formalism presents a method that takes into account clustering effects in granular systems. The theory is based on the assumption that multipolar interactions higher than the dipolar decay so rapidly as a function of interparticle distance that only multipolar interactions between particles within each scattering unit need to be considered in calculating the effective medium dielectric constant.

References

1. These two models are described elsewhere. The original papers are J.C.Maxwell-Garnett. Philos.Trans.R.Soc.London **203**, 385 (1904); D.A. Bruggeman. Ann. Phys. (Leipzig) **24**, 636 (1935).

2. M.H. Lindsay, M.Y. Lin, D.A. Weitz, P. Sheng, Z. Cheng, R. Klein, and P. Meakin. Faraday Disc. Chem. Soc. **83**, 153 (1987).

3. Z. Chen and P. Sheng. Phys.Rev.B **39**, 9816 (1989).
4. V.V.Varadan, V. Bringi, and V.K. Varadan. Phys. Rev.D **19**, 2480 (1979).
5. P.C. Waterman. Phys. Rev. D **3**, 825 (1971).
6. B. Peterson and S. Ström. Phys. Rev.D **8**, 3661 (1973).
7. L.Cruz,L.F. Fonseca,and M.Gómez. Phys.Rev.B **40**, 7491 (1989).

Part VI

Properties of Thin Films

Solar Energy Materials: Survey and Some Examples

C.G. Granqvist

Physics Department, Chalmers University of Technology and
University of Gothenburg, S-41296 Gothenburg, Sweden

This paper introduces materials for energy efficiency and solar energy utilization and discusses some current trends for basic research and development. Most of the materials involve thin surface coatings. Brief overviews are given for solar absorber surfaces, transparent infrared reflectors and transparent conductors, large-area chromogenics for transmittance control in "smart windows", and transparent convection-suppressing materials, whereas solar cell materials are not included. The paper treats a few examples of specific coatings that are presently being investigated; data are given for angular-selective transmittance through porous Cr films with oblique columnar microstructure, transparent and conducting non-stoichiometric SnO_2 films, and chromogenic effects in Li-intercalated VO_2 films.

1. Introduction

The limited availability of fossil and nuclear fuel, and their environmental impact, have led to a growing awareness of the importance of renewable energy sources. Political considerations and incidental market fluctuations may have short-term effects but will not offset the tendency that solar energy materials are going to play an ever-increasing role both in the industrialized and the lessdeveloped countries. Given this situation, research and development on solar energy materials is sure to be of growing importance [1].

Modern technology gives a multitude of options for manmade collectors of solar energy and for energy-efficient passive design in architecture [2-4]. Among the collectors, one can distinguish between those utilizing thermal conversion ("solar collectors") and quantum conversion ("solar cells"). The pertinent "solar energy materials" have properties tailored specifically according to the requirements set by the spectral content and intensity of the solar radiation [2]; most of them involve thin films or surface treatment in one way or another. In this paper we exclude materials for quantum conversion from the discussion.

Section 2 below sets the scene for solar energy materials by introducing the solar irradiance spectrum and its relation to luminous and thermal spectra. Section 3 gives brief *overviews* over materials categories for which vigorous research and development activities are going on, with consecutive presentations of solar absorber surfaces, transparent infrared reflectors and transparent conductors, large-

Springer Proceedings in Physics, Volume 62
Surface Science Eds.: F.A. Ponce and M. Cardona

area chromogenic materials and devices [5], and transparent convection-suppressing materials. Sections 4-6 then give more in-depth expositions of three *specific* types of coatings for which active research is currently done in the author's laboratory; the discussions regard angular-selective transmittance through porous Cr films with oblique columnar microstructure, transparent and conducting non-stoichiometric SnO_2 films, and chromogenic effects in Li-intercalated VO_2 films.

2. Natural Radiation

Solar energy materials are designed to take advantage of the natural radiation in our environment. There are four radiative properties that need to be introduced. *Thermal radiation* from a material is represented by a blackbody spectrum multiplied by a numerical factor - the emittance - which is less than unity. In general, the emittance is wavelength dependent. Thermal radiation lies in the $2 < \lambda < 100$ μm wavelength range for temperatures of practical interest. Extraterrestrial *solar radiation*, on the other hand, is confined to the $0.25 < \lambda < 3$ μm interval, so that there is almost no overlap with the spectra for thermal radiation. Most energy-related applications take place at ground level, and hence the *atmospheric absorption* is of interest. In clear weather, most of the solar radiation can be transmitted, and, furthermore, there is an "atmospheric window" allowing transmittance of thermal radiation in the $8 < \lambda < 13$ μm band. The atmospheric absorptance is strongly dependent on meteorological conditions. Finally, the spectral *sensitivity of the human eye* is limited to the $0.4 < \lambda < 0.7$ μm range.

The different types of ambient radiation are spectrally selective, i.e., confined to well-defined and often non-overlapping wavelength regions. This is of major significance for the desired properties of solar energy materials, and by adequate design one can achieve the following:

- High solar absorptance or transmittance can be combined with low thermal emittance (i.e., low heat transfer) and accompanying high electrical conductivity. Such properties are useful for solar absorber surfaces, low emittance windows, and transparent front electrodes on solar cells.

- High luminous transmittance can be combined with rejection of infrared solar radiation. These properties are desired for "solar control" windows.

- Varying meteorological and climatic conditions can be compensated for by chromogenic materials, characterized by a dynamic throughput of radiant energy. These properties will be used in future "smart windows" [5].

- Transmission of energy through the "atmospheric window" can be used for passive cooling [6]. Among the many conceivable applications we note food preservation and condensation irrigation in arid regions.

3. Survey of Solar Energy Materials

3.1 Solar Absorber Surfaces

A solar collector is a device which absorbs solar radiation, converts it to thermal energy, and delivers the thermal energy to a heat-transfer medium. Energy efficiency, i.e., minimized losses associated with the energy transfer, can be achieved by using suitable materials in the components of the solar collector, and by integrating these components into a well-designed device. The standard flat-plate solar collector includes a spectrally selective absorber surface under a cover glass. Spectral selectivity here means high absorptance (low reflectance) at $0.3 < \lambda < 3$ µm and low emittance (high reflectance) at $3 < \lambda < 100$ µm. Much research and development was conducted during the later half of the 1970's and the early 1980's on coatings and surface treatments yielding such properties. This work is reviewed in [7] and [8]. Today, durability issues are in focus [9].

It is possible to exploit several different design options and physical mechanisms in order to create a selectively solar-absorbing surface. The most straight-forward one is to use a material whose *intrinsic radiative properties* have the desired kind of spectral selectivity. Generally speaking, this approach has not been very fruitful, but data on certain transition metal diborides, and on some other compounds, indicate that intrinsically selective materials do exist.

Semiconductor-metal tandems can give the required spectral selectivity by absorbing short-wavelength radiation in a semiconductor whose bandgap is ~ 0.6 eV and having low thermal emittance due to the underlying metal. The useful semiconductors have undesirably large refractive indices, which give high reflection losses, and hence it is necessary to apply an antireflection coating which is effective in the solar range. Work on silicon-based designs is particularly well known.

Multilayer coatings of the type dielectric/metal/dielectric/ ... can be tailored so that they become efficient selective solar absorbers. It is straight-forward to compute the optical properties, which facilitates design optimization. Coatings with the dielectric being Al_2O_3 and the metal being Mo have good properties; the coatings are readily produced by large scale vacuum coating technology.

Metal-dielectric composite coatings consist of very fine metal particles in a dielectric host. The ensuing optical properties can be intermediate between those of the metal and of the dielectric. The coatings have to be backed by a metal. The metal-dielectric concept offers a high degree of flexibility, and the solar selectivity can be optimized with regard to the choice of the constituents, coating thickness, particle concentration (which can be graded), and the size, shape and orientation of the particles. Effective medium theory can be used to quantitatively model the optical properties [10]. The solar absorptance can be boosted by use of a suitable substrate and by applying an antireflection coating. A variety of techniques for producing the coatings - some of which are suitable for large areas - are well established.

Textured surfaces can produce a high solar absorptance by multiple reflections of the incident radiation against dendrites which are ~ 2 μm apart. The long-wave thermal emission, on the other hand, is rather unaffected by this treatment. Dendritic surfaces can be produced by judiciously chosen deposition or etching techniques.

The final concept considered here involves a *selectively solar transmitting coating on a blackbody-like absorber*. The absorber can be chosen among materials with proven long-term durability (such as black enamel), and the coating can be a heavily doped oxide semiconductor (for example SnO_2:F). We return to coatings of this type below.

3.2 Transparent Infrared Reflectors and Transparent Conductors

Surface coatings that are transparent at $0.3 < \lambda < 3$ μm and reflecting at $3 < \lambda < 100$ μm can be used in low emittance windows, and surface coatings that are transparent at $0.4 < \lambda < 0.7$ μm and reflecting at $0.7 < \lambda < 3$ μm can be used in solar control windows. Materials with high infrared reflectance are electrically conducting and hence of interest as transparent electrodes in a variety of applications, including solar cells. The required solar selectivity can be obtained with noble-metal based as well as with doped oxide semiconductor based coatings.

Thin *noble-metal films* can combine short-wavelength transmittance (up to ~ 50 %) with high long-wavelength reflectance [11]. By embedding the metal between high-refractive-index dielectric layers one can use antireflection to maximize the transmittance in a desired wavelength range. Current research and development considers techniques to produce thinner continuous noble metal layers than those now used as well as techniques to combine spectral selectivity with a pronounced angular dependence of the transmittance.

Doped oxide semiconductor films offer an alternative to the noble-metal based films. The semiconductor must have a wide bandgap, so that it allows good transmission in the luminous and solar ranges. Further, it must allow doping to a level that makes the material metallic and hence infrared reflecting and electrically conducting. Most of the materials that are known to be useful are oxides based on Zn, Cd, In, and Sn and alloys of these. The required doping is often achieved by the addition of a foreign element; particularly good properties have been obtained with SnO_2:F [12], In_2O_3:Sn [13], and ZnO:Al [14]. When the doping is sufficient, the "impurity" atoms are ionized and the free electrons form an electron gas, whose properties are limited by the unavoidable scattering against the ionized "impurities". Further details on the attainable optical properties are given in [13]. Another possibility is to provide doping via a moderate oxygen deficiency. If prepared properly, the above mentioned coatings can be virtually nonabsorbing for luminous and solar radiation. A specific and important advantage of the doped oxide semiconductors is their excellent chemical and mechanical durability. The commercially produced large-area coatings do not yet match the theoretical limits.

3.3 Chromogenic Materials

Chromogenic materials are characterized by their ability to change the throughput of radiant energy in accordance with dynamic needs. They are expected to be of large importance for future "smart windows" [5]. Chromogenics is subject to much current research and development, both in academia and in industry.

There are several different types of chromogenic materials. The most well known of these is *photochromic* glass, whose luminous absorptance is increased when subjected to ultraviolet irradiation. Such glass is widely used in ophthalmics. Certain novel polymers (spirooxazines) have similar properties.

Thermochromic thin films can produce a decrease of their transmittance when a certain "critical" temperature τ_c is exceeded. Thermochromism can be used for automatic temperature control in buildings, provided that τ_c is close to a comfort temperature. Vanadium-oxide-based coatings show thermochromism associated with a reversible metal-insulator transition. Bulk VO_2 crystals have $\tau_c \approx 68°$ C, which clearly is undesirably high. The transition temperature can be depressed in $W_xV_{1-x}O_2$; the latter materials can be prepared as thin films. Fluorination has the added benefit of enhancing the transmittance at temperatures below τ_c [15].

Electrochromic-based multilayer coatings give possibilities to obtain very flexible control of the radiative throughput in "smart windows". The coating includes five layers backed by a glass plate or positioned between two glass plates in a laminate configuration. The outermost layers are transparent electrical conductors, for example of In_2O_3:Sn. One of these is in contact with the optically active electrochromic layer, which can be of an inorganic transition metal oxide based on W, Ni, Co, Mo, Ti, Ir, etc., or of one of several possible organic materials. The other transparent conductor is in contact with an "ion storage", which can be either optically passive (for example V_2O_5) or coloured/ bleached in synchronization with the colouration/bleaching of the base electrochromic layer. The intermediate layer, finally, is of an ion conducting solid material; it can be either a suitable inorganic thin layer or a polymeric layer. For the latter option, one can combine ionic conductivity with adhesiveness so that the overall design can comprise two glass plates, each having a two-layer coating, laminated together by the ion conductor [16]. When a voltage (< 2 volts) is applied between the transparent electrical conductors, ions (H^+, Li^+,...) can be moved from the ion storage, via the ion conductor, and inserted into the electrochromic layer. The change in the optical properties can be between widely separated extrema, and occurs gradually and reversibly.

Liquid-crystal-based chromogenic materials offer several possibilities for transmittance control. These materials are not discussed here, though.

3.4. Transparent Convection-suppressing Materials

Transparent convection-suppressing (i.e., insulating) materials are of considerable interest for solar collectors, energy-efficient windows and skylights, innovative wall claddings, etc. The materials can be organized into four groups: (1) thin flexible polymer foils; (2) polymer honeycomb materials; (3) bubbles, foams and fibres, and; (4) inorganic microporous materials, especially silica aerogels [17]. Materials (1) and (4) can be almost invisible to the eye and allow, in principle, very high solar energy throughput. Materials (2) and (3) cause strong scattering.

Aerogels are of particular interest for transparent insulation. Such materials can be obtained by supercritical drying of colloidal silica gel. The ensuing substance consists of silica particles of size ~1 nm interconnected so that a loosely packed structure with pore sizes of 1 to 100 nm is formed. The porosity can be up to 97 %. Silica aerogel can be prepared both as transparent tiles and as a translucent granular material.

4. Example One: Angular-selective Transmittance Through Obliquely Evaporated Cr Films [18,19]

Most view windows should have high luminous transmittance along a near-horizontal line-of-sight, whereas it may be advantageous to have a low transmittance for lines-of-sight that form large angles to the horizon so that overheating and glare are minimized. For *vertical* windows this calls for coatings whose transmittance falls off monotonically with increasing angle θ to the surface normal. Metal-based three-layer coatings (cf. Sec. 3.2) can show a rather strong angular dependent transmittance. Even more pronounced angular dependent transmittance can be achieved in a five-layer coating containing two metal layers [18]. For *inclined* windows - such as windscreens and rear windows in cars and glass louvres in buildings - it is generally an advantage to have optical properties that are angular selective. Angular selectivity means that the optical properties, usually the transmittance, are different for equal angles on either side of the surface normal, i.e., for $+\theta$ and $-\theta$.

Angular selectivity may emerge when a light beam passes the boundary between two optically different media, provided that at least one of these is characterized by an optical axis that deviates from the surface normal. This situation is illustrated in Fig. 1 where a collimated light beam is incident onto a substrate with a coating represented by identical inclined cylindrical columns. The optical properties are conveniently represented with regard to a vector a in the surface plane. Now one can describe the incident light by its polar angle θ and azimuthal angle ϕ and choose a so that $T(\theta, \phi = 90°) = T(\theta, \phi = 270°)$. Other orientations of the light beam yield

$$T(\theta,\phi) \neq T(\theta,\phi + 180°). \qquad (1)$$

In general, the difference between the transmittance values in the inequality is largest at $\phi = 0$, i.e., for light incident in the plane spanned by a and the surface

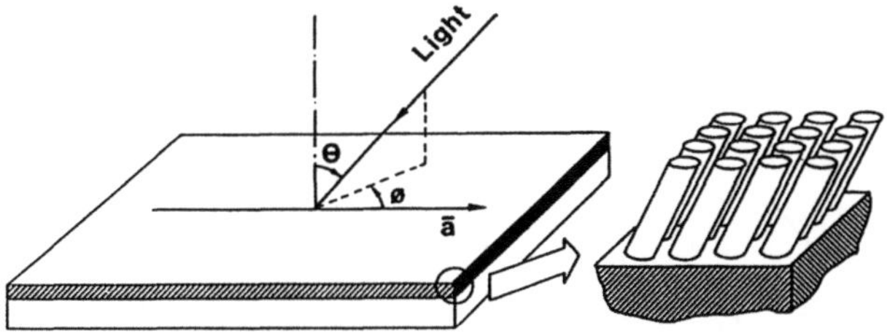

Fig. 1 Left-hand part defines the geometry for a light beam incident onto a coating of a uniaxial material. Right-hand part shows a schematic model for an oblique columnar microstructure. From [18,19].

normal. This particular configuration leads to a simple criterion for angular selectivity, which can be written for s and p polarization as [18,19]

$$T_s(\theta) = T_s(-\theta), \quad (2)$$

$$T_p(\theta) \neq T_p(-\theta). \quad (3)$$

Here the sign convention $+\theta$ ($-\theta$) denotes light having a propagation vector with a component opposite (parallel) to *a*.

Coatings with inclined columnar microstructure can be made by oblique angle vacuum deposition, using evaporation or sputtering, as well as by special etching techniques. The relation between deposition angle α and column orientation β is often given by the "tangent rule" [20]

$$\tan\beta = (1/2)\tan\alpha. \quad (4)$$

The general validity of this "rule" is questionable, though [21].

Recently, we prepared Cr coatings on glass by oblique angle evaporation with $\alpha < 80°$ [18,19]. The structure was analyzed by micro-fractography. Spectral transmittance was measured for $-70 < \theta < 70°$. Figure 2 shows angular-dependent transmittance at $\lambda = 0.5$ µm for s- and p-polarized light and for unpolarized (u) light. The latter quantity was obtained from

$$T_u(\theta) = [T_s(\theta) + T_p(\theta)]/2. \quad (5)$$

It appears that $T_s(\theta)$ is symmetric around $\theta = 0$ and peaked at normal incidence, whereas $T_p(\theta)$ varies in a more irregular and interesting manner and is strongly peaked at $\theta = +60°$. The quantity of most importance for energy-related applications is T_u, which increases monotonically from ~18 % at $\theta = -60°$ to ~29 % at $\theta = +60°$. An analogous variation exists for the luminous and solar transmittance. The optical data are fully consistent with a theory [18,22] built on effective-medium concepts and generalized Fresnel relations.

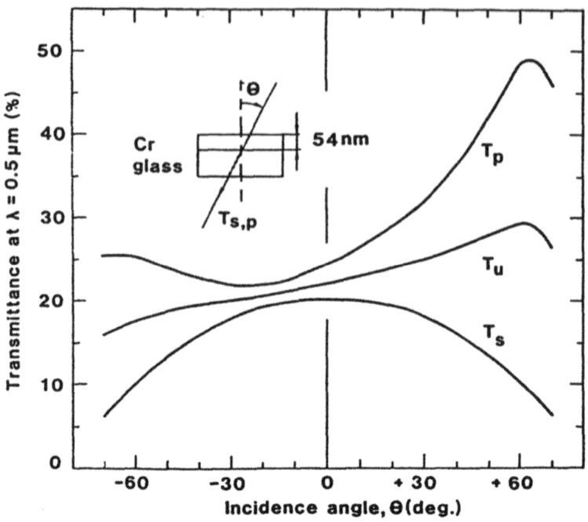

Fig. 2 Angular-dependent transmittance at $\lambda = 0.5$ μm measured as sketched in the inset. Data are given for s-polarized, p-polarized and unpolarized light. From [18,19].

5. Example Two: Transparent and Conducting SnO_x Films [23-26].

As discussed in Sec. 3.2, heavily doped wide bandgap semiconductors such as SnO_2:F, In_2O_3:Sn and ZnO:Al have important applications as energy efficient window coatings. Their optical properties can be understood in terms of an effective-mass model for n-doped semiconductors well above the Mott critical density [13]; the doping results from *singly* ionized impurities. Non-stoichiometric tin oxide, denoted SnO_x, can be heavily doped by *doubly* ionized oxygen vacancies [27], and it is not obvious that the theoretical model for SnO_2:F, for example, can be successfully extended to SnO_x. The study briefly described here shows that, indeed, SnO_x can be understood in terms of the earlier theory [25]. Another, more practical, reason for studying SnO_x is that coatings of this material can be made by high-rate magnetron sputtering onto temperature-sensitive substrates [23], wheres SnO_2:F requires substrates that are heated to high temperatures.

In this work, SnO_x films were produced by reactive RF magnetron sputtering [23,24]. The process parameters for thin film deposition were optimized by correlating the RF power (P_{rf}), the sputter gas pressure, and the O_2/Ar gas flow ratio (Γ) with optical and electrical measurements. It was found that Γ, in particular, had to be very accurately controlled in order to get optimized film properties. Quantitatively, we could accomplish a luminous transmittance of 75±1.5 %, a luminous absorptance of 9±2.5 %, and a DC resistivity of ~3×10^{-3} Ω cm at low P_{rf} and ~10^{-2} Ω cm at high P_{rf}. The sputter rate was approximately proportional to P_{rf} and could be as large as ~2.85 nm/s. Solid curves in Fig. 3 indicate the spectral transmittance and reflectance we could reach in

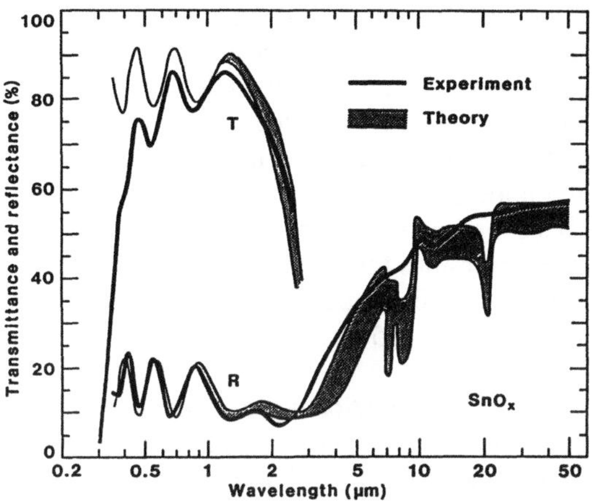

Fig. 3 Solid curves indicate spectral transmittance T and reflectance R of a 0.344-μm-thick SnO_x film sputter deposited onto glass with $P_{rf} = 10$ W. Shaded band refers to a calculation for a SnO_x film represented by the measured thickness and the parameters given in the main text. From [25].

a SnO_x film with minimum resistivity. Further data on such films were obtained by electron microscopy, Hall effect measurements, and Mössbauer spectrometry [26].

We now turn to the theoretical model for the electromagnetic properties of SnO_x.The frequency dependent dielectric function $\varepsilon(\omega)$ is obtained as a sum of additive contributions due to free carriers (electrons), valence electrons, and phonons. Well away from the semiconductor bandgap (in the ultraviolet) and phonon resonances (in the mid-thermal range) one can write

$$\varepsilon(\omega) = \varepsilon_\infty + i/[\varepsilon_o\, \omega\rho(\omega)]\ , \tag{6}$$

where ε_∞ is the high-frequency dielectric constant of SnO_x, ε_o is the permittivity of free space, and $\rho(\omega)$ is the complex dynamic resistivity due to free electrons. For the case of electrons scattered against Coulomb-like ion potentials, and taking the semiconductor to be non-polar, one obtains [28]

$$\rho(\omega) = i\,\frac{Z^2 N_i}{6\pi^2 \varepsilon_o n_e^2 \omega} \int_o^\infty k^2\, dk \left(\frac{1}{\varepsilon^{eg}(\mathbf{k},\omega)} - \frac{1}{\varepsilon^{eg}(\mathbf{k},0)} - i\,\frac{\omega}{\varepsilon_o \omega_p^2} \right., \tag{7}$$

where Z is the charge of the ions, N_i is their density, n_e is the free-electron density, ε^{eg} is the dielectric function of the free-electron gas, and ω_p is the plasma frequency. The latter quantity is

$$\omega_p{}^2 = n_e\, e^2 / \varepsilon_o\, \varepsilon_\infty\, m_c{}^*, \tag{8}$$

with e being the electronic charge and m_c^* the effective conduction band mass. Quantitative data for the optical properties of SnO_x were computed [25] presuming that doubly ionized oxygen vacancies in the SnO_2 lattice serve as donors [27]. Thus we put $Z = 2$ and $N_i = n_c/2$ in Eq. (7). ε^{eg} was computed for a degenerate electron gas using the Random Phase Approximation as extended to include exchange effects. According to the literature, one has $\varepsilon_\infty \approx 4$ and $m_c^*/m_o \approx 0.39$, where m_o is the free-electron mass. Once $\varepsilon(\omega)$ was fully specified, spectral optical properties were computed from Fresnel's formulas for a thin film on a thick substrate. The substrate was represented by the dielectric function for glass at $\lambda < 2.5$ µm and for sputter-deposited SiO_2 at $\lambda > 2.5$ µm.

The shaded band in Fig.3 shows results from a calculation using experimentally determined values of the relevant parameters, as given above, and taking the free-electron density to lie in the range 1.0 to 1.4 x 10^{20} cm^{-3}. This choice is consistent with the measured resistivity, and the interval is wide enough to account also for possible errors in film thickness determination, etc. The main result is that theory and experiment are in very good agreement. The fact that the experimental transmittance drops increasingly below the theoretical prediction as one approaches the shortest wavelengths is due to absorption in the glass substrate as well as to some residual absorption - perhaps due to SnO-like inclusions - that are not accounted for by the theory. The lack of detailed agreement for the phonon-induced absorption features in the thermal infrared is not unexpected since the glass composition deviates from SiO_2.

6. Example Three: Chromogenic Effects in Li_xVO_2 [29]

Vanadium-oxide-based materials have several properties of potential interest for energy efficient fenestration. Thus thermochromic switching between a transparent low-temperature state and a less transparent high-temperature state is possible in thin films based on VO_2, as discussed in Sec.3.3. The critical temperature at which the transition takes place can be changed by the incorporation of dopants. Li-doping is an interesting, yet unexplored, possibility that can be accomplished by electrochemical means. This technology also allowed us to investigate - as far as we know for the first time - the electrochromism of Li_xVO_2.

The Li_xVO_2 films were prepared by a two-step predure. First, VO_2 films were made by reactive RF magnetron sputtering of V followed by annealing posttreatment [30]. Secondly, Li ions were inserted by cyclic voltammetry in an electrochemical cell containing $LiClO_4$ [16]. Cyclic voltammograms gave clear evidence for intercalation/deintercalation processes and were also useful for assessing Li contents.

Left-hand part of Fig. 4 shows spectral transmittance at ambient temperature for a 170-nm-thick VO_2 film, in initial state and after electrochemical lithiation with the voltammetric cycle interrupted so that three magnitudes of the lithium content were

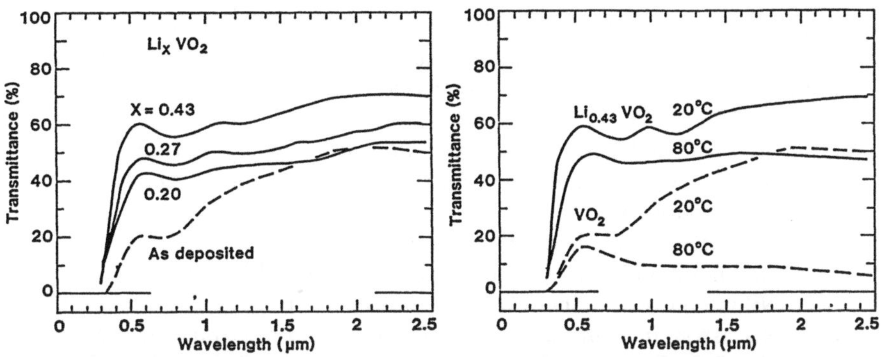

Fig. 4 Spectral transmittance for a 170-nm-thick Li_xVO_2 film with different lithium contents and at different temperatures. Left-hand part shows data for $0 < x < 0.47$ and room temperature. Right-hand part shows data for two x-values and two temperatures. From [29].

obtained. It is seen that the intercalation of Li induces a large increase of the transmittance. This effect prevails over the full wavelength range and is most pronounced for visible light. The transmittance goes up monotonically with increasing lithiation. The optical modulation is reversible, and hence the Li_xVO_2 film shows *electrochromism* with anodic colouration.

Right-hand part of Fig. 4 reports on *thermochromism* and shows spectral transmittance for the 170-nm-thick VO_2 film in initial state and after full lithiation so that $Li_{0.43}VO_2$ was obtained. The measurements were taken at room temperature and at ~80°C, i.e., below and above the critical temperature for thermochromic switching of VO_2. The unlithiated film shows a strongly temperature-dependent transmittance, particularly in the infrared, and the transmittance is lowered at enhanced temperature as desired for automatic temperature control in a thermochromic "smart window". The results are in good agreement with earlier data [30]. The more transparent $Li_{0.43}VO_2$ film also displays some thermochromism and a suppression of transmittance at elevated temperature, but the modulation of the optical properties is smaller than for VO_2 both in relative and absolute terms.

7. References

1. World Commission on Environment and Development, *Our Common Future* (Oxford University Press, Oxford, U.K., 1987).
2. C.G. Granqvist, *Spectrally Selective Surfaces for Heating and Cooling Applications*, SPIE Opt. Engr. Press, Bellingham, USA, 1989).
3. C.M. Lampert, in *Workshop on Materials Science and the Physics of Non-conventional Energy Sources*, edited by G. Furlan, D. Nobili, A.M. Sayigh and B.O. Seraphin (World Scientific, Singapore, 1989), p. 45.
4. C.G. Granqvist, in *Energy and the Environment into the 1990s*, edited by A.A.M. Sayigh (Pergamon, Oxford, U.K., 1990), Vol. 3, p. 1465.

5. C.M. Lampert and C.G. Granqvist, editors, *Large-area Chromogenics: Materials and Devices for Transmittance Control* (SPIE Opt. Engr. Press, Bellingham, USA, 1990); C.G. Granqvist, Crit. Rev. Solid State Mater. Sci. **16**, 291 (1990).
6. C.G. Granqvist and T.S. Eriksson, in *Materials Science for Solar Energy Conversion Systems*, edited by C.G. Granqvist (Pergamon, Oxford, U.K., 1991), to be published.
7. W.F. Bogaerts and C.M. Lampert, J. Mater. Sci. **18**, 2847 (1983).
8. G.A. Niklasson and C.G. Granqvist, J. Mater. Sci. **18**, 3475 (1983).
9. G.A. Niklasson, Proc. Soc. Photo-Opt. Instrum. Engr. **1272**, 250 (1990).
10. G.A. Niklasson, in *Materials Science for Solar Energy Conversion Systems*, edited by C.G. Granqvist (Pergamon, Oxford, U.K.,1991), to be published.
11. G.B. Smith, GA. Niklasson, J.S.E.M. Svensson and C.G.Granqvist, J. Appl. Phys. **59**, 571 (1986).
12. H. Haitjema, J.J.P. Elich and C.J. Hoogendorn, Solar Energy Mater. **18**, 283 (1989).
13. I. Hamberg and C.G. Granqvist, J. Appl. Phys. **60**, R123 (1986).
14. Z.-Č. Jin, I. Hamberg and C.G. Granqvist, J. Appl. Phys. **64**, 5117 (1988).
15. K.A. Khan, G.A. Niklasson and C.G. Granqvist, J. Appl. Phys. **64**, 3327 (1988); K.A. Khan and C.G. Granqvist, Appl. Phys. Lett. **55**, 4 (1989).
16. A.M. Andersson, C.G. Granqvist and J.R. Stevens, Appl. Opt. **28**, 3295 (1989).
17. V. Wittwer and W. Platzer, Proc. Soc. Photo-Opt. Instrum. Engr. **1272**, 284 (1990).
18. G. Mbise, G.B. Smith, G.A. Niklasson and C.G. Granqvist, Proc. Soc. Photo-Opt. Instrum. Engr. **1149**, 170 (1989).
19. G. Mbise, G.B. Smith, G.A. Niklasson and C.G. Granqvist, Appl. Phys. Lett. **54**, 987 (1989).
20. H.J. Leamy, G.H. Gilmer and A.G. Dirks, in *Current Topics in Materials Science*, edited by E. Kaldis (North-Holland, Amsterdam, The Netherlands, 1980), Vol. 6, p. 309.
21. J. Krug and P. Meakin, Phys. Rev. A **40**, 2064 (1989); P. Meakin and J. Krug, Europhys. Lett. **11**, 7 (1990).
22. G.B. Smith, Opt. Commun. **71**, 279 (1989); Appl. Opt. **29**, 3685 (1990).
23. B. Stjerna and C.G. Granqvist, Appl. Opt. **29**, 447 (1990).
24. B. Stjerna and C.G. Granqvist, Solar Energy Mater. **20**, 225 (1990).
25. B. Stjerna and C.G. Granqvist,Appl. Phys. Lett. **57**, 1989 (1990).
26. B. Stjerna, C.G. Granqvist, A. Seidel and L. Häggström, J. Appl. Phys. **68**, 6241 (1990).
27. Z.M. Jarzebsky and J.P. Marton, J. Electrochem. Soc. **123**, 199c, 299c, 333c (1976).
28. E. Gerlach, J. Phys. C **19**, 4585 (1986).
29. M.S.R. Khan, K.A. Khan, W. Estrada and C.G. Granqvist, J. Appl. Phys. **69**, 3231 (1991).
30. S.M. Babulanam, T.S. Eriksson, G.A. Niklasson and C.G. Granqvist, Solar Energy Mater. **16**, 347 (1987).

The Optical Response of Composites at Low Filling Fractions: A New Diagrammatic Summation

R.G. Barrera[1], *C. Noguez*[2], *and E. Anda*[3]

[1]Instituto de Física, Universidad Nacional Autónoma de México, Apdo. Postal 20-364, 01000 México, D.F., Mexico

[2]Facultad de Ciencias, Universidad Nacional Autónoma de México, Circuito Exterior, 044510 México, D.F., Mexico

[3]Instituto de Física, Universidade Federal Fluminense, 24210 Niteroi, RJ, Brazil

Abstract. We extend a previously developed diagrammatic formalism for the calculation of the effective dielectric response of composites, prepared as a collection of small spherical inclusions embedded in an otherwise homogeneous matrix. This is done within the long wavelength, dipolar approximation for a low filling fraction of spheres. We propose a new diagrammatic approximation and we compare our results with recently reported numerical simulations.

1. Introduction

The electromagnetic response of composite media has attracted the attention of many investigators since the pioneering work of JC Maxwell more than a century ago. In his Treatise[1], JC Maxwell poses the problem, and advances an approximate solution, of calculating the effective dc conductivity of a conductor with a well-defined volume fraction of small insulating inclusions. The main difficulty in this problem is to find a proper way of averaging the fields and currents generated by the presence of the inclusions. Furthermore, as the volume fraction of the inclusions increases, the system goes through a metal-insulator transition which nowdays is being treated with percolation theory[2]. A well documented review of the historical development of this problem can be found in the work of Landauer[3]. Here we will be interested not in the dc response but rather in the electromagnetic response of the composite at finite frequencies. We will consider an homogeneous matrix filled with identical spherical inclusions with radius much less than the wavelength of the electromagnetic radiation. The effective dielectric response of this system, as a function of frequency and filling fraction, was first calculated, within the mean field approximation (MFA), by JC Maxwell Garnett[4] as early as 1904. His result becomes completely equivalent to the Clausius-Mossotti or Lorentz-Lorenz relation[5], which links the dielectric response of a fluid with the polarizability and density of its molecules, if the molecules are regarded as polarizable spheres embedded in vacuum.

In the MFA one assumes that in the presence of a long-wavelength external electric field all the spheres acquire exactly the same induced dipole moment which is taken equal to the average dipole moment and is calculated self-consistently. Therefore any improvement upon MFA has to include, in some way or another, the effect of the fluctuations around the average of the induced dipole moments. If the spheres were arranged in a periodic lattice, then every sphere would have exactly the same surroundings and all of them would acquire, in the long wavelength limits, the same induced dipole moment: there would be no fluctuations. In this case, the MFA yields an exact result[5] and we conclude that it is the disorder in the location of the spheres the source of the dipolar fluctuations.

The problem of considering the effects of disorder and consequently the effects of dipolar fluctuations in the dielectric response of a composite has also a long history. There have been many diffe-

Springer Proceedings in Physics, Volume 62
Surface Science Eds.: F.A. Ponce and M. Cardona

rent types of approaches to this problem: density expansions[6], linked cluster expansions[7], perturbation expansions[8], integral equations[9], multiple-scattering methods[10], intuitive arguments[11,12], diagrammatic theories[13], location of bounds[14], numerical simulations[15,16], etc. In order to decide the benefits of each of these approaches, one has to compare their results with the experimental ones. Due to the fact that the experiments performed up to now do not resemble properly the models used in the theoretical work, the comparison between theory and experiment has been a painful process. The preparation of homogeneous and isotropic samples, with a well-defined filling fraction of identical spheres with radius in the nanometer range, has not been an easy task. Problems like particle clustering, a distribution of shapes and sizes, and an anomalous high density of dislocations in the small particles have obscured a clear interpretation of the effects of disorder in the optical experiments. On the other hand, as pointed out by several authors,[9,12] beyond MFA the effective dielectric response of a composite depends not only on the filling fraction of the spheres but also on the structure of their two-and three-particle distribution functions. In other words, different types of disorder will lead to different results. Now, since most of the experimentalist do not report the actual distribution functions of the inclusions in their samples, this yields to another source of confusion; this might explain some of the discrepancies[17] found in experiments performed in differently prepared samples.

In this work, we reformulate a diagrammatic approach reported earlier,[13] for the calculation of the effective dielectric response of a composite prepared as mixture of identical spheres embedded in an otherwise homogeneous matrix. This formalism is valid in the low-density regime where all the m-th particle distribution functions of the spheres can be approximated by unsymmetrized products of two-particle distribution functions. After setting up the formalism, we extend a previously performed diagrammatic summation[13] by including an infinite set of diagrams which should be important at low densities. Then we compare our results with the only "experiments" that, we believe, will give the fairest possible comparison: the numerical simulations recently performed by Cichocki and Felderhof[16] for a collection of Drude spheres within the dipolar approximation. The structure of the paper is as follows: in section II we develop the theoretical framework and section III is devoted to results, comments and conclusions.

II. Formalism

Lets consider an homogeneous and isotropic ensemble of $N \gg 1$ spheres of radius a and dielectric function ϵ_m embedded in a host medium with dielectric function ϵ_h. The system is in the presence of a space- and time-dependent external electric field which oscillates with frequency ω and wave-vector $\mathbf{q}$. Furthermore, we assume that $qa \ll 1$ thus the induced interaction between the spheres can be taken in the quasi-static limit. The local electric field induces an effective dipole $\mathbf{p}_i$ on the i-th sphere given by

$$\mathbf{p}_i(\omega) = \alpha(\omega)[\mathbf{E}_i^\circ + \sum_j \overleftrightarrow{\mathbf{t}}_{ij} \cdot \mathbf{p}_j(\omega)], \tag{1}$$

where $\mathbf{E}_i^\circ$ is the electric field induced in the medium at $\mathbf{R}_i$ in the absence of the spheres, $\alpha(\omega) = a^3[\epsilon_m(\omega) - \epsilon_h(\omega)]/[\epsilon_m(\omega) + 2\epsilon_h(\omega)]$ is the effective polarizability of an isolated sphere in the medium and

$$\overleftrightarrow{\mathbf{t}}_{ij} = (1 - \delta_{ij})\nabla_i\nabla_j(1/R_{ij}) \tag{2}$$

is the dipole-dipole interaction tensor in the quasi-static limit. Here $R_{ij} \equiv |\mathbf{R}_i - \mathbf{R}_j|$ and δ_{ij} is the Kronecker delta.

The polarization is then defined as the average dipole moment per unit volume and it can be related to the effective dielectric response ϵ_{eff} of the system through[12]

$$\frac{\epsilon_h(\omega)}{\epsilon_{eff}(\omega)} = 1 - 4\pi\epsilon_h(\omega)\chi^{ex,\ell}(q \to 0, \omega), \tag{3a}$$

where $\overleftrightarrow{\chi}^{ex}(\mathbf{q},\omega)$ is the external susceptibility defined by

$$n < \mathbf{p} > (\mathbf{q},\omega) = \overleftrightarrow{\chi}^{ex}(\mathbf{q},\omega) \cdot \mathbf{E}^{ex}(\mathbf{q},\omega), \tag{3b}$$

and the superscript ℓ denotes longitudinal projection. Here $\mathbf{E}^{ex}(\mathbf{q},\omega)$ and $n < \mathbf{p} > (\mathbf{q},\omega)$ are the Fourier transforms of the external field and the average polarization field, respectively, n is the number density of spheres and $<>$ means ensemble average. The appearance of the longitudinal projection, is only a matter of convenience which takes advantage of the fact that the $q \to 0$ limit of either the longitudinal or transverse response coincide. The limiting process $(q \to 0)$ is necessary in order to get around the evaluation of a few non-convergent integrals.[12]

We consider that the system is excited by a longitudinal external field of the form $\mathbf{E}^{ex} = \hat{q}E^{ex}e^{i(\mathbf{q}\cdot\mathbf{r}-\omega t)}$ and we rewrite Eq. (1) as

$$\mathbf{P}_i = \alpha(\mathbf{E}_L + \sum_j \Delta\overleftrightarrow{\mathbf{T}}_j \cdot \mathbf{P}_j), \tag{4}$$

where $\mathbf{E}_L = \hat{q}E^{ex}/\epsilon_h + N < \overleftrightarrow{\mathbf{T}} > \cdot < \mathbf{P} >$ is the Lorentz field. Here

$$\mathbf{P}_i = \mathbf{p}_i e^{-\mathbf{q}\cdot\mathbf{R}_i} \qquad \text{and} \qquad \overleftrightarrow{\mathbf{T}}_{ij} = \overleftrightarrow{\mathbf{t}}_{ij} e^{-i\mathbf{q}\cdot(\mathbf{R}_i-\mathbf{R}_j)} \tag{5}$$

are so defined in order to get rid of trivial exponential factors, and N is the total number of spheres.

The formal solution of Eq. (4) is

$$\mathbf{P}_i = \alpha \sum_j (\overleftrightarrow{\mathbf{V}}^{-1})_{ij} \cdot \mathbf{E}_L, \tag{6a}$$

where

$$\overleftrightarrow{\mathbf{V}}_{ij} = \mathbf{1}\delta_{ij} - \alpha\Delta\mathbf{T}_{ij} \tag{6b}$$

and $\mathbf{1}$ is the unit matrix. We now define the Lorentz susceptibility as

$$n < \mathbf{P} > (q,\omega) = \overleftrightarrow{\chi}^{L}(q,\omega) \cdot \mathbf{E}^{L}(q,\omega), \tag{7}$$

and it can be easily shown that the effective dielectric response ϵ_{eff} is given by

$$\epsilon_{eff} = \frac{1 + 2f\hat{\alpha}^*}{1 - f\hat{\alpha}^*}, \tag{8a}$$

where $f = n4\pi a^3/3$ is the volume fraction of spheres, $\hat{\alpha}^* = \alpha^*/a^3$ and

$$n\alpha^* = \chi^{\ell}_L(q \to 0,\omega) = lim_{q\to 0} < \sum_j (\overleftrightarrow{\mathbf{V}}^{-1})^{\ell}_{ij} > . \tag{8b}$$

Eq. (8a) is an exact equation and it has the same functional form as the Maxwell Garnett formula[4] (or Clausius-Mossotti relation[5]), except that the bare polarizability α is replaced by a dressed polarizability α^* which is proportional to the dipolar response of the sphere to the Lorentz field rather than to the local field.

According to Eq. (8b) the calculation of α^* requires the evaluation of the ensemble average of the inverse of matrix $\overleftrightarrow{\mathbf{V}}_{ij}$, defined in Eq. (6b), which in the thermodynamic limit becomes an infinite matrix with stochastic elements. This is obviously a complicated problem. We perform instead a series representation of the inverse of $\overleftrightarrow{\mathbf{V}}_{ij}$ in powers of $\alpha\Delta\overleftrightarrow{\mathbf{T}}_{ij}$, that is

$$\sum_j (\overleftrightarrow{\mathbf{V}}^{-1})_{ij} = 1 + \alpha \sum_j \Delta\overleftrightarrow{\mathbf{T}}_{ij} + \alpha^2 \sum_{jk} \Delta\overleftrightarrow{\mathbf{T}}_{ik} \cdot \Delta\overleftrightarrow{\mathbf{T}}_{kj} + ..., \tag{9}$$

we then take an ensemble average assuming that in the low-density regime the s-particle distribution function can be factored as unsymmetrized sequential products of two-particle distribution functions, that is

$$\rho^{(s)}(\mathbf{R}_1,\ldots,\mathbf{R}_s) = \prod_{\substack{ij \\ (j=i+1)}} \rho^{(2)}(R_{ij}). \tag{10}$$

The end result is a series representation of α^* which can cast in a diagrammatic form as

$$\xi \equiv \frac{\alpha^*}{\alpha} = \tag{11}$$

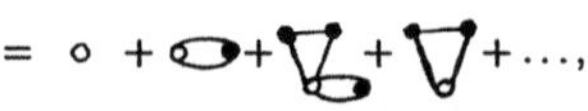

where each diagram in this series is irreducible; this means that it cannot be split into two independent diagrams by cutting a single line. The precise definition of each diagram is given in Ref. 13; here we will only say that in order to draw a diagram one cannot lift the pencil from the paper, that each line carries a factor α, each black dot carries a factor n, the other factor is an integral over the coordinates of the particles depicted by black dots, being the white dot the reference sphere. The integrand contains the longitudinal projection of the scalar product of r tensors T_{ij} and the s-particle distribution functions where r is the total number of lines and s is the total number of dots (blacks and white).

In this work we propose the following diagrammatic approximation:

$$\xi = \tag{12a}$$

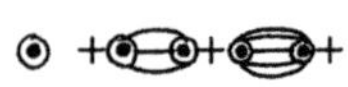

where

$$\Delta = \tag{12b}$$

and

$$\eta = \tag{12c}$$

This is an extension of the renormalized polarizability theory (RPT) developed in Ref. 13, which only considered

$$\xi = \Delta \quad \text{and} \quad \eta = \tag{13}$$

Here we are including diagrams that, we believe, should be important in the low-density regime because they take full account of the interaction between only two renormalized dots. For example, if we replace the renormalized dots by unrenormalized ones, we then recover the two-particle-linked-cluster-expansion results of Felderhof, Ford and Cohen[18].

The solution of the system of diagrammatic equations given by Eq. (12) yields to

$$\xi = \Delta + \frac{1}{3} f \hat{\alpha} \Delta^2 log\left(\frac{64 - \hat{\alpha}^2\Delta^2}{64 - 4\hat{\alpha}^2\Delta^2}\right) \tag{14a}$$

and

$$\Delta^{-1} = 1 - \frac{1}{3} f \hat{\alpha}\sqrt{\Delta}\, log \frac{(4+\hat{\alpha}\sqrt{\Delta})(8+\hat{\alpha}\sqrt{\Delta})}{(4-\hat{\alpha}\sqrt{\Delta})(8-\hat{\alpha}\sqrt{\Delta})}, \tag{14b}$$

which have to be solved self-consistently. Here $\hat{\alpha} \equiv \alpha/a^3$.

III. Results and Discussion

We present our results in terms of the Bergman's spectral representation of the effective dielectric function. It has been shown[19] that ϵ_{eff} can be written as

$$\epsilon_{eff} = 1 - f \int_0^1 \frac{g(u)}{t-u} du, \tag{15a}$$

where

$$t = \frac{1}{1 - \epsilon_m/\epsilon_h} \quad . \tag{15b}$$

The spectral representation is a normal-mode representation in a normal-mode variable u, where the frequency of the modes is determined by the pole location ($t = u$) and $g(u)$ is their strength. The main advantage of this representation is that $g(u)$ does not depend on the physical nature of the elements which constitute the composite but only the geometrical location of the spheres.

In Fig. 1 we show the spectral function $g(u)$, as a function of u, for volume fractions of 0.1, 0.2 and 0.3, calculated with Eq. (14) (solid line), with RPT as defined in Eq. (13) (broken line) and the "experimental" results of Cichocki and Felderhof[16] (dotted line). For the two-particle distribution function, here we used a simple step function $\theta(R_{12} - 2a)$ which should be valid in the low density regime. We can see that for $f = 0.1$ we obtain an excellent agreement with "experiment" and the difference between these new results and RPT clearly demonstrates the importance of the additional class of diagrams contained in the present approximation. For $f = 0.2$ the agreement between theory and experiment is not so good and our new results lie now something in between the "experiment" and RPT. Finally for $f = 0.3$ our new results resemble very much RPT but the agreement with experiment is far from being good. Therefore, for such high filling fractions the effect of the additional class of diagrams is indeed negligible. These corroborates our earlier assertion about the importance of this class of diagrams in the low-density regime. The disagreement between theory and "experiment", at higher

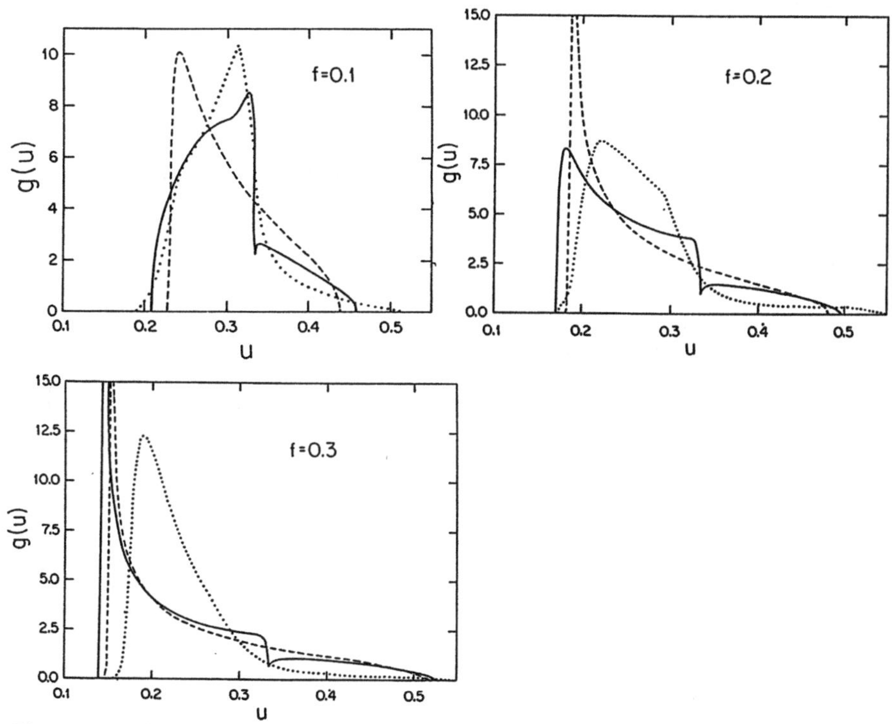

Fig 1. The spectral function $g(u)$ as a function of u for volume fractions of 0.1, 0.2 and 0.3, calculated with Eq. (14) (solid line), with RPT as defined in Eq. (13) (broken line) and the "experimental" results of Cichocki and Felderhof (dotted line).

volume fractions, can be explained by noticing that in this regime the factorization of the distribution functions, as given by Eq. (10), is no longer valid. Also, at high filling fractions, $\rho^{(2)}(R_{12})$ is not a simple step function any more and better approximations, something like Percus-Yevick[20], should be used. Work along these lines is now in progress and it will be reported elsewhere.

Acknowledgements

We acknowledge stimulating discussions with Marcelo del Castillo, Luis Mochán and Guillermo Monsivais. The financial support of Dirección General de Asuntos del Personal Académico of the National University of Mexico, through grant IN-01-4689-UNAM, is also acknowledged.

References

1. J.C. Maxwell, *A Treatise on Electricity and Magnetism*, Vol. 1, (Reprint: Dover, New York, 1954) Sec. 314, p. 440.

2. See for example: D. Staufer, Phys. Rep. **54**, 1 (1979).

3. R. Landauer in *Electrical Transport and Optical Properties of Inhomogeneous Media*, edited by J.C. Garland and D.B. Tanner, AIP Conference Proceedings Number 40 (American Institute of Physics, New York, 1978) p. 2.

4. J.C. M. Garnett, Philos Trans. R. Soc. London **302**, 385 (1904).

5. See for example: J.D. Jackson, *Classical Electrodynamics*, second Edition (J. Wiley and Sons, New York, 1975) pp. 152-155.

6. J.G. Kirkwood, J. Chem. Phys. **4**, 592 (1936); J. de Boer, F. Van der Maesen, and C.A. Ten Seldam, Physica (Utrecht) **19**, 265 (1953); J.D. Ramshaw *ibid.* **62**, 1 (1972); B.R.A. Nijboer, Physica A **80**, 398 (1975).

7. A. Isihara and R.V. Hanks, J. Chem. Phys. **36**, 433 (1962); M.S. Wertheim, Mol. Phys. **25**, 211 (1973); M.S. Wertheim Ann. Rev. Phys. Chem. **30**, 471 (1979); B.U. Felderhof, G.W. Ford and E.G.D. Cohen, J. Stat. Phys. **28**, 135 (1982); **28**, 649 (1982).

8. L.D. Landau and E.M. Lifshitz, *Electrodynamics of Continuous Media* (Pergamon Press, Oxford, 1960) pp. 45-47.

9. V.A. Davis and L. Schwartz, Phys. Rev. B **31**, 5155 (1985); **33**, 6627 (1986); L. Tsang and J.A. Kong, J. Appl. Phys. **53**, 7162 (1982).

10. J.E. Gubernatis, in *Electrical Transport and Optical Properties of Inhomogeneous Media*, edited by J.C. Garland and D.B. Tanner, AIP Conference Proceedings Number 40 (American Institute of Physics, New York, 1978) p. 84; M. Gómez, L. Fonseca, G. Rodríguez, A. Velázquez and L. Cruz, Phys. Rev. B **32**, 3429 (1985).

11. H. Looyenga, Physica (Utrecht) **31**, 401 (1965); A. Liebsch and P. Villaseñor, Phys. Rev. B **29**, 6907 (1984); A. Bittar, S. Berthier, and J. Lafait, J. Phys. (Paris) **45**, 623 (1984); G.A. Niklasson, Solar Energy Mater. **17**; 217 (1988); P. Sheng, Phys. Rev. Lett. **45**, 60 (1980).

12. R.G. Barrera, G. Monsivais and W.L. Mochán, Phys. Rev. B **38**, 5371 (1988).

13. R.G. Barrera, G. Monsivais, W.L. Mochán and E. Anda, Phys. Rev. B **39**, 9998 (1989).

14. D.J. Bergman, Phys. Rev. Lett. **44**, 1285 (1980); G.W. Milton, Appl. Phys. **37**, 300 (1980).

15. Y. Kantor and D.J. Bergman, J. Phys. C. **15**, 2033 (1982); S. Kumar and R.I. Cukier, J. Phys. Chem. **93**, 4334 (1989).

16. B. Cichocki and B.V. Felderhof, J. Chem. Phys. **90**, 4960 (1989).

17. W.J. Kaiser, E.M. Logothetis and L.W. Wenger, J. Phys C. **18**, L837 (1985).

18. B.U. Felderhof, G.W. Ford and E.G.D. Cohen, J. Stat. Phys. **28**, 135 (1982).

19. J. Korringa, Geophysics **49**, 1760 (1984); D.J. Bergman, Ann. Phys. (N.Y.) **138** 78 (1982).

20. See for example: M.S. Wertheim, Phys. Rev. Lett., **10**, 321 (1963).

Determination of Impurity Content in SnO_2 Thin Films Using Nuclear Reactions

R. Asomoza[1], *A. Maldonado*[1], *J. Rickards*[2], *E.P. Zironi*[2], *M.H. Farías*[3], *and L. Cota-Araiza*[3]

[1]Dpto. de Ingeniería Eléctrica, CINVESTAV-IPN,
Apdo. Postal 14-740, 07000 México, D.F., Mexico
[2]Instituto de Física, Universidad Nacional Autónoma de México,
Apdo. Postal 20-364, 01000 México, D.F., Mexico
[3]Instituto de Física, Universidad Nacional Autónoma de México,
Laboratorio de Ensenada, Apdo. Postal 2681, 22800 Ensenada,
Baja California, Mexico

ABSTRACT. The determination of the dopant concentration in semiconductors is important to understand their transport properties. In the case of SnO_2:F thin films, prepared by spray pyrolysis, the usual surface analysis methods (Auger, ESCA) do not give reliable values of the fluorine content, however the resonant nuclear reaction technique has been applied successfully to determine the fluorine concentration. From the intensity of gamma rays coming from the nuclear reaction $^{19}F(p,\alpha\gamma)^{16}O$, the concentration as well as the distribution of the fluorine atoms in the sample can be obtained. The results have shown a uniform distribution of fluorine within the sample as well as no diffusion into the silicon substrate.

1. INTRODUCTION

The need to improve the characteristics of a-Si:H solar cells makes it necessary to obtain high quality semitransparent contacts. One compound used frequently as such a contact is SnO_2 which shows good chemical stability and, when conveniently doped, low resistivity and high transparency [1,2,3]. Moreover, tin oxide is replacing the use of ITO films because it shows a more stable interface [4,5].

Several techniques are used to prepare tin oxide films, among which the spray pyrolysis technique [6,7,8] gives low cost and high quality tin oxide thin films. Using this technique we have obtained films over large areas with excellent uniformity and reproducibilty [8]. Doping was achieved by adding fluorine atoms during the growth. The films were characterized by measuring their electrical, optical and structural properties. In order to correlate these properties to the fluorine content, it is desirable to determine it. The nuclear reaction method, which shall be described in detail later, proved to give, not only a figure proportional to the fluorine content, even at low concentrations, but also its profile within the sample. It also showed that the interface SnO_2:F/substrate is rather stable.

2. EXPERIMENTAL

Fluorine doped SnO_2 thin films were obtained by spraying a solution of $SnCl_4$-$5H_2O$, 0.2 molar concentration in ethyl alcohol containing HF, onto glass and silicon substrates. The HF concentration, C_{HF}, was varied in the range of 2 to 8 weight %. The substrates were kept at constant temperature, T_s, by means of a liquid tin bath, three

Springer Proceedings in Physics, Volume 62
Surface Science Eds.: F.A. Ponce and M. Cardona

inches in diameter. T_s was varied in the range from 300 to 450 °C. The carrier gas was nitrogen with a flux of 4 l/min. The films were grown to a thickness within the range of 70 nm to 600 nm at a deposition rate of about 20 A/sec. The resistivity was measured with a conventional setup using a four point probe and the optical transmission was performed using a Perkin-Elmer Lambda 3 spectrophotometer in the range from 300 to 850 nm.

The fluorine content was measured using Auger electron spectroscopy (AES) and the resonant nuclear reaction method (RNR) [9,10].

The Auger spectra were obtained with a JEOL JAMP-30 Auger microprobe, working with E_p = 5 keV and I_p = 300 nA. In this spectrometer the electron beam impinges on the sample almost parallel to the surface, at 90° with respect to the CMA. The AES technique is widely used and will not be described here [11].

We will describe the RNR method briefly here: the sample was bombarded with a proton beam having an energy close to an isolated resonance in a reaction with the element to be detected. In the present case the nuclear reaction $^{19}F(p,\alpha\gamma)^{16}O$ was used. This reaction shows a resonance at 340 keV with a high cross section, 2.4 keV wide. The proton beam from the 700 kV Van der Graaff accelerator from the Instituto de Física was used. The 6.14 MeV gamma and two escape peaks from the reaction were detected by means of a 10.2 x 10.2 cm NaI(Tl) scintillation detector, placed as close as possible to the target.

To obtain the excitation curve, the energy of the incident protons was increased in steps, about 1 to 2 keV, starting with a value slightly below the resonance and the emitted gamma rays produced in each step were counted. If the fluorine atoms were located at the surface, the excitation curve would only show a narrow resonance at 340 keV, if on the contrary they were located at a certain depth within the sample, it would be necessary to increase the bombarding energy in order to compensate for the energy loss between the surface and the position of the fluorine atoms. Therefore, the excitation curve obtained is directly related to the fluorine concentration and the energy scale of the excitation curve can be transformed to a depth scale. Typical values were 20 μC per point at a current of 100 nA. The proton energy at a given depth was obtained from the formula of Montenegro et al [12] and the Bragg rule was used for adding stopping cross sections of the different elements in these compound targets. The stopping power for protons in tin and oxygen calculated in this manner are very similar to those obtained from tables like Ziegler's [13]. The calculated excitation curves were obtained by integrating these quantities over the appropriate energies and depth intervals. These results were directly compared to the experimental curves.

From the excitation curve measured in this way, a depth profile for the fluorine concentration was obtained [14].

3. RESULTS AND DISCUSSION

It is known that different mechanisms are responsible for the high electrical conductivity of tin oxide films: a) intentional doping; b) deviations from stoichiometry and c) residual impurities [15,16].

Fluorine is known to act as a donor impurity in tin oxide [2,3]. It has the advantage of increasing its conductivity without much degrading its optical properties. However, the addition of fluorine atoms beyond a certain limit may modify the crystalline structure of this oxide. Therefore it is important to determine its concentration in the films as well as a possible segregation at the grain boundaries or at the tin oxide/substrate interface.

In order to determine more precisely the role of fluorine on the electrical and optical properties of the films, a systematic study

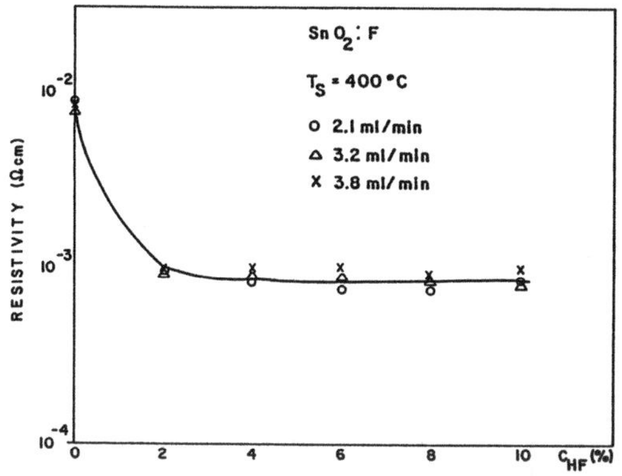

Figure 1- Resistivity as a function of HF concentration in the departure solution (C_{HF}) of fluorine doped tin oxide film prepared at a substrate temperature of 400 °C. The symbols refer to different rates of solution.

was carried out varying the concentration of HF in the preparation solution as well as the substrate temperature. These two parameters were identified as the most important in modifying the electrical properties [17].

Figure 1 shows that the addition of 2 % HF to the departure solution produces a decrease of the film resistivity of about one order of magnitude. A further increase in C_{HF}, up to 8 % produces only small changes in the resistivity. On the contrary, the substrate temperature has a more drastic effect on the electrical resistivity, as can be seen in figure 2. Films prepared with C_{HF}= 8 % show a decrease in the resistivity of four orders of magnitude as T_s increases from 300 to 450 °C, which has a minimum at around 400 °C whose value is below 10^{-3} Ωcm. Figure 3 shows, however, that the transmittance in the range 350-850 nm, is much better in samples prepared at T_s= 450 °C. We have thus selected C_{HF} = 8 % and T_S = 450 °C as the deposition conditions for our semitransparent electrodes.

Concerning the determination of the fluorine content by using Auger spectroscopy, it has been reported that the electron beam used for the excitation can induce fluorine desorption from the surface, making the measurements unreliable [4,18].

In order to reduce the damage produced by the electron beam on the sample, and therefore the fluorine desorption, the JEOL JAMP-30 spectrometer mentioned above, where the electron beam is nearly parallel to the surface, was used. This geometry reduces the damage produced on the sample. The Auger spectrum is shown in figure 4a). This spectrum corresponds to a highly doped sample having C_{HF} = 50 % It is to be noted that even though a fluorine signal is detected, the calculated concentration does not corresponds to the expected one, as can be seen in figure 4b). However, the depth profile shown in this figure indicates a rather uniform distribution within the sample.

In contrast to the Auger results, the gamma ray signal originated in the nuclear reaction, mentioned above, is very strong, even at low C_{HF} values. Figure 5 shows typical excitation curves obtained in samples prepared with C_{HF} = 2 and 8 % and T_S = 400 °C. The continuous line is just to guide the eye. We can observe different trends in this figure: a) the gamma ray count is well above the background in both curves; b) there is a uniform

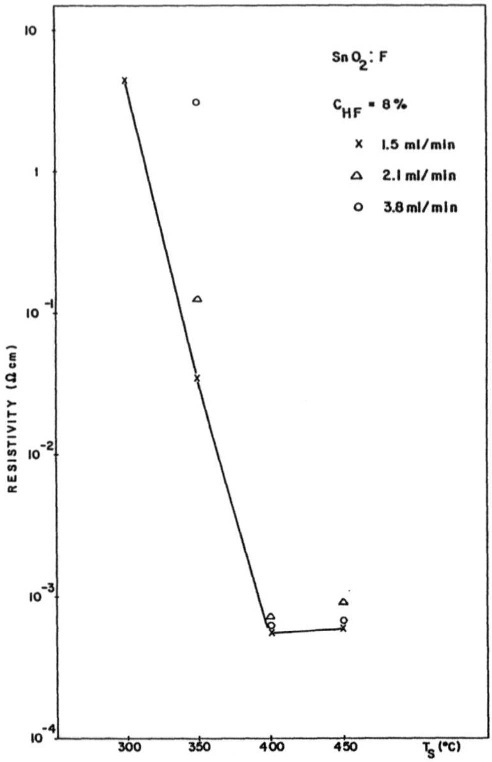

Figure 2- Resistivity as a function of substrate temperature for a fluorine doped tin oxide sample with C_{HF} = 8 %. In this figure the symbols refer to different rates of the deposition solution.

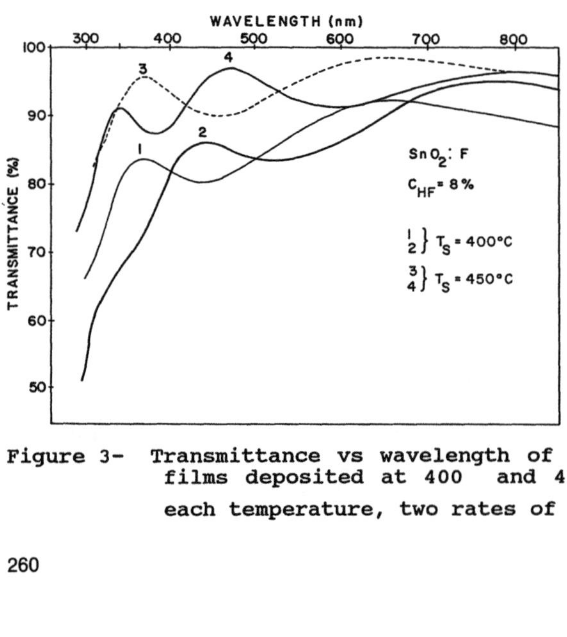

Figure 3- Transmittance vs wavelength of fluorine doped tin oxide films deposited at 400 and 450 °C, and C_{HF} = 8 %. For each temperature, two rates of the solution were used.

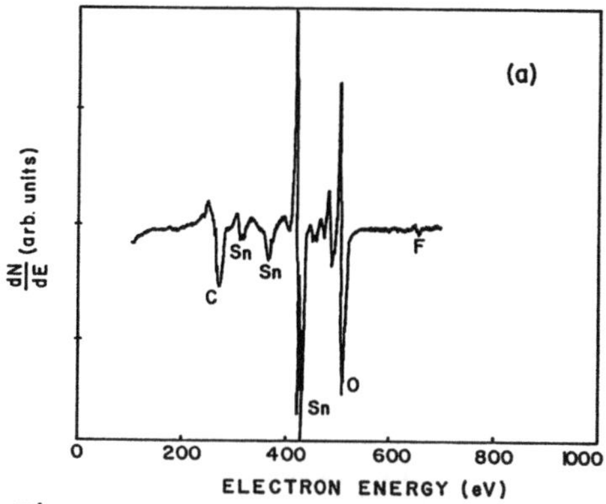

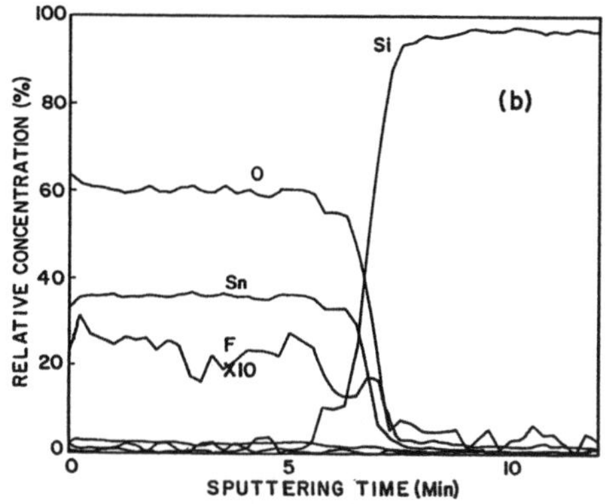

Figure 4- Auger electron spectrum obtained from a fluorine doped tin oxide sample prepared at 450°C and C_{HF} = 50 %; a) original surface; b) depth profile giving the relative concentration of the elements detected as a function of the sputtering time.

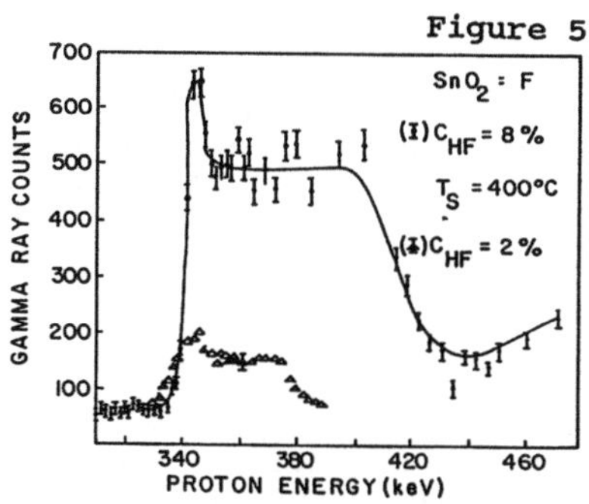

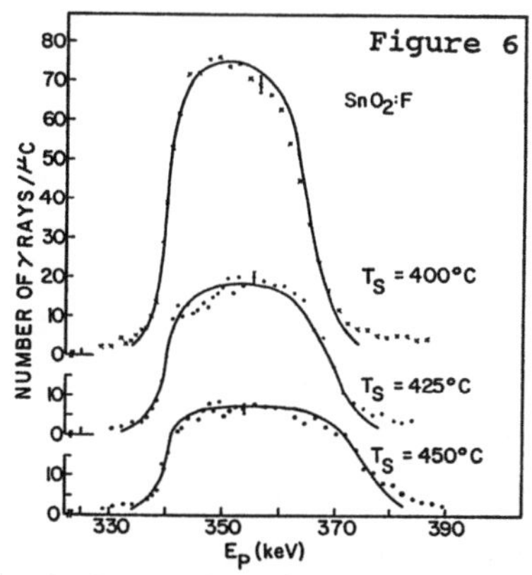

Figure 5- Excitation curve obtained from fluorine doped tin oxide samples with C_{HF} = 8 and 2 %. The gamma ray counts, which are proportional to the fluorine concentration are plotted as function of the incident proton energy.

Figure 6- Experimental and calculated excitation curves measured on samples grown at different substrate temperatures and a constant C_{HF} = 8 %. A loss of fluorine is observed as the deposition temperature is increased.

distribution of fluorine into the sample; c) there is a higher concentration at the surface and d) the tin oxide/substrate is well defined.

From these curves we can obtain the average fluorine concentration, its distribution in depth and the thicknesses of the films.

The rapid decrease of the gamma ray counts in the sample/substrate interface indicates that it is quite abrupt. Annealing treatments are under way to determine the stability of this interface at high annealing temperatures. This is important if the films are going to be used as a part of electronic devices.

With these results in mind, we prepared a batch of samples at a constant HF concentration of 8 % and the substrate temperature in the range 400 to 450 °C. The results are shown in figure 6. In this

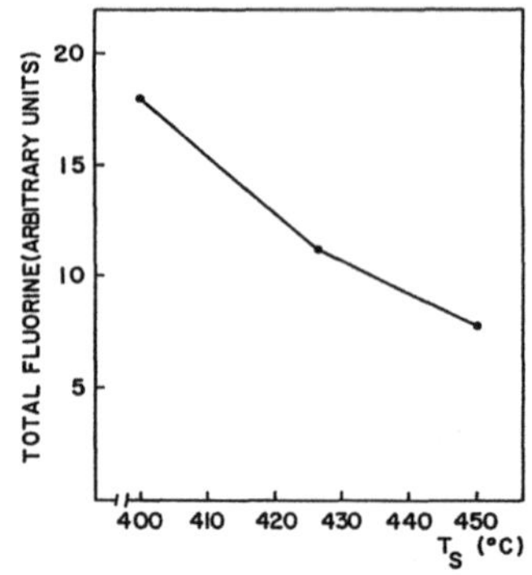

Figure 7- Total fluorine concentration in the films of figure 6, calculated from the area under the excitation curve, plotted as a function of the substrate temperature.

figure, the solid lines were calculated assuming stoichiometric tin oxide and adjusting the ordinate with an arbitrary constant. Taking a value of 6.95 g/cm^3 for the density of tin oxide, from the FWHM of each distribution, the thicknesses of the samples were obtained. The values found agreed within an average of 13 % with the values found by ellipsometry.

The total fluorine concentrations of the films were calculated by integrating the excitation curves of figure 6 and are shown in figure 7, as a function of the substrate temperature. In this figure, the ordinate is proportional to the fluorine concentration. In order to obtain a true value it is necessary to determine the efficiency of the counting system, which can be accomplished by using a calibrated sample, in our case we used polycrystalline LiF. It can be seen that the fluorine concentration diminishes as T_s increases, even though the thicknesses, determined as mentioned above, showed an increase in the same temperature range.

The RNR method has shown that the incorporation of fluorine is more effective at low deposition temperatures, for a constant departure concentration. The increase in the transmittance of samples prepared at 450 °C is compatible with a smaller fluorine content in the films. However, the electrical resistivity is much less sensitive in the temperature range from 400 to 450°C to the change found in the fluorine concentration. The rapid variation of the resistivity with the increase of T_s shown in figure 2 seems to be due more to a structural change than to the incorporation of fluorine atoms to the tin oxide. Indeed, SnO_2 is known to be amorphous when prepared below 300 °C, having a high resistivity [4,6]. As the substrate temperature increases, in the range between 300 and 400 °C, the tin oxide becomes a mixture of two polycrystalline phases. Selected area diffraction showed that these phases were Sn_3O_4 and SnO_2[19]. At higher substrate temperatures, the main phase was identified to be SnO_2 [19], which has a high resistivity when it approaches stoichiometric SnO_2. One drawback of the RNR method is that the size of the proton beam is rather large for studying grain boundary segregation , it is about 2 mm^2, therefore only average values can be obtained over this large area. In consequence, even though a surface accumulation of fluorine was detected in some samples it was impossible to conclude anything about the grain boundaries.

4. CONCLUSION

We have studied SnO_2 thin films prepared by the spray pyrolysis method. The layers showed excellent adherence and stability. In the best cases, their electrical resistivity was lower than 10^{-3} Ωcm and its transparency was higher than 90 % in the visible region.

The Resonant Nuclear Reaction method proved to be very sensitive to the presence of fluorine atoms. We were able to calculate the fluorine concentration, its distribution within the films as well as a surface segregation in some samples. One advantage of this method over the usual surface spectroscopies is that it is non-destructive. The results found are in qualitative agreement with the observed electrical and optical properties. We have applied these films as semitransparent electrodes in a-Si:H solar cells.

References

1- G. Haacke, Ann. Rev. Mater. Sci., 1(1977)73.

2- J.C. Manifacier, L. Szepessy, J.F. Bresse M. Perotin and R. Stuck, Mat. Res. Bull., 14(1979)109.

3- J.C. Manifacier, L. Szepessy, J.F. Bresse M. Perotin and R. Stuck, Mat. Res. Bull., 14(1979)163.

4- M. Fantini and I. Torriani, Thin Solid Films, 138(1986)255.

5- A Antonaia, S. Aprea and P.T. Menna, 21st. IEEE Photovoltaic Specialists Conference, Kissimmee, Florida (1990).

6- J. Sanz Maudes and T. Rodríguez , Thin Solid Films 69(1980)183

7- G. Mavrodiev, M. Gajdarddziska and N. Novkovski, Thin Solid Films,113(1984)93.

8- A.T. Silver, Master Science thesis, Electr. Eng. Dept. CINVESTAV IPN, Mêxico, 1985 (unpublished)

9- G. Amsel, J.P. Nadai, E. D'Artemare, D. David, E. Girard and J. Moulin, Nucl. Instr. and Methods 92(1971)481.

10- E.P. Zironi, J. Rickards, A. Maldonado and R. Asomoza, Nucl. Instr.and Methods B45(1990)115.

11- see for example "Auger Electron Spectroscopy" A. Joshi, L.E. Davis and P.W. Palmberg in Methods of Surface Analysis, Ed. A.W. Czanderna, Elsevier Sci. Publ. Co. (1975)p. 159.

12- E.C. Montenegro, S.A. Cruz and C. Vargas-Aburto-Phys. Lett., A92 (1982)195.

13- J.F. Ziegler, Handbook of Stopping Cross Sections for Energetic Ions in all Elements, Pergamon Press, 1980.

14- F. de Anda, J. Rickards, E.P. Zironi, A. Maldonado and R. Asomoza, Proccedings of the VIII National Congress on Surfaces and Interfaces, SMCSV, Guanajuato, Gto., Mexico (1988)p.150.

15- C.A. Vincent, J. Electrochem. Soc., 119(1972)515.

16- C.A. Vincent and D.C. Weston, J. Electrochem. Soc.,119(1972)518.

17- A.T. Silver, A. Maldonado and R. Asomoza Technical Report A53, Electrical Engineering Department, CINVESTAV, Mexico, (1989)

18- S.V.Pepper, J.Appl.Phys.,45(1974)2947.

19- D.R. Acosta, R. Asomoza and A. Maldonado, Proceedings of the XII International Congress for Electron Microscopy, San Francisco, CA (1990).

6. CONCLUSION

Electrochromic dc Sputtered Nickel-Oxide-Based Films: Optical Structural, and Electrochemical Characterization

W. Estrada[1], *A.M. Andersson*[2], *C.G. Granqvist*[2], *A. Gorenstein*[3], *and F. Decker*[3]

[1]Universidad Nacional de Ingeniería, Facultad de Ciencias, Lima, Peru
[2]Physics Department, Chalmers University of Technology and University of Gothenburg, S-41296 Gothenburg, Sweden
[3]DFA/IFGW, UNICAMP, CP 6165, 13081 Campinas, SP, Brazil

Abstract. NiO_x films were made by dc magnetron sputtering of Ni in O_2. The grain size was ~10 nm, and the crystal structure was cubic. Subsequent electrochemical treatment in KOH established electrochromism. The material was studied by cyclic voltammetry, in situ measurements of optical transmittance and mechanical stress, spectral infrared reflectometry, and spectrophotometric measurements in the 0.35-2.5 μm range. Electrochemical data showed that electrochromic bleaching was associated with proton insertion. P-polarized infrared reflectance showed OH stretching vibrations representative of "free" OH for the bleached state and OH in the presence of hydrogen bonds for the coloured state. The luminous and solar transmittance could be varied between 80% and 20% and between 74% and 24%, respectively. The electrochromism is produced by absorption modulation.

1. Introduction

Nickel-oxide-based thin films were produced by sputtering onto glass followed by electrochemical treatment. Structural, compositional, electrochemical, optical and mechanical properties were investigated with the aim of elucidating parameters of importance for using these films in electrochromic "smart windows".

Electrochromic materials are characterized by a reversible and persistent change of the optical properties under the action of an applied electric field. Thin films of these materials enable a dynamic control of the throughput of radiant energy, and it follows that they can be used on windows in order to achieve balanced lighting and air conditioning levels. These functions are of obvious importance for energy-efficient architecture. Other applications of electrochromics-based systems are in anti-dazzling devices for vehicles and in high-contrast nonemissive information displays. A thorough review of electrochromic materials and systems, and their uses in advanced fenestration technology, is given in ref. 1. Numerous inorganic and organic substances exhibit electrochromism [1,2]. Nickel-oxide-based materials are of particular interest and combine a number of favourable properties, viz. a large span in luminous and solar transmittance between fully bleached and fully coloured states, good durability, low materials cost, and possibilities to manufacture by reactive dc magnetron sputtering which is an attractive technique for large-area high-rate deposition.

Springer Proceedings in Physics, Volume 62
Surface Science Eds.: F.A. Ponce and M. Cardona

The organization of this paper is as follows: Section 2 covers sample production by reactive dc magnetron sputtering onto glass coated with In_2O_3:Sn (i.e., ITO), for a subsequent conversion of the as-deposited absorbing films into electrochromic NiO_xH_y by electrochemical treatment in aqueous KOH. It also reports data taken by transmission electron microscopy (TEM). Section 3 reports data of electrochemical characterization, spectrophotometric measurements and infrared reflection spectroscopy (IRS) for ion-intercalated nickel oxide. Section 4, finally, summarizes the main results.

2. Sputter-Deposited Nickel-Oxide Films

2.1 Sample Preparation

Nickel oxide films were produced by reactive dc magnetron sputtering in a versatile deposition system [3]. After evacuation oxygen was introduced and sputtering was conducted from a nickel metal target at a constant discharge power. Further details on the thin film deposition are given in ref. 4.

Two kinds of substrates were employed: Most of the studies used 1 mm thick glass coated with a transparent and electrically conducting ITO layer having a resistance/square of ~10 Ω. Measurements of mechanical stress employed 0.15 -mm-thick microscopy cover glasses coated with ITO, using a procedure in ref. 5, so that their resistance/square was ≤ 200 Ω. The thickness of the nickel oxide films was between 0.1 and 0.2 μm. A typical deposition rate was 0.1 nm/s.

2.2 Transmission Electron Microscopy

Transmission electron microscopy was used to investigate the microstructure of as-deposited nickel oxide films sputtered onto carbon-covered on grids. Micrographs showed a granular structure with individual grains appearing most clearly as bright spots on a dark-field image [4]. The average grain diameter was about 10 nm.

The TEM was used in diffraction mode in order to investigate the crystal structure of the films. Most features of the patterns were consistent with a cubic NiO (bunsenite) structure. The lattice parameter was 0.42 nm, which is the expected bulk value. Evidence was found for a fiber texture in the (001)-direction [4]. The diffraction patterns also showed some additional spots, between the (200) and (220) rings, consistent with (210)-diffraction. Since (210)-diffraction is forbidden for cubic NiO this result suggests the existence of two phases in the sputter deposited nickel oxide films [6].

3. Ion-intercalated Nickel-Oxide-Based Films

3.1 Cyclic Voltammetry and Mechanical Stress Related to Electrochromism

Ion intercalation in as-sputtered nickel oxide films was accomplished by electrochemical techniques. The basic phenomena and processes can be illustrated by combining cyclic voltammetry with measurements of mechanical stress and monochromatic optical transmittance. In the experimental set up for these measurements, a nickel-oxide-coated substrate was put in 0.1 M

KOH electrolyte in an electrochemical cell which also contained a Pt counter electrode and a saturated calomel electrode (SCE) as potential reference. The substrate was rigidly fixed at its upper end, which extended somewhat above the electrolyte. A laser beam at 632.8 nm wavelength was sent via a beam-splitter onto the lower end of the substrate. Bending of the sample could be recorded as a deflection of the reflected laser beam on a position sensitive photodetector. The bending is directly related to the stress in the films. The same system allowed transmittance measurements. Also ex-situ stress measurements could be made. This novel experimental set up is described in some detail in ref. 7.

Figure 1 shows data taken during potentiodynamic cycling from -0.8 V to +0.6 V vs. SCE of a 0.2 μm thick NiO_xH_y film. The measurements started with an as-deposited sample at an open circuit potential of +0.2 V, and the initial scan was anodic. The curves trace the property evolution during the first two complete cycles and also give results under stable conditions reached after 50 cycles. It is observed that the stress varies during each cycle, accompained by an increase both of the transmittance variation between coloured and bleached states, and the cathodic and anodic charge involved in each cycle. Figure 1a reports on the current density. Irreversible electrochemical reactions during the initial cycling are manifested as unbalanced charge during the anodic and cathodic excursions. Considering that the electrochemical reaction corresponds to a simultaneous ion and electron insertion (extraction), and supposing that ions are monovalent cations, an upper limit for the number of inserted/extracted ions of nickel-oxide-based film at each cycle can be calculated. If furthermore it is considered that the NiO molecular volume approximately corresponds to 6×10^{22} molecules/cm^3 [8], the non-reversible charge insertion corresponds approximately to a 5% increase of the total number of Ni atoms in the film.

Figure 1b shows the relative stress evolution during cycling. The level of zero stress in the NiO_xH_y film was fixed by ex-situ measurements of the radius of curvature for samples with a bare ITO coating as well as with nickel-oxide-based overlayers in as-deposited and electrochemically treated states. All of the films were found to be under compressive stress. The stress condition was about the same for the bare ITO coating as after overlaying it with as-deposited nickel oxide. This state is taken to correspond to zero stress in the NiO_xH_y film, and compressive and tensile stress evolution are signified by the radius of curvature being diminished or enlarged. It is seen in Fig. 1b that the stress becomes increasingly tensile upon initial cycling. At negative potential the NiO_xH_y film is in a compressive state, which gives evidence for ion insertion; a positive potential corresponds to tensile stress, indicating ion extraction.

Figure 1c shows the transmittance evolution for 632.8 nm wavelength. The as-deposited film has a low transmittance. Electrochemical treatment is found to produce almost stable electrochromism after 50 cycles. The transmittance values at -0.8 and +0.6 V were ~85% and ~40%, respectively. An even larger span in transmittance could have been obtained by increasing the potentiodynamic range.

Two further experiments, complementary to the stress study reported above, were performed in order to identify the ion involved in the insertion/extraction process. In the first of these, a sample of the earlier described type was overcoated with a ~0.1 μm thick Pd film by sputtering and then subjected to

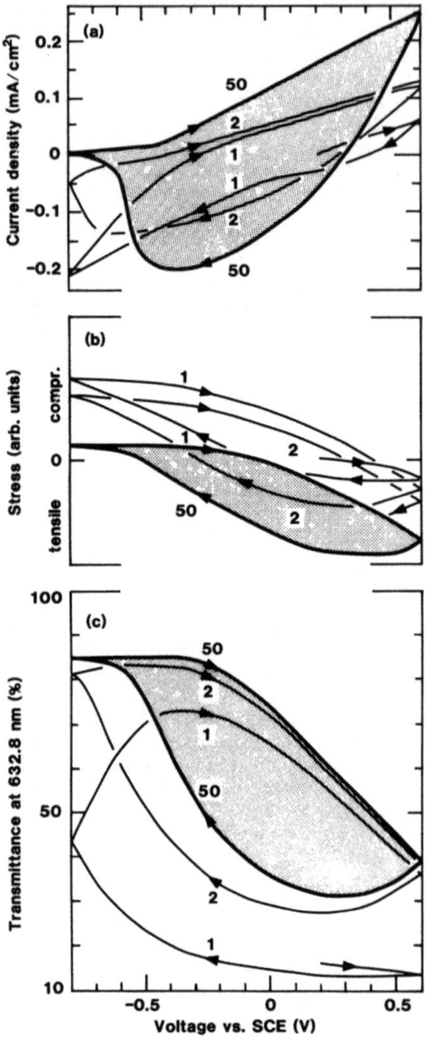

Fig. 1 Current density, relative mechanical stress, and transmittance at 632.8 nm during potentiodynamic cycling of a 0.2-μm-thick nickel-oxide-based sputtered film on ITO coated glass. The cycle number is given, and arrows indicate potential scan direction. Shaded regions represent stable electrochemical conditions reached after 50 cycles. The transmittance refers to the system (cell+electrolyte+sample), with the 100% transmittance value taken to be the reading for (cell+electrolyte).

potentiodynamic cycling as described in this section. In this case the electrochromism was followed by reflection of a He-Ne laser beam incident onto the uncoated side of the substrate. It was found that the sample showed similiar electrochromic behaviour as in the absence of Pd. Bleaching was also obtained if the sample, overcoated with Pd, was located in a vacuum chamber and then exposed to hydrogen gas. However, a subsequent colouring state was not possible to obtain in this latter experiment. It is well known that Pd acts as a filter for hydrogen, and hence the results of the two above-mentioned experiments give strong evidence in favour of H^+ being the mobile species during electrochromism.

Several conclusions can be drawn from the data:

Ions are inserted into the film in the cathodic cycle. The volume of the film then is increased, giving rise to a compression of the electrochromic layer. In the anodic cycle, ions are extracted from the film back into the electrolyte, and the stress becomes tensile.

- The first cycles are not completely reversible. Hence a net number of ions are incorporated into the film.

- The electrochemically stabilized situation corresponds to reversible insertion/extraction of ions at each cycle.

- The experiments give conclusive evidence that the mobile ions are protons.

The latter of these points proves that the nickel-oxide-based films adequately can be represented by NiO_xH_y.

3.2 Spectral Optical Properties

An experiment for studying the spectral optical properties of electrochromic nickel-oxide-based films was performed as follows: The specimen was placed in 1M KOH electrolyte and run through a number of colour-bleach cycles by applying alternately +0.1 and -0.1 mA between the working electrode and the counter electrode. The polarity was reversed each 150th second, i.e. the cycling time was 5 minutes. When a desired number of cycles was completed, the sample was withdrawn from the electrolyte, rinsed in distilled water, blown dry with filtered air, placed in the sample compartment of a spectrophotometer, and measured. After the measurements, the sample could be put back into the electrolyte and run through more colour-bleach cycles. The cycling was interrupted with the sample either in fully bleached state, or in partially or fully coloured state. The coloured states were specified by an amount of extracted charge Q_{ex} per unit area. Spectral normal transmittance T and near-normal reflectance R were measured in the $0.35 < \lambda < 2.5$ μm wavelength interval by using a double-beam spectrophotometer.

Figure 2 shows transmittance and reflectance spectra for a sample with 0.12 μm of ITO covered with 0.14 μm of NiO_xH_y. The sample was run through 2000 colour-bleach cycles prior to the

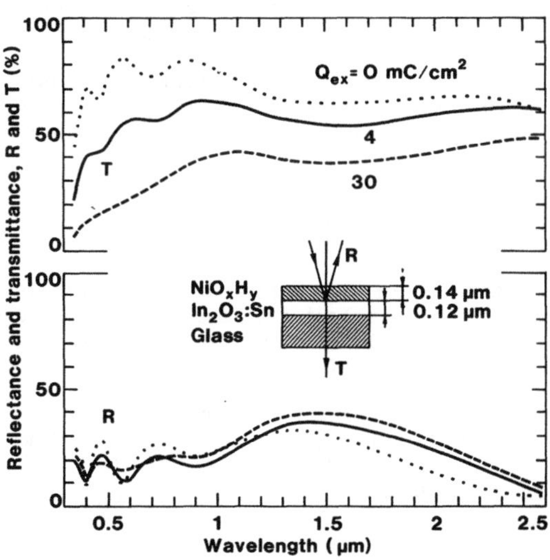

Fig. 2 Spectral normal transmittance (upper part) and spectral reflectance at 10° angle of incidence (lower part) for a sample including an electrochromic NiO_xH_y layer. Dotted, solid, and dashed curves refer to different amounts of extracted charge per unit area Q_{ex}. The same designation is used for transmittance and reflectance data. The sample configuration is shown in the inset.

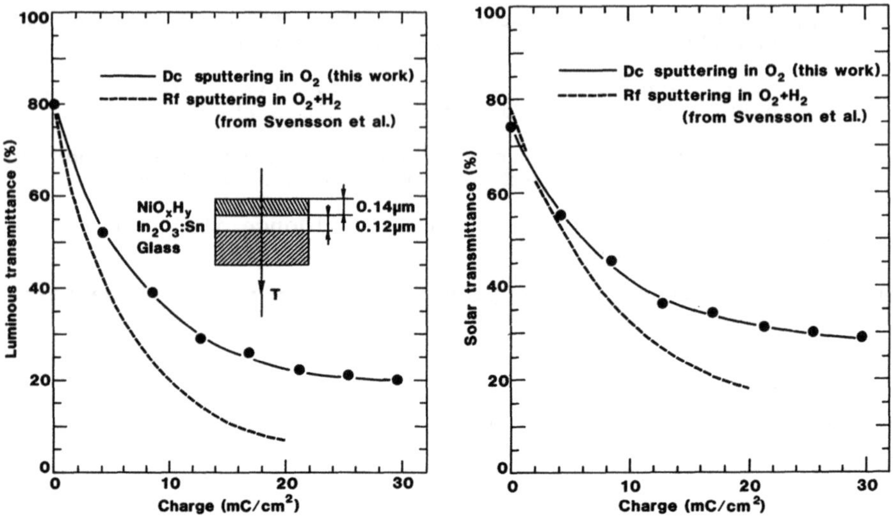

Fig. 3 Integrated luminous (left) and solar (right) transmittance for a sample including an electrochromic NiO_xH_y layer with different amounts of extracted charge per unit area. Dots represent evaluations for the sample sketched in the inset. Solid curve was drawn only to guide the eye. Dashed curve denotes data from ref. 11 on electrochromic NiO_xH_y layers made by rf magnetron sputtering of Ni in O_2+ H_2.

optical measurements. Dotted and dashed curves correspond to fully bleached and coloured samples, respectively. The solid curves in Fig. 2 represent results for intermediate colouration obtained with Q_{ex} = 4 mC/cm^2. In this state, T is ~46% at 0.5 μm wavelength; the corresponding values are ~17% and ~74% for fully coloured and bleached samples, respectively. The reflectance spectra, given in the lower part of Fig. 2 are similar irrespective of colouration level. This proves that electrochromic NiO_xH_y exhibits absorptance modulation rather than reflectance modulation.

The usefulness of NiO_xH_y -based coatings on smart windows is conveniently assessed through suitable integrated optical properties. The luminous (lum) and solar (sol) transmittance are defined by

$$T_{lum(sol)} = \int d\lambda\ \phi_{lum(sol)}(\lambda)\ T(\lambda) \Big/ \int d\lambda\ \phi_{lum(sol)}(\lambda)\ .$$

Here ϕ_{lum} denotes the standard luminous efficiency function for photopic vision [9] and ϕ_{sol} is the solar irradiance spectrum for one air mass [10]. ϕ_{lum} is confined to the 0.4 < λ < 0.7 μm wavelenght interval and is peaked at λ = 0.55 μm. ϕ_{sol} extends over the 0.3 < λ < 3 μm range.

Figure 3 shows T_{sol} and T_{lum} as a function of Q_{ex} for a sample with 0.14 μm of NiO_xH_y and 0.12 μm of ITO. Some spectral transmittance data for this sample were given in Fig. 2. It is seen that T_{lum} goes from 80% to 20% when Q_{ex} increases from 0 to 30 mC/cm^2. The corresponding variation of T_{sol} is from 74% to 29% .

The transmittance in the coloured state seems to be dependent on the technique for fabricating the electrochromic NiO_xH_y-based layers. The present coatings, made by dc magnetron sputtering of Ni in O_2, did not yield as low magnitudes of T_{lum} and T_{sol} as found in earlier studied coatings made by rf magnetron sputtering of Ni in O_2+H_2 [11]. The difference is clearly seen by comparing the solid and dashed curves in Fig. 3.

Nevertheless, the optical modulation is substantial for both types of layers, which clearly indicates that electrochromic NiO_xH_y-based coatings are potentially useful for regulating the throughput of visible light and solar energy in smart windows.

3.3 Infrared Reflectance Spectroscopy

Structural data for coloured and bleached nickel-oxide-based thin films can be inferred from infrared reflectance spectra, and hence such data are useful for determining the physical mechanisms underlying the electrochromism.

Figure 4 shows IR-reflectance in the 4000 to 2000 cm^{-1} range for a 0.15-μm-thick film in fully bleached state. After the first potentiodynamic half-cycle, a broad absorption is located at ~ 3500 cm^{-1}. This feature is typical for OH stretching vibrations when hydrogen bonding is present [12]. Cycling makes the reflectance increase somewhat and, more important, leads to the evolution of a sharp absorption band at 3620 cm^{-1}. This latter absorption is characteristic [12] of stretching vibrations in "free" OH, i.e. in the absence of hydrogen bonding. This result gives clear evidence that the electrochemically bleached state is associated with the occurrence of "free" OH.

Figure 5 gives further information on the relation between electrochromism and hydrogen bonding by showing reflectance in the bleached, intermediate and coloured states for the same sample, and in the same spectral range, as in Fig. 4. Following the arguments given above, electrochromic colouration is associated with formation of hydrogen bonds. The data are consistent with electrochromism according to the reaction:

$$Ni(OH)_2 \underset{\text{bleach}}{\overset{\text{colour}}{\rightleftharpoons}} Ni\,OOH + H^+ + e^- .$$

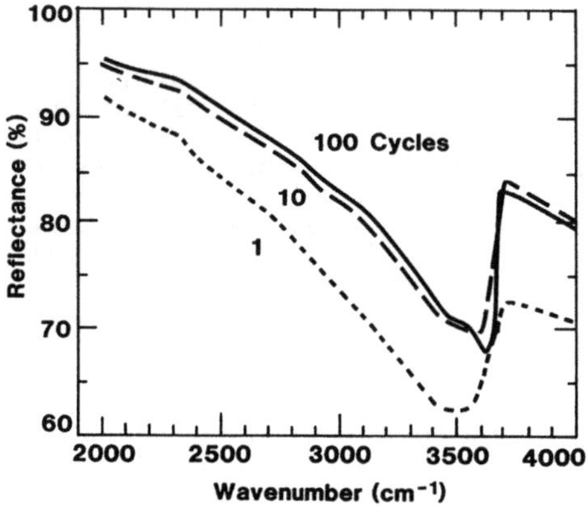

Fig. 4 Spectral infrared reflectance measured at 60° angle of incidence and p-polarized light for a 0.15-μm-thick bleached electrochromic NiO_xH_y film on ITO-coated glass. The curves curves refer to different numbers of potentiodynamic cycles.

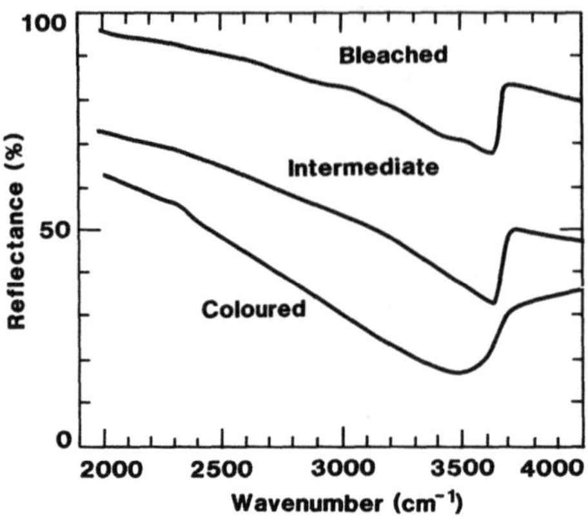

Fig. 5 Spectral infrared reflectance measured at 60° angle of incidence and p-polarized light for a 0.15-µm-thick electrochromic NiO_xH_y film on ITO-coated glass. The curves refer to different levels of colouration.

The far-infrared absorption was studied further for the 0.15-µm-thick electrochromic NiO_xH_y film in fully bleached and coloured states. Absorptance feature was noticed at [6] ~ 590 cm^{-1}; it was weaker than those around ~ 3500 cm^{-1}. This band can be understood as a consequence of LO-phonons of Ni-O vibrations [13].

Finally it should be noted that water does not seem to be involved in the electrochromism, since neither its typical bending vibration at ~1600 cm^{-1} nor its torsional modes at ~530 cm^{-1} were found in the infrared spectra [12-14].

4. Conclusions

We prepared non-stoichiometric nickel oxide films by reactive dc magnetron sputtering onto ITO-coated glass. These films were subsequently transformed into electrochromic NiO_xH_y by electrochemical treatment in KOH solution. As-deposited films were absorbing, presumably due to the presence of an oxygen-rich phase dispersed in NiO. The grain size was ~10 nm, and the crystal structure was cubic (bunsenite). Electrochemical cycling gave reversible insertion and extraction of ions. Combined measurements of mechanical stress, monochromatic transmittance, and electrochemical parameters gave clear evidence for electrochromic colouring being associated with ion extraction, and a separate experiment verified that the mobile species were protons.

Spectral infrared p-polarized reflectance measurements showed unambiguous absorption due to OH stretching vibrations, and it was found that the bleached state was associated with "free" OH and the coloured state with OH in the presence of hydrogen bonds. From spectrophotometric data it was deduced that luminous and solar transmittance could be varied between 80% and 20% and between 74% and 29%, respectively. The results in this study provide a number of elements in a theory of the electrochromism of nickel-oxide-based materials, but a detailed theory must await further work.

Acknowledgement. Two of us (A.M.A.and W.E.) wish to thank the International Science Program, Uppsala University, Sweden, for scholarships and travel grants. A.G. acknowledges the

International Center for Theoretical Physics (ICTP, Trieste, Italy) and FAPESD (S.P., Brazil) for Visiting Scientist Fellowships. Work at Chalmers was financially supported by grants from Swedish Natural Science Research Council, the National Swedish Board for Technical Development, and the Swedish Board for Building Research. Work in Campinas was financially supported by FINEP and FAPESP.

References

1. C.M. Lampert and C.G. Granqvist, editors, Large-area Chromogenics: Materials and Devices for Transmittance Control (SPIE Opt. Engr. Press, Bellingham, USA, 1990)

2. M.K. Carpenter and D.A. Corrigan, editors, Proc. Symp. Electrochromic Mater.(The Electrochem. Soc., Pennington, USA, 1990), Vol. 90-2.

3. T.S. Eriksson and C.G. Granqvist, J. Appl. Phys. 60, 2081 (1986)

4. W. Estrada, A.M. Andersson and C.G. Granqvist, J. Appl. Phys. 64, 3678(1988)

5. S.J. Jiang and C.G. Granqvist, Proc. SPIE 562, 129 (1985)

6. W. Estrada, Ph.D. Thesis, Facultad de Ciencias, Universidad Nacional de Ingenieria, Lima, Peru, 1990

7. J. Scarminio, S.N. Sahu and F. Decker, J. Phys. E 22, 755(1989)

8. Z.M. Jarzebski, "Oxide Semiconductors", Pergamon Press, Oxford Vol. 4 (1973)

9. G. Wyszecki and W.S Stiles, Color Science, 2nd edition (Wiley, New York, 1982), p.256

10. M.P. Thekaekara, in Solar Energy Engineering, edited by A.M. Sayigh (Academic, New York, 1977)

11. J.S.E.M. Svensson and C.G. Granqvist, Appl. Phys. Lett. 49, 1566(1986)

12 F.P. Kober, J. Electrochem. Soc.112, 1064(1965); 114, 215(1967)

13. S. Mochizuki, Phys. Stat. Sol. B 126, 105(1984)

14. P.J. Lucchesi and W.A. Glasson, J. Am. Chem. Soc. 78, 1347(1956)

International Center for Theoretical Physics (ICTP, Trieste, [illegible]

Photoluminescence Characterization of the Crystalline Quality in Intrinsic GaAs Epitaxial Layers

G. Torres-Delgado, J.G. Mendoza-Alvarez, and B.E. Zendejas

Departamento de Física, Centro de Investigación y de Estudios Avanzados del IPN, Apdo. Postal 14-740, 07000 México, D.F., Mexico

Abstract. The influence of melt supercooling during the growth of GaAs intrinsic epitaxial layers by Liquid Phase Epitaxy on the crystalline properties of these layers has been studied using low temperature Photoluminescence spectroscopy. From the layer surface morphology, as is well know, a supercooling of ~3°C gives the best layers. From our measurements we show that the best PL spectra are obtained for layers grown under near-equilibrium conditions, even when these layers show a terrace-like aspect. In this way a compromise between the surface quality required and the bulk crystalline quality should be established in order to choose the adequate growth parameters. We also show that, probably due to an increase in As vacancies as supercooling increases, the density of antisite defects Ga_{As} in the bulk of the layer increases as the supercooling parameter Δ increases.

1. Introduction

The actual development in the field of optoelectronic devices has been based on the optimization of III-V semiconductor epitaxial layer growth. In particular, the GaAs/AlGaAs system continues to be fundamental for near-infrared emission devices. One crucial step in the development of GaAs/AlGaAs multilayer structure is the achievement of a good homoepitaxial growth to obtain layers with defect densities lower than those available in commercial GaAs substrates. In this respect, the study of defects induced by the growth conditions and/or by the substrate used, has a particular importance in terms of the goal of getting the best possible GaAs epitaxial layers. Usually the surface morphology of the layer is used as a measure of the growth quality and, for example, a "cross hatch" has been attributed to the presence of misfit dislocations at the interface, or even that its origin could be in substrate defects [1]. Optical characterization techniques, like Photoluminescence (PL) and Raman spectroscopies have been used to study the quality of, for example, very pure GaAs layers grown by MBE [2], and MOCVD [3]; or the influence of shallow acceptor defects on the PL bound-exciton line [4].

Springer Proceedings in Physics, Volume 62
Surface Science Eds.: F.A. Ponce and M. Cardona

In this work we present results on the use of low temperature PL spectroscopy to study the crystalline quality of intrinsic GaAs layers grown by the Liquid Phase Epitaxy technique under different initial growth temperatures that correspond to a departure, at different levels, of the equilibrium conditions. We show that from the optical point of view the best layers are those grown at very near-equilibrium conditions even when the surface morphology presents terrace formation, which do not appear when the growth is done with supercooling conditions. In this way we show that a compromise should be established between surface morphology and PL properties to get the best intrinsic GaAs layers. From the PL results we also conclude that supercooling favors the formation of antisite defects Ga_{As}, probably as the result of an increase in the formation of As vacancies.

2. Experimental Details

Intrinsic GaAs epitaxial layers were grown using the Liquid Phase Epitaxy (LPE) technique using a conventional horizontal system. As substrates, (100) GaAs single crystals doped with Si ($n \sim 2 \times 10^{18}$ cm^{-3}) from Laser Diode Inc. were used. All the epitaxial layers were grown after a 4-hour baking time for the melt in a high purity Pd-purified H_2 atmosphere. All the samples were grown under the same conditions, except for the temperature at which the cooling was initiated, T_i, and the temperature at which growth started, T_{ig}, as shown in Table 1. In this Table we also show the layer thickness obtained for each GaAs layer. The cooling rate, R, was kept constant at approximately 0.24 ± 0.02 °C/min, and the growth time was always 10 minutes; except for sample CEI110 which took 25 min. to grow. In total, we analyzed 11 different samples grown in the temperature range T_{ig}: 788.7 - 804.0 °C. It should be observed from Table 1 that the temperature difference between T_i and T_{ig} was always around 3°C. The weights used to prepare the growth solutions were calculated from the solubility curve for the binary system Ga-As [5], for a saturation temperature of around 800°C. As shown in Table 1, the weights used in this work were: W_{Ga}= 2.00001±0.00004 gr, and W_{GaAs}= 98.58±0.04 mg. Based on the expressions for the layer thickness resulting from the solution of the one-dimensional diffusion equation for the cases of equilibrium and super-cooling LPE variations, we have obtained that the real saturation temperature is around 803.5°C.

Surface morphology was studied through optical microscopy. Low temperature photoluminescence (PL) spectra were measured with a conventional system. As the exciting source, the 4880 Å Argon laser line was shined on the sample contained in a closed-cycle helium cryostat capable of controlling temperatures in the range 10-300°K. The typical laser power densities used for recording the spectra were of the order of 1 watt/cm^2. The lumines-

Table 1. Growth parameters for the intrinsic GaAs epitaxial layers grown in the LPE system. W_{Ga} and W_{GaAs} are the weights which were used to prepare the melt. R is the cooling rate. In the first column the label identification for each sample is presented. The second column is the temperature at which the cooling ramp was switched on, T_i. In the third column it is shown the temperature at which substrate and melt are bring in contact. The measured thickness is presented in the fourth column.

$W_{Ga} = 2.00001 \pm 0.00004$ gr

$W_{GaAs} = 92.58 \pm 0.04$ mg

$R = 0.24 \pm 0.02$ °C/MIN

CEI	T_i(°C)	T_{ig}(°C)	d(±0.15μm)
100	791.7	788.7	1.6
99	793.6	790.5	2.3
98	795.0	791.9	4.1
97	797.7	794.7	6.9
103	800.2	797.0	5.8
104	802.1	799.1	4.4
105	803.7	800.7	3.4
111	805.5	802.0	2.5
106	806.0	803.0	2.0
110	806.5	803.0	5.1
108	807.0	804.0	1.5

cence emitted by the sample was focused on the entrance slit of a Jobin-Ivon double monochromator model HRD-1, and at the exit slit signal was detected by a Hamamatsu R636 GaAs:Cs photomultiplier tube. Phase sensitive techniques were used for signal detection which were implemented using a SRS Lock-in amplifier, model SR530, interfaced with a PC computer for data processing.

3. Results and Discussion

We have already mentioned that the saturation temperature for our experimental growth parameters is around 803.5°C; this implies that for the first eight samples listed in Table 1, the growth takes place under the supercooling situation, with the supercooling parameter, Δ, changing from ~1.5°C for sample CEI111 up to ~8.8°C for sample CEI97. Also, since the temperature at which the cooling starts is approximately 3°C higher than T_{ig}, it should be pointed out that for samples CEI: 100, 99, 98, 97, 103, and 104, the melt is already supercooled when the cooling ramp is switched on.

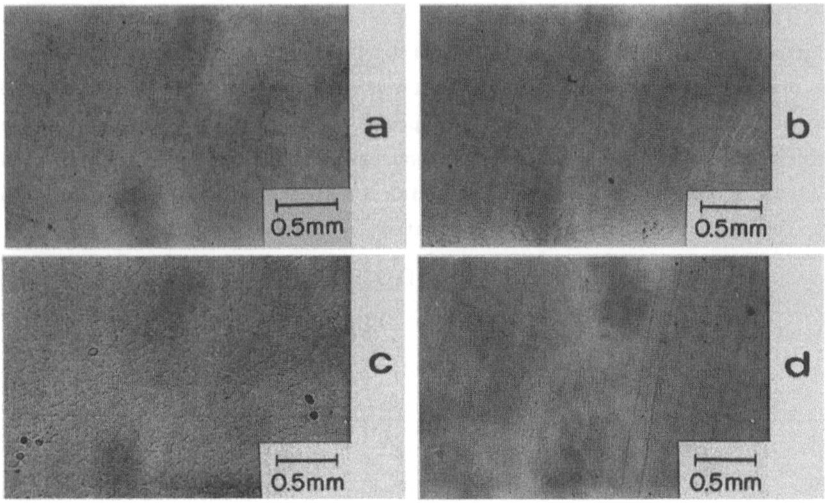

Fig. 1. Optical microscope photographs taken on samples: a) CEI106, showing terraces; b) CEI110 and c) CEI108, with small terraces; and d) CEI105 showing a smooth surface.

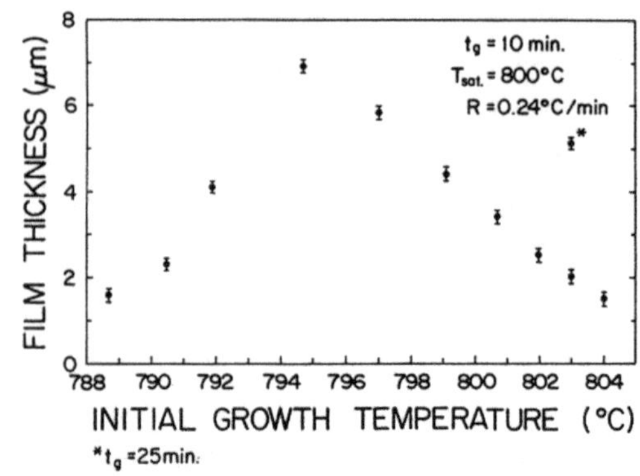

Fig. 2. Thickness of the GaAs layer as a function of the initial growth temperature. Note the linear relationship between d and Δ for the temperature range: 796-804 °C.

Surface morphology for the set of layers, as observed under the optical microscope, is shown in figure 1a) to 1d). Fig. 1a corresponds to sample CEI106, grown at near-equilibrium conditions, and we can observe the presence of terraces which are undesirable for optoelectronic device purposes. For samples CEI110 and CEI108, shown in figs. 1b and 1c respectively, the terraces appear smaller and closer to each other. Fig 1d corresponds to sample CEI105 for which the supercooling parameter was around 3°C and shows an uniform surface aspect without terraces, as expected from what we already know about the effect of supercooling on the surface morphology [6].

In figure 2 the layer thicknesses, with corresponding error bars for the different GaAs samples, are shown in terms of the initial growth

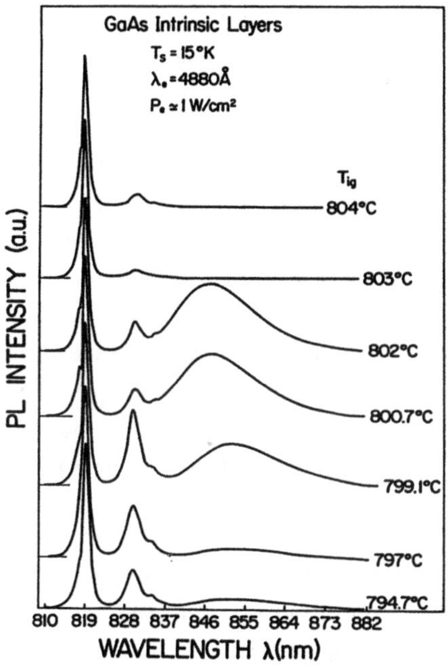

Fig. 3. Photoluminescence spectra for samples grown at different T_{ig}'s. Note the abrupt appearance of the low energy broad band for samples grown at $T_{ig} \leq 802$ °C.

temperature T_{ig}. Saturation temperature corresponds to a point located between layers CEI106 and CEI108. Sample CEI110 corresponds to a layer with a growth time of 25 min. As can be observed in this figure, the thickness increases as T_{ig} decreases; this is because for supercooling growth, the total thickness is given in terms of the sum of the equilibrium and step-cooling contributions [7]:

$$d = \frac{1}{m} \sqrt[2]{\frac{D}{\pi}} \left[4 \, \Delta T \, t^{1/2} + \frac{8 \, R \, t^{3/2}}{3} \right]$$

where D is the diffusion coefficient, t is the growth time, m is the slope of the liquidus curve in the Ga-As phase diagram at the saturation temperature, and R is the cooling rate. From the above equation we see that d increases linearly with Δ. This situation is in good agreement with the experimental results of fig.2 for samples CEI106 down to CEI103. The thickness for sample CEI97 is slightly smaller than expected, probably because some precipitation started in the melt previous to its contact with the GaAs substrate. For samples CEI: 98, 99 and 100, the thickness drops because the supercooling is so high that a severe precipitation occurs and therefore we have a two-phase mechanism of growth from solution.

The low temperature PL spectra measured at 15 °K for the samples CEI: 108, 106, 111, 105, 104, 103 and 97 are shown in figure 3 normalized to the exciton peak. It is observed that there are three well-defined regions for all the spectra. These spectra are dominated by the excitonic emission at higher energies with the peak around 8190 Å; depending on the sample, this band presents some structure related to transitions involving excitons bounded to different types of residual impurities. From its energy position, the excitonic peak corresponds to an acceptor-bound -exciton, (X,A), where the acceptor is probably carbon at As sites, since carbon is the most probable residual impurity. For all the spectra shown in fig.3 the FWHM is smaller than 2 meV, an evidence of the very good crystalline quality of our samples [8]. In the intermediate energy range we can observe a band centered at around 8300 Å which corresponds to a conduction band-to-acceptor transition [9], the acceptor being carbon in As sites: C_{As}, (e,C_{As}). For T_{ig}'s below the saturation temperature, the intensity of this band, compared to the exciton band intensity, increases as T_{ig} decreases down to the sample CEI104, and then decreases for samples CEI: 103 and 97. The shoulder present in some spectra at the low energy-side of the carbon-related band corresponds to the transition: (e,Si_{As}) [9]. At the lowest energies, for wavelengths above 8370 Å, a broad band is observed in the spectra corresponding to samples grown with supercooling parameters greater than 1°C. The maximum of this broad band seems to shift to lower energies as T_{ig} decreases. It should be noted that for those samples grown near equilibrium conditions, the broad low-energy band is absent, and the spectrum is strongly dominated by the excitonic recombination band.

In order to analyze more carefully the PL spectra for the low energy region, a magnification of this region for samples CEI: 111, 105, 104, 103 and 97 is shown in figure 4. Here we see that the broad band in samples 111, 105 and 104 is really a superposition of two bands, and that the lower-energy band increases in intensity and dominates for samples CEI: 103 and 97. The band at the lower energy has its peak at ~8560 Å (1.448 eV). In agreement with van de Vent et al [10], we propose that this band comes from donor-to-acceptor recombination transitions involving an acceptor level due to antisite defects: Ga_{As}, the origin of this defect related to the V_{As} defect. Therefore, we observe that as the supercooling increases, the antisite defect presence increases, probably because of an increase in arsenic vacancies when the melt is supercooled. The band at the high energy side of the broad-band emission has its maximum located at around 8475 Å (1.463 eV), and its intensity decreases as supercooling increases, as observed in fig.3. The possibility that this broad band could come from substrate emission is discarded because in such a case the intensity of the emission would increase as the thickness decreases; however, the PL spectra for samples CEI106 and CEI108 do not present such an emission band even when the samples are thinner than sample CEI111 (see fig.3). At present we

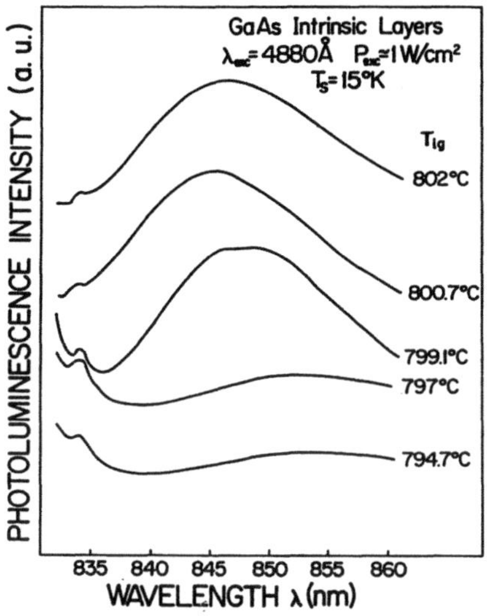

Fig. 4. Magnification of the low energy side of the PL spectra shown in Fig.3 to remark the details in the broad PL band related to emission involving defects. Two broad bands are present with maxima at ~8475 Å, and ~8560 Å.

have not been able to identify the corresponding transition and other measurements are in progress for its identification. The donor-to-acceptor nature of the broad-band transition has been confirmed from its dependence with the laser intensity: it shifts to lower energies as the laser intensity decreases, at a rate of ~5 meV per intensity decade.

Since for the sample CEI111 the broad-energy band is abruptly shown, as seen in fig.3, and the relative intensity of this band, compared to the excitonic emission, decreases as the thickness increases for the other samples, one could be tempted to explain such a broad-band as due to some kind of interface defect whose concentration decreases as layer thickness increases. To explore this possibility we grew a GaAs layer under near-equilibrium conditions, similar to those used for sample CEI106 but increasing the growth time up to 25 min. to get a thicker layer; this corresponds to sample CEI110 with a d~5 μm. PL spectra for these two samples are shown in figure 5, where we observe that there is a slight modification in the excitonic peak in terms of its structure and that the emission bands from the carbon and silicon acceptors increase in intensity for the thicker layer, a result expected because a higher growth time favors impurity incorporation. But the most relevant result is that the broad-band at the lower energies is absent also in the thicker sample; this allow us to conclude that the defect-related band does not depend on the thickness or growth time of the layer but on the initial growth conditions, mainly in relation with the departure from the equilibrium conditions.

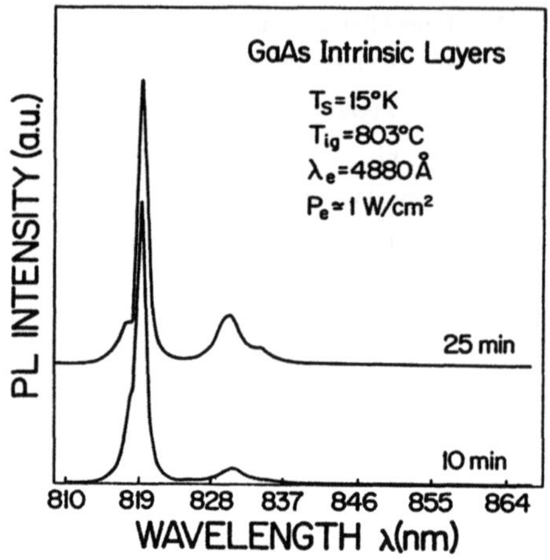

Fig. 5. Photoluminescence spectra for samples CEI106 and CEI110, both grown at near-equilibrium conditions, but with growth times of 10 and 25 min, respectively.

4. Conclusions

In summary, in this work we have shown that even when the supercooling growth produces GaAs epitaxial layers with the best surface morphology, as was shown in fig. 1 and has been long recognized [5], the photoluminescence properties of such layers are not the best ones. In fact they show an important broad recombination band at lower energies which appears abruptly as soon as the growth conditions depart from the ideal near-equilibrium ones. The origin of this band is related to the presence of bulk defects in the layer. This in fact means that one should establish a compromise between the bulk crystalline quality and the surface morphology of the layer required for a specific application. From the PL point of view, the best GaAs layers are those grown at near-equilibrium conditions, showing PL spectra dominated by an intense excitonic-recombination peak.

From our PL results we have also found that, probably as a result of an increase in the As vacancies density when the supercooling parameter Δ increases, the density of antisite defects of the type Ga_{As} (gallium at arsenic sites) also increases with Δ as seen in the corresponding PL spectra. These show an increase in the intensity of the broad band centered at ~1.448 eV when the samples are grown with higher values of Δ. Photoreflectance spectroscopy measurements are under way to get more information about the nature of this defect-related low energy PL broad band. These results will be published elsewhere.

5. References

[1] J.E. Cunningham, T.H.Chiu, A.Ourmazd, J.Shah, and W.T.Tsang, J. Appl. Phys. **60**, 4165 (1986)

[2] H. Temkin and J.C.M. Hwang, Appl. Phys. Lett. **42**, 178 (1983)

[3] M. Razeghi, F.Omnes, J. Nagle, M. Defour, O. Acher, and P. Bove, Appl. Phys. Lett. **55**, 1677 (1989)

[4] I. Szafranek, M.A. Plano, M.J. McCollum, S.A. Stockman, S.L. Jackson, K.Y. Cheng, and G.E. Stillman, J. Appl. Phys. **68**, 741 (1990)

[5] M.B. Panish, J. Cryst. Growth **27**, 6 (1974)

[6] Nobuyuki Toyoda, Minoru Mihara, and Tohru Hara, Appl. Phys. Lett. **27**, 625 (1975)

[7] J.J. Hsieh, J. Cryst. Growth **27**, 49 (1974)

[8] T. Bryskiewicz, M. Bugajski, j. Lagowski and H.C. Gatos, J. Cryst. Growth **85**, 136 (1987)

[9] D.W. Kisker, H. Tews, and W. Rehm, J. Appl. Phys. **54**, 1332 (1983)

[10] J. van de Ven, W.J.A.M. Hartman, and L.J. Gilling, J. Appl. Phys. **60**, 3735 (1986)

Optical Properties of Thin Films of Polymerized Acetylene Deposited by dc and rf Glow Discharge

J.H. Dias da Silva[1], *M.P. Cantão*[2], *J.I. Cisneros*[2], *C.S. Lambert*[2], *M.A. Bica de Moraes*[2], *and R.P. Mota*[3]

[1]Departamento de Física, UNESP, 17033 Bauru, SP, Brazil
[2]IFGW UNICAMP, 13081 Campinas, SP, Brazil
[3]DFQ, FEG, UNESP, 12500 Guaratinguetá, SP, Brazil

1 Introduction

By glow discharge polymerization [1], organic thin films having promising important technologycal applications [2,3,4,5,6] can be obtained. These materials can be obtained in a range of physical and chemical properties, depending on the starting monomer and on the deposition parameters.

In this work we report a study of the optical properties of polymerized acetylene films obtained by radiofrequency (RF) and direct current (DC) discharges.

The films were obtained by a process described elsewhere [7] and were deposited on quartz, glass, single-crystal silicon and KCl substrates.

Infrared, visible and ultraviolet transmission analyses were performed by a JASCO IR 700 (5000 - 400 cm^{-1})and a PERKIN ELMER LAMBDA 9 (54000 - 3125 cm^{-1})spectrometers.

The hydrogen content (C–H bond density) of the films has been determined by the integrated absorption of the C–H stretching bands at 2950 cm^{-1}, as the method described by Couderc and Catherine [8].

2 Results and Discussion

Table 1 summarizes the data for the DC and RF glow discharge polymerized acetylene films.

Various a–C:H film studies show a dependence of the absorption coefficient, α, with photon energy, E, described by the Tauc expression [9], $\sqrt{\alpha E} = B(E - Eg)$ where B is the density of states and Eg is the optical gap.

The Eg values are obtained by extrapolating the dashed lines (fig.1) plotted in function of the cathode voltage for the DC polymerized films. Figure 2 shows a schematic diagram of the optical gap shift for both DC and RF polymerized films.

There is an optical gap widening that may be due to the hydrogen amount in the material. However, the gap shift for the RF polymerized samples were lesser than the DC ones.

Table 1 - Summary of results

DC DISCHARGE

Cathode Voltage (V)	Pressure C2H2 (mtorr)	Pressure Argon (mtorr)	Optical Gap (eV)	Integrated Absorption (cm-1)	Thickness (μm)
700	8	60	2.28	67	0.420
1000	8	60	2.08	64	0.470
1300	8	60	1.90	40	0.785
1600	8	60	1.58	29	0.838
1900	8	60	1.44	21	0.700
2100	8	60	1.25	18	0.811
2500	8	60	1.05	11	0.451
3100	8	60	0.40	12	0.489

RF DISCHARGE

Power (W)	Pressure C2H2 (mtorr)	Optical Gap (eV)	Integrated Absorption (cm-1)	Thickness (μm)
8	240	2.15	49	1.512
13	240	2.07	51	1.757
19	240	2.05	66	1.376
13	130	2.03	47	0.593
13	370	2.06	64	3.658
13	1050	2.80	53	1.692

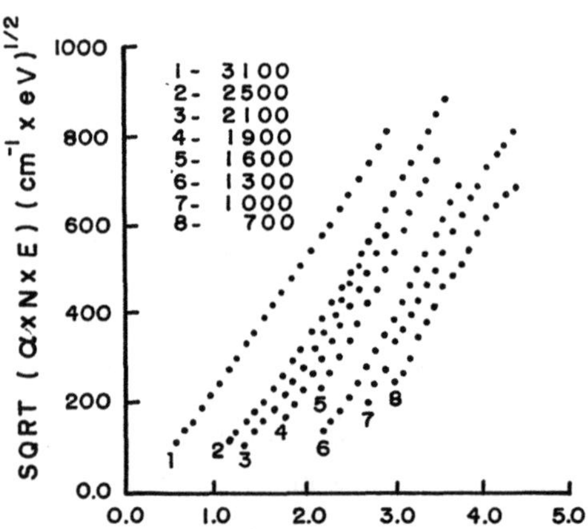

Fig. 1 - Optical gap - C_2H_2 + Ar, DC glow discharge polymerization.

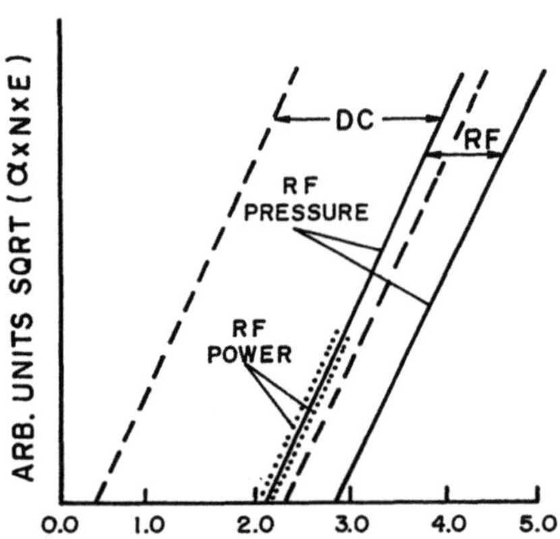

Fig. 2 Schematic diagram - Optical gap - DC and RF glow discharge polymerization.

3 Conclusions

For the films synthetized in DC glow discharge, we have found that the energy gap decreases with increasing cathode voltage and varies from 0.4 to 2.28 eV for cathode voltage ranging from 3100 to 700 V. Graphytic carbon is present when depositions are made at high cathode voltages and can be attributed to sputtering of the carbon-covered eletrode [10].

In both DC and RF glow discharges, the optical gap increased with decreasing hydrogen content as measured by the C–H stretch integrated absorption.

[1] H. Yasuda (1981). J. Polym. Sci.: Macromol. Rev. 16, 199

[2] M. Hori, H. Yamada, T. Yoneda, S. Morita and S. Hattori (1987) J. Electrochem. Soc., 134, 707

[3] P. K. Tien, G. Smolinsky and R. J. Martin (1972) Appl. Opt. 11, 637

[4] Y. Osada and M. Takase (1985) J. Polym. Sci.: Polym. Chem. Ed. 23, 2425

[5] P. Schreiber, M. R. Wertheimer and A. M. Wrobel (1980) Thin Solid Films, 72, 487

[6] N. Inagaky, K. Suzuki and K. Nejigaki (1983) J. Polym. Sci.: Polym. Let. Ed. 21, 353

[7] M. P. Cantão, C. S. Lambert, J. H. Dias da Silva, R. P. Mota, J. I. Cisneros and M. A. Bica de Moraes (1990) Proceedings of XIII Encontro Nacional de Física da Matéria Condensada. Caxambu, Brazil

[8] P. Couderc and V. Catherine (1987) Thin Solid Films, 146, 93

[9] J. Tauc, R. Grigorovici and A. Vancu (1966) Phys. Stat. Sol. 15, 627

[10] R. Parissari and M. A. Bica de Moraes Rev. Fis. Apl. Instr. (to be published)

3 Conclusions

Incoherent Light Assisted $CuInSe_2$ Thin Film Processing

H. Galindo[1], *J.M. Martin*[1], *A.B. Vincent*[1], *and L.D. Laude*[2]

[1]Universidad de Los Andes, Facultad de Ciencias, Mérida, Edo. Mérida, 5101, Venezuela

[2]Laboratoire de Physique de l'Etat Solide, Université de l'Etat, 23 Av. Maistriau, B-7000 Mons, Belgium

Abstract. $CuInSe_2$ thin films were prepared by stoichiometric evaporation of each of its constituent elements on a glass substrate, and subsequent processing by a non-coherent light source. Several films were produced, changing one parameter at a time in the process. The optical and structural properties of the obtained films were then analyzed, using optical absorption and Raman scattering spectroscopies. The results revealed that this Incoherent Light Assisted technique for semiconducting thin films is a promising and inexpensive alternative.

1. Introduction

The ternary compound $CuInSe_2$ has attracted much attention in the last decade for its growing potential as a photovoltaic cell material [1]. The preparation of this compound in large areas, with optimum optical and electrical characteristics, is still an open issue. Techniques like flash evaporation, sputtering, MBE, spray pyrolysis and co-evaporation have been largely used for the preparation of $CuInSe_2$ thin films [2].

Recently, Laser Induced Synthesis (LIS) has been successfully used for the formation of ternary compounds. This procedure is based on the interaction of a laser beam with a film of the evaporated elements disposed on sandwich layers [3]. In this way, good quality semiconductor films have been achieved with similar properties as those obtained with more elaborated techniques.

In this work, we present an experimental study of an alternative process to LIS, using incoherent light instead, to synthesize the compound.

2. Experimental

Films with 250 nm thickness were formed by evaporation (pressure at 10^{-6} torr) of layers of copper, indium and selenium in a stoichiometric proportion of 1:1:2, on a glass substrate. Their thickness was controlled by a quartz crystal monitor with a 7% precision. The obtained films were subsequently processed on a system which simulates the effect of a commercial rapid thermal processor (RTP) with controlled temperature.

After this, each sample was characterized by optical absorption and Raman spectroscopies, in the usual way, as reported in previous works [4-7].

Springer Proceedings in Physics, Volume 62
Surface Science Eds.: F.A. Ponce and M. Cardona

3. Results

Figure 1 shows the optical absorption spectra of three of the morphologically best samples, which were produced at 200°C, 250°C and 300°C, exposed during 60 seconds.

For the sample treated at 250°C, the spectrum clearly indicates the formation of an optical absorption edge located at about 1.06 eV, which does not appear in the other two samples. These data suggest that the synthesis of the ternary compound occurs at just under 250°C with, however, the presence of small amounts of In-Se binary compounds. The possible formation of InSe or In_2Se_3 gives place to untreated metallic copper in the obtained film, creating localized states which increase optical absorption in the infrared region.

Figure 2 shows the corresponding Raman spectra. These have been used only to identify the presence of the ternary compound and/or its binary phases. The γ-In_2Se_3 single crystal vibrational spectrum [8,9] presents, in the long wavenumber region, bands at 178, 203, 220 and 225 cm^{-1}, associated to the transverse optical phonons dominated by the In-Se bonds, and in the low energy side, bands at 54, 74, 85 and 97 cm^{-1}. On the other hand, InSe is usually characterized by its Raman peaks at 115, 178 and 199 cm^{-1} [10], and In_4Se_3 is clearly identified by the structures at 104, 131 and 150 cm^{-1}. This confirms the existence of the above referred phases. Some characteristic peaks of amorphous selenium (113 cm^{-1}) and copper oxide (105 and 145 cm^{-1}) are also observed.

In the spectrum corresponding to the sample treated at 250°C, some extra peaks are observed at 77, 170, 179, 197 and 215 cm^{-1} which we assign to the E^5, B^1, B^2_2, A^1_2, B^1_2 and E^1 modes of $CuInSe_2$ reported by Newmann [11] and Gan et al. [12]. These results are summarized in table 1 which indicates that the sample treated at 250°C effectively forms the $CuInSe_2$ compound, in presence of several contaminants.

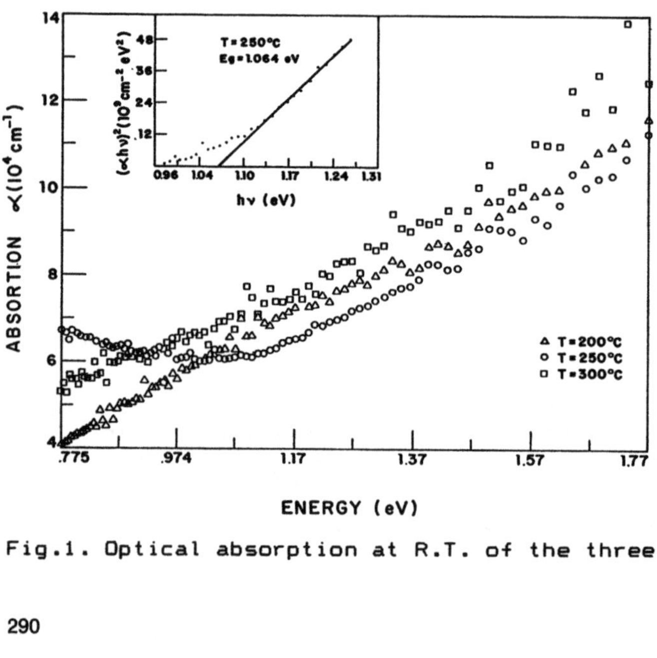

Fig.1. Optical absorption at R.T. of the three specimens.

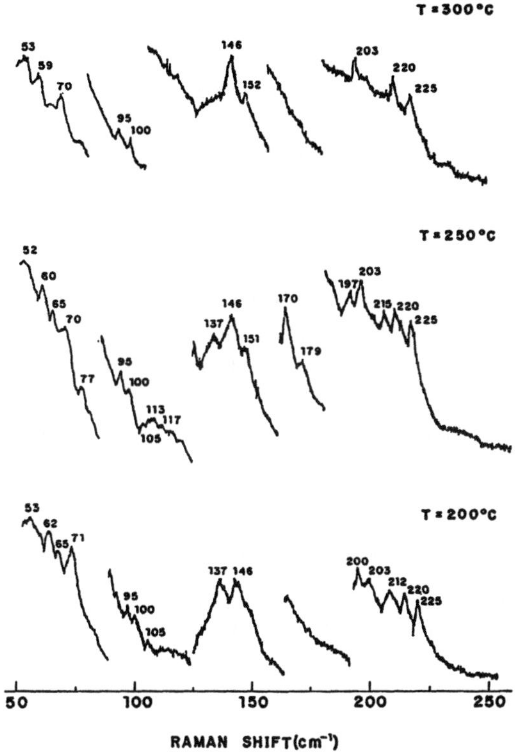

Fig. 2. Raman scattering of the three specimens.

Table 1. Summary of Raman peaks for the sample treated at 250°C.

Observed Modes (cm^{-1})	Comparation with the reported values of the vibrational modes of $CuInSe_2$ and related binary phases.					
	$CuInSe_2$ Ref. 11.12 (cm^{-1})	InSe Ref.10 (cm^{-1})	γ-In_2Se_3 Ref.8.9 (cm^{-1})	In_4Se_3 Ref.10 (cm^{-1})	a-Se Ref.7 (cm^{-1})	Cu_2O Ref.7 (cm^{-1})
52			54			
60	61(E^6)					
65	64(B_2^3)					
70			74			
77	78 (E^5)					
95			97			
100				104		
107						105
113					113	
117		115				
137				131		
146						145
151				150		
170	165(B_1^2)					
179	178(B_2^2)	178	178			
197	196(A_2^1)					
203		199	203			
215	213/214(E^1+B_2^1)					
220			220			
225		227	224			

4. Discussion

The sequential evaporation of thin layers of In, Cu and Se induces the spontaneous formation of In-Cu alloys and binary compounds of $Cu_{2-x}Se$. Furthermore, the thermal treatment reaches its maximum temperature in about 10 seconds; during this time, interdifusion of the elemental constituents is induced, producing primary reactions in which the binary phases are formed; this situation reduces the possibility of ternary formation [13]. When Se, with a melting point of 217°C, diffuses through the Cu layer, it forms some $Cu_{2-x}Se$ compounds which are not stable and/or with very short life times, resulting that a large amount of Se gets through the Cu layer without reacting, reaching the In layer with no problem. The In as well, with a melting point of 156.6°C, diffuses through the Cu layer, reaching the Se atoms with which they may react.

The thermal treatment at 250°C, however, permits diffusion of the atoms of the elemental constituents resulting, via a solid-solid reaction, in the ternary compound formation, along with some of the binary phases mentioned above.

The band gap of this film treated at 250°C was calculated from the straight line of the $(\alpha h\nu)^2$ v.s. $h\nu$ plot (inset fig. 1). This gave an optical gap of 1.064 ± 0.006 eV, very close to the single crystal value of 1.04 eV [5]. Rather high sub-band gap absorption coefficients were observed for films presenting second phase sections at the film surface. Similar results were previously reported on chemically deposited films [13].

5. Conclusions

Layers of In-Cu-Se in 1:1:2 atomic proportion, thermally evaporated onto a glass substrate, were processed using an incoherent light source. The optical absorption and Raman scattering data revealed that the compound synthesis occurs with the treatment at 250°C. Small amounts of secondary phases are, however, present. Comparing this technique with laser induced synthesis (LIS), we conclude that increasing the temperature rate at the beginning of the process, we should be able to produce $CuInSe_2$ thin films on large areas, with a single phase. The ability to obtain $CuInSe_2$ with this technique and the possibility to improve it, leads us to believe that solar cell quality material should be achieved.

Acknowledgments

We are thankful to CDCHT-ULA and Conicit-Venezuela for financing research projects under which this work was carried out. We also thank the Research Agreement between Belgium and Venezuela for financial assistance. This work was also supported by the Ministerio de Energía y Minas.

References

1 .- L.L. Kazmerski. Intern. Mat. Rev., 34, pp. 185-210 (1989).
2 .- See the special issue of $CuInSe_2$ in Solar Cells, 16, pp. 131-207 (1986).

3 .- L.D. Laude. Prog. in Cryst. Growth and Characterization, 10, 141 (1984).
4 .- D. Barreto, J. Luengo, Y. De Vita and N.V. Joshi, Applied Optics, 26, 5280 (1987).
5 .- N.V. Joshi in Photoconductivity, Art, Science and Technology, Marcel Dekker, New York, 1990.
6 .- H. Galindo, F. Hanus, M.C. Joliet, A.B. Vincent and L.D. Laude in Thin Films and Small Particles, ed. M. Cardona, World Scientific, 207 (1988).
7 .- H. Galindo, M.C. Joliet, F. Hanus, L.D. Laude and N.V. Joshi. Thin Solid Films, 170, 227 (1989).
8 .- K. Kambas, C. Julien, M. Jouanne, A. Likforman and M. Guittard. Phys. Stat. Sol. (b), 124, K105 (1984).
9 .- K. Kambas and C. Julien. Mater. Res. Bull., 12, 1573 (1982).
10.- J.Y. Emery, L. Brahim-otsmane, M. Jouanne, C. Julien and M. Balkanski. Mat. Sci. and Eng., B3, 13 (1989).
11.- H. Newmann. Solar Cells, 16, 399 (1986).
12.- J.N. Gan, J. Tauc, V.G. Lambrecht and M. Robbins. Phys. Rev. B, 13, 3610 (1976).
13.- J. Szot and V. Prinz. J. Appl. Phys., 66, 6077 (1989).

A Study of the Dispersive Behavior of an Anisotropic Gold Film on Mica

J.M. Siqueiros[1], *R. Machorro*[1], *J. Valenzuela*[1], *L. Morales*[1], *and L.E. Regalado*[2]

[1]Instituto de Física, Universidad Nacional Autónoma de México, Laboratorio de Ensenada, Apdo. Postal 2681, 22800 Ensenada, Baja California, Mexico

[2]Centro de Investigación en Física de la Universidad de Sonora, Apdo. Postal H-8, Hermosillo Sonora, Mexico

Abstract. A gold film was vacuum deposited on air-cleaved mica at an oblique angle, producing thus a sample with anisotropic structure. The anisotropic dispersive behavior was analyzed using the attenuated total reflection (ATR) technique for different orientations of the sample, and its surface structure was obtained by tunneling electron microscopy (STM).

1. Introduction

A gold film was vacuum deposited on air-cleaved mica at an oblique angle, producing thus a sample with anisotropic structure. The anisotropic dispersive behavior was analyzed using the attenuated total reflection (ATR) technique for different orientations of the sample, and its surface structure was obtained by tunneling electron microscopy (STM). We also performed spectral transmittance measurements and, using a control sample, Auger electron spectroscopy (AES) to analyze the Au/mica interface. The Au/mica system has been widely studied using different surface spectroscopy techniques [1–5]. The addition of scanning tunneling microscopy as a new technique for surface analysis has opened up a whole new field and the comparison of the results it produces with those obtained by other methods may prove to be very enlightening. In this particular work we intend to correlate the dispersive behavior of an anisotropic gold film on mica studied by ATR with its atomic structure as obtained by STM.

2. The Experiment and Characterization of the Sample

2.1 Sample Preparation

We started by air cleaving a mica substrate, first with a sharp knife and, finally, with a piece of tape until a surface free of defects was obtained. The gold film was deposited at a pressure of 2×10^{-6} Torr on a substrate hanging at an angle of 20° with respect to the line joining the boat to the substrate. The deposition rate was 2 Å/s. We expected a columnar growth of the film with the row of columns closest to the evaporation source projecting a shadow on the following rows (Fig. 1).

Springer Proceedings in Physics, Volume 62
Surface Science Eds.: F.A. Ponce and M. Cardona

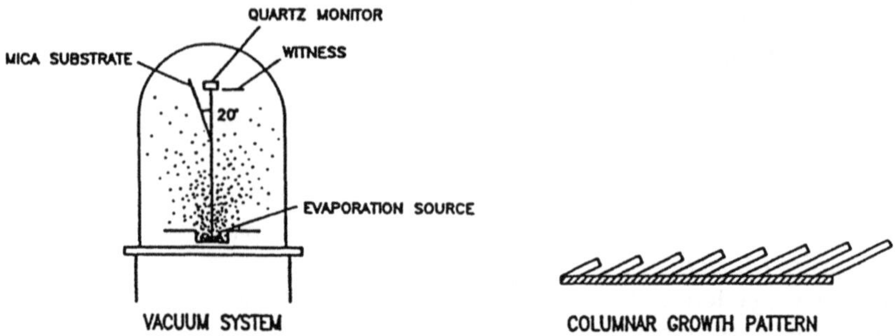

Fig. 1 Experimental setup for the deposition of the gold film and the expected growth pattern

2.2 Optical Measurements

1) ATR measurements were made using the experimental arrangement shown in Fig. 2, where we generated a set of curves of reflection against wave vector with the wavelength as parameter (Fig. 3).
2) Ellipsometric measurements were also performed to try to resolve a possible difference in the refractive index between isotropic and anisotropic gold films.
3) Transmittance measurements were made in the 200 – 2600 nm region using

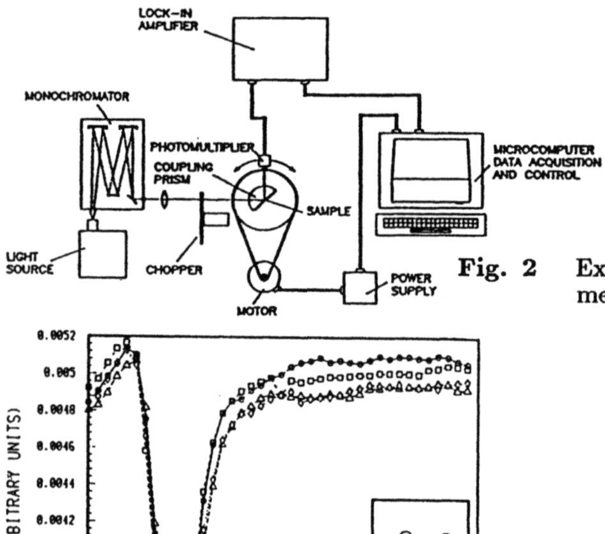

Fig. 2 Experimental setup for RTA measurements

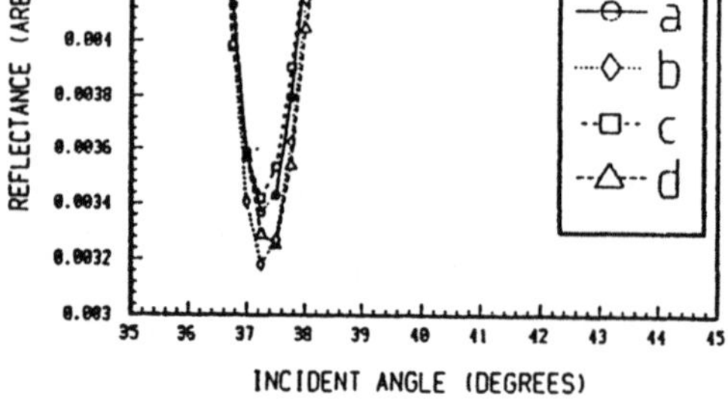

Fig. 3
RTA curves for four different positions of the Au/mica sample for $\lambda = 770$ nm

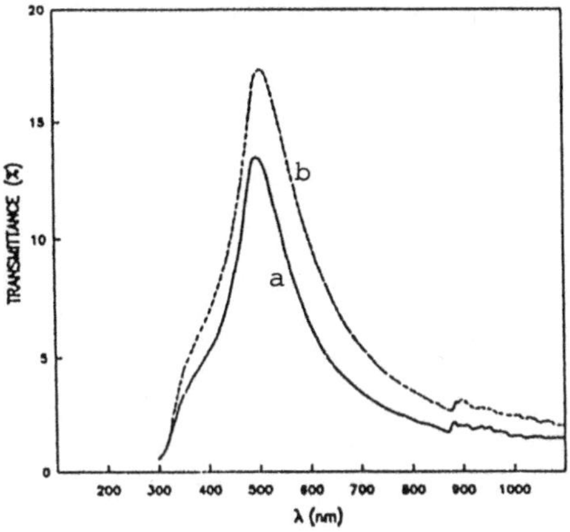

Fig. 4 Transmittance spectra (visible only), for the Au/mica sample at the thick end (a) and at the thin end (b), where the scale is 0 to 50%

a PE 330 Spectrophotometer. By illuminating different zones of the sample we could see that the gold film was wedged shaped, with the thickest part closest to the evaporation source (Fig. 4).

2.3 Transmission Electron Microscopy

TEM was performed on one portion of the sample specially prepared for this purpose in order to estimate the grain size of the gold film (Fig. 5a).

2.4 Scanning Tunneling Microscopy

STM scans were performed at different scales and on different samples. In Fig. 5b, the grain structure may be observed and compared with the TEM results. Figure 6a shows the atomic structure for a normally deposited film and Fig. 6b the grain structure for an obliquely deposited one.

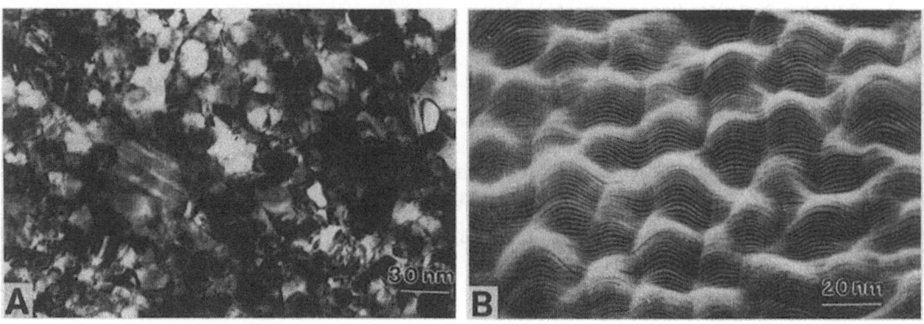

Fig. 5 (a) A TEM micrograph of the gold film and (b) a STM generated image at the same scale

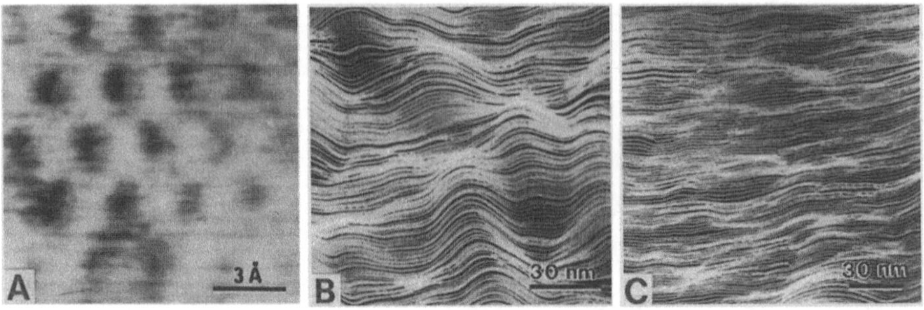

Fig. 6 (a) The atomic structure of the (111) face of the gold film. (b) Rows parallel to the wedge can be observed. (c) A blow up of a smooth region appearing beside the rougher area shown in (b)

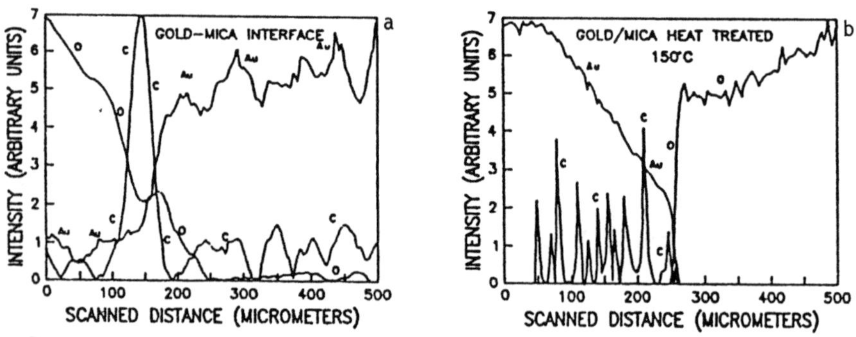

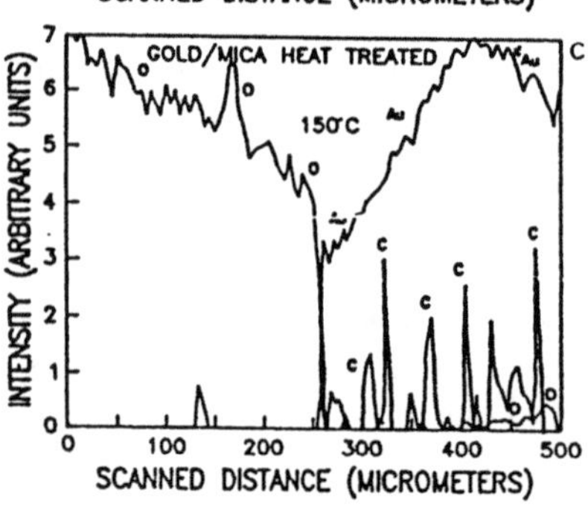

Fig. 7
Auger spectra of the Au/mica interface. (a) Sample as deposited; (b), (c) after a 4 h heat treatment at 150°C

2.5 Auger Electron Spectroscopy

AES was performed on the Au/mica interface by carving an extended crater on the gold film of a control sample. We obtained interesting information regarding the presence of different elements and their dynamics in that zone of the sample (Fig. 7).

3. Results and Discussion

Figure 3 shows a set of RTA curves (reflected intensity vs incident angle) obtained from our sample using p-polarized light for a wavelength of 770 nm. Curve a (b) corresponds to the **E** field perpendicular to the wedge direction, thin side up (down) and curve c (d) to the **E** vector parallel (antiparallel) to the wedge direction. As can be seen, there are only small intensity differences between the four curves and they could be attributed to the differences in thickness of the illuminated areas and not necessarily to a change in the optical properties. The plasmon resonance peak remained unaltered within the resolution of our system (one eighth of a degree). From the ellipsometric measurements we obtained a value for the refractive index of the gold film with the wedge parallel to the plane of incidence of the ellipsometer. We give here also, for comparison, the value reported by *Palik* [6] for 546 nm:

$$N = 0.404 - i2.506 \qquad \text{(our sample)}$$
$$N = 0.402 - i2.540 \qquad \text{(Palik).}$$

From the transmittance measurements presented in Fig. 4 and the value given by the quartz monitor for the film thickness (603 Å), we found a difference in thickness from one side to the other of 22%, approximately. The TEM micrographs, Fig. 5, show that the film consists of randomly oriented crystalline grains with sizes varying from 10 to 35 nm, the same value as obtained with the STM.

The analysis with the STM shows, at the atomic level, the typical structure of a gold (111) surface (Fig. 6), where the distance between minima was 3.1 + 0.3 Å, which is in reasonable agreement with the value reported in the literature of 2.88 Å for the volume.

At a larger scale, scanning a 250 × 250 nm area, there is a corrugated region where we can observe rows running parallel to the wedge direction (Fig. 6b). The average corrugation height is of the order of 50 Å. Beside this region, there is another where the corrugation at this scale practically disappeared (Fig. 6c). We do not have an explanation for this result, except for the fact that we are working in air and the surfaces of the substrate and the gold film cannot be free of contamination.

This last statement is supported by the results of the AES analysis where strong oxygen and carbon signals were present in the spectra of the surface of the film (as deposited). These signals disappeared in the first minute of ion bombardment but the carbon signal reappeared when we reached the Au/mica interface. When we heated the sample at 150°C for 4 h, the strong carbon signal at the interface turned into smaller localized signals distributed along the film thickness, suggesting a diffusion of the carbon to the grain boundaries (Fig. 7). Another feature that can be observed from these spectra is the absence of diffusion of gold into the mica and vice versa. The poor adherence of the gold film to the mica substrate may be related to this observation.

Acknowledgements. We acknowledge the support given by F. Ruiz (TEM) and G. Vilchis (photography) for the realization of this work. We also thank Dr. L. Cota for the Auger analysis.

References

1. V.M. Hallmark et al.: Phys. Rev. Lett. **59**, 2879 (1987)
2. H. Poppa et al.: J. Vac. Sci. Tech. **8**, 471 (1971)
3. H. Poppa et al.: Nucl. Instrum. Methods **102**, 521 (1972)
4. H.S. Kim et al.: J. Vac. Sci. Technol. A **8**, 314 (1990)
5. G.K. Binnig, H. Rohrer, Ch. Gerber, E. Stoll: Surf. Sci. **144**, 321 (1984)
6. E.D. Palik (ed.): *Handbook of Optical Constants of Solids* (Academic, Orlando, FL 1985)

A Non-homogeneous Thin Film Model and the Evaluation of Its Properties by Ellipsometric Methods

Y. Torres and A. Plata

Universidad Industrial de Santander, Departamento de Física,
Laboratorio de Optica y Tratamiento de Señales,
A.A. 678, Bucaramanga, Colombia

Abstract. A material produced by ion implantation or by chemical or mechanical action reveals a variable refraction index over its surface. This non-homogeneous thin film can be analyzed by ellipsometric method[1].

A model based on the generalization of Scandonne-Ballerine formulas and the variational method for estratified homogeneous thin films is proposed. Finally, the results are presented for effective complex index determination.

1. Introduction

Non-homogeneous thin films are characterized by variations of optical properties with thickness. Examples of these films can be the result of differents actions on the surface of substrate.

The Scandonne-Ballerine (S-B) method[2,3,4] applied to null ellipsometer measurements is proposed for the evaluation of optical properties of those films.

2. Direct problem

The non-homogeneous thin film is considered as a superposition of homogeneous thin films. The generalized reflection coefficient has the recurrence formula:

$$R_{j,0} = \frac{r_{j+1,j} + R_{j-1,0}\, e^{-j2\delta_j}}{1 + r_{j-1,j}\, R_{j-1,0}\, e^{-j2\delta_j}}\,, \tag{1}$$

$$\delta_j = 2\pi \left(\frac{d_j}{\lambda}\right) N_j \sin(\varphi_j)\,.$$

Where for the jth film: r, is the Fresnel coefficient; d, the film thickness and φ is the angle of incidence.

This coefficient evaluated for the S-B equation by the variational method has the form $R = R_o - \delta R$. Where δR is the non-homogeneous contribution.

For the hipothesis of a transparent external medium ($k=0$) and a thin film with variable refraction coefficient N(z), this coefficient can be expressed by

$$N(z) = N_{substrate} + \Delta N(z)\,. \tag{2}$$

Springer Proceedings in Physics, Volume 62
Surface Science Eds.: F.A. Ponce and M. Cardona

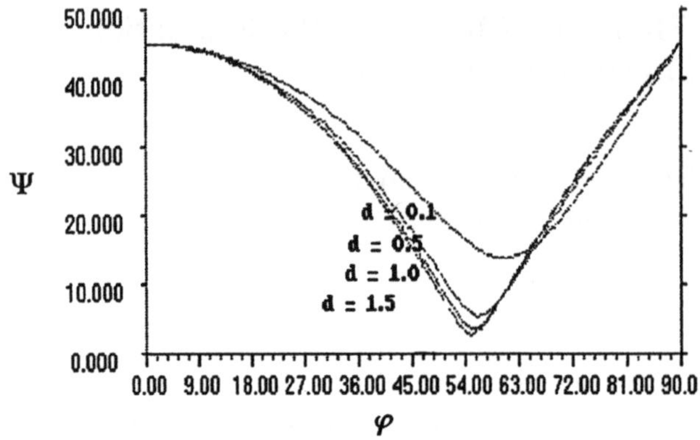

Fig. 1. Numerical simulation of the model for different values of thickness d in μm.

Where $\Delta N(z)$ is the total variation of refraction coefficient of the thin film related to substrate (for $z \rightarrow \infty$, $N(z) \rightarrow N_{substrate}$).

The contribution of the jth film is deduced in terms of the difference between the reflection coefficients of thin film with equivalent thickness Z_j and a thin film with thickness $Z_j + \Delta Z_j$. The Fresnel coefficients are calculated in approximations up to $(\Delta N_{equivalent})^2$. Finally, the contribution of the jth layer is

$$\delta R_0^{P,S} = -j\, 2\, k_{subs}\, A_0^{P,S}\, \Delta n_j\, \Delta z_j\, R_0^{P,S}\, e^{-j2\delta_0} \,.$$

The total contribution is obtained by summation of elementary contributions. This total contribution decreases with wave penetration in the surface and depends of the wavelength, of the optical properties of external medium and the gradient of refraction index of the non-homogeneous thin film.

The final form of the relative reflection coefficient is

$$\rho = \rho_0 \left(1 - j2k_{subs}\, A_0 \int_0^{\infty} \Delta_j\, e^{-j2\delta_0}\, dz\right), \tag{3}$$

$$A_0 = \frac{2\, N_0 N_1 \sin(\varphi_{ext})}{N_0^2 \cos^2(\varphi_{ext}) + N_1^2 \sin^2(\varphi_{ext})} \,.$$

Fast inspection of above equation points out the reflection coefficient as the Fourier transform of the variation of the index of refraction.

Figure 1 is a numerical simulation of the model for non-homegeneous thin films with linear variation of the index of refraction from $N_{substrate}$ = 1.63-j0.02 to $N_{maximum}$ = 1.96-j0.02, corresponding to glace doped with plumb.

3. Inverse Problem

The following optimization function allows a correct choice of the variation of the reflection index[6,7]:

$$S = \sum_j \left[(\delta\Delta_j)^2 + (\delta\Psi_j)^2 \right] \frac{A_j^{-2}}{\sum_j A_j^{-2}},$$

where j is the number of experimental measurements and the quadratic expressions are the difference between the theoretical and experimental results.

4. References

[1]R.M.A. Azzam and N.M. Bashara, *Ellipsometry and polarized light* (North-Holland Publishing Co., Amsterdam, 1977).
[2]A.B. Rshanov, *Ellipsometry: Research method of surfaces*, in Russian (Nauka, Novosibirsk, 1983).
[3]A.B. Rshanov, *Ellipsometry: Theory, methods and applications*, in Russian (Nauka, Novosibirsk, 1987).
[4]A.B. Rshanov, *Software and algorithms for the solution of problems in ellipsometry*, in Russian (Nauka, Novosibirsk, 1980).
[5]C.A.P. Garnier and Y. Torres, II Encontro Latinoamericano sobre Laser e suas Aplicaçoes (Niteroi R.J., Brasil, 1986).
[6]C.A.P. Garnier y Y. Torres, Memorias XII Congreso Nacional de Física (Popayán, Colombia, 1987), p. 30.
[7]C.A.P. Garnier y Y. Torres, Memorias XII Congreso Nacional de Física (Popayán, Colombia, 1987), p. 39.

Computer Aided Ellipsometry Applied to Thin Films

Y. Torres, A. Plata, and C.A.P. Garnier

Universidad Industrial de Santander, Departamento de Física,
Laboratorio de Optica y Tratamiento de Señales,
A.A. 678, Bucaramanga, Colombia

Abstract. The association of computer and interactive software to an ellipsometer[1,2,3] allows efficient evaluation of electromagnetic properties of thin films. The interactive software has been developed in quick-C language for IBM-AT or compatible computers.

Two problems of ellipsometry have been designed to enable the user work in a flexible interactive manner: the simulation of thin film properties for different models and the solution of inverse problem for experimental data.

1. Introduction

Development of an interactive software by mathematical formalism and facilities for interpretations of the experimental data and numerical simulations is desirable for real time applications[4]. Additionally, this software is very useful if it can be used as tutorial.

2. The CAE system

The Computer Aided Ellipsometry system written in C language allows the computer to perform the numerical evaluation of direct and inverse problems for a null ellipsometer[1,2]. The menu structure of CAE system, figure 1, is divided into four submenus: The fundamental theory of ellipsometry, the different models of thin

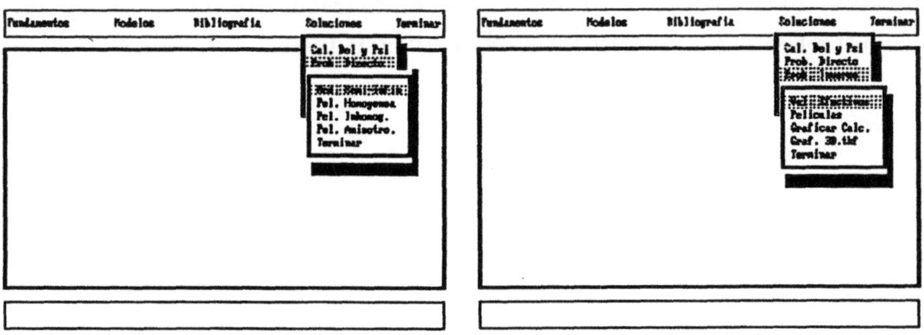

Fig. 1. Two screen examples of submenu options in the CAE system.

Springer Proceedings in Physics, Volume 62
Surface Science Eds.: F.A. Ponce and M. Cardona

films, one database for management of references in this specific subject and the direct or inverse problem evaluation.

The values of delta and psi are calculated from the A, P and C experimental data for the PCSA configuration of the null ellipsometer and the Jones matrix formalism[1,5]. Graphical representations of delta and psi allow efficient evaluation of experimental values.

3. Direct problem

This problem evaluates the reflection coefficient for homogeneous thin films (Drude method[1,2]) and for non-homogeneous thin films[6]. The case of anisotropic thin films estimates the generalized Fresnel coefficients.

4. Inverse problem

The numerical inversion optimizes the error function[6] in terms of the difference between the theoretical (direct problem) and experimental results. 3-D representations aid to user for a rapid local determination of the minimum in two steps.

5. References

[1]R.M.A. Azzam and N.M. Bashara, *Ellipsometry and polarized light* (North-Holland Publishing Co., Amsterdam, 1977).
[2]A.B. Rshanov, *Ellipsometry: Research method of surfaces*, in Russian (Nauka, Novosibirsk, 1983).
[3]A.B. Rshanov, *Ellipsometry: Theory, methods and applications,* in Russian (Nauka, Novosibirsk, 1987).
[4]A.B. Rshanov, *Software and algorithms for the solution of problems in ellipsometry*, in Russian (Nauka, Novosibirsk, 1980).
[5]C.A.P. Garnier y Y. Torres, I Encuent.Nal. Optica (Med., Colombia, 1987), p. 100.
[6]Y. Torres and A. Plata, Proceedings of SLAFS-6 (Cusco, Peru, 1990), Submitted.

Electrical Resistance in Hydrogenated Nb Thin Films

D.E. Azofeifa and N. Clark

Escuela de Física, Universidad de Costa Rica, San José, Costa Rica

Abstract. The electrical resistivity of Nb films (50 to 100 nm thick) has been studied as a function of hydrogen pressure. The Nb films were covered with a small Pd overlayer. The resistivity change reaches saturation at about 15 Torr with a maximum increase of 8%.

1. Introduction

Research in metal-hydrogen systems has been motivated by the desire to have fundamental understanding of the nature of these interesting systems and by the desire to explore and develop technological applications [1]. The present report is part of an on-going project of our group to study the transport properties of hydrogenated metal films [2] and it is motivated by the interesting behavior found in film samples different from that of bulk [3].

Nb has been the subject of many studies because of its very large capacity to absorb hydrogen (up to 160 at.%), which makes it a good model substance in the practical aspects of storage. Also, abnormally high capacity of absorption in Nb films has been reported [4].

2. Experimental Procedure

The films were prepared evaporating Nb (99.5% pure) at $5x10^{-6}$ Torr and room temperature, onto glass substrates. Sample thickness was determined using a quartz crystal monitor. Hydrogen pressure was measured with a mercury manometer. Electrical resistance, R, measurements were made using the 4 point method. In order to minimize oxidation and to enhance hydrogen intake, all samples had a small (approx.3 nm) Pd overlayer.

Each sample was submitted to several hydrogenation cycles, which consisted in admitting hydrogen gas into the chamber and, once R became stable, the pressure was

Springer Proceedings in Physics, Volume 62
Surface Science Eds.: F.A. Ponce and M. Cardona

further increased. After several increments the chamber was evacuated and the process repeated.

3. Results and Conclusions

The film resistance decreased after the first hydrogenation cycles, however it became stable after the fifth or sixth cycle (see Fig.1). This may be explained by two mechanisms: a hydrogen induced ordering occurring in the lattice inside the grains [5], and/ or, a relaxation of the strain fields on, or near, the grain boundaries due to the hydrogen diffusing along them [6]. The second mechanism may explain also why the magnitude of this change was thickness dependent, being larger for thinner samples (19% for the 50 nm film and 9% for the 100 nm films as seen in Fig. 1).

All cycle-stabilized samples showed an increase in R as a function of increasing hydrogen pressure. This change became negligible from 15 Torr up to the tested pressures (see Fig. 2). The maximum resistance increase was approx-

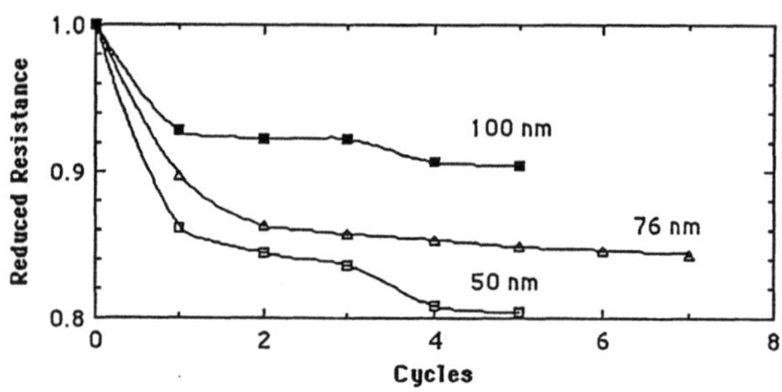

Fig. 1. Resistance of the samples after each cycle, normalized to the resistance before hydrogen is introduced in the system. Lines are only an aid to the eye.

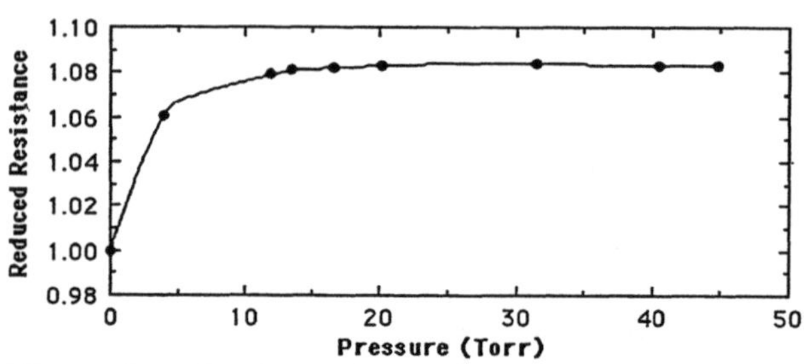

Fig. 2. Reduced resistance (normalized to the initial resistance) of a 76 nm film which has reached cycle stability, as a function of hydrogen pressure. Note the steep change up to 11 Torr and the saturation after 15 Torr.

imately 8% and it did not show a measurable thickness dependence, in the studied range. This independence means that the change is due to the absorption of hydrogen in the films, which become saturated at approximately 15 Torr.

Finally it is interesting to note the sensitivity of the resistance due to the presence of hydrogen between 0 and 11 Torr. Changes as small as 1 Torr induced measurable R changes. This raises the possibility of using Nb films as a hydrogen detector in the 0 to 10 Torr range. This possibility is being further explored.

References

1. As example of applications see: J. Topler and K. Feucht - Z. für Physikalische Chemie **163**,1451(1989); S.Suda - Z. für Physikalische Chemie **163**, 1463(1989).
2. D.E.Azofeifa and N.Clark - Z.für Physikalische Chemie **163**, 621 (1989).
3. M.Lee and R.Glosser - J.Appl.Phys. **57**, 5236 (1985); M. Nicolas and J.P.Burger - Z.für Physikalische Chemie **163**, 67 (1989).
4. S.Moehlecke, C.F. Palatnik and A.T.Myron Strongin. - Phys. Rev B **131**, 6804 (1985).
5. A.V.Irodova, V.P.Glazkov, V.A.Somenkov, I.V.Kurchatov, V.E. Antonov and E.G.Ponyatovsky - Z.für Physikalische Chemie **163**, 53 (1989).
6. R.P.Volkova L.S. Palatnik and A. T. Pugachev. - Sov.Phys.Dokl. **26**, 695 (1981).

Interdiffusion of Cu-In Films Studied by the Resistometric Method

N. Clark and D.E. Azofeifa

Escuela de Física, Universidad de Costa Rica, San José, Costa Rica

Abstract: The inter diffusion of Cu-In bilayers is studied by the resistometric method. In each case one of the films is at least 8 times thicker than the other, and the total thickness is in the order of 140 nm. The difussion of Cu in In lowers the resistance and the diffusion of In in Cu raises it. The diffusion constant, at 295 K, is calculated by means of a simple model of the diffusion of a thin layer of material deposited on a thicker film. From the results it is concluded that the dominant mechanism is grain boundary diffusion.

1. Introduction

Diffusion in thin films differs from that in bulk because of the increased surface-to-volume ratio and the presence of grain structures in the films [1,6]. In this work the changes in the electrical resistance, R, due to diffusion are studied for two-layer Cu-In films, in which one of the components is substantially (at least 8 times) thicker than the other.

2. Experimental Procedure and Results

Cu films were deposited at 10^{-6} Torr on glass substrates. Once R had become stable an In film was deposited as an overlayer. All measurements were done at 295±2 K.

The most interesting feature in the behavior of R after the deposition of the In film (see figure 1) is that it decreases when the In film is thick and the Cu is thin, while it increases in the opposite case. The behavior changes from one type to the other as the ratios of the thicknesses varies.

The reason for the distinct changes is found in the different type of diffusion that is dominant in each case: lattice diffusion (LD) which increases R, and grain boundary diffusion (GBD) which tends to decrease it [2].

It is also known [1] that LD dominates at higher temperatures, $T>0.5T_m$, while GBD dominates at low

Springer Proceedings in Physics, Volume 62
Surface Science Eds.: F.A. Ponce and M. Cardona

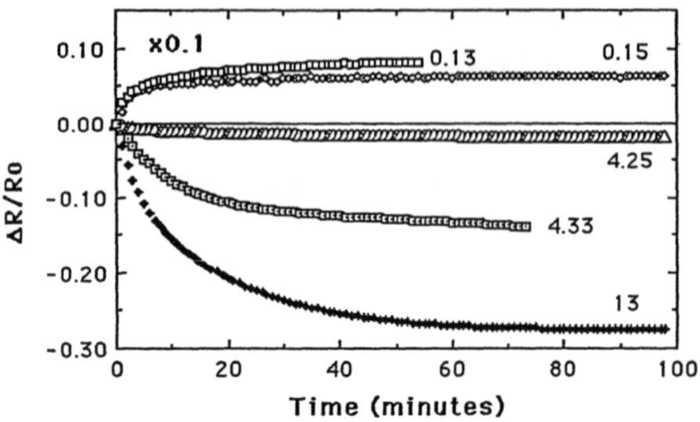

Fig.1 Resistance change after deposition of the second layer. Ro is the initial resistance of the bilayer. The numbers on the different curves are the different ratios of the In-film thickness to the Cu-film thickness. Values of the 0.13 and 0.15 curves have been multiplied by 10.

temperatures, $T<0.3T_m$, T_m is the melting temperature of the host metal. At room temperature we clearly have both regimes because the relations are $0.22T_m$ for Cu and $0.64\ T_m$ for In [3].

3.Theory and Discussion

Diffusion in thin films can be studied in relation to the resistivity, ρ, by assuming a linear behavior of ρ as a function of the diffusing substance concentration [4]. This is

$$\rho(z,t) = \rho_0 + 100\ \alpha\ C(z,t) \qquad (1)$$

where t is time, z the space coordinate along the direction of diffusion, C(z,t) the concentration at a given position and time, which can be calculated by solving the diffusion equation [5] with boundary conditions of the bilayer. ρ is the resistivity of the bilayer at t=0 and α the change on ρ induced by 1% of the diffusing substance.

The resistance per unit length of the bilayer is given by

$$R(t)^{-1}=\int_0^{\Delta} \frac{a\ dz}{\rho_0+100\ \alpha\ C(z,t)}\ , \qquad (2)$$

where Δ is the thickness of the bilayer and a the thickness of the thicker film. Then, by defining the integral

$$I(t) = \Delta^{-1}\int_0^1 \frac{dz}{1+\gamma y(z,t)} ,$$

where $\gamma = 100\ \alpha\ C_F/\rho$, ($C_F$ being the final concentration); $y = C(z,t)/C_F$; and $\tau = D\ t/\ \Delta^2$ (D being the diffusion constant). Equation (2) transforms to $R(t) = R_0/I(\tau)$ where $I(\tau)$ can be calculated numerically as a function of τ with γ as parameter.

The comparison to the experimental data is done through the relation

$$\phi(t) = \frac{R(t) - R_0}{R_f - R_0} , \qquad (3)$$

which as a function of γ and τ is written as

$$\phi(\tau) = \frac{1}{\gamma}\left[\frac{1}{I(\tau)} - 1\right] . \qquad (4)$$

The actual comparison is done by curve fitting (4) to (3) evaluated with the experimental data. This allows to find the proper γ and the relation between τ and t, which in turn allows the calculation of the diffusion coefficient at room temperature.

When applied to the Cu-In bilayers described above the results are: 1.26×10^{-14} cm^2/s for the grain boundary diffusion coefficient of Cu in In and 1.32×10^{-14} cm^2/s for the diffusion coefficient of In in Cu. The former value is in good agreement with the expected results [1]. The latter is two orders of magnitude higher than expected, this indicates that not only LD is taking place but also the faster GBD is occurring [6]. However both results show that this simple method is capable of finding D from the resistance change in bilayers.

References:

1. D. Gupta - In Diffusion Defect Data. Solid State Data A Defect Diffusion Forum A59, 137 (1988).
2. R. P. Volkova, L.S.Palatnik and A.T.Pugachev - Sov. Phys. Dokl. 26, 695 (1981).
3. CRC Handbook of Chemistry and Physics - CRC PRESS, Inc. Florida(1979).
4. S.Ceresara, T.Federighi and F.Pieragostini - Phys. Stat. Sol 16 439 (1966).
5. B.S.Bokshtein - Difusión en Metales, Sec.5 - Edit. MIR (1978).
6- A. Wagendristel - Phys. Stat. Sol.(a) 13 131 (1975).

Structure and Optical Absorption of $Li_yV_2O_5$ Thin Films

A. Talledo[1], *A.M. Andersson*[2], *and C.G. Granqvist*[2]

[1]Universidad Nacional de Ingeniería, Facultad de Ciencias, Apartado 1301, Lima, Peru

[2]Physics Department, Chalmers University of Technology and University of Gothenburg, S-41296 Gothenburg, Sweden

Films of lithiated vanadium pentoxide ($Li_yV_2O_5$) are of interest for ion storage in electrochromic "smart windows" and in lithium battery technology. In this paper we study optical absorption of $Li_yV_2O_5$ films in the $0.3 < \lambda < 2.5$ µm wavelength range, encompassing visible light and solar radiation, in relation to the microstructure.

V_2O_5 films were prepared by dc magnetron sputtering from a V target in Ar + 8 % O_2 onto glass precoated with transparent and conducting In_2O_3:Sn layers. Lithiation was accomplished in 0.25 M $LiClO_4$ in propylene carbonate according to

$$yLi^+ + ye^- + V_2O_5 \rightleftharpoons Li_yV_2O_5 \ . \qquad (1)$$

The magnitude of y was determined from the inserted charge. Details of the sample fabrication are given in [1].

X-ray diffractometry on 0.2-µm-thick films sputtered onto unheated substrates indicated a nanocrystalline (n-X) structure with a single peak due to (001) planes in orthorhombic V_2O_5.

Similar measurements for films sputtered onto substrates heated to 180°C indicated a polycrystalline (p-X) structure with peaks from the (200), (001) and (400) planes. Lithiation was accompanied by shifts in the diffraction peaks.

Spectral optical properties were studied by spectrophotometry. Figure 1 shows spectral absorptance for 0.15-µm-thick n-X and p-X films of $Li_yV_2O_5$ with different y's. For n-X films the spectra show several peaks at $\lambda > 0.5$ µm whose details depend on y. There is qualitative agreement with some earlier work that ascribed such peaks to V^{4+} ions in a distorted lattice [2]. The absorption edge, due to O2p → V3d transitions, appears to be blue-shifted in proportion with y. It is conceivable that quantum confinement effects are associated with bandgap features. For p-X films, the bandgap is smaller than for n-X films and shows no strong dependence on y. A well-defined absorption peak is observed at $\lambda = 1.0$ µm. At $y = 0$, this absorption is unambiguously connected with oxygen vacancies in the V_2O_5 lattice - specifically with electrons localized at V atoms adjacent to a vacancy and leading to the formation of V^{4+} [3]. At $y \neq 0$, the electron insertion indicated in reaction (1) accounts for the observed enhancement of the absorption at 1.0 µm.

Springer Proceedings in Physics, Volume 62
Surface Science Eds.: F.A. Ponce and M. Cardona

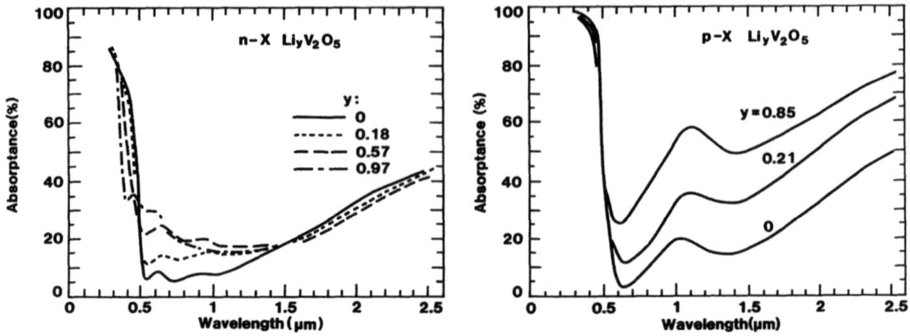

Figure 1. Spectral absorptance for nanocrystalline (n-X; left-hand part) and polycrystalline (p-X; right-hand part) films of $Li_yV_2O_5$ with 0.15 µm thickness and the shown y's.

The p-X films also display broad-band infrared absorption. This may be associated with 3d electrons of V^{4+} ions with freedom to hop along the crystallographic b-direction in orthorhombic V_2O_5. The In_2O_3:Sn layer also contributes some absorption.

A detailed discussion of the optical properties of $Li_yV_2O_5$ films with n-X and p-X structure is given in [4].

References

1. A.Talledo, A.M. Andersson and C.G. Granqvist, J. Mater. Res. **5**, 1253 (1990).
2. F.P. Koffyberg and N.J. Koziol, J. Appl. Phys. **47**, 4701 (1976).
3. P. Clauws and J. Vennik, Phys. Stat. Sol. **B66**, 553 (1974).
4. A. Talledo, A.M. Andersson and C.G. Granqvist, J. Appl. Phys. **69**, 3261 (1991).

Part VII

Semiconductors

Phonons in Semiconductor Superlattices

M. Cardona

Max-Planck-Institut für Festkörperforschung,
Heisenbergstr. 1, W-7000 Stuttgart 80, Fed. Rep. of Germany

Abstract. The lattice dynamics of superlattices made out of tetrahedrally coordinated semiconductors (Ge-Si, GaAs-AlAs) is discussed with emphasis on the modes which can be observed by Raman and/or infrared spectroscopy. These phonons are classified into folded acoustic, confined optic, and interface modes. Particular attention is paid to the interface modes and the relation between their macroscopic and microscopic theoretical treatments. It is shown that interface modes are not new modes of polar superlattices but they arise from the conventional confined ones of odd order (especially $m = 1$) when their wave vector is tilted with respect to the growth axis.

1. Introduction

Optical spectroscopies (Raman, infrared, nonlinear spectroscopies such as hyper-Raman or CARS) yield only very limited information concerning phonons in conventional crystalline solids. This limitation arises from the $\vec{k}$-conservation selection rule: since the wavelength of light is very large compared with characteristic atomic dimensions (i.e., lattice constants), its wave vector has nearly zero magnitude compared with the dimensions of the Brillouin zone (BZ). Hence only phonons with $\vec{k}$-vectors very close to the center of the BZ (i.e., the Γ-point) can be observed with optical spectroscopies. This is in sharp contrast with neutron or high resolution electron energy loss spectroscopies (EELS) where, because of the short wavelength of the probing particles, phonons throughout the whole BZ can be investigated.

We discuss in this paper superlattices made out of tetrahedral semiconductors with the diamond (germanium, silicon) or zincblende (GaAs, AlAs) bulk structures. These bulk materials have two atoms per primitive cell (PC) which leads to three acoustic phonons (observable with Brillouin spectroscopy) and three optic phonons (observable with Raman spectroscopy) near the Γ-point of the BZ. In the diamond structure the two atoms in the PC are identical (the material is nonpolar) and the optic phonons, threefold degenerate at Γ, are not ir-active. In zincblende, however, the two atoms are different. Electron transfer occurs from one to the other (e.g. from Ga to As in GaAs) and the bonding becomes ionic (also called polar). The inversion symmetry of the diamond lattice is lost and the optic phonons become ir (and hyper-Raman) active. As a result of this activity these phonons split for $\vec{k}$-vectors very close to Γ (but not closer than $2\pi/\lambda$ where λ is the reststrahlen wavelength, i.e.,

Springer Proceedings in Physics, Volume 62
Surface Science Eds.: F.A. Ponce and M. Cardona

the wavelength which correspond to the TO-phonon frequency in the medium) into one longitudinal, LO, (which vibrates along $\vec{k}$) and two transverse (TO, vibrate perpendicular to $\vec{k}$). While only the latter are ir-active, both LO and TO are Raman (and hyper-Raman) active. A review of Raman scattering by phonons can be found in [1].

The sharp restriction imposed on optical spectroscopies by $\vec{k}$-conservation can be partially lifted by destroying, perturbing, or decimating the translational lattice. Destruction of translational invariance is achieved by converting the materials into their amorphous form while a less drastic perturbation of that invariance is achieved when impurities are added to the material (e.g. through doping) or when two similar materials are alloyed (e.g. Ge-Si mixed crystals). A better controlable way of reducing translational invariance is to form a superlattice (SL) composed of n_1 monolayers of material A alternating with n_2 of material B. This results in a drastic reduction (decimation) of the number of translations under which the crystal remains invariant: those which translate between an atom in layer A and one in B are no longer symmetry operations. With n_1 GaAs and n_2 AlAs monolayers, the PC is increased to contain n_1+n_2 molecules and, correspondingly, the size of the BZ is reduced by the factor n_1+n_2 (the so-called mini-Brillouin zone or MBZ). The increase in the number of atoms per PC results in an increase of the number of phonons at the Γ point (into a total of $6(n_1+n_2)$), of which only three are acoustic, with a corresponding increase in the number of phonons accessible to optical spectroscopies. Since the superlattice phonons bear a close relationship to those of the bulk materials, the method leads to considerable experimental information about the lattice dynamics of the bulk constituents which can otherwise be only obtained by inelastic neutron scattering (the latter, however, requires considerably more experimental effort and yields poorer accuracy and resolution than Raman spectroscopy). Recent reviews of the topic can be found in [2–6].

Superlattices for vibrational studies are mostly grown by molecular beam epitaxy (MBE) although other vapor phase techniques such as metal-organic chemical vapor deposition (MO-CVD) are also used. Sometimes superlattice-like structures (polytypes) appear in nature or can be grown by conventional crystal growth techniques under carefully controlled conditions (e.g. SiC which appears in more than one hundred polytypes [7]). With MBE (and also MO-CVD) tetrahedral semiconductor superlattices grow most easily and perfectly with the layers perpendicular to the [001] direction. Recently, however, other directions, such as [111] [8], [110] [9,10], [012] [11], [211] [12], have been grown and investigated by optical spectroscopies. Growth along the various crystal directions produces superlattice structure with different symmetry properties, as described by membership to a certain space group. The most striking variety is obtained for Ge_mSi_n: for growth along [100] six different groups obtain, depending on the values of n_1 and n_2 [13] : if one of them is even the SL has a center of inversion, otherwise not, if one of them is odd the SL is tetragonal, otherwise orthorhombic. Superlattices grown along low symmetry directions have been discussed in [14]. Table I lists the space groups which result in the GaAs/AlAs case. In fact Table I applies to all AB/CD superlattice systems

TABLE I: Symmetry properties of $(GaAs)_{n_1}/(AlAs)_{n_2}$ superlattices grown along various directions of practical interest. The space groups are given in both Schönfliess and International notation. Note that the [110] $n_1 = n_2 = 1$ case is equal to [001] $n_1 = n_2 = 1$.

growth direction	crystal system	n_1, n_2	space group Schönfliess	space group Intern.
[001]	tetragonal	$n_1 + n_2$ even	D_{2d}^5	$P4m2$
		$n_1 + n_2$ odd	D_{2d}^9	$I\bar{4}m2$
[111]	trigonal	$n_1 + n_2 = \dot{3}$	C_{3v}^1	$P\ 3m$
	rhombohedral	$n_1 + n_2 \neq \dot{3}$	C_{3v}^5	$R\ 3m$
[110]	orthorhombic	$n_1 = n_2 = 1$	D_{2d}^5	$P4m2$
		n_1, n_2 even	C_{2v}^7	$Pmn2_1$
		n_1, n_2 odd	C_{2v}^1	$Pmm2$
		$n_1 + n_2$ odd	C_{2v}^{20}	$Imm2$
[211]	monoclinic	$n_1 + n_2$ odd	C_s^3	$B11m$
		$n_1 + n_2$ even	C_s^1	$P11m$
[012]	monoclinic	$n_1 + n_2$ odd	C_2^3	$B112$
		$n_1 + n_2$ even	C_2^1	$P112$

sharing a common cation (A = C) or anion (B = D), while if neither is common, e.g. InAs/GaSb, the symmetry is usually lower for several growth directions which have been realized in practice.

It is interesting to note that the point groups of Table I fall into two categories, those which include either a reflection plane (C_s, C_{2v}) or a twofold axis (D_{2d}, C_{2v}, C_2) perpendicular to the growth direction and those which do not (C_s, C_{3v}) [8]. In the former case the eigenvectors at Γ of the mini-BZ must be odd or even with respect to those operations (i.e., sums of either sines or cosines of $q\bar{z}$, with $\bar{z}$ the direction of growth), in the latter the parity is mixed. In the former case important selection rules for optical transitions can be derived. Similar conditions apply to $Ge_{n_1}Si_{n_2}$ with the additional feature that the inversion can also be a symmetry operation (if either n_1 or n_2 are even). In this case Raman allowed modes are ir-forbidden and ir-allowed ones Raman forbidden. Note that even when "parity" along the axis does not exist, the effects of the parity violation are small [8,15].

A problem one is often confronted with in superlattices is that of mismatch between the lattice constants of the constituents. This mismatch is rather small in the GaAs-AlAs systems, but it reaches 4% for Ge-Si. Large mismatches cannot be accommodated beyond a certain layer thickness: mismatch dislocations are formed and the built-in strain is lost. Fortunately the effects of strain, both uniaxial and hydrostatic, have been investigated in many of the bulk constituents [16,17] and can be carried over to the SL's. Conversely, one can, from the phonon frequencies measured in SL's by Raman scattering, determine the details of the mismatch strain, which is also affected by the substrate on which the SL is grown. Strained layer SL's are of practical interest since the strain can change the details of the band structure, e.g. transforming indirect gap materials into direct ones [18]. It should also be possible to grow unstable bulk materials, by utilising the interface energy produced by strain. Such is the case for grey tin, only stable below 13 C in the bulk but to much higher temperatures when grown as a film on a tetrahedral substrate such as InSb or CdTe.

2. Folded Acoustic Modes

The acoustic branches of the phonon dispersion relations of the bulk constituents have a common point ($\omega = 0$) at $\vec{k} = 0$. In order to grow epitaxially upon one other, the physical parameters (lattice constants, sound velocities, ...) must be similar. Hence there is a large frequency region starting at $\omega = 0$, in which the dispersion relations of the acoustic modes of both materials have common eigenvalues (for different $\vec{k}$'s). In this region one can approximate the superlattice by an average material with a speed of sound:

$$\bar{v} = \left(v_1^{-1} d_1 + v_2^{-1} d_2\right)^{-1} \cdot (d_1 + d_2) \tag{1}$$

where d_1 and d_2 are the thicknesses of constituents 1 and 2. $\bar{v}$ describes sound propagation along the superlattice axis and can be derived by simply adding the propagation time through each constituent. The effect of the superlattice on the average medium can be represented by a small modulation of v when going from medium 1 to 2 which can be treated by perturbation theory.

In order to represent the phonons of the superlattice in the MBZ we first "fold" the bulk dispersion relation by translating the pieces which lie outside the MBZ by reciprocal lattice vectors of the superlattice. In the Debye model (elastic approximation) a folded straight line is obtained for longitudinal (LA) and transverse (TA) phonons (note that the modes are only pure longitudinal and transverse for propagation along [001] and [111] growth axes. This holds also true for the [110] case in Ge/Si superlattices. For the GaAs/AlAs [110] case the modes are also pure TA and LA within the elastic approximation valid at long wavelengths. At shorter wavelengths L-T mixing occurs [9]).

Each of the folded acoustic branches for $(GaAs)_{n_1}/(AlAs)_{n_2}$ has thus $(n_1 + n_2 - 1)$ additional states at $\vec{k} = 0$ (Γ), which have become "optical" modes of the SL and are, in principle, accessible to optical spectroscopies. In the average $\bar{v}$-approximation most of these modes are twofold degenerate as a result of folding (nondegenerate are only the highest which may be at Γ or at the MBZ-edge). The superlattice modulation of v leads to the splitting of these degeneracies. In the elastic approximation the folded acoustic bands can be calculated by means of elasticity theory with the appropriate boundary conditions for stress and strain. This was done many years ago by Rytov [19] for the "macroscopic" case of propagation of seismic waves in stratified media.

For short period superlattices it is possible to obtain the dispersion relations of the acoustic (and optic) modes by diagonalizing the microscopic dynamical matrix instead of using macroscopic elasticity theory. Such calculations have appeared for many superlattice systems. We show in Fig. 1 dispersion relations calculated for Ge_4Si_4 [001] superlattices with $\vec{k}$ along the growth direction [15]. For these calculations the planar force constants of the bulk materials were used [22]. The results of Fig. 1 are similar to those given in [20,21].

The acoustic phonon bands of Fig. 1 are folded $(n_1 + n_2)/2 = 4$ times. For the LA phonons (between 0 and 9 THz) the nearly linear folded behavior expected for the elastic continuum model is followed rather well in the whole range. For the TA case, however, considerable deviations arise: the upper two folded modes are nearly dispersionless, a fact which arises from the well-known anomalous dispersion of the TA branches in the bulk [22]. Moreover, the two TA branches, degenerate in the bulk and in the SL within the elastic model, split in the microscopic treatment thus reflecting the orthorhombic symmetry (space group D^1_{2h}). This splitting has yet to be observed experimentally.

Figure 1 also displays the splitting of the "optical" (folded acoustic) modes at Γ ($A_g - B_{1u}$) and at the edge of the MBZ induced by the superlattice modulation. Note that the modes at Γ (and also at the zone edge) show the effects of the parity operation (center of inversion) mentioned in Sec. 1: these folded acoustic "optical" modes are either odd (B_{1u}) or even (A_g) with respect to the inversion; the A_g modes are Raman active while the B_{1u} ones are symmetry-wise ir-allowed. Their ir activity, however, arises from the rather weak ionicity of the Ge-Si bonds at the interfaces [23] and is hardly expected to be observable, the less so the larger the period of the superlattice. Note that, for the usual configuration involving backscattering along the superlattice axis, only the LA (A_g) modes are Raman allowed: coupling to the B_{2g}, B_{3g} modes requires that either the incident or the scattered polarization be parallel to that axis.

For small period superlattices (such as that of Fig. 1) the $\vec{k}$-vector transferred in backscattering is still small compared with the MBZ dimensions. Hence only one (A_g) of the $A_g - B_{1u}$ doublet components should be seen. For larger periods the $\vec{k}$-vector can have values in the middle or even close to the edge of the MBZ (the latter happens typically for $d = d_1 + d_2 \simeq 40$ nm). For periods of the order of 10 nm, a doublet is seen which just reflects the folding of the dispersion relation and not the splitting at Γ. The eigenvectors can then be expressed as linear combinations of their odd and even counterparts at Γ (multiplied by the appropriate Bloch factor) with nearly equal weights. Thus, doublets are seen in the backscattering spectra (see Fig. 2) with components of nearly equal intensities. In forward scattering along the growth axis the $\vec{k}$-transfer is nearly negligible and the parity selection rule is observed [24].

The coupling to the folded acoustic modes in Raman scattering takes place mainly through the superlattice induced modulation in the photoelastic constants [2]. Hence the same selection rules apply as for Brillouin scattering in the bulk [25]. We have already mentioned that for the [001] growth direction only LA modes can be seen when the incident and scattered direction are also [001]. For other growth directions the TA modes are, in principle, observable. We can see in Fig. 2, for a [012] SL, doublets which correspond to LA and TA modes, although in this case those modes are not pure longitudinal and transverse but mixed (hence QT and QL labels, Q stands for "quasi").

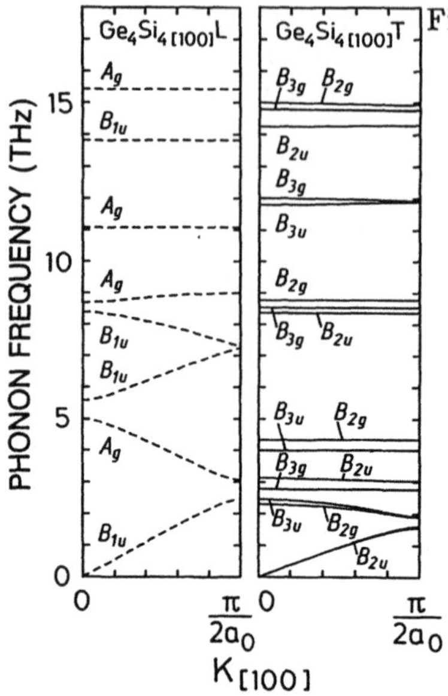

Fig. 1

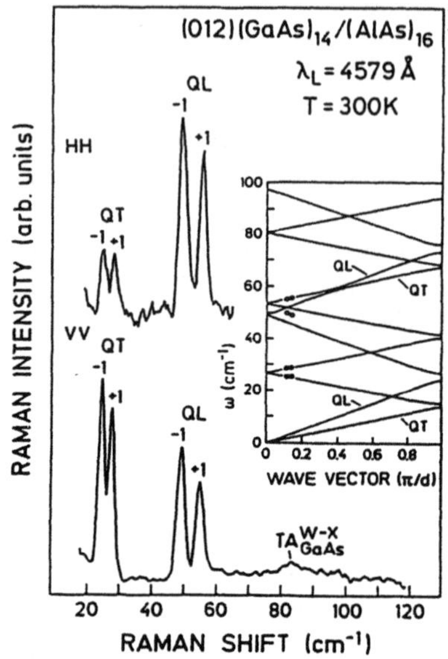

Fig. 2

Fig. 1 Phonon dispersion relations of Ge_4Si_4 [001] superlattices along the growth direction as calculated in [15]. Left panel: longitudinal, right panel: transverse modes.

Fig. 2 Folded acoustic phonons of a $(GaAs)_{14}/(AlAs)_{16}$ superlattices grown along the low symmetry [012] direction. The two doublets correspond to mixed TA-LA modes [11].

We conclude by mentioning that the L-T mixing in low-symmetry SL's induces gaps in the interior of the BZ when the folded QT branch crosses the unfolded QL (or at higher crossings) [26]. Such gaps (and L-T mixing) also arises in high symmetry SL's for $\vec{k}$ tilted with respect to the high symmetry growth axis. The latter gaps have been observed by performing phonon spectroscopy with superconducting tunnel diodes [27].

3. Optical Modes

We have just shown that the acoustic bands of the bulk components of SL's overlap considerably in frequency. This leads to folded acoustic phonons with gaps at the center and the edges of the MBZ. The situation is rather different for the optic modes, where usually (e.g. for Ge/Si, GaAs/AlAs) no overlap exists. Under these conditions it is not possible to obtain propagating optic modes (except maybe for very short period SL's) since they can only propagate in one of the constituents. Modes are then confined to either set of constituent layers. The vibrations of the individual layers of one materials do not interact with each other in the case of non-polar constituents (except for very short periods). Such interaction is blocked by the separating layers of the other constituent which act as "phonon barriers". Hence no dispersion along the direction of growth takes place for the optic phonons of the SL (see the three upper longitudinal modes in Fig. 1). Note, however, that the optic modes of Ge overlap with the LA's of Si. Nevertheless, due to the different nature of the acoustic and optic eigenvectors, this does not affect significantly their confinement, provided the period is not too short [20].

In polar superlattices large range electrostatic fields exist, which enable the vibrations to "communicate" with each other. These lead to dispersive modes which are usually referred to as interfaces (IF) modes. Both cases are discussed below.

3.1 Optical Modes: Nonpolar Constituents

We discuss first the high-symmetry case of [001] growth using $Ge_{n_1}Si_{n_2}$ as an example, with a period of length $d = d_1 + d_2$, consisting of n_1 layers of Ge (thickness $d_1 = n_1 a/2$) and n_2 of Si (thickness $d_2 = n_2 a/2$, where a is the distance between second neighbor planes, equal to one-half of the cubic lattice constant). To a first approximation we assume that the vibrations of each constituent are confined to its respective layers and thus vanish at the boundary between them. This approximation is rather good provided n_1 and n_2 are not too small (e.g. n_1, $n_2 \gtrsim 4$), as has been shown by many microscopic calculations. It leads to nearest neighbor planes vibrating against each other with an amplitude modulated by an envelope function which vanishes at the edge of the layer. Due to the existence of either an inversion center (n_1 or n_2 odd) or, in its absence, a twofold axis perpendicular to [001], this envelope must be odd or even with respect to the "parity" operation. The vibrations are thus standing waves of the bulk (obtained as a linear combination of running waves with wave vectors $\pm\vec{k}$). Their frequencies can be read off the dispersion relations of the bulk for the effective wave vectors along [001]:

$$q = \frac{\pi}{d_{1,2}} m, \quad m = 1, 2, \ldots \ . \tag{2}$$

For m = odd, the envelope function is $\cos qz$ (origin of z in the center of the layer) and corresponds to vibrations of odd parity. For m = even, the envelope is $\sin qz$ and the vibrations are even.

Although (2) yields good eigenfrequencies and eigenvectors for large period superlattices, it can be slightly improved to better represent the low period case [28]. This is done by keeping in mind that from the microscopic point of view, the envelope should not vanish at the layer boundaries (which have no microscopic reality anyhow, since they cut through empty space) but at the nearest layer of the nonvibrating atoms. Equation (2) is thus modified to

$$q = \frac{2\pi}{a\,(n_{1,2} + \delta)} \tag{3}$$

where δ is a parameter, of the order of one, which can be either regarded as

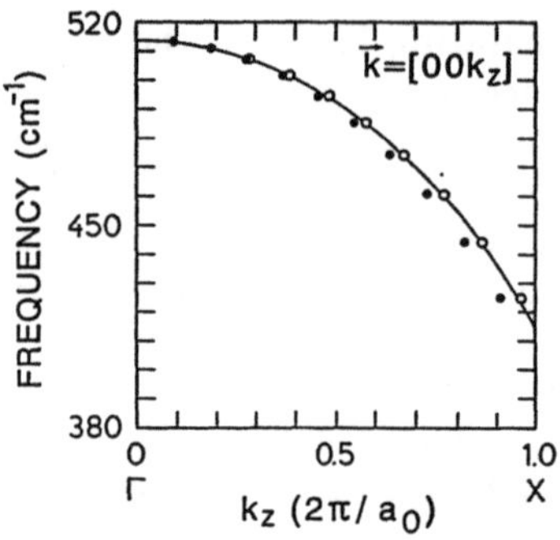

Fig. 3 Calculated LO frequencies of a $Si_{20}Ge_4$ [001] superlattice plotted vs. the nominal wavevector of (3) with $\delta = 2$ (circles, [20]) and $\delta = 1$ (dots) together with the bulk dispersion relation (line).

adjustable, (so as to fit best experimental data or microscopic calculations) or fixed, so that π/q corresponds to the spacing between the first non-vibrating planes outside the layer under consideration. With the latter criterion we find $\delta = 1$ for Ge/Si grown along [001].

As an illustration we show in Fig. 3 a comparison between the bulk dispersion relation of Si along [001] and the corresponding frequencies calculated for a [001] Ge_4/Si_{20} superlattice plotted vs the q's obtained with (3) [20]. The dots were obtained by the authors of [20] for $\delta = 2$; they deviate considerably from the bulk dispersion relation, especially for large q's. The circles, however, plotted by us, with the more physical value $\delta = 1$ (see above) fall exactly on the bulk dispersion relation.

Equation (3) can, and has, also been used to describe envelope functions in superlattices of lower symmetry. The case of [110] growth has been treated in [20], where it has been shown that best agreement between calculated frequencies and bulk dispersion relations is obtained for $\delta = 1$. In this case LO and TO modes also do not mix and parity applies, i.e., the envelope functions are $\cos qz$ (m odd) and $\sin qz$ (m even). The case of [111]-grown SL's has been treated in [15]. The strict separation between odd and even modes still applies, but parity is only a good quantum number, if either n_1 or n_2 are even (inversion center). In the absence of parity, sine and cosine envelope functions mix, especially for small period superlattices. This has been demonstrated in [15] for a Ge_1/Si_5 superlattice. The [111] growth direction leads to two distinct configurations when n_1 (or n_2) is even: the first Ge-Ge bond in the n_1 (or Si-Si in n_2) layer can lie along the growth direction (type A) or along the other $< 1\bar{1}1 >$ directions (type B). Using the argument given above, we find $\delta = 1/2$ for the A and $\delta = 3/2$ for the B configuration and $\delta = 1$ in the case of n_1 (or n_2) odd. A survey of the frequencies calculated with (3) (referred to as nominal wave vector), using the δ's just given, for a few relatively short period superlattices and those obtained by diagonalization of the full dynamical matrix of the SL, is shown in Table II. The agreement is quire remarkable, especially in view of the rather short period ($n_1 + n_2 \leq 8$) of these superlattices.

TABLE II: Calculated frequencies of $Ge_{n_1}Si_{n_2}$ superlattices at $\vec{k}=0$ compared to those obtained from bulk data with the nominal wave vector (NWV) for confined modes (see text). The NWV's can be easily determined (see Eq. (3)). Frequencies are expressed in THz.

Super-lattice	Polari-zation	Mode	Superlattice frequency	NWV frequency	m
$Ge_2Si_{2[001]}$	longitudinal	A_g	14.37	14.6[a]	1
	longitudinal	A_g	8.06	8.4[b]	1
$Ge_2Si_{2[111]A}$	longitudinal	A_{1g}	14.27	13.4[a]	1
	longitudinal	A_{1g}	8.07	7.8[b]	1
$Ge_2Si_{2[111]B}$	longitudinal	A_{1g}	13.91	14.5[a]	1
	transverse	E_g	15.30	15.0[a]	1
	transverse	E_g	8.95	8.75[b]	1
$Ge_4Si_{4[001]}$	longitudinal	A_g	15.44	15.4[a]	1
	longitudinal	B_{1u}	13.82	13.8[a]	2
	longitudinal	A_g	8.69	8.9[b]	1
	transverse	$\frac{B_{3g}+B_{2g}}{2}$	$\frac{15.00+14.79}{2}=14.90$	14.7[a]	1
	transverse	$\frac{B_{3g}+B_{2g}}{2}$	$\frac{8.77+8.54}{2}=8.65$	8.6[b]	1
$Ge_4Si_{4[111]A}$	longitudinal	A_{1g}	15.13	15.0[a]	1
	longitudinal	A_{2u}	13.60	13.0[a]	2
	longitudinal	A_{1g}	8.49	8.7[b]	1
	transverse	E_g	15.23	15.15[a]	1
	transverse	E_g	8.84	8.8[b]	1
$Ge_4Si_{4[111]B}$	longitudinal	A_{1g}	15.10	15.2[a]	1
	longitudinal	A_{2u}	13.16	13.8[a]	2
	longitudinal	A_{1g}	8.58	8.85[b]	1
	transverse	E_g	15.35	15.2[a]	1
	transverse	E_u	15.01	15.0[a]	2
	transverse	E_g	8.97	8.85[b]	1
	transverse	E_u	8.75	8.7[b]	2
$Ge_5Si_{1[001]}$	longitudinal	B_2	8.87	9.0[b]	1
	longitudinal	A_1	8.50	8.45[b]	2
	transverse	E	8.77	8.7[b]	1
	transverse	E	8.30	8.3[b]	2
$Ge_5Si_{1[111]}$	longitudinal	A_1	8.75	8.9[b]	1
	longitudinal	A_1	8.21	8.2[b]	2
	transverse	E	8.97	8.9[b]	1
	transverse	E	8.71	8.7[b]	2

[a] Si-confined mode [b] Ge-confined mode.

3.2 Polar Materials: Macroscopic Theory

Infrared-active modes of polar materials are accompanied by electrostatic fields, which may induce coupling of confined modes across the barriers and thus dispersion along the growth direction. These modes can be represented macroscopically by a dielectric function $\epsilon(\omega)$:

$$\epsilon(\omega) = \epsilon_\infty \frac{\omega_{\rm LO}^2 - \omega^2}{\omega_{\rm TO}^2 - \omega^2} \tag{4}$$

where $\omega_{\rm LO}$ and $\omega_{\rm TO}$ are the longitudinal ($\epsilon(\omega_{\rm LO}) = 0$) and transverse ($\epsilon(\omega_{\rm TO}) = \infty$) frequencies, respectively. They may be chosen to depend on $\vec{k}$, in which case $\epsilon(\omega) = 0$ and $\epsilon(\omega) = \infty$ yield the dispersion relations of LO and TO modes.

We present next a macroscopic treatment of the optic modes, based on the $\epsilon(\omega)$ of the two materials in a way similar to that used for acoustic modes in Sec. 2, which was based on the elastic constants and densities of both media. For simplicity we neglect electromagnetic propagation effects (i.e. retardation) and treat the problem as a purely electrostatic one. This implies that $k \gg 2\pi/\lambda$, where λ is the reststrahlen wavelength. This condition certainly applies for Raman backscattering, but not for forward scattering (if $k \simeq 2\pi/\lambda$). In the latter cases polariton effects appear.

From the absence of charges and retardation [29], one concludes that the electric field $\vec{E}$ and the electric displacement $\vec{D}$ ($\vec{D} = \epsilon\vec{E}$) must fulfill [5]

$$\vec{\nabla} \times \vec{E} = 0\,, \tag{5a}$$
$$\vec{\nabla} \cdot \vec{D} = 0\,. \tag{5b}$$

These equations are simultaneously fulfilled in the bulk materials either for $\epsilon = 0$ ($\vec{D} = 0$, LO modes) or for $\epsilon = \infty$ ($\vec{E} = 0$, TO modes). However, other solutions are possible in SL's.

Equation (5a) implies that $\vec{E}$ (or $\vec{D}$) derives from a scalar (or vector) potential ϕ (or $\vec{A}$). Replacing $\vec{D} = \epsilon\vec{E}$ in (5b) and $\vec{E} = -\vec{\nabla}\phi$, we obtain

$$\nabla \cdot \epsilon \nabla \phi = 0 \tag{6}$$

where ϵ is discontinous at the interface. We can solve (6) within each medium:

$$\epsilon_{1,2} \nabla^2 \phi = 0 \tag{7}$$

and treat the discontinuity of ϵ by means of boundary conditions at the interfaces. Due to the translational invariance perpendicular to z, ϕ must fulfill a Bloch condition. In the continuum approximation (implicit in the use of $\epsilon(\omega)$) this leads to an in-plane dependence of $\phi \sim e^{ikx}$, where k is the magnitude of the in-plane wave vector, and x is a distance in the plane. Let us consider the possibility of confined modes arising from (7). Confinement in medium 1 implies that in medium 2 ϕ must be constant, since $E_2 = D_2 = 0$. The continuity of E_x and the mechanical displacement $\vec{u} \sim (\vec{D} - \vec{E})$ at the boundary requires ϕ to be continuous and $d\phi/dz = 0$. The continuity of D_z is automatically fulfilled if $\epsilon_1 = 0$ [3]. We express ϕ_1 as

$$\phi_1(x,z) = e^{ikx}\varphi_1(z) \quad . \tag{8}$$

The continuity conditions thus require, for $x \neq 0$, that φ_1 and $d\varphi_1/dz$ vanish at the boundary. If we decompose φ_1 in Fourier series (sines and cosines) these conditions require mixing of various components, a fact which invalidates the treatment in 3.1 since different components correspond to different eigenfrequencies ω. The macroscopic treatment can only be pursued if we either assume $k = 0$ (in this case ϕ need not vanish at the interface) or if we neglect the dispersion of the bulk optical frequencies. For $k = 0$ we obtain

$$\phi_1 \sim \begin{array}{ll} \sin m\frac{\pi}{d_1}z, & m \text{ odd} \quad , \\ \cos m\frac{\pi}{d_1}z, & m \text{ even} \quad . \end{array} \tag{9}$$

This potential leads to an envelope of the atomic displacements $u_z \sim E_z \sim d\phi/dz$:

$$u_{1z} \sim \begin{array}{ll} \cos m\frac{\pi}{d_1}z, & m \text{ odd} \quad , \\ \sin m\frac{\pi}{d_1}z, & m \text{ even} \; , \end{array} \tag{10}$$

equivalent to those found in 3.1 and corresponding to the LO bulk frequencies obtained for the q's of (2). Note that if we take the expressions (9) to be those of the vector potential component A_y, we obtain (10) for the TO displacements along the x direction. From Maxwell's equation we derive for $\omega \to 0$ (no retardation):

$$\epsilon^{-1}\vec{\nabla} \times \vec{\nabla} \times \vec{A} = -\epsilon^{-1}\nabla^2\vec{A} = 0 \tag{11}$$

which is satisfied if $\epsilon = \infty$, i.e. for the bulk transverse frequencies.

We have already mentioned that it is not possible to fulfill both electrical and mechanical boundary conditions simultaneously with a single function of the type (9) when multiplied by the Bloch factor e^{ikx} i.e., for $k \neq 0$ (8). Modified functions φ, which fulfill both $\varphi = 0$ and $d\varphi/dz = 0$ at the boundaries, have been recently proposed [30,31]. They are

$$\varphi_1(z) = \begin{array}{ll} \cos(m\pi z/d_1) - (-1)^{m/2} & m = 2,4,6 \\ \sin(\mu_m \pi z/d_1) + C_n(z/d_1) & m = 3,5,7 \end{array} \tag{12}$$

where $\mu_3 = 2.86$, $\mu_5 = 4.91$, $\mu_7 = 6.95$, tending to m for large m, and C_n assumes values which tend to ± 2 ($C_3 = 1.95$, $C_5 = -1.98$, $C_7 = 2$). These functions are shown in Fig. 4 together with the corresponding $d\varphi/dz \sim E_z \sim u_z$ for $m = 2,3,4,5$. It can be seen in Fig. 4 and inspection of (12) confirms that the u_z's are the same as those obtained above (10) for m even. They differ by a constant offset from them for m odd, although this difference is only significant for $m = 3$. Eq. (12) can similarly be applied to the TO modes, by replacing ϕ by the vector potential $\vec{A}$. The u_z envelopes obtained for m odd differ somewhat from those of the simple sine functions of the nonpolar case in that their pitch is changed slightly (by the difference between μ_m and n) and they are shifted vertically by an amount proportional:

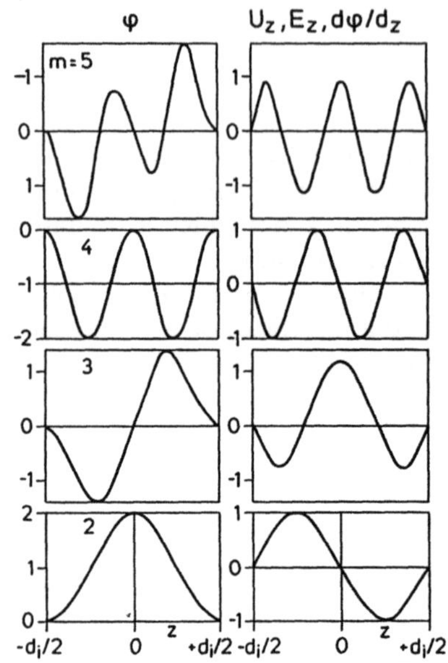

Fig. 4 Potential $\varphi(z)$, which represents longitudinal confined modes in polar superlattices, according to (12). Also, $d\varphi/dz$ which represent the electric field (E_z) and mechanical displacement (u_z) along z.

$$\frac{C_n}{\mu_m \pi} \simeq \pm \frac{2}{m\pi} \tag{13}$$

which tends to zero for large m and amounts to only 0.22 for $m = 3$. Even in this extreme case the offset is small. These differences are not very significant; and furthermore they should not be taken too seriously, since they have been derived under the assumption of no dispersion of ω vs. $\vec{k}$ in the bulk. This requirement arises from the C_n term in (13), i.e., the offset (13) which, when expanded in Fourier series, contains components of all odd m's and thus can only be mixed if the q_m-dispersion is neglected. These conclusions are confirmed by the microscopic treatment of the ir-active modes (see Sec. 3.3). It is found that constant offsets in an otherwise sinusidal envelope function correspond to long range electrostatic fields, i.e., to so-called interface modes which have an off-axis $\vec{k}$-vector.

The discussion above has revealed several important points. First of all, there is no effect of the long range electrostatic fields for m even. This could have already been inferred from the fact that the envelope functions are odd and the corresponding total eigenvectors are even, with respect to the parity operations (i.e., two two-fold axes perpendicular to z). Hence the modes are ir-inactive (ir-activity requires vector behavior, i.e., odd eigenvectors). Only ir-active modes, such as found for m = odd, are accompanied by long-range fields. The treatment of the m-even modes is the same as for nonpolar materials since they are nonpolar modes.

We note that the $m = 1$ envelope is missing in (12) since no corresponding solutions for μ_1 and C_1 can be found. This case must be treated separately and will lead to the so-called interface phonons. The $m = 1$ mode, found to be exactly given by the envelope of (10) for $k = 0$ (on-axis propagation), evolves into an interface mode when $\vec{k}$ is tipped away from the z-axis (see below).

3.3 Interface Modes: Macroscopic Theory

We note that the number of phonon modes must be the same in a polar as in a nonpolar lattice with the same number of atoms. In an attempt to search for the $m = 1$ modes, missing in (12), we consider the possibility of fulfilling (7) for $\epsilon_{1,2} \neq 0$ by requiring $\nabla^2\phi = 0$, with the appropriate boundary conditions. We neglect, for the time being, the mechanical boundary conditions and impose only the continuity of E_x and D_z. We find for the case of a single interface between media 1 ($z > 0$) and 2 ($z < 0$) the potential

$$\phi_{1,2} = e^{ik\bar{x}} e^{\mp kz} \tag{14}$$

where $-(+)$ applies to medium 1 (2). The boundary condition for E_x is automatically fulfilled by (14). That for D_z leads to the well known secular equation for interface excitations

$$\eta(\omega) = \frac{\epsilon_1(\omega)}{\epsilon_2(\omega)} = -1 \quad . \tag{15}$$

Equation (15) has two solutions, one for a frequency between ω_{TO} and ω_{LO} of medium 1 and the other between ω_{TO} and ω_{LO} of medium 2. For a superlattice we must construct the corresponding potential as a linear combination of ϕ's of the type (14) at a 1-2 and at the nearest 2-1 interface. These potentials must then be carried over to subsequent layers s by multiplying them by Bloch factors $\exp(isdq)$, where s is an integer which denotes the separation between layer s and that taken as origin. Imposing electrostatic boundary conditions at both interfaces we reach the secular equation [32]

$$\cos qd = \cosh(kd_1)\cosh(kd_2) + \frac{\eta(\omega)^2 + 1}{2\eta(\omega)} \sinh(kd_1)\sinh(kd_2) \tag{16}$$

where $\eta(\omega)$ is given in (15). Typical examples of the dispersion relations $\omega(q,k)$ obtained with (16) are shown in Fig. 3 of [32].

It is easy to show that the solutions $\omega(q,k)$ of (16) are singular for $q, k \to 0$, i.e., they depend on the angle $\Theta = \arctan(k/q)$ of the wave vector with the axis of growth. For $q, k \to 0$ we find from (16)

$$\langle\epsilon\rangle \langle\epsilon^{-1}\rangle = -\frac{d^2}{d_1^2 + d_2^2} \cot\Theta \tag{17}$$

where $\langle\rangle$ denotes the linear average over the two media, i.e. $\langle\epsilon\rangle = (d_1\epsilon_1 + d_2\epsilon_2)/d$. For $\Theta \to 0$ (17) yields

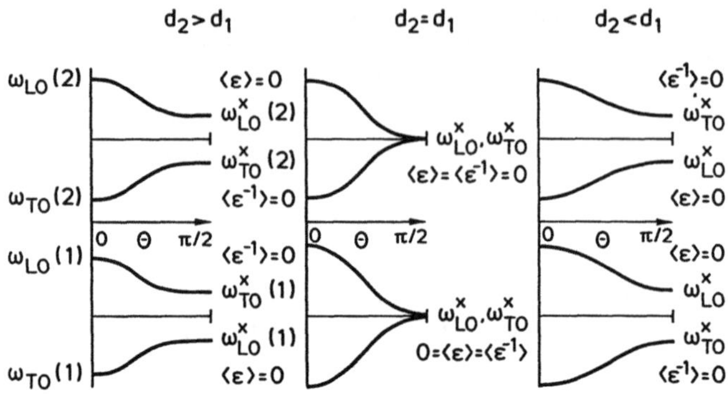

Fig. 5 Angular dispersion of long wavelength interface modes (i.e., the missing $m = 1$ of Fig. 4) for superlattices with two polar constituents.

$$\langle\epsilon\rangle = \infty \quad \text{or} \quad \langle\epsilon^{-1}\rangle = \infty \quad . \tag{18}$$

The solutions of (18) are $\omega_{\mathrm{TO},1}$, $\omega_{\mathrm{TO},2}$, $\omega_{\mathrm{LO},1}$, $\omega_{\mathrm{LO},2}$, i.e., the longitudinal and transverse modes of the bulk as already found in Sec. 3.2. For $\Theta \to \pi/2$, i.e., for in-plane propagation, (17) yields

$$\langle\epsilon\rangle = 0 \quad \text{or} \quad \langle\epsilon^{-1}\rangle = 0 \quad . \tag{19}$$

The solutions of (19) can be interpreted as the longitudinal and transverse frequencies of an effective medium of average dielectric function $\langle\epsilon\rangle$ and $\langle\epsilon^{-1}\rangle$, respectively. There are two such frequencies, each lying between ω_{TO} and ω_{LO} of the corresponding medium. For $d_1 = d_2$ both conditions (19) lead to the same frequency ($\omega_{\mathrm{TO}} = \omega_{\mathrm{LO}}$) which is found by solving

$$\eta(\omega) = \frac{\epsilon_1(\omega)}{\epsilon_2(\omega)} = -1 \tag{20}$$

i.e. we recover the secular equation (15) for interface modes.

The angular dispersion just described for interface modes in the long-wavelength limit is illustrated in Fig. 5 for three cases, $d_2 > d_1$, $d_2 = d_1$, and $d_2 < d_1$. It is easy to understand the decrease in LO-TO splitting for $\Theta = \pi/2$ as compared with the bulk splitting found for $\Theta = 0$. For $\Theta = 0$ the LO mode of medium 1 is upshifted by the electrostatic restoring force produced by the charges induced on the layer boundaries. The same situation as for the bulk obtains. For $\Theta = \pi/2$, i.e. in-plane propagation, the polarization charges are induced on the sides of the slabs. One must thus average the contribution of medium 1 and medium 2 to these restoring forces and one obtains $\langle\epsilon\rangle = 0$ for the LO modes. This leads to a smaller LO-TO splitting than in the bulk: for the 1-like vibration, medium 2 does not contribute to the electrostatic restoring force and vice versa. In the long wavelength $\Theta = \pi/2$ limit, the electric fields E and displacements D, and thus the mechanical displacements also, are constant over each layer. For the LO (TO) mode u_x (u_z), which is proportional to E (D) are discontinuous at the interfaces since either D (LO-mode) or E (TO-mode) is discontinuous. The discontinuity of u is unphysical, resulting from having neglected the mechanical boundary condition in the macroscopic treatment. As a result of the short range nature of the mechanical forces, it is expected that this boundary condition will lead to a rounding off of the constant u_x (u_z) towards the layer boundaries (see Fig. 6).

3.4 Polar Materials: Microscopic Theory

The correct treatment of the mechanical and electrostatic boundary conditions requires microscopic lattice dynamical calculations. An attempt at such microscopic calculations was made in [29]. It is, however, too simple to properly account for the microscopic effect, e.g., it neglects frequency dispersion in the bulk.

Several microscopic calculations have been performed for polar superlattices [8,9,11,33]. We present in Fig. 6 the atomic displacements obtained for $\Theta = \pi/2$ long wavelength modes of a [111] $(GaAs)_9/(AlAs)_{10}$ superlattice [8]. The mode labeled $IF_{\parallel}$ (Fig. 6a) represents the LO-like mode (in-plane vibration). We recognize in medium 1 the flat envelope function discussed in 3.3 and the rounding off towards the layer boundaries so as to fulfill the mechanical boundary condition. Note that the vibrational amplitude in medium 2 is not zero. In Fig. 6b, we show a transverse "interface" mode (note that the "interface" designation is misleading, these modes are stronger in the middle of the layer than at the interfaces), which is strongly mixed with an $m = 3$ "confined" mode ($IF_{\perp}$-LO_3) thus yielding the vertically offset sinusoidal pattern discussed in Sec. 3.3. Its mixed partner $IF_{\perp}$-LO_3 is displayed in Fig. 6c.

The singular behavior of the m-odd modes for $\vec{k} \rightarrow 0$, i.e., their dependence on Θ, as calculated microscopically for the $(GaAs)_9/(AlAs)_{10}$ [111] superlattices [8], is displayed in Fig. 7. We note that m-even modes are indeed nondispersive, a conclusion already reached in Sec. 3.2. For small Θ the LO_1 and TO_1 modes are strongly dispersive. The latter displays in Fig. 6 for $\Theta = \pi/2$, the characteristic envelope function which corresponds to the "missing" $m = 1$ mode in Sec. 3.3. The LO_3 mode, with weak dispersion for $\Theta \approx 0$, as corresponds to an $m = 3$ mode of the macroscopic theory, mixes strongly with the LO_1 mode for increasing Θ and exhibits an anticrossing with it (related to the fact that both have the same symmetry; no anticrossing with the m-even modes, whose symmetry is different, occurs). For $\Theta = \pi/2$, the roles of the LO_1 and LO_3 modes reverse as they both become strongly mixed, i.e., superpositions of $m = 3$ oscillations and a vertical offset (Figs. 6b and c). The higher frequency mode is, however, for $\Theta = \pi/2$, more ($m = 3$)-like than the lower frequency one. Note that for $\Theta = \pi/2$ the high frequency interface modes are TO-like ($IF_{\perp}$) while the lower frequency ones have LO character, in accordance with the macroscopic results of Fig. 5 ($n_1 < n_2$).

Figure 8 displays the envelope functions of the $m = 1$ ($\Theta = 0$) and $m = 2$ (any Θ) GaAs-like confined modes of the [111] $(GaAs)_{10}/(AlAs)_{11}$ SL as calculated microscopically. They exhibit clear sinusoidal character (Eq. (10)),

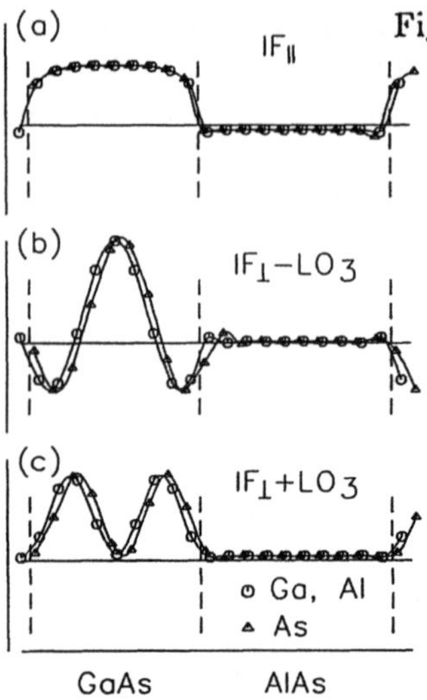

Fig. 6

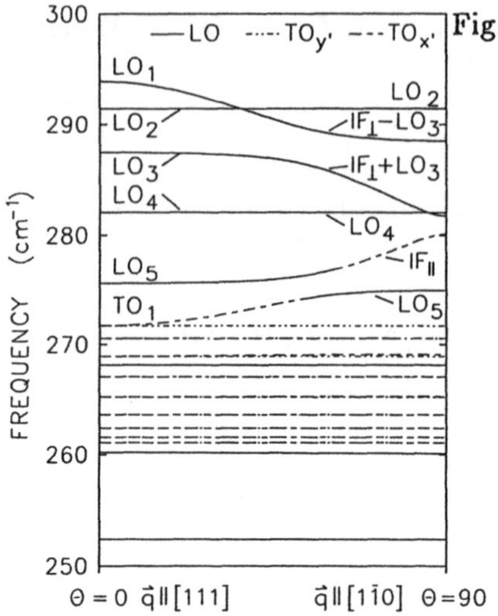

Fig. 7

Fig. 6 Microscopic envelope functions of interface-like modes in a $(GaAs)_9/(AlAs)_{10}$ [111] superlattice [8].

Fig. 7 Angular dispersion of the long wavelength modes of the superlattice of Fig. 6 [8], as obtained from microscopic lattice dynamical calculations [8].

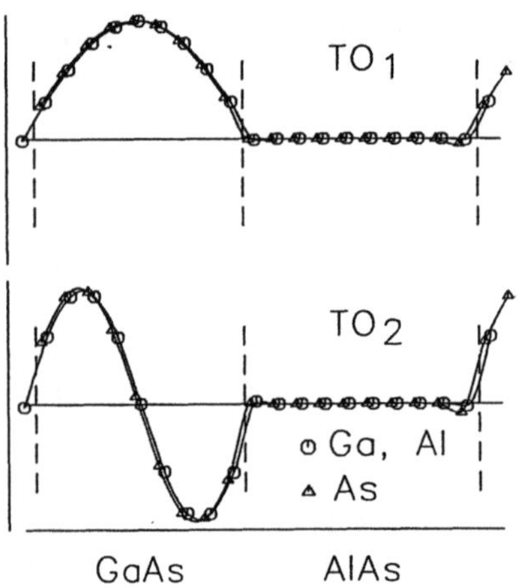

Fig. 8 Envelope function of confined modes of the superlattice of Fig. 6 as obtained from microscopic calculations [8].

as well as slight penetration into the AlAs layer. It is interesting to note that those modes do not strictly show parity behavior. The deviation, although small, can be appreciated for the As vibration and results from the lack of two-fold axis or reflection plane perpendicular to [111]. The vibrations of either Ga or Al, however, exhibit parity-like behavior, a fact which has been shown

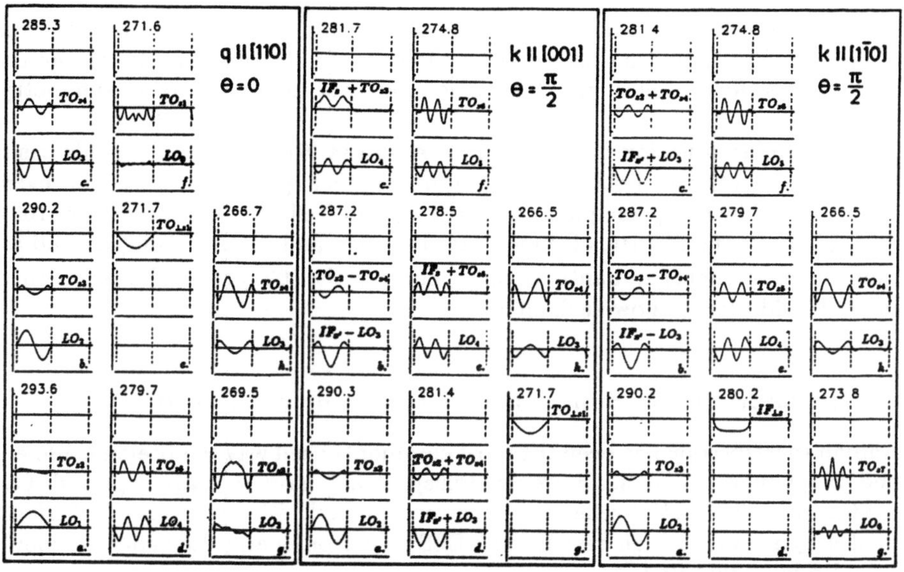

Fig. 9 Microscopically obtained envelope functions of long wavelength modes in a (GaAs)/(AlAs) [110] superlattice for axial ($\Theta = 0$) and transverse ($\Theta = \pi/2$) propagation, showing, in the latter case, typical interface effects [9].

to arise from a supersymmetry which results from having assumed equal force constants and effective charges for bulk GaAs and AlAs [8]. Deviations from symmetry-antisymmetric behavior would not be expected to be large, even if the force constants were slightly different.

Similar results have been obtained for [110] and [012] superlattices [9,10]. In these cases, however, bulk LO and TO modes can mix in the superlattice. In Fig. 9 we show the envelope functions calculated for a [110] $(GaAs)_{13}/(AlAs)_{14}$ SL [9]. Note that the LO mode mixes with the $(TO)_z$ mode (polarized along [001]). Despite these modes having opposite bulk parities (with respect to [001] two-fold axis), they display the same parity in the SL when m is odd for one of them and even for the other. The odd-even mixture is clearly seen in the envelope functions of Fig. 9. Note that in this case one must distinguish between the two different propagating directions perpendicular to [110] for $\Theta = \pi/2$: $k \parallel [001]$ and $k \parallel [1\bar{1}0]$, since they are not symmetry equivalent. The flat envelope function with rounded-off edges characteristic of IF-modes, is displayed by the mode near 280 cm^{-1} for $k \parallel [1\bar{1}0]$. Other interface-like modes exhibit the mixture of this shape with a sinusoidal function, also found in connection with IF modes mixed with $m \geq 3$, odd (e.g. those near 281 and 287 cm^{-1} modes for $k \parallel [1\bar{1}0]$). For $\vec{k} \parallel [1\bar{1}0]$ and $\Theta = \pi/2$, all modes are ir-active, through either the transverse or the longitudinal component, since m odd and even mix: all LO-TO_z modes anticross vs. Θ. For $\vec{k} \parallel [1\bar{1}0]$, however, the infrared activity lies for m (LO) even (m (TO) odd) along the [001] direction, which is perpendicular to the plane defined by $\vec{k}$ and the direction of growth. These modes thus exhibit no long range electrostatic effects and do not anticross for $0 < \Theta < \pi/2$, a situation like that found for higher symmetry superlattices.

A similar situation is found for GaAs/AlAs [012] SL's [9]. In this case, however, the LO and both TO modes mix, with appropriate m = odd, even switchings. The two principal $\vec{k}$-planes in which IF-effects can be seen are (100) and (02$\bar{1}$). In the former, all modes are ir-active and m (LO) odd, as well as even, show IF-effects and anticrossings. For $\vec{k}$ in the (02$\bar{1}$) plane, however, only m (LO) odd planes show such effects.

4. Optical Modes in Superlattices: Experimental Results

Raman spectroscopy has been the most commonly used way to observe confined and IF phonons in SL's. However, Pusep et al. have recently reported TO modes by ir differential reflection spectroscopy [34]. Observations of interface modes by means of high resolution EELS (electron energy loss spectroscopy) has also been reported [35].

4.1 Confined Modes

Detailed experimental investigations of ω vs. m for confined modes in Ge/Si superlattices have been mostly performed at the Walter-Schottky Institute [36]. A large number of papers have been devoted to confined modes in SL's made out of III-V and II-VI semiconductors. Since we cannot review them all, the reader is referred to [2–9,11]. Recent Raman work for [001] GaAs/AlAs SL's [37] has yielded data for GaAs confined phonons with m up to 8 and AlAs-like ones with m up to 4, and compared the measured frequencies with the best available bulk dispersion relations. We reproduce these results in Fig. 10. The data for the AlAs-like phonons allow us to assess the accuracy of three lattice dynamical calculations available for the bulk material shown in Fig. 10b: no neutron scattering results are available for AlAs since large enough single crystals cannot be grown. It can be seen in Fig. 10b that the "ab initio" calculations of Baroni et al. [38] give the best agreement with experiment.

We recall that for [001] SL's only LO-like modes can be observed in backscattering (TO-modes are observable for other growth directions [8,9,11]). In this case it is easy to distinguish between m-odd (B_2) symmetry and m-even (A_1) modes. They are both Raman allowed by symmetry, although B_2 requires the $z(x,y)\bar{z}$ backscattering configuration, while A_1 is observed for $z(x,x)\bar{z}$ and $z(y,y)\bar{z}$ (the letter outside the brackets indicate propagation, those inside the polarization directions of incident and scattered light). Note that the $z(x,x)\bar{z}$, $z(y,y)\bar{z}$ configurations lead to dipole forbidden scattering in the bulk which nevertheless is made quadrupole allowed by the electrostatic Fröhlich interaction [25]. The scattering becomes dipole allowed in the superlattice since the "quadrupole-inducing" wave vector transfer has to be replaced by the superlattice – given $q = m\pi/d_{1,2}$. The striking m odd/even selection rule just

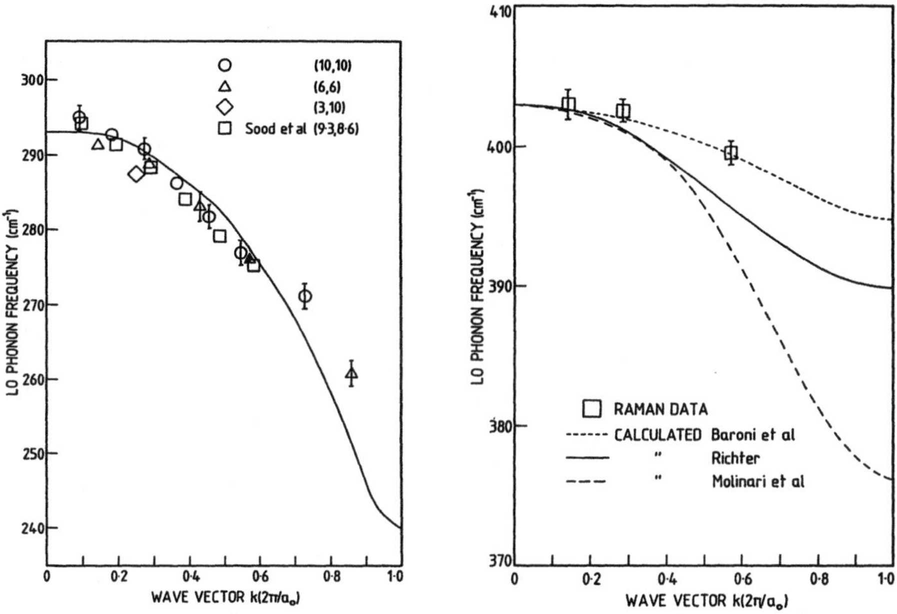

Fig. 10 Bulk LO dispersion relations of GaAs and AlAs obtained using (3) from measurements on [001] superlattices [37] compared with calculated bulk dispersion relations [38].

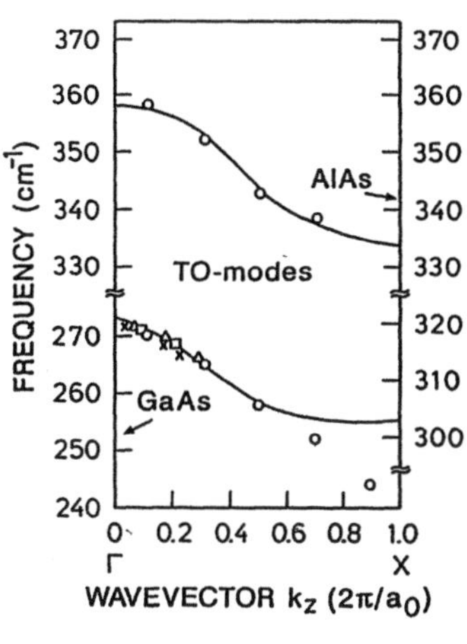

Fig. 11 Some as Fig. 10 but for TO modes. The points were obtained from ir spectra [34].

mentioned for the polarization configurations is easily observed [39] and can be helpful in determining which order m a particular peak corresponds to.

TO modes of [001] GaAs/AlAs SL's (both GaAs and AlAs-like m odd) have been recently observed by differential ir reflectivity [34] and also by Raman scattering (under extremely resonant conditions, only GaAs-like, all m's) [39]. The frequencies obtained are plotted in Fig. 11 and compared with existing lattice dynamical calculations.

4.2 Interface Modes

We have already mentioned the observation of interface modes of GaAs/AlAs superlattices by means of EELS [35]. These modes appear as a broad band between the LO and TO frequencies of both bulk GaAs and AlAs.

Interface modes are also usually observed in Raman backscattering. This is somewhat surprising, since in backscattering configuration one should only couple to phonons with $\Theta = 0$ (confined modes). It is usually assumed that the transverse k required for coupling to interface phonons is supplied by defects, possibly related to interface roughness. In support of this, it has been observed that the strength of interface modes varies in a strongly nonlinear way with the laser power (see Fig. 12) [40], in contrast to the linear power dependence of allowed confined modes. It is believed that light excited electron-hole pairs screen out the disorder potential and thus decrease forbidden coupling to the IF-modes.

In Fig. 2 of [32] the Raman spectra of AlAs-like IF modes are displayed for three superlattices (A: $d_1 = d_2 = 2$ nm; B: $d_1 = 2$ nm; $d_2 = 6$ nm; and C: $d_1 = 6$ nm, $d_2 = 2$ nm). In case A a broad peak centered between the LO and TO bulk frequency is seen. For case B the main peak is shifted towards LO, with a weaker shoulder close to TO, while in case C a situation opposite to that of case B obtains. It is believed that this difference is related to the LO-TO switching seen in Fig. 5 for $d_1 \gtrless d_2$.

A calculation of these spectra, as induced by either neutral or ionized randomly distributed point defects, has been recently performed [41]. Typical results are reproduced in Fig. 13 where it is seen that the neutral defect

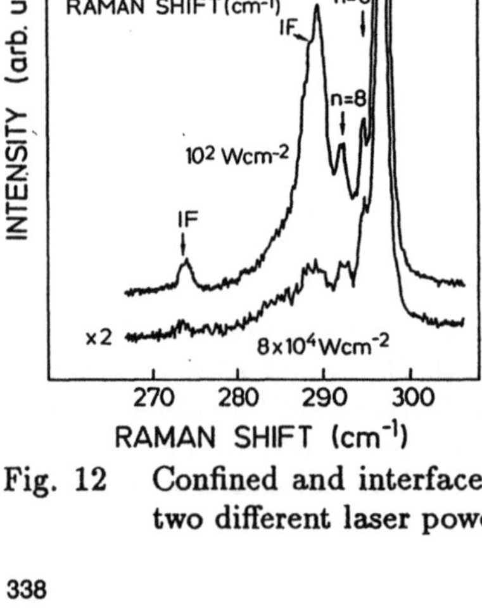

Fig. 12 Confined and interface modes of (GaAs)/(AlAs) superlattice for two different laser power densities [40].

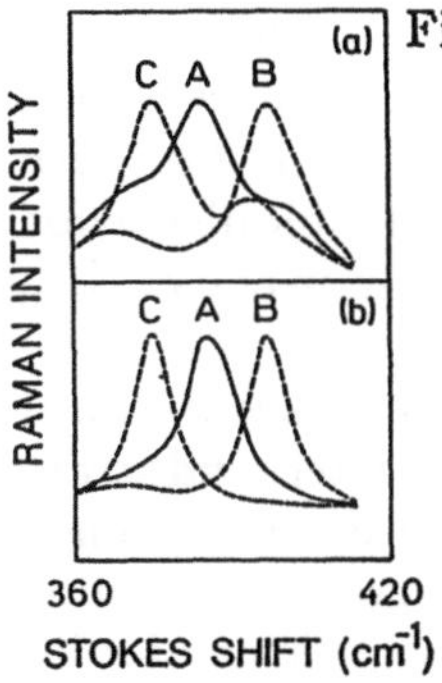

Fig. 13

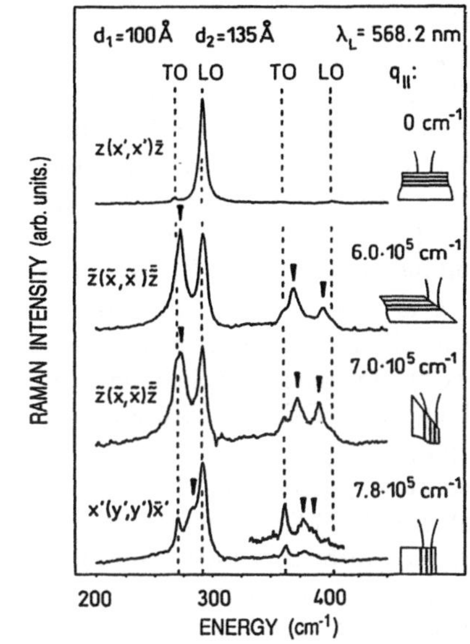

Fig. 14

Fig. 13 Calculated backscattering by interface modes in GaAs/AlAs superlattices under the presence of charged (b) or neutral (a) impurities. A: (20/20Å), B: (20/60Å), C: (60/20Å) [41].

Fig. 14 Dispersion of interface modes in a GaAs/AlAs superlattice (arrows) obtained in backscattering as illustrated by the diagrams on the right [42].

case reproduces better the shoulder observed experimentally for cases B and C. This contradicts, however, the interpretation of the power dependence of the data in Fig. 12, discussed earlier, where the scattering was suggested to be due to charged defects, which are screened out by the light induced carriers. Obviously further work, including a calculation of scattering induced by interface roughness, is required to clarify the matter.

The observations just mentioned do not allow for a determination of IF frequencies vs wave vector $\vec{k}$. Recent work [42], involving scattering on the side of a GaAs/AlAs superlattice (Fig. 14), has provided data on such a dispersion and has shown that it agrees with the results of the macroscopic calculation of Sec. 3.3. Coupling to IF modes of well defined $\vec{k}$ has also been achieved by backscattering studies on a superlattice, which had a vertical 16 nm-period grating deposited on top [43].

4.3 Multiphonon Spectra

The Raman spectra of GaAs/AlAs SL's show, especially near resonance with the lowest direct excitations, a great wealth of multiphonon peaks. Overtones and combinations of phonons near Γ (mini-BZ), mainly confined but sometimes also IF, are observed [44–47]. The fact that only modes near Γ, instead of the whole density of two phonon states, is observed suggests that we are dealing here with iterated scattering processes via first order electron-phonon (most likely Fröhlich) interaction. Processes in which deformation-potential-electron-two-phonon interaction is at work yield much broader, density-of-states-like spectra [25]. It has been noted that for not too short period superlattices only overtones and combinations of modes with m even are observed in Raman scattering [47]. This has been attributed to the angular dispersion found for m odd phonons (Figs. 5 and 7), which tends to smear out the multiphonon structures ($\vec{k}$ is close to zero but can have any direction). For short period superlattices, IF multiphonon peaks are also observed [45]. This has been attributed to the formation of electron minibands with non-negligible dispersion along z which favor Fröhlich interaction induced scattering by ir-active (i.e., IF) modes [45].

The scattering processes described in this article are resonant for laser (ω_L) or scattered (ω_S) frequencies close to strong excitonic (or interband) transitions. The ω_S resonances are usually dominant, especially in multiphonon processes [45]: double resonances, in which both ω_L and ω_S agree with exciton frequencies, can be induced in SL's as a result of the light-heavy hole band splitting. By appropriate choice of layer thickness this splitting can be made nearly equal to the expected Raman shift. For two-phonon scattering the intermediate state reached after the emission of the first phonon can also be resonant and strong triple resonances result [47]. Under these conditions crossed-polarized scattering, e.g., $z(x,y)\bar{z}$, induced by Fröhlich interaction becomes dominant [30,47]. This is a seemingly paradoxical situation since Fröhlich interaction usually favors parallel-polarized scattering [25]. It can, however, be simply understood when the symmetries of the split value band states are properly taken into account [47,48].

I would like to thank A. Shields for a very careful reading of the manuscript.

References

[1] *Light Scattering in Solids*, Vols. I – VI, edited by M. Cardona and G. Güntherodt (Springer Verlag, Heidelberg, 1975 – 1991).

[2] B. Jusserand and M. Cardona, Ref. [1] Vol. IV, 1989.

[3] M. Cardona, Superlatt. and Microstr. 7, 183 (1990).

[4] M. Cardona, in *Lectures on Surface Sciences* (SLAFS 1986) G.R. Castro and M. Cardona eds. (Springer Verlag, Heidelberg, 1987), p. 2.

[5] M.V. Klein, in *Raman Spectroscopy: Sixty Years of Vibrational Structure and Spectra* (Elsevier, Amsterdam, 1989), p. 203.

[6] J. Menéndez, J. Luminesc. **44**, 285 (1989).

[7] D.W. Feldmann, J.H. Parker, Jr., W.I. Choyke, and L. Patrick, Phys. Rev. **173**, 787 (1968); S. Nakashima and K. Takara, Phys. Rev. B **40**, 6339 (1989).

[8] Z.V. Popović, M. Cardona, E. Richter, D. Strauch, L. Tapfer, and K. Ploog, Phys. Rev. B **41**, 5904 (1990).

[9] Z.V. Popović, M. Cardona, E. Richter, D. Strauch, L. Tapfer, and K. Ploog, Phys. Rev. B **40**, 3040 (1989).

[10] E. Friess, H. Brugger, K. Eberl, G. Krötz, and G. Abstreiter, Solid State Commun. **69**, 899 (1989).

[11] Z.V. Popović, M. Cardona, E. Richter, D. Strauch, L. Tapfer, and K. Ploog, Phys. Rev. B, in print.

[12] S. Subbanna, H. Kroemer, and J.L. Merz, J. Appl. Phys. **59**, 488 (1986).

[13] P. Molinàs i Mata, M.I. Alonso, and M. Cardona, Solid State Commun. **74**, 347 (1990).

[14] M. Cardona, in *Proceedings of the NATO Symposium on Light Scattering in Semiconductors Microstructures*, Mt. Toremblant 1990, ed. by J. Lockwood and J. Young (Plenum Press, New York, 1991).

[15] P. Molinàs i Mata and M. Cardona, Superlatt. and Microstr., in press.

[16] B.A. Weinstein and R. Zallen, Ref. [1], Vol. IV and references therein.

[17] E. Anastassakis, A. Cantarero, and M. Cardona, Phys. Rev. B **41**, 7529 (1990).

[18] *Strained-Layer Superlattices Physics*, T. P. Pearsall ed. (Academic Press, Boston, 1990).

[19] S.M. Rytov, Akust. Zh. **2**, 71 (1956). [Sov. Phys. Acoust. **2**, 68 (1956)].

[20] R.A. Ghambari, J.D. White, G. Fasol, C.J. Gibbings, and C.G. Tuppen, Phys. Rev. B **42**, 7033 (1990); see also R.A. Ghambari and G. Fasol, Solid State Commun. **70**, 1025 (1989).

[21] Z. Jian, Z. Kaiming, and Xie Xide, Phys. Rev. B **41**, 12862 (1990).

[22] P. Molinàs i Mata and M. Cardona, Phys. Rev. B, in press.

[23] U. Schmid, N.E. Christensen, and M. Cardona, Phys. Rev. B **41**, 5919 (1990).

[24] B. Jusserand, F. Alexandre, J. Dubard, and D. Paquet, Phys. Rev. B **33**, 2897 (1986); P. Santos, L. Ley, J. Mebert, O. Koblinger, Phys. Rev. B **36**, 4858 (1987).

[25] M. Cardona in Ref. [1], Vol. II (1983).

[26] F. Calle, M. Cardona, E. Richter, and D. Strauch, Solid State Commun. **72**, 1153 (1989).

[27] O. Koblinger, J. Mebert, E. Dittrich, S. Dottinger, W. Eisenmenger, P.V. Santos, and L. Ley, Phys. Rev. B **35**, 9372 (1987).

[28] A. Fasolino, E. Molinari, and K. Kunc, Phys. Rev. Lett. **56**, 1751 (1986); B. Jusserand and D. Paquet, ibid **56**, 1752 (1986).

[29] R. Enderlein, Phys. Status Solidi (b) **150**, 85 (1988).

[30] K. Huang, B. Zhu, and H. Tang, Phys. Rev. B **41**, 5825 (1990).

[31] K. Huang and B. Zhu, Phys. Rev. B **38**, 13377 (1988).

[32] A.K. Sood, J. Menéndez, M. Cardona, and K. Ploog, Phys. Rev. Lett. **54**, 2115 (1985).

[33] S.F. Ren, H. Chu, and Y.C. Chang, Phys. Rev. B **37**, 8899 (1988).

[34] Iu.A. Pusep, A.F. Milekhin, M.P. Sinyukov, K. Ploog, and A.I. Toporov, to be published. See also G. Scarmacio, L. Tapfer, W. König, K. Ploog, E. Molinari, and S. Baroni, to be published.

[35] P. Lambin, J.P. Vigneron, A.A. Lucas, P.A. Thiry, M. Liehr, J.J. Pireaux, R. Caudano, and T.J. Kuech, Phys. Rev. Lett. **56**, 1227 (1986).

[36] E. Friess, K. Eberl, U. Menczingar, and G. Abstreiter, Solid State Commun. **73**, 203 (1990).

[37] D.J. Mowbray, M. Cardona, and K. Ploog, Phys. Rev. B, in press.

[38] S. Baroni, P. Gianozzi, and E. Molinari, Phys. Rev. B **41**, 3870 (1990).

[39] A.K. Sood, J. Menéndez, and M. Cardona, Phys. Rev. Lett. **54**, 2110 (1985).

[40] G. Ambrazevicius, M. Cardona, R. Merlin, and K. Ploog, Solid State Commun. **65**, 1035 (1988).

[41] F. Herzel, D. Suisky, J. Röseler, and R. Enderlein, *Proceedings of the 20th International Conference on the Physics of Semiconductors*, E. Anastassakis and S. Pantelides, eds. (World Scientific, Singapore, 1991).

[42] A. Huber, T. Egeler, W. Ettmüller, H. Rothfritz, G. Tränkle, and G. Abstreiter, *Proceedings of the Int. Conf. on Superl. and Microstr.*, Berlin, 1990, to be published.

[43] H. Fuchs, C.H. Grein, C. Thomsen, M. Cardona, W.L. Hansen, E.E. Haller, and K. Itoh, Phys. Rev. B, in press.

[44] A.K. Sood, J. Menéndez, M. Cardona, and K. Ploog, Phys. Rev. B **32**, 1412 (1985).

[45] D.J. Mowbray and M. Cardona, to be published.

[46] M.H. Maynadier, E. Finkman, M.D. Sturge, J.M. Worlock, and M.C. Tamargo, Phys. Rev. B **35**, 2517 (1987); A.M. Brodin, M.Ya. Valakh, V.I. Gavrilenko, M.P. Lisitsa, A.P. Litvinchuk, V.G. Litovchenko, and K. Ploog, JETP Lett. **51**, 178 (1990); M.Ya. Valakh, A.A. Klochikhin, and A.P. Litvinchuk, Sov. Phys. Solid State **26**, 1558 (1985).

[47] A. Alexandrou, M. Cardona, and K. Ploog, Phys. Rev. B **38**, 2196 (1988).

[48] M. Cardona and C. Trallero-Giner, Phys. Rev. B, in press.

Dispersive Transient Charge Carrier Transport in Polycrystalline Films of CdTe

F. Sánchez-Sinencio[1], *J.M. Figueroa*[1;*;+], *R. Ramírez-Bon*[1;@], *O. Zelaya*[1], *G.A. González de la Cruz*[1], *J.G. Mendoza*[1], *G. Contreras-Puente*[2], *and A. Díaz-Góngora*[2]

[1]Departamento de Física, Centro de Investigación y de Estudios Avanzados del IPN, Apdo. Postal 14-740, 07000 México, D.F., Mexico
[2]Departamento de Física, Escuela Superior de Física y Matemáticas del IPN, Apdo. Postal 75-702, 07738 México, D.F., Mexico

Abstract. The time-of-flight of electrons and holes has been measured laterally in intrinsic polycrystalline CdTe films. The time dependent charge transport results can be interpreted as manifestations of dispersive (non-Gaussian) transient transport. This transport is similar to that observed in amorphous semiconductors. Many types of trapping states in the polycrystalline films may generate disorder. Hole and electron transit times (t_T) have been observed in the same sample indicating that the CdTe material is of high quality. Two experimental results give evidence of dispersive transport: 1) The carrier current transients have a power-law behavior, as follows $I(t) \sim t^{-(1-\alpha)}$ for $t < t_T$ and $I(t) \sim t^{-(1+\alpha)}$ for $t > t_T$. 2) The experimental field dependence of electron transit time is nonlinear. All carrier current transients studied in this work were measured at room temperature.

1. Introduction

The understanding of electronic transport in solids is being constructed through several steps. The big initial step was given with the theory of a perfect periodic lattice, followed by the development of the theoretical and experimental knowledge from dispersive (non-Gaussian) transport [1-3] in amorphous materials. However, there is an intermediate case between a perfect lattice and an atomically disordered solid,

---

*) Also at Escuela Superior de Física y Matemáticas del IPN, México, D.F.
+) Scholarship COFAA-IPN.
@) Permanent address: Centro de Investigación en Física, Universidad de Sonora, Hermosillo, Sonora, Mexico.

Springer Proceedings in Physics, Volume 62
Surface Science Eds.: F.A. Ponce and M. Cardona

that is the case of a polycrystalline material. Electrical transport phenomena at semiconductor grain boundaries has been mainly studied in the steady state and for time dependent excitations (frequency modulated experiments) [4,5]. Several theoretical models, most of them based on the idea that the grain boundaries have an inherent space charge region due to the interface, have been proposed in order to explain the experimental results. These models lead to an electronic transport controlled by potential barriers (double Schottky barrier). These barriers can be easily built in doped material. However for an intrinsic semiconductor, the bands will be flat and in this case the barrier potential does not have any influence on the electronic transport. On the other hand, for the intrinsic material the experimental characterization of the electronic transport by steady state measurements, is a severe problem. Hall effect, conductivity and I-V characteristic measurements do not readily lead to meaningful conclusions; almost nothing is known about charge transport in an intrinsic polycristalline semiconductor like CdTe.

In this work, we have used the time-of-flight technique in order to measure the transit time in polycrystalline CdTe films; i.e. the time spent by a charge carrier to travel between two lateral electrodes. The experiments give less ambiguous results than do more conventional steady-state techniques, which always involve, by necessity, long-time averages. The transient charge carrier transport has a dispersive character.

2. Experimental

The transit time was measured in polycrystalline 10 mm x 12 mm x 30-100 μm CdTe samples grown, on 7059 Corning glass substrates, by the Close-Spaced Vapor Transport (CSVT) technique [6]. The CSVT technique developed by Nicoll [7], is a convenient method for growing semiconductor films. The method is simple and allows one to obtain large crystalline grains. This method has been successfully used in the growing of thin films of different semiconductor materials such as Ge [7,9], Si [10], III-V compounds [7,8,11-16], II-VI compounds [6,17-24] and ternary compounds [25,26]. Different authors [27,28] have reported large grain size (up to 70 μm) in CdTe films, grown on glass substrates. Grain size and crystallite orientation depend on the Ar-pressure and substrate temperature [29]. As-grown samples were successively washed in acetone, ethanol, and distilled water,

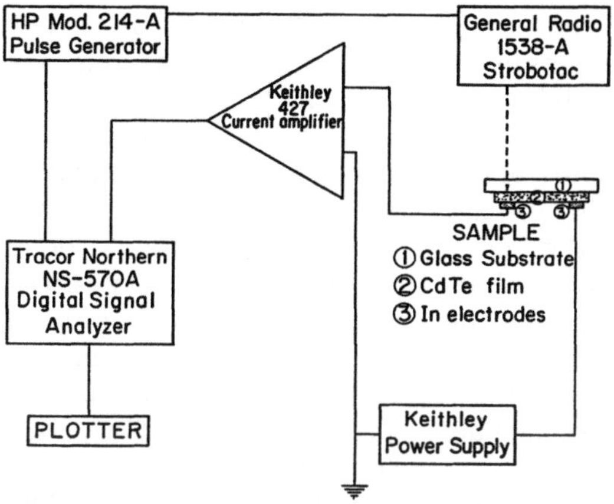

Fig. 1. Experimental arrangement used in the time-of-flight experiments performed laterally in intrinsic polycrystalline CdTe films.

followed by immersion (10-15 sec.) in a 0.5% bromo-methanol solution and rinsed with ethanol and distilled water. Two indium electrodes of 1x10 mm were evaporated on each sample; the distance between electrode varies from 0.2 to 10 mm.

The experimental arrangement used in the time-of-flight experiments is illustrated in the block diagram in Fig. 1. The excitation light source is a stroboscopic lamp with a 10-15 μsec pulse. This pulse duration is roughly two orders of magnitude shorter than most of the transit times measured in the samples. The light is passed through a rectangular opening of 1x10 mm and travels along the glass and the CdTe film, in spatial coincidence with one of the electrodes. Most of the light is absorbed in the top few hundred angstroms of the film. In order to study the importance of the uniform hole-electron pairs generation through the whole thickness of the sample, in some experiments the light was filtered with a 1000 Angstrom CdTe film, as a filter. A pulse generator permits manual control of the stroboscopic light and also serves to trigger the digital signal analyzer. A bias voltage, given by a power supply connected directly to one of the sample electrodes, allows the selection of holes or electrons as the drifting charge inside the sample. The transient current is amplified by a high-gain wide bandwidth amplifier that also insures that the response time of the measurement circuit be shorter than the transit time. On the other hand, the d.c. resistivity measured in the samples leads to relaxation time of the order of 10^{-5} sec. which is shorter than the measured transient time. From this last result, we can assume the existence of two different materials, possibly the intragrain and the intergrain regimes in the films, one with resistivity of about 10^7 Ωcm and other with much higher resistivity. In most of the samples, it was possible to observe hole and electron carrier transit times using an applied field in the range 1-100 V/cm. This is a good indication that the carrier lifetime against deep trapping is not a problem and that the high resistivity CdTe polycrystalline material grown as described above, is of high quality. The digital signal analyzer is used as a signal averager of the observed transient currents. In this way, a true signal component can be extracted from a repetitive, but not necessarily periodic, signal function composed of the desired information component plus an undesirable random noise component. As sweeps are accumulated, the value of the noise component will tend to increase proportionally to the square root of the number of sweeps, while the desired signal component will increase in a directly linear fashion. The degree of signal-to-noise enhancement thus increases as more signal averaging sweeps are performed. In our measurements the number of sweeps varied from 3 to 36 depending on the severity of the trapping in the sample. In some samples electric polarization can be a problem, and in those cases illumination of the sample with white light and without any applied field is required between consecutive sweeps.

3. Experimental results

Figure 2a shows electron current transients plotted on a linear scale for three different applied bias: 1.5 V, 2.5 V and 3.0 V. The currents were arbitrarily shifted along the I-axis, for comparison of the results. Notice that in all current transients is possible to define a transit time, and this is indicated by an arrow on each curve. These transit times are more evident when the same data shown in Fig. 2a are replotted as log I versus log t. In Fig. 2b the current decay follows approximately a power law in time with a gradual change to a faster power law after the transit time. On the other hand, there exists a second knee in all the current transients plotted in Fig. 2b. One possible explanation for this result is the inhomogeneity in applied field along the film. Under this circumstance the electrons travel in some regions faster than in others; this situation is probably due to charge polarization in the sample. However, it is important to note that any of the observed

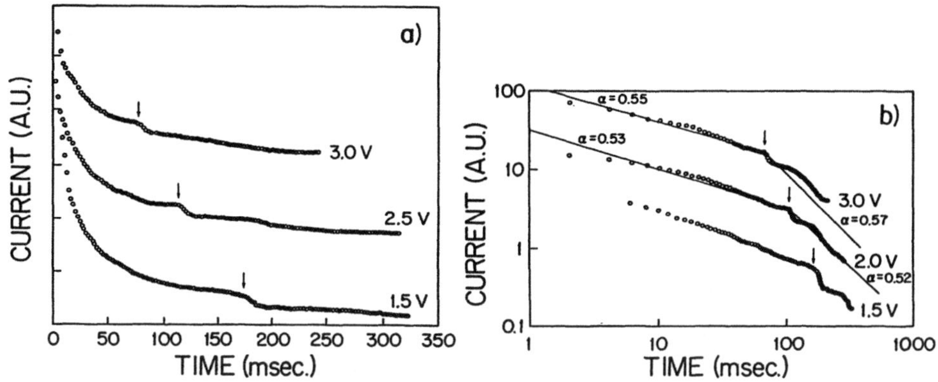

Fig. 2.a) Electron photocurrent transients measured at three different applied voltages: 1.5 V, 2.5 V and 3.0 V; L = 1.5 cm. b) The same electron photocurrent data of Fig. 2a in units log I versus log t. The arrows indicate the transit time for each case.

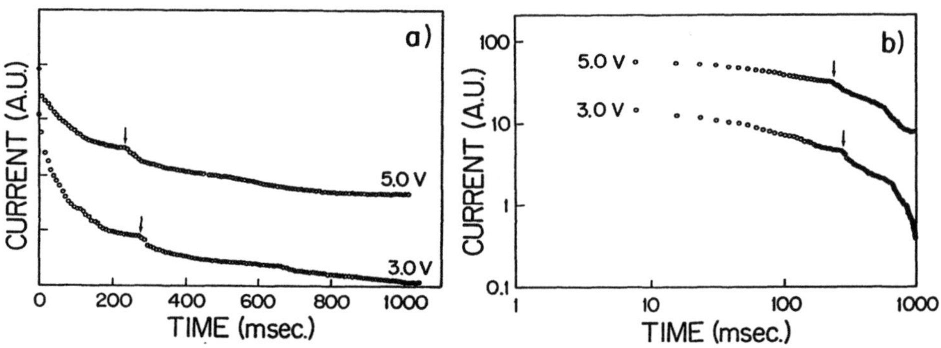

Fig. 3. a) Hole photocurrent transients measured with 3.0 V (bottom) and 5.0 V (top); b) The same hole photocurrent data of Fig. 3a in units log I versus log t. The arrows indicate the transit time for each case.

electron transit times have a nonlinear electric field dependence.

Figure 3a shows hole current transients plotted on a linear scale for two different applied bias: 3.0 V and 5.0 V. The Currents were arbitrarily shifted along the I-axis, for comparison of results. Hole current transients show similar dispersive transport phenomena; it is possible to define a transit time and these are more evident when the same data are replotted as log I versus log t. Here there also exists a second knee on both current transients and the explanation may be the same, i.e. space charge. Note that the transit

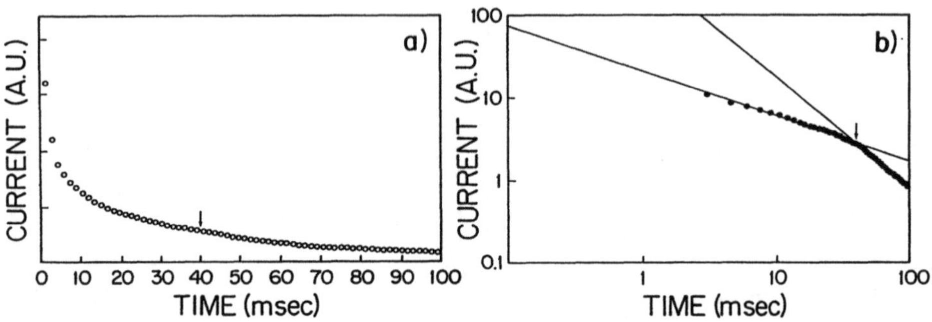

Fig. 4. a) Electron photocurrent transient measured with 3.0 V. The transit time indicated by the arrow was determined from the log plots shown in Fig. 4b. b) The same electron photocurrent data as plotted in Fig. 4a. replotted on a logarithmic scale.

time dependence on electric field is weaker than for electron transport.

Some other samples studied in this work shown the characteristic current transient where it is not possible to define a transit time on a linear plot of I vs. t. This case is illustrated in Fig. 4a. However, when the same data are replotted on a log scale the transit time is easily defined as is evident in Fig. 4b.

4. Discussion of results and conclusions

Scher and Montroll [1] developed a stochastic transport model which describes the motion of the drifting charge inside disordered semiconductors as a carrier dispersive (non Gaussian) transport process. Tiedje and Rose [30] gave a physical interpretation, for the dispersive charge transport, based on a progressive thermalization of electrons in an exponential distribution of traps. The principal results of both theories are essentially the same and we will use them on the discussion of our experimental results. In the theory, the current decay at short times ($t < t_T$) has the form $t^{\alpha-1}$ and at long times ($t > t_T$), the form $t^{-\alpha-1}$. Approximately the same value of α ($\sim$ 0.54) can fit the rate of current decay at short times and also at long times, for the 2.5 and 3.0 V cases shown in Fig. 2b. The best fitting correspond to the straight lines drawn in this figure. In the case where an applied voltage of 1.5 V was used, we get the best fitting with a smaller α ($\sim$ 0.30). Probably, in this last case, the polarization field is more relevant than in the two previous cases. The transit time has an electric field dependence as $E^{-1/\alpha}$, thus a drift mobility defined by $\mu_d = L^2/Vt_T$ has the power-law dependence $E^{1/\alpha-1}$. The electron transit times for 3.0 and 2.5 V, which are clearly evident as a knee in Fig. 2b and the resulting drift mobility have a power-law field dependence that is consistent with the theory with $\alpha=0.51$. The hole current decay has the same power-law behavior, however there is not a single value of α able to fit both short and long times regimes. Cases like this have been reported by other authors [2]. Time of flight measurements on holes show also an electric field dependence, as can be seen in Fig. 3b. However, in this case the resulting drift mobility is not a superlinear function of the applied electric field.

In conclusion, from our preliminary results we can say that the transient charge carrier transport in CdTe polycrystalline films has a dispersive (non Gaussian) behavior. Assuming detrapping times of carriers in deep states for which E > kT, given by $\omega_o^{-1}\exp(E/kT)$, where ω_o is the phonon frequency, the electron time of flight experiment shows that the trapping state distribution is exponential from 0.46 to 0.63 eV below the bottom of the conduction band. Additional experiments where the temperature dependence of the current transients are studied over longer times, will be necessary in order to get a complete picture of the trapping states in the energy gap of CdTe. We have also studied the dependence on the grain size in the films of the electron and hole mobility, and the results will be published elsewhere [31].

Acknowledgements. This work was partially supported by Consejo Nacional de Ciencia y Tecnología CONACyT-México. Dirección General de Investigación Científica y Superación Académica (SEP-México).

References

1. Harvey Scher and Elliot W. Montroll, Phys. Rev. B12, 2455 (1975).
2. G. Pfister and H. Scher, Adv. in Phys. 27, 747 (1987).
3. T. Tiedje, in Semiconductors and Semimetals, Vol. 21, Part C, Academic Press, Inc. (1984), Chap. 6, Pg. 207.07.
4. Lawrence L. Kasmerski, in "Polycrystalline and Amorphous Thin Films and Devices", Ed. by Lawrence L. Kasmerski, Academic Press, Inc. (1980), Chap. 3, pg. 59.

5. F. Greuter and G. Blatter, Semicond. Sci. Technol. 5, 111, (1990).
6. C. Menezes, C. Fortman, and S.J. Casey, J: Electrochem. Soc. 132, 709 (1985).5).
7. F.H. Nicoll, J. Electrochem. Soc. 110, 1165, (1963).
8. P.H. Robinson, RCA Rev. 24, 574 (1963).
9. R.F. Tramposh, J. Electrochem. Soc. 116, 654 (1963).
10. J.E. May, J. Electrochem. Soc. 112, 710 (1965).
11. G.E. Gottlieb and J.F. Corboy, RCA Rev. 24, 585 (1963).
12. E. Sirtl, J. Phys. Chem. Solids 24, 1285 (1963).
13. H. Flicker, B. Goldstein, and P.A. Hoss, J. Appl. Phys. 35, 2959 (1964).
14. R.G. Schulze, J. Appl. Phys. 37, 4295 (1966).
15. N. Isawa, Jpn. J. Appl. Phys. 7, 81 (1968).
16. O. Igarashi, J. Appl. Phys. 41, 3190 (1970).
17. O. Igarashi, Jpn. J. Appl. Phys. 8, 642 (1969).
18. W.H. Strehlow, J. Appl. Phys. 42, 4035 (1970).
19. J. Saraie, M. Akiyama, and T.Tanaka, Jpn. J. Appl. Phys. 11, 1758 (1972).
20. A. Yoshikawa, R. Kondo, and Y. Sakai, Jpn. J. Appl. Phys. 12, 1096 (1973).
21. K. Mitchell, A.L. Fahrenbruch, and R.H. Bube, J. Vac. Sci. Technol. 12, 909 (1975).5).
22. T.C. Anthony. A.L. Fahrenbruch, and R.H. Bube, J. Vac. Sci. Technol. A2, 1296 (1984).4).
23. C.A. Menezes, F. Sánchez-Sinencio, and A. Sosa, Sol. Energ. Mater. 11, 401 (1985).
24. T.C. Anthony, A.L. Fahrenbruch, M.G. Peters, and R.H. Bube, J. Appl. Phys. 57, 400 (1985).
25. O.N. Tufte and E.L. Stelzer, J. Appl. Phys. 40, 4559 (1969).
26. G. Cogen-Solal and Y. Riant, Appl. Phys. Lett. 19, 436 (1971).
27. A.L. Fahrenbruch, V. Vasilchenko, F. Buch, K. Mitchell, and R.H. Bube, Appl. Phys. Letters 25, 605 (1974).
28. T.L. Chu, S.S. Chu, F. Firszt, H.A. Naseem, and R. Stawski, J. Appl. Phys. 58, 1345 (1985).5).
29. O. Zelaya, F. Sánchez-Sinencio, J.G. Mendoza-Alvarez, M.H. Farías, L. Cota-Araiza and G. Hirata-Flores, J. Appl. Phys. 63, 410 (1988).
30. T. Tiedje and A. Rose, Solid State Comm. 37, 49 (1980).
31. J.M. Figueroa, F. Sánchez-Sinencio,J.G. Mendoza-Alvarez, O. Zelaya, G. Contreras-Puente and A. Diaz-Góngora, to be published in J.Crystal Growth.

Fabrication and Theoretical Simulation of $Cu(In,Ga)Se_2$/(ZnCd)S Thin Film Solar Cells

G. Gordillo

Departamento de Física, Universidad Nacional Bogotá, Colombia

Abstract. The quaternary compound $Cu(In,Ga)Se_2$ presents great flexibility for the design of efficient single solar cells or stacked junctions, because it can be prepared with a bandgap in the range of 1.0 – 1.7 eV.

Measurements of the spectral quantum efficiency of $Cu(In,Ga)Se_2$(ZnCd)S heterojunctions demonstrated a strong influence of reducing/oxidizing post-deposition treatment on the red and infrared response of the cells. Theoretical calculations of the quantum efficiency of $Cu(In,Ga)Se_2$/(ZnCd)S solar cells were carried out on the basis of a simulation of the heterojunction performance. From these results, information about the parameters influencing the quantum efficiency of the cells during reducing/oxidizing treatments was obtained.

1. Introduction

Solar cells based on polycrystalline $CuInSe_2$ thin films have reached efficiencies around 14% [1]. The specific material properties of $CuInSe_2$ have made it possible to realize the most efficient thin-film devices to date. By gradual substitution of Ga for In, the optical bandgap can be increased from 1.04 to 1.68 eV.

An increment of the gallium content leads to an increase of the open circuit voltage of the $Cu(In,Ga)Se_2$ based cells, because the open circuit voltage of the devices corresponds to about half of the energy bandgap of the absorber.

The grain size of polycrystalline $Cu(In,Ga)Se_2$ thin films is of the order of the minority carrier diffusion length [2], therefore the performance of the $Cu(In,Ga)Se_2$ based solar cells depends not only on the bulk material but also on grain boundaries. Since the properties of the grain boundaries are influenced by post-deposition treatments, these can be used to increase the cell efficiency and to obtain additional information concerning the physics of the device.

2. Fabrication of the $Cu(In,Ga)Se_2$/(ZnCd)S Solar Cells

$Cu(In,Ga)Se_2$/(ZnCd)S thin-film solar cells with the structure shown in Fig. 1 were prepared by evaporation in high vacuum.

The $Cu(In,Ga)Se_2$ films with a thickness of 2 – 3 μm and a Ga content in the range 30 – 40% were deposited on molybdenum coated glass substrates using spatially separated effusion sources [3]. During the evaporation, the Cu rate was varied by approximately 20%, in order to obtain $Cu(In,Ga)Se_2$ layers

Springer Proceedings in Physics, Volume 62
Surface Science Eds.: F.A. Ponce and M. Cardona

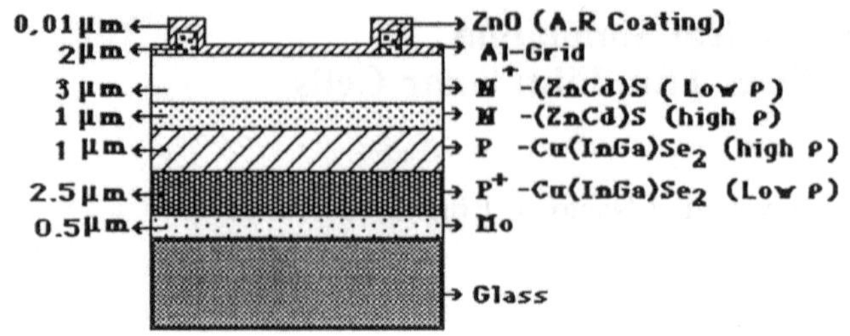

Fig. 1 Structure of the $Cu(In,Ga)Se_2/(ZnCd)S$ solar cells

with high and low conductivity. The heterojunction was formed by evaporation of (Zn,Cd)S on the absorber layer in a different vacuum system using a fixed temperature coaxial effusion source [4]. Ga-doped (ZnCd)S layers with a thickness of 1 – 3 μm and Zn concentration in the range 10 – 40%, were used in the fabrication of the solar cells.

Sputtered ZnO layers were used to reduce the sheet resistance in the case of highly resistive window layers and additionally as a partial antireflection coating.

3. Influence of Post-Deposition Treatment

For cells based on evaporated $CuInSe_2$ films, annealing in air at a temperature around 200°C results in optimum efficiency [5]. Oxygen increases the effective net acceptor concentration of the films and eliminates the formation of a buried $CuInSe_2$ homojunction [6]. In $Cu(In,Ga)Se_2$ based solar cells, annealing in air usually increases their open circuit voltage but the efficiency is often found to decrease because the carrier collection deteriorates. On the other hand, reducing treatments like dipping in hydrazine can significantly increase the filling factor and efficiency [7]. Figure 2 illustrates the influence exerted on the spectral response by a reducing treatment following by an air anneal. After hydrazine treatment, the quantum efficiency increases, in particular for red and infrared light, where the carriers are generated at a greater distance from the junction and the transport properties are more important.

The behavior of the $Cu(In,Ga)Se_2$ based solar cells after oxidizing/reducing treatments can be related to a modification of: effective doping (and

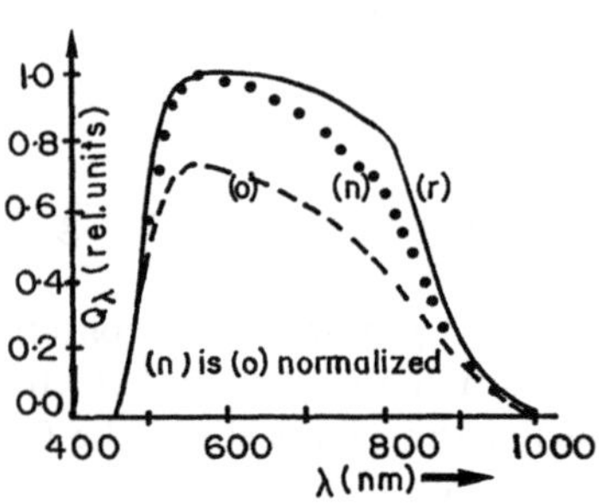

Fig. 2 Spectral response after air anneal (o) and after hydrazine treatment (r)

therefore junction field), absorption coefficient and minority carrier diffusion length, caused by post-deposition treatment. It is, however, difficult to obtain reliable experimental evidence of the variation of these parameters caused by oxidizing/reducing post-deposition treatments, and the corresponding influence on the spectral response.

4. Theoretical Calculation of the Quantum Efficiency of the $Cu(In,Ga)Se_2/(ZnCd)S$ Heterojunction

Theoretical calculation of the quantum efficiency of the $Cu(In,Ga)Se_2/(ZnCd)S$ heterojunction was carried out in order to obtain information about the influence of the electric field, diffusion length, absorption coefficient and width of the space charge region on the quantum efficiency of the heterojunction. The quantum efficiency was calculated including only the collection of carriers generated in the $Cu(In,Ga)Se_2$ layer. The (ZnCd)S layer does not contribute to the quantum efficiency because the majority of minority carriers are generated near the (ZnCd)S surface and they do not reach the interface, due to the short diffusion length.

The calculation of the internal quantum efficiency Q was made considering that:

- Within the space charge region (SCR) there exists an electric field which is assumed constant. Therefore the carrier transport inside the space charge region is caused by diffusion and drift mechanisms.

- Outside the space charge region (SCR) there is no electric field. Therefore the carrier transport outside the space charge region is caused only by a diffusion mechanism.

The internal quantum efficiency (Q) can be calculated by using

$$Q(\lambda) = Q_1(\lambda) + Q_2(\lambda), \tag{1}$$

$$Q_1(\lambda) = \left(\frac{J_{n1}(\lambda)}{qF(\lambda\,[1 - R(\lambda)]}\right) \mathrm{ICF}, \tag{2}$$

$$Q_2(\lambda) = \left(\frac{J_{n2}(\lambda)}{qF(\lambda\,[1 - R(\lambda)]}\right) \mathrm{ICF}, \tag{3}$$

where J_{n1} is the photocurrent density due to electrons generated outside of the SCR,

J_{n2} is the photocurrent density due to electrons generated inside the SCR,

$F(\lambda)$ is the incident photon flux,

$R(\lambda)$ is the reflection coefficient in the (ZnCd)S surface, and

ICF is the interface collection factor.

In general the photocurrent of the heterojunction is given by the sum of the diffusion and drift components:

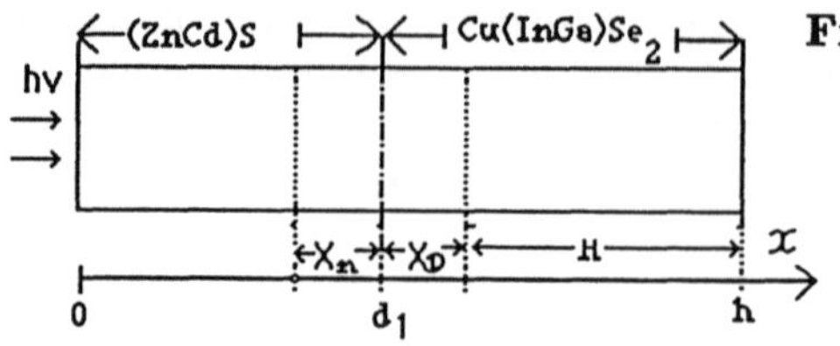

Fig. 3 Structure of the Cu(In,Ga) Se_2/(ZnCd)S heterojunction

$$J_n = nq\mu_n E + qD_n(dn/dx), \tag{4}$$

where μ_n is the mobility of electrons inside the SCR,
E is the electric field within the SCR, and
n is the minority carrier density.

The density of minority carriers generated inside and outside the space charge region of the $Cu(Ga,In)Se_2$ layer can be obtained from the solution of the transport equation. Under low level injection conditions, the transport equation of minority carriers is

$$D_n\left(\frac{d^2n_p}{dx^2}\right) + \mu_n Eq\left(\frac{dn_p}{dx}\right) + G - \frac{(n_p - n_{po})}{T_n} = 0 \tag{5}$$

for carriers generated inside the SCR and

$$D_n\left(\frac{d^2n_p}{dx^2}\right) + G - \frac{(n_p - n_{po})}{T_n} = 0 \tag{6}$$

for carriers generated outside the SCR, where G is the generation rate of electron-hole pairs.

The structure of the heterojunction shown in Fig. 3 will be considered in the solution of the transport equations (5) and (6).

The general solution of (5) is

$$n_p = S_1\cosh(y/L) + S_2\sinh(y/L) - C\exp(-A_2y) + n_{po}, \tag{7}$$

where

$$\begin{aligned} C &= (A_2T_nN)/A, \\ N &= F(1-R)e^{-A_1d_1}e^{-A_2x_p}, \\ A &= A_2^2L_2^2 - 1, \quad L = (D_nT_n)^{1/2} \end{aligned} \tag{8}$$

A_1, A_2 are the absorption coefficients of (ZnCd)S and $Cu(In,Ga)S_2$, respectively.

The constants S_1 and S_2 can be obtained by using the following boundary conditions:

$$S(n_p - n_{po}) = -D_nd(n_p - n_{po})/dy, \quad \text{for } y = H \text{ with } H = h - (d_1 + X_p), \tag{9}$$

$$n_p - n_{po} = J_n/qV_{sl}, \quad \text{for } y = 0, \tag{10}$$

$$S_1 = C + (kJ_n)/qV_{sl}, \tag{11}$$

$$S_2 = -\left\{S_1 B + C\exp\left(-A_2 H\right)\left[L A_2 - T\right]\right\}/D, \tag{12}$$

where

$$\begin{aligned} B &= T\left[\cosh(H/L) + \sinh(H/L)\right], \quad T = STn/L \\ D &= T\left[\sinh(H/L) + \cosh(H/L)\right] \end{aligned} \tag{13}$$

The photocurrent due to electrons collected at $y = 0$ is

$$J_{n1} = \frac{qL^2 C V_{sl}\left[A_2 L D - B - \exp\left(-A_2 H\right)\left(A_2 L - T\right)\right]}{\left[D L V_{sl} T_n + B L_2\right]}. \tag{14}$$

Outside the space charge region, the transport equation is

$$\frac{d^2(n_p)}{dz^2} + \left(\frac{qE}{kt}\right)\frac{dn_p}{dz} - \frac{(n_p - n_{p_0})}{L^2} = \left(-A_2/D_n\right) N_1 \exp\left(-A_2 z\right), \tag{15}$$

where

$$N_1 = F(1-R)\exp(-A_1 Z) \quad \text{and} \quad Z = x - d_1. \tag{16}$$

The general solution of (15) is

$$n_p = \frac{S_3 \exp\left[-\left(F+G\right)z\right] + S_4 \exp\left[-\left(F-G\right)z\right] - A_2 N_1 \exp\left(-A_2 z\right)}{D_n G^2\left(b^2 - 1\right)}, \tag{17}$$

where

$$F = \left(\frac{q}{2kT}\right)E, \quad G = \left(F^2 + \frac{1}{L^2}\right)^{1/2}, \quad b = \frac{(A_2 - F)}{G}. \tag{18}$$

The constants S_3 and S_4 can be calculated using the following boundary conditions:

$$n_p - n_{p_0} = 0, \qquad \text{for } z = 0, \tag{19}$$

$$n_p - n_{p_0} = J_n/qV_{sl}, \qquad \text{for } z = X_p, \tag{20}$$

$$S_3 = C_1 - K, \tag{21}$$

$$S_4 = \left(\frac{J_n}{(qV_{sl})} + K_1 B_1\right)/C_2. \tag{22}$$

The photocurrent due to electrons collected in $z = 0$ is

$$J_{n2} = qD_n C_1\left[P_1 + B_2\left(P_2 - P_1\right)/C_2 + \left(B_1/C_2\right)\left(P_2 - P_1\right) + A_2\right], \tag{23}$$

where

$$B_2 = \frac{M}{T_n}\frac{\left[G^2\left(b^2-1\right)L^4 \exp\left(-A_2 x_p\right)\right]}{A}, \tag{24}$$

$$M = T_n \frac{\left[A_2 L D - B - \exp\left(-A_2 H\right)\left(A_2 L - T\right)\right]}{\left(T_n D L V_{sl} + B L^2\right)}. \tag{25}$$

Replacing (14) in (2) we obtain

$$Q_1 = \mathrm{ICF} \times MT_n L_2 V_{sl} \left[A_2 \exp\left(-A_1 d_1\right) \exp\left(-A_2 x_p\right)\right] / A \tag{26}$$

Replacing (23) in (3), we obtain

$$Q_2 = \mathrm{ICF} \times \left[A_2 \exp\left(-A_1 d_1\right)\right] \times \\ \times \left[P_1 + A_2 + \frac{B_1\left(P_2 - P_1\right)}{C_2} + \frac{B_2\left(P_2 - P_1\right)}{C_2}\right] / G^2\left(b^2 - 1\right). \tag{27}$$

4. Results

In this section we report the results concerning the calculation of the quantum efficiency of Cu(In,Ga)Se_2/(ZnCd)S solar cells. We used the values $V_{sl} = 10^6$ cm/s, $T_n = 10^{-9}$ s, $S = 10^3$ cm/s obtained from the literature. The values of A_1 used in the calculation, were obtained experimentally [8], whereas for the values of A_2, the relation $A_2 = R_2\left(h\nu - E_g\right)^{1/2}$ was used, which reproduces the experimental measurements of A_2 very well [9].

Figures 4a,b and 5a–c illustrate the influence of absorption coefficient A_2, space charge region width x_0, internal electric field E_0, minority carrier diffusion length L, and bias voltage V on the quantum efficiency of Cu(In,Ga)Se_2/(ZnCd)S solar cells.

The results indicate that all the above-mentioned parameters influence the magnitude of the Q, but only A_2 and L influence the form of the curves of $Q = f(\lambda)$.

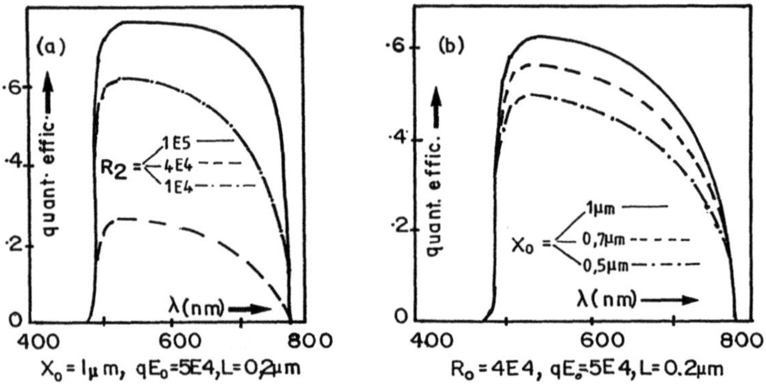

Fig. 4 Quantum efficiency of Cu(In,Ga)Se_2/(ZnCd)S solar cells as a function of **(a)** absorption coefficient of the Cu(In,Ga)Se_2 layer, and **(b)** width of the space charge region

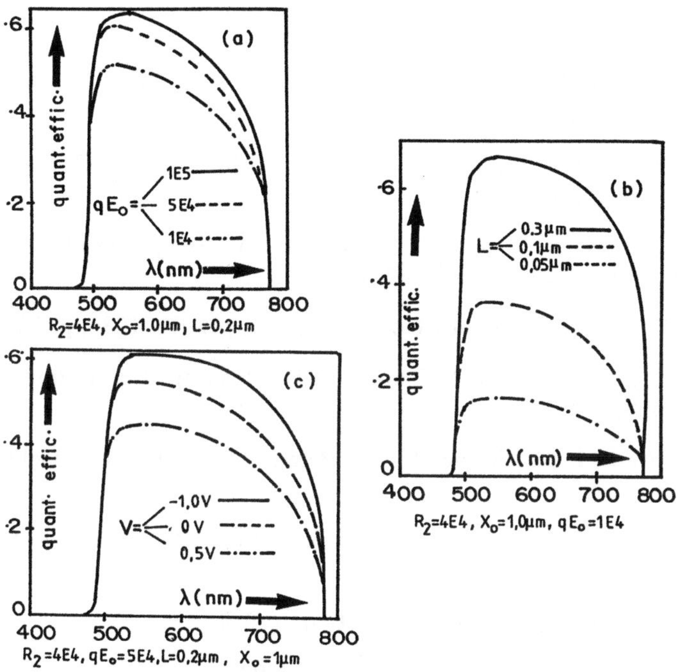

Fig. 5 Quantum efficiency of $Cu(In,Ga)Se_2/(ZnCd)S$ solar cells as a function of **(a)** internal electric field, **(b)** minority carrier diffusion length, and **(c)** bias voltage

5. Conclusions

Evaporated $Cu(In,Ga)Se_2/(ZnCd)S$ thin-film solar cells with efficiencies greater than 7% were fabricated. Reducing treatments with hydrazine after cell fabrication lead to an increment of the conversion efficiency. Experimental measurements of the quantum efficiency of $Cu(In,Ga)Se_2/(ZnCd)S$ solar cells have indicated that oxidizing/reducing post-deposition treatments influence the red and infrared response of the cells, especially under bias voltage. From theoretical calculation of the quantum efficiency it was possible to explain the behavior of these cells by assuming that the absorption coefficient of the absorber layer, minority carrier diffusion length, internal electric field, and width of the SCR are modified during the oxidizing/reducing treatments. An oxidizing process causes a decrease of the diffusion length and absorption coefficient of the absorber layer. In this way the bias voltage influences the form of the $Q = f(\lambda)$ curves, especially in the red and infrared region.

Acknowledgements. This work was supported by COLCIENCIAS, TWAS, University of Stuttgart and by the National University of Colombia.

References

1. K. Mitchell, C. Eberspacher, J. Emer, D. Pier: Proc. 20th IEEE Photov. Spec. Conf., Las Vegas, 1988

2. R. Klenk, R. Menner, D. Schmid, D. Cahen, H.W. Schock: 9th E.C. Photov. Sol. Energy Conf., Freiburg, 1989

3. B. Dimmler, H. Dietrich, R. Menner, H.W. Schock: Int. Symp. Trends and New Appl. Thin Films, Strasbourg, 1987, p. 103

4. G. Gordillo, H.W. Schock: Proc. 5th Int. Solar Forum, Berlin 1984, p. 786

5. R.W. Birkmire, L.C. Di Netta, P.A. Laswell, J.D. Meaking, J.E. Phillips: Solar Cells 16 (1986)

6. R.J. Matson, R. Nouti, R.K. Ahrenkiel, R.C. Powell, D. Cahen: Solar Cells 16 (1986)

7. R. Klenk, R.H. Mauch, R. Menner, H.W. Schock: 20th IEEE Photov. Spec. Conf., Las Vegas, 1988

8. G. Gordillo: PhD Thesis, University of Stuttgart (1984)

9. R. Klenk: Diplomarbeit, University of Stuttgart (1987)

Photoelectric Response of Thin Films for Solar Cells

A. Valera

Universidad Nacional de Ingeniería, Laboratorio de Física,
Av. Túpac Amaru s/n, Apartado 1301, Lima, Peru

Abstract. The spectral dependence of the photoelectric response of metal/a-Si:H/metal photovoltaic configurations in the energy range 0.75 – 2.2 eV is used to study the energy level diagram of the device.

Significant information is recovered from the spectra with reverse/forward bias and illumination. The results are interpreted as arising from alternative generation processes. We can identify internal photoemission, optical band-to-band absorption and localized-to-extended-state absorption.

1. Introduction

In the last few years progress has been made in investigating the photoelectrical properties of amorphous semiconductors by using new measuring techniques. In addition to potential applications as large area solar cells, Schottky barrier diodes made of hydrogenated amorphous silicon (a-Si:H) are important tools for studying this material. Photoconductivity spectra are commonly used to obtain information on sub-bandgap absorption in thin films of a-Si:H [1], where direct optical transmission experiments fail.

In this contribution we outline a procedure, dependent upon a few assumptions, which can be used to quantitatively elucidate the photoelectric response of metal/a-Si:H/metal systems from measurements of photoconductivity.

2. Material

Intrinsic a-Si:H films were prepared by the decomposition of silane in a DC glow discharge system directly onto glass and over half-metallized substrates; the former were used to measure the photoresponse of the a-Si:H film alone, and the latter were complemented with transparent metallic dots in order to elaborate metal/a-Si:H/metal configurations. The metallic dots were from Pd, Cr, Ag, and Al and formed the front contacts. The back contacts were Ag and Cr.

3. Experimental

The spectral dependence of the samples was measured in the energy range 0.75 – 2.2 eV using a two-beam experimental setup with chopped monochromatic light. The resulting AC photocurrent was detected by a lock-in technique. Complementary DC measurements were also performed.

Springer Proceedings in Physics, Volume 62
Surface Science Eds.: F.A. Ponce and M. Cardona

The photoconductivity response was measured as the number of charge carriers liberated per photon (quantum yield) that contributes effectively to the total current (factor G in our spectra).

From transmission spectra we determined the thickness (2 to 3 μm) and energy gap $E_g = 1.77$ eV of the a-Si:H samples.

4. Results and Discussion

In the photocurrent measurements on our metal/a-Si:H/metal probes we obtain bulk and surface contributions. Figure 1 shows the photocurrent of an a-Si:H sample measured in a coplanar configuration, this is mainly the bulk contribution of our multilayer systems. In order to elucidate the surface contributions we first find the level at which the reference (bulk spectrum) is located. After this calibration the bulk contribution can be subtracted and the surface processes identified.

Figure 2 shows the photocurrent obtained from a Pd/a-Si:H/Cr probe with -0.4 V bias voltage. After calibration we identified two surface contributions. Fitting them to photocurrent spectra gave

$$G_1 = 577(h\nu - 0.875)^3, \quad (1)$$
$$G_2 = 1.6 \times 10^5 (h\nu)^{-1}(h\nu - 1.63)^2. \quad (2)$$

G_1 is associated with internal photoemission from Pd to a-Si:H through a thin intermediate layer (process B, Fig. 3). This process has been reported also by

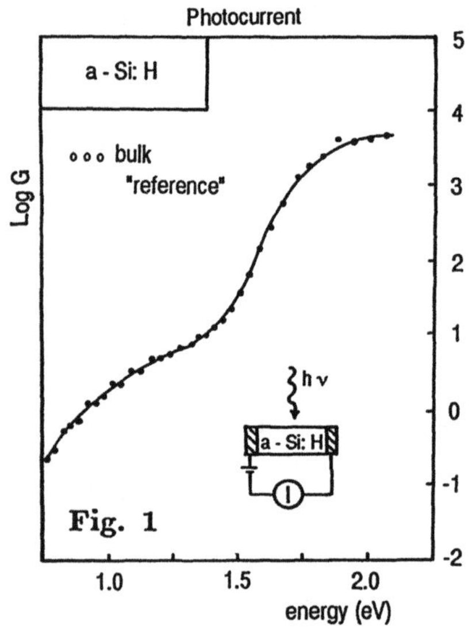

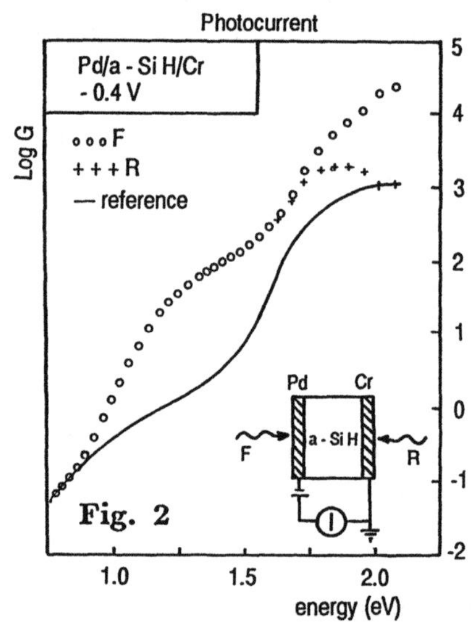

Fig. 1 Photocurrent spectrum of an a-Si:H sample, measured in a coplanar configuration (bulk contribution)

Fig. 2 Photocurrent spectrum of Pd/a-Si:H/Cr with -0.4 V bias voltage for frontal (F) and back (R) illumination

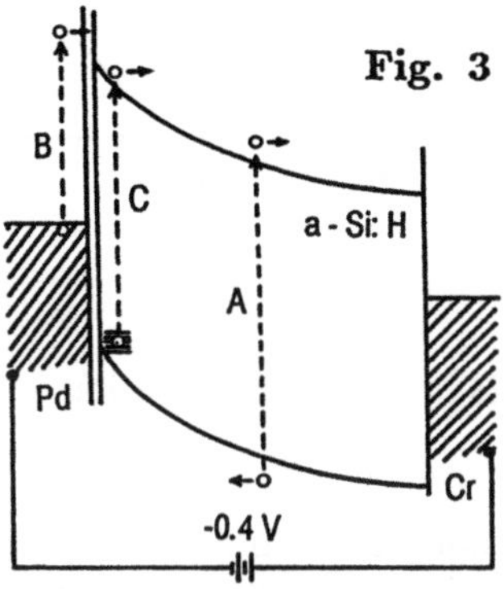

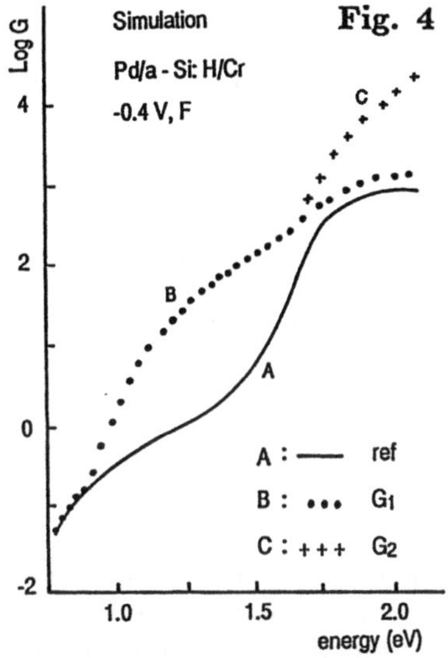

Fig. 3 Schematic of the optical transitions that contribute to the photoresponse of Pd/a-Si:H/Cr with −0.4 V bias voltage. A, band-to-band absorption; B, internal photoemission; C, localized-to-extended-state absorption

Fig. 4 Simulation of the photocurrent measurement on Pd/a-Si:H/Cr with −0.4 V bias voltage and frontal (F) illumination by consideration of the processes A, B, and C of Fig. 3

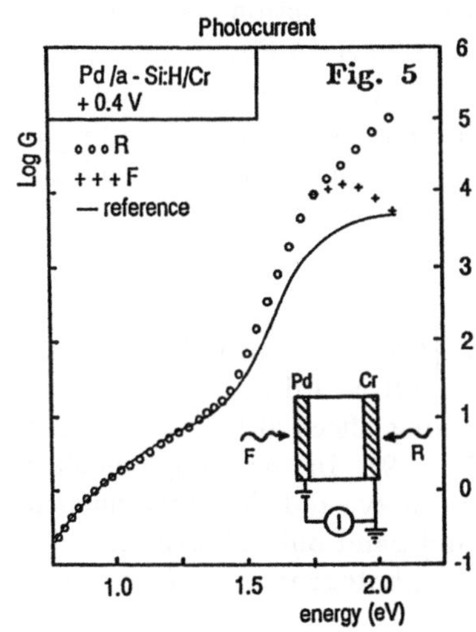

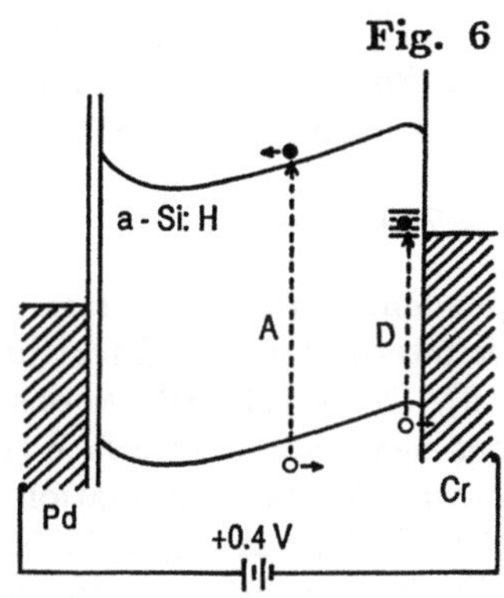

Fig. 5 Photocurrent spectrum of Pd/a-Si:H/Cr with +0.4 V bias voltage for frontal (F) and back (R) illumination

Fig. 6 Schematic of the optical processes that contribute to the photoresponse of Pd/a-Si:H/Cr with +0.4 V bias voltage. A, band-to-band absorption; D, localized-to-extended-state absorption

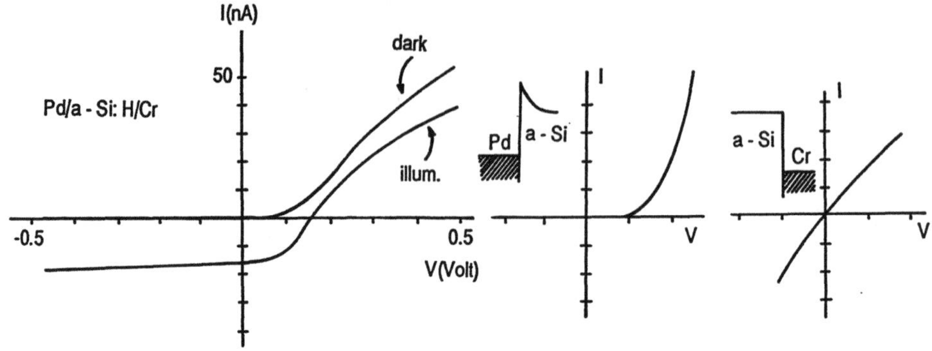

Fig. 7 *I-V* measurements on Pd/a-Si:H/Cr and its border decomposition

Yamamoto [2]. G_2 obeys Tauc's law [3], which implies subband-band absorption in the a-Si:H zone near the Pd interface, where the main contribution probably arises from the hole transport (process C, Fig. 3). The simulation used to obtain G_1 and G_2 agreed very well with experimental data, as is apparent from Fig. 4.

Figure 5 shows the photocurrent spectrum from Pd/a-Si:H/Cr with a bias of +0.4 V. We obtain only one surface contribution at 1.35 eV, the origin of which is located in the a-Si:H zone near the Cr interface. We conclude that this generation arises from states near to the conduction band (process D, Fig. 6) and that the main transport arises from holes. We observe no internal photoemission from the Cr side, which implies that the junction is ohmic. This interpretation is supported by *I-V* measurements shown in Fig. 7.

The photocurrent spectrum of Pd/a-Si:H/Cr with +0.2 V bias and back illumination (Fig. 8) shows a different behavior. The generation according to (1), arising from the Pd internal photoemission, disappears and a quadratic enhancement evolves at 1.0 eV. The dependence now fits

$$G_3 = 240(h\nu - 1.00)^2. \qquad (3)$$

This dependence follows the classical Fowler rule [4] for the internal photoemission. It is interpreted as a direct jump of electrons from Pd to a-Si:H (process E, Fig. 9). Near 1.35 eV we observe a quenching effect arising again from transitions near the Cr interface (process D, Fig. 9). In the complementary spectrum with frontal illumination (Fig. 10) we observe additionally a quenching effect at 2 eV, starting at 1.75 eV where band-band bulk absorption takes place. Finally there is again an increase due the photogeneration (process C, Fig. 9) on the front side.

On the basis of the present measurements, and our results on other systems, we are able to obtain the conclusions summarized in Fig. 11.

Fig. 8

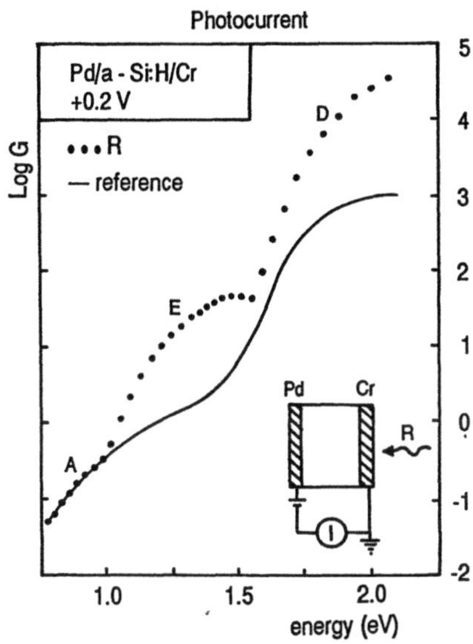

Fig. 9

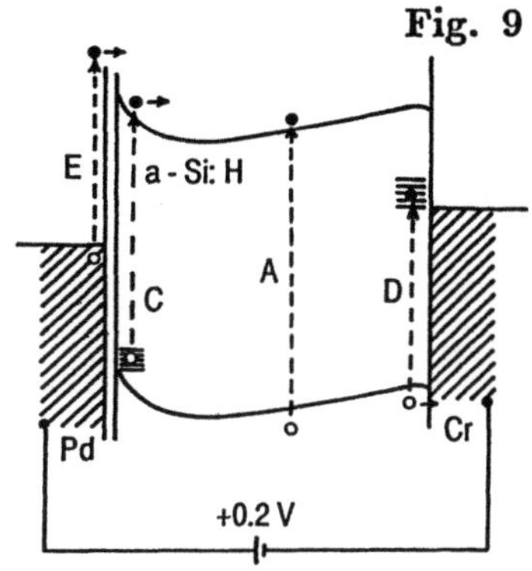

Fig. 8 Photocurrent spectrum of Pd/a-Si:H/Cr with +0.2 V bias voltage and back illumination. Enhancements A, E, and D are due to processes shown in Fig. 9

Fig. 9 Schematic of the processes that contribute to the photoresponse of Pd/a-Si:H/Cr with +0.2 V bias voltage

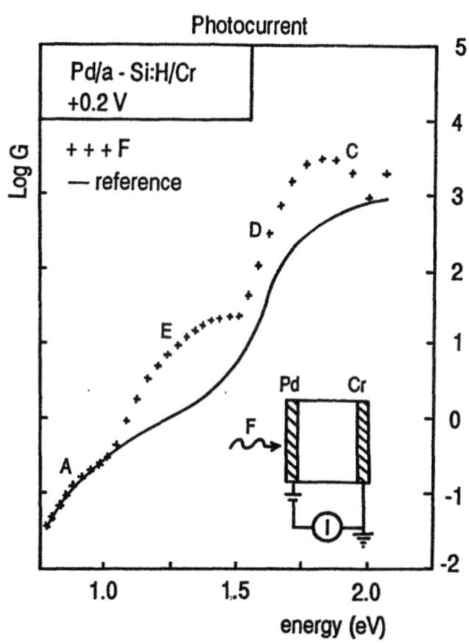

Fig. 10 Photocurrent spectrum of Pd/a-Si:H/Cr with +0.2 V bias voltage and frontal illumination. Enhancements A, E, D, and C are explained in Fig. 9

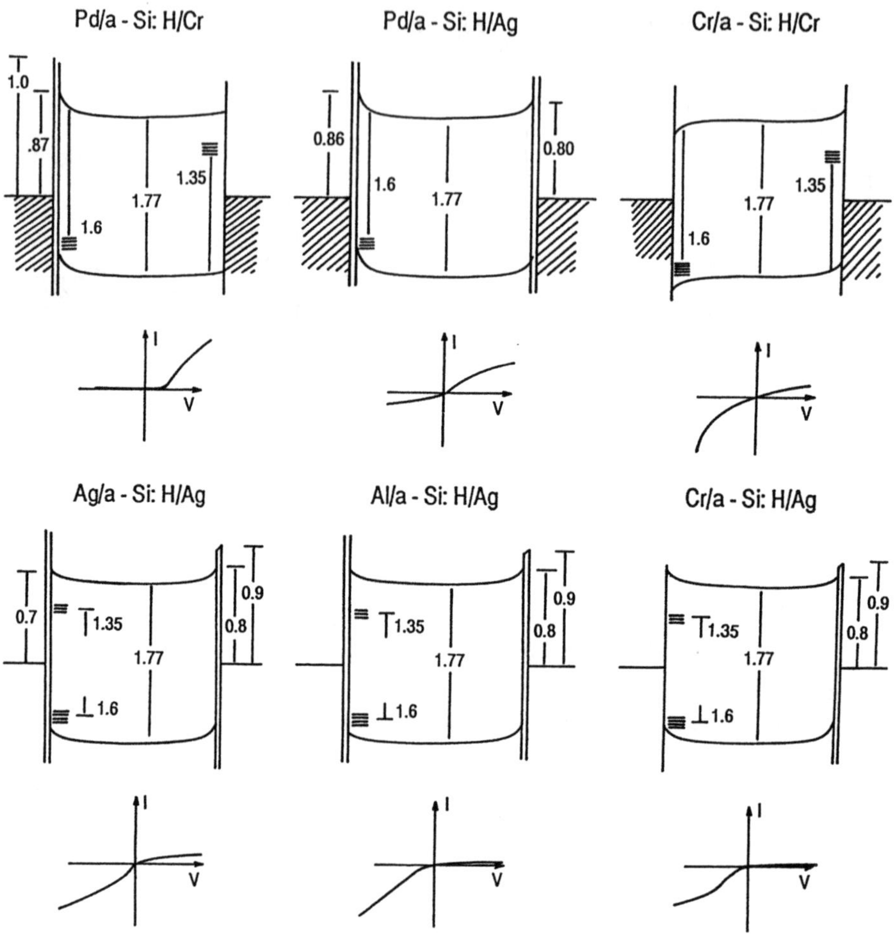

Fig. 11 Summary of the conclusions obtained for some metal/a-Si:H/metal configurations obtained from photocurrent measurements. The energy values are given in eV

5. Conclusions

We have shown that photocurrent measurements give quantitative information on processes and gap states involved in the photoresponse of metal/a-Si:H/metal configurations. The main assumption which was employed is that the surface radiation processes and the bulk radiation processes can be separated.

Acknowledgments. It is a pleasure to thank my colleagues Christoph Nebel and Helga Weller for the facilities given and stimulating discussions. Carlos Paucarchuco deserves special thanks for undertaking the DC measurements and the computer evaluation task.

The work was partially supported by the BRD under a DAAD fellowship.

References

1. J. Kocka: J. Non-Cryst. Solids **90**, 91 (1987)
2. T. Yamamoto et al.: Jpn. J. Appl. Phys. **20**, Suppl. 20-2, 185 (1981)
3. J. Tauc: In *Optical Properties of Semiconductors*, ed. by F. Abeles (North-Holland, Amsterdam 1972), p. 279
4. R.J. Nemanich: in *Semiconductors and Semimetals*, Vol. 21C, ed. by J.I. Pankove (Academic, New York 1984), p. 385

Characterization of Palladium Contacts to a-Si:H and a-Si:N:H

M.G. da Silva and S.S. Camargo, Jr.

Laboratorio de Estudos de Materials e Interfaces, PEMM/COPPE,
Universidade Federal do Rio de Janeiro, Caixa Postal 68505,
CEP 21945, Rio de Janeiro, RJ, Brazil

Abstract. Palladium contacts to hydrogenated amorphous silicon and hydrogenated amorphous silicon-nitrogen alloys were characterized by current versus voltage measurements and the internal photoemission technique. Results show that the incorporation of small amounts of nitrogen to the semiconductor increases the diode ideality factor and decreases the contact resistance. Thermal annealing of the samples reduces the former while increases the later. For larger amounts of nitrogen an increase in the barrier height is also observed. Possible explanations for these effects are discussed.

1. Introduction

The physics of metal-semiconductor Schottky barriers has received considerable interest over the last decades due to their fundamental and technological aspects [1]. The use of some very simple techniques like measuring current versus voltage characteristics allows the determination of the fundamental parameters of the barrier, but their relation to structural properties of the contacts may be much more complex.

Metal contacts to hydrogenated amorphous silicon have been studied mainly due to their potential applications in solar cells. In spite of the large amount of work already done, a conclusive relation between the electrical characteristics and structural properties of these contacts is still lacking. Pd/a-Si:H contacts are one of the most investigated both by structural and electrical characterization. In this case, the observed reduction of the recombination current upon annealing was associated with the formation of a silicide layer which would consume the near surface defective region and/or to the formation of a more laterally uniform interface [2].

In this paper the results of the characterization of Pd/a-Si:H and Pd/a-Si:N:H Schottky diodes are reported and possible explanations for the observed effects are discussed.

2. Experimental

Hydrogenated amorphous silicon (a-Si:H) and silicon nitrogen alloy (a-Si:N:H) Schottky diodes were produced by glow discharge decomposition of silane and silane-ammonia mixtures, respectively, onto nickel-chromium and indium-tin oxide covered Corning glass substrates. In order to achieve good ohmic back contacts thin (30 nm) a-Si:H layers doped with 1% of phosphine in the gas phase were deposited and followed by

Springer Proceedings in Physics, Volume 62
Surface Science Eds.: F.A. Ponce and M. Cardona

thick (600 to 2000 nm) nominally intrinsic layers. Finally, a number of 12 mm^2 palladium dots with thickness ranging from 10 to 50 nm were deposited to form the diodes. Thermal annealing experiments were performed in vacuum at a temperature of 150°C.

3. Results and Discussion

In case of Pd/a-Si:H contacts the obtained IxV curves for low forward bias voltages show a typical diode exponential behavior from which the contact ideality factor can be determined. For higher voltages, a deviation from this behavior due to the series resistance of the diode is observed. This deviation is larger for films with higher nitrogen contents thus impeding the ideality factor determination. For this reason IxV measurements were performed only for the films with the lowest nitrogen concentrations corresponding to a few atomic percent in the film as determined by AES.

Figure 1 shows the variation of the ideality factor as a function of thermal annealing time for a-Si:H and a-Si:N:H diodes. Two main effects can be observed, namely the reduction of the ideality factor upon annealing and the fact that, except for the as prepared samples which may yield unreproducible results, nitrogen incorporation to the semiconductor increases the ideality factor. Contact resistance, defined as the inverse of the slope of the IxV curve at zero bias, measurements also show that changes in the current conduction mechanisms are likely to occur. Figure 2 shows its change with annealing time and its dramatic decrease due to nitrogen incorporation.

In order to understand these effects one must consider the possible current transport mechanisms through the barrier [1]. Thermionic emission/diffusion over the barrier and recombination conduction mechanisms have been already observed in case of metal/a-Si:H (intrinsic) Schottky contacts [2]. On the other hand, tunneling through the barrier was only observed for doped films [3]. Hole injection could, in principle, also occur but due to the small hole diffusion

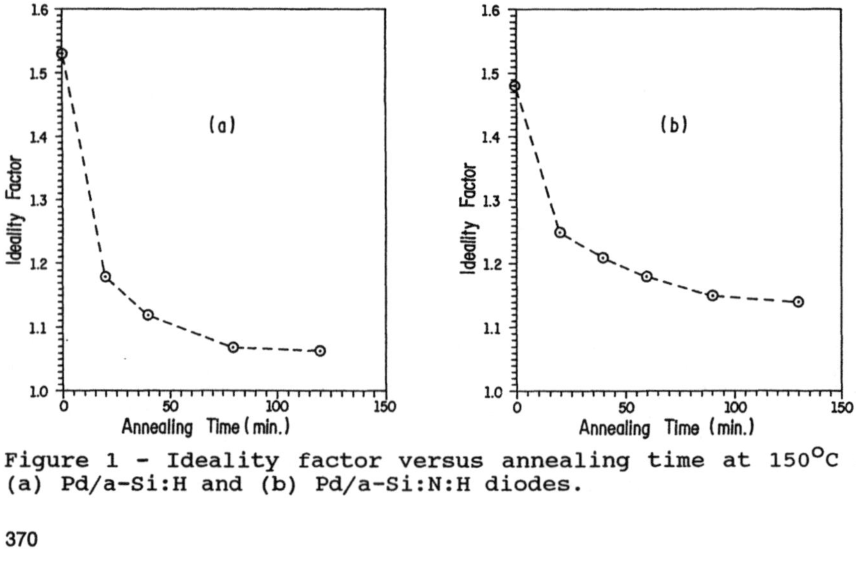

Figure 1 - Ideality factor versus annealing time at 150°C for (a) Pd/a-Si:H and (b) Pd/a-Si:N:H diodes.

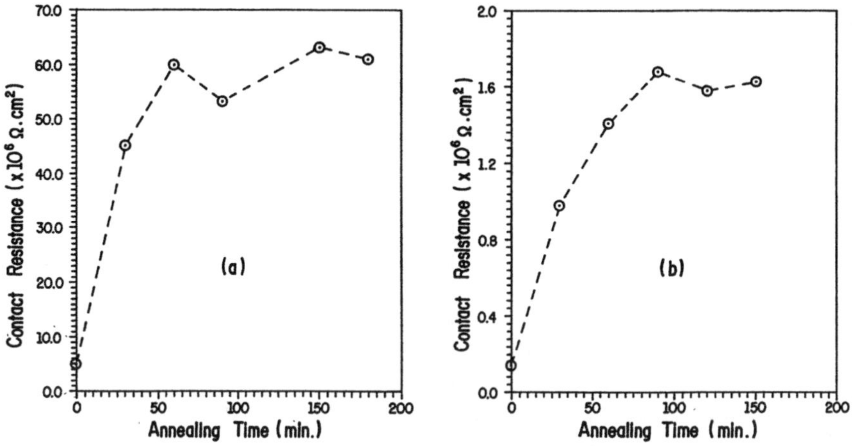

Figure 2 - Contact resistance versus annealing time at 150°C for (a) Pd/a-Si:H and (b) Pd/a-Si:N:H diodes.

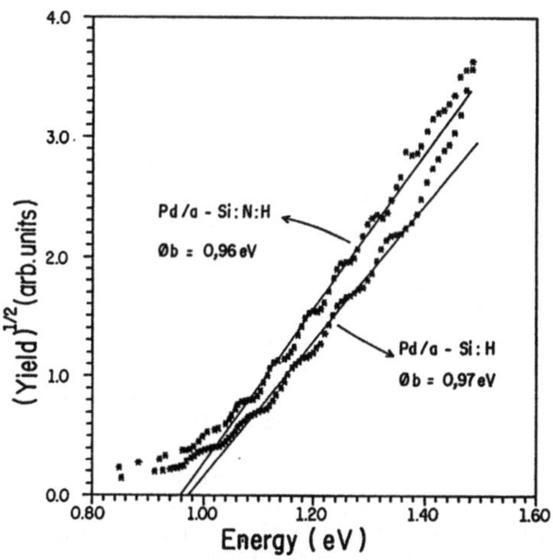

Figure 3 - Square root of the photoelectric yield versus photon energy for Pd contacts on (a) a-Si:H and (b) a-Si:N:H.

coefficient in this type of material its contribution to the total current is certainly negligible.

In order to clarify this problem barrier height measurements were done by the internal photoemission technique. As shown in figure 3 the obtained barrier heights for a-Si:H and a-Si:N:H only differ by an amount of the order of the experimental error. In addition, it was found that these values are not affected by thermal annealing. So, one can conclude in a first approximation that the thermionic emission/diffusion mechanism is not affected either by the incorporation of nitrogen impurities or by thermal annealing. Therefore, the observed effects may be attributed to the change of the carrier recombination in the depletion region and/or tunneling through the barrier conduction mechanisms.

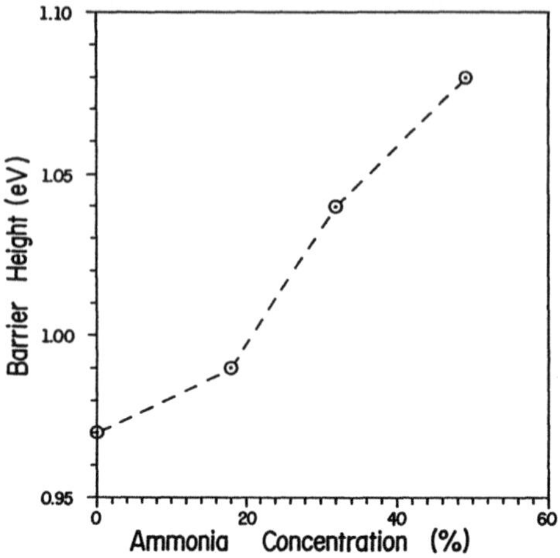

Figure 4 - Barrier heights versus ammonia concentration in the gas for Pd/a-Si:N:H diodes.

Two main reasons may be given for the change of the above current conduction mechanisms. The first and more important is the change in the defect density within the mobility gap which increases upon nitrogen incorporation [4] and may be reduced by thermal annealing. The second is doping which may also occur for the low nitrogen concentrations used here.

Other authors attributed the effect of thermal annealing to the growth of a silicide layer which would consume the near-surface defective region or to the formation of a more laterally uniform interface. However, these authors have observed a reduction in the ideality factor at temperatures lower than the silicide formation temperature for Pt/a-Si:H contacts [5], revealing that other mechanisms may be involved. In addition, preliminary results for Pd/a-Si:H diodes under light soaking showed that the effects of thermal annealing can be reversed by illumination. Therefore, we suggest that hydrogen redistribution inside the material, as already observed for Pt/a-Si:H contacts [6], and the consequent passivation of defects may play an important role, although some influence of silicide formation and the consequent translation of the metal/semiconductor interface cannot be directly discarded. Unfortunately, Pd silicide formation and hydrogen diffusion in a-Si:H possess similar activation energies, thus making their distinction difficult in our experiments.

Finally, the effect of incorporation of larger amounts of nitrogen on the barrier height was investigated. Figure 4 shows that the barrier height increases with nitrogen concentration what may be related to the increase of the mobility gap. Further experiments to clarify this point are under way.

Acknowledgements - The financial support of Brazilian (CNPq, Finep) and German (VW Foundation) institutions is gratefully acknowledged.

4. References

[1] E.H.Roderick and R.H.Williams, "Metal-Semiconductor Contacts", Second edition, Clarendon Press, Oxford, 1988.

[2] M.J.Thompson, R.J.Nemanich and C.C.Tsai, Surf. Sci. 132, 250 (1983).

[3] A.Madan, W.Czubatj, J.Yang, M.S.Shur and M.P.Shaw, Appl. Phys. Lett. 40, 234 (1982).

[4] M.Meaudre and R.Meaudre, Phil. Mag. B 55, 417 (1987).

[5] R.J.Nemanich, M.J.Thompson, W.B.Jackson, C.C.Tsai and B.L.Stafford, J. Vac. Sci. Tech. B 1, 519 (1983).

[6] W.Beyer, C.E.Gatts, J.Herion, W.Losch and H.Wagner, J. Non-Cryst. Sol. 97&98, 951 (1987).

Chemical Homogeneity and Charge Transfer in Amorphous Si-N Alloys

M.M. Guraya[1], *H. Ascolani*[1], *G. Zampieri*[1], *J.I. Cisneros*[2], *J.H. Dias da Silva*[2], and *M.P. Cantão*[2]

[1]Centro Atómico Bariloche, 8400 Bariloche, Argentina
[2]UNICAMP, CP 6165, 13081 Campinas, SP, Brazil

This work completes a study of amorphous SiN_x:H thin films with $0 \leq x < 1.5$, in which we have combined XPS, EELS, and optical measurements to determine the bonding structure. /1/ We analyze here the lineshapes and the chemical shifts of the Si-2p and N-1s peaks.

The main results are shown in Table 1. The N content x was determined from the ratio of the intensities of the N-1s and Si-2p peaks. We have used two parameters to characterize the lineshapes: The full width at half maximum, W, and an asymmetry factor defined as follows

$$A = \frac{\Delta H - \Delta L}{\Delta H + \Delta L} , \qquad (1)$$

where ΔH and ΔL are the half width at half maximum towards the high- and low-binding energy sides of the peak, respectively.

Table 1: Results of the measurements.

x	Si-2p BE eV	Si-2p W eV	Si-2p A %	N-1s BE eV	N-1s W eV	N-1s A %	<BE(Si-2p)> eV
0.0	99.2	1.85	+4.3	-	-	-	99.25
0.36	99.6	2.80	+9.8	397.2	1.95	+1.5	99.8
0.49	100.2	3.00	+2.7	397.5	2.00	0.0	100.3
0.87	101.3	3.00	-13.9	397.5	2.15	+1.4	100.8
1.11	101.4	2.45	-3.2	397.6	2.15	+1.2	101.4
1.35	101.9	2.40	-1.7	398.0	2.15	+1.2	101.87

We have referred all the binding energies (BE) to that of the Ar-$2p_{3/2}$ level, to which we have assigned the arbitrary value BE = 242 eV. With this referencing we avoid peak shifts due to charging of the sample and/or shifts of the Fermi level in the band gap.

The lineshape of the Si-2p peak varies strongly as the N content is increased. The peak first broadens asymmetrically towards high binding energies, then towards low binding energies, and finally narrows and becomes rather symmetric again. A significant shift of the maximum from 99.2 to 101.9 eV is observed. In contrast, the N-1s peak remains practically unchanged.

We can use the Si-2p lineshape to analyze the chemical homogeneity of the films. In a mixture model the material is thought as composed of microclusters of pure Si and silicon nitride; the corresponding Si-2p spectrum can be simulated by combining the Si-2p spectra of pure Si and pure silicon nitride in the proportion determined by the total N content of the film. Two synthetic spectra, produced by combining the corresponding "pure" spectra of amorphous and crystalline materials, are shown in fig. 1 together with the actually measured Si-2p spectrum. Similar comparisons have been made for all the films. It is clear that the mixture model is com-

Springer Proceedings in Physics, Volume 62
Surface Science Eds.: F.A. Ponce and M. Cardona

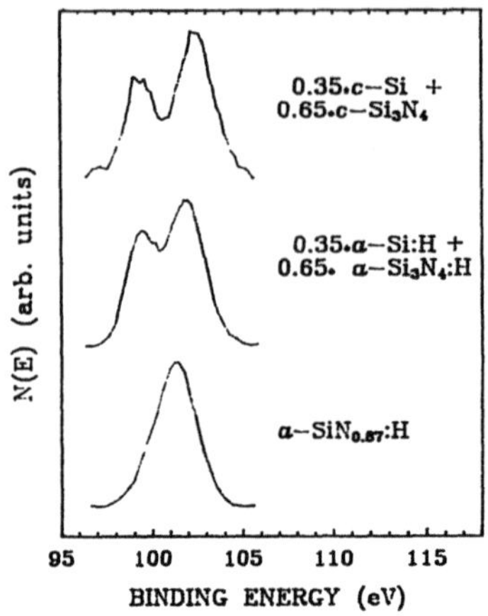

Figure 1: Synthetic spectra simulating segregation and the actually measured spectrum of Si-2p for the sample with x = 0.87.

pletely inadequate to describe the structure of the films. A random bonding model, in which Si-(Si_kN_{4-k}) units occur on a statistical basis, offers a much better description.

The chemical shift of the Si-2p peak is due to the replacement of homopolar Si-Si bonds by heteropolar Si-N bonds; each Si-N bond removes charge from the Si atoms thereby producing an increase in the binding energy of the Si-2p level. If the charge transfer per Si-N bond is independent of the number of N atoms bound to the Si atom, then the mean binding energy must be proportional to the mean number of N atoms bound to a Si atom

$$\langle BE(Si\text{-}2p)\rangle = BE_o + \Delta E_{Si-N} \cdot \frac{(Si\text{-}N)}{[Si]} \qquad (2)$$

where BE_o is the binding energy of Si-2p in a-Si:H, ΔE_{Si-N} is the chemical shift per Si-N bond, and (Si-N) and [Si] are the densities of Si-N bonds and Si atoms, respectively. Using the values of <BE(Si-2p)> of Table 1 and (Si-N) and [Si] from ref. 1, the linear regression yields ΔE_{Si-N} = 0.78 eV, in excellent agreement with a derivation by another method./2/ Converting the chemical shift into charge transfer with the scale factor deduced by Grunthaner et al,/3/ 2.2 eV/electron, we obtain a charge transfer of 0.35e per Si-N bond.

References:

/1/ M.M. Guraya, H. Ascolani, G. Zampieri, J.I. Cisneros, J.H. Dias da Silva, and M.P. Cantão, Bonding structure of amorphous SiN_x:H films, in these proceedings.
/2/ R. Kärcher, L. Ley, and R.L. Johnson, Phys. Rev. B 30, 1896 (1984).
/3/ F.J. Grunthaner, P.J. Grunthaner, R.P. Vasquez, B.F. Lewis, J.Maserjian, and A. Madhukar, Phys. Rev. Lett. 43, 1683 (1979).

Photocurrent Oscillations in a-SiC:H Double Barrier Devices Exhibiting Negative Differential Resistance

M.P. Carreño and I. Pereyra

Laboratorio de Microelectronica, Escola Politécnica da Universidade de São Paulo, Caixa Postal 8174, CEP 01051, São Paulo, Brazil

Abstract. Recently we have reported the existence of voltage controlled negative differential resistance (NDR) in a-Si:H/a-SiC:H double barrier devices. The NDR results were attributed to sequential tunnelling phenomena. Here we report a new interesting feature of those structures, namely, voltage sweep rate dependent photocurrent oscillations. Moreover, the dark current of previously illuminated samples also exhibits oscillations. The correlation of these oscillations with negative resistance and sequential tunnelling is analyzed.

1. Introduction

Lately, many works have reported quantum size effects in the electric transport through double barrier devices (DBD) [1,2], but just bumps in the I vs.V curves have been observed. Pereyra et al., however, reported also the observation of negative differential resistance (NDR) in a-Si:H/a-SiC:H double barrier structures [3]. They pointed out some peculiar features, particularly the dependence of the peak to valley ratio on the applied voltage sweep rate, which was attributed to trapping effects [4]. In this work we describe the existence of current and photocurent oscillations in double barrier devices, which under particular conditions of sweep rate, illumination and temperature evolve into negative resistance peaks.

2. Results and Discussion

In this work, the current-voltage and photocurrent-voltage characteristics of DBDs with different barrier heights, are analyzed for various temperatures. First, we study the I vs. V dark characteristics of two samples with different barrier heights, which exhibit negative differential resistance (NDR). For the shallow well device (~150 meV) NDR is observed in the 100 K up to 170 K temperature range for an applied voltage varying from 0.2 to 0.3 volts. The deeper well structure

Springer Proceedings in Physics, Volume 62
Surface Science Eds.: F.A. Ponce and M. Cardona

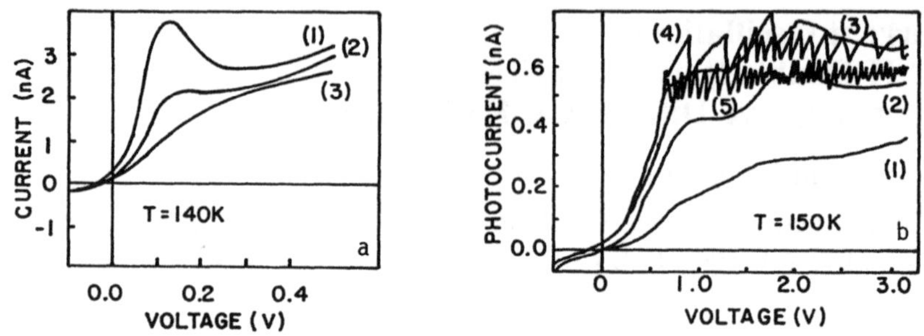

Fig.1 (a) dark current-voltage for the 150 meV deep well at (1) 10 V/s, (2) 125 mV/s and (3) 80 mV/s sweep rates. (b) Photocurrent vs. voltage for 400 meV at (1) 40 mV/s, (2) 200 mV/s, (3) 250 mV/s, (4) 500 mV/s and (5) 10 V/s, deep well.

(400 meV) shows NDR in the 100 K up to 260 K temperature range, for an applied voltage varying from 0.6 to 1.2 V approximately. In both cases the sweeping rate of the applied voltage was 10 V/s. For both samples, only one NDR peak is observed, although for the deeper well device the NDR region is wider. These results were attributed to a sequential tunnelling phenomenon, eventhough theoretical estimates predict two peaks for the deeper well. As it was pointed out previously [4], the high resistivity of the films difficults the determination of the electric field distribution along the devices. To overcome this problem, photocurrent experiments were performed. Also, in order to obtain almost stationary state conditions, the curves were traced under lower sweep rates. Fig.1b shows the photocurrent vs. voltage characteristic for a 400 meV deep well device illuminated with 10 mW/cm^2 white light at different sweep rates. Now, for some sweep rates, two NDR regions are observed, as expected from the theory. Moreover, for fast sweeps photocurrent oscillations occur. Decreasing the sweep rate makes the oscillations evolve to NDR peaks which for even slower sweeps disappear and only bumps in the characteristics remain. This rate dependence is also observed in the dark characteristics as shown in Fig.1a for the 150 meV deep well device. Eventhough this would be an unexpected result for a pure sequential tunnelling effect, it should not be surprising when dealing with transport in amorphous material with high density of scattering centers and traps [4]. Prior exposure to light affected the oscillations (persistent photocurrent). In Fig.2a, the dark current of a device (400 meV deep well) which was previously exposed for 2, 3 and 4 minutes to 10 mW/cm^2 of white light are shown. It can be observed that for 2 minutes exposure a NDR peak appears but for 3 and 4 minutes exposure,

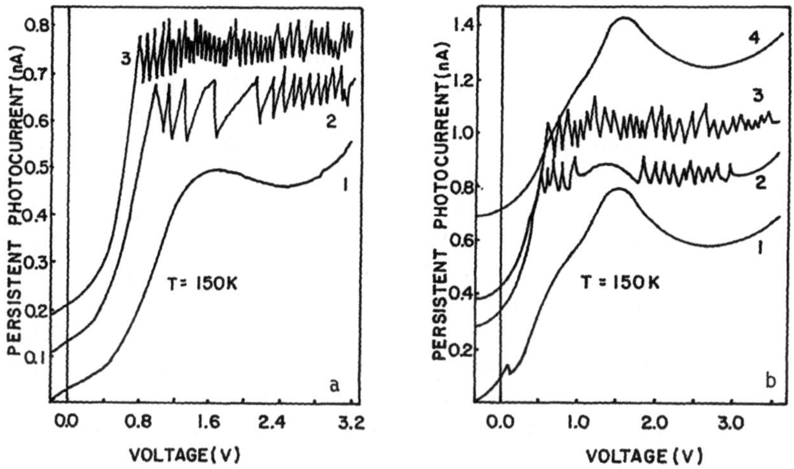

Fig.2 (a) Persistent photocurrent for 2, 3 and 4 minutes of previous illumination (curves 1, 2 and 3 respectively) at a fixed 10 V/s sweep rate. (b) Persistent photocurrent for diferent sweep rates at a fixed previuos illumination: (1) 10 V/s, (2) 200 mV/s, (3) 125 mV/s and (4) 40 mV/s. In both cases the values of photocurrent are valid only for curve (1), the other curves were dislocated to visualize better.

the NDR gives place to current oscillations whose frequency increases with the previous light exposure. Fig.2b shows the dark current characteristic of a sample which was previously exposed for 1 minute to the same light intensity but for 4 different sweep rates. It is observed that for the higher rate (10 V/s) a well defined NDR peak appears (curve 1). For a slightly lower rate, the NDR peak disappears and two groups of oscillations separated by a smooth peak show up (curve 2). For the next lower rate, only oscillations of higher frequency are observed (curve 3) and finally, for the slowest sweep rate a NDR peak appears again (curve 4). Temperature also influences the photocurrent oscillations. Increasing temperatures make the oscillations evolve to NDR peaks and further increases smear the peaks, remaining only bumps in the curves.

The reported occurrence of negative differential resistance (NDR) in amorphous double barrier devices is a complex phenomenon which depends on several parameters. This fact makes difficult the observation and reproduction of the phenomena, being much more frequent the observation of oscillations or simply bumps in the I vs. V curves. At a first glance the oscillations could be attributed to instabilities in the measurement circuit due to the NDR. The sweep rates at which they are observed depend on the resonance frequency of the circuit which at the same time depends on the illumination and temperature of the device. In this way the presence of oscillations would indicate the existence of NDR whose direct observation is difficulted because the other parameters (illumination, sweep rate, temperature) do not assume the proper values. On the other hand, the existence of bumps in the curves does not imply necessarily the ocurrence of sequential tunnelling since simple contact effects can produce the same features, as we proposed previously [5].

Summarizing, the occurrence of NDR in amorphous silicon DBDs is a rather critical phenomenon whose observation depends on specific conditions of parameters as temperature, illumination, voltage sweep rate, among others. When those conditions are not met, oscillations of the current and photocurrent are observed. In this way the presence of current oscillations would imply in the existence of NDR.

The question of the origin of the NDR still remains unresolved. Eventhough sequential tunnelling is a possible explanation, the results suggest that this phenomenon may be related to others, particularly transport through traps which can mask the effect almost completely.

3. References

1. S. Miyasaki, Y. Ihara, M. Hirose, Phys. Rev. Lett., vol.59(1987) pp.125.
2. Y.L. Jiang and H.L. Hwang, J.J. of Appl. Phys.,Vol.27, No.12 (1988)L2434- L2437.
3. I. Pereyra, M.P. Carreño, R.K. Onmori, C.A. Sassaki, A.M. Andrade and F. Alvarez, J. of Non-Cryst. Sol., 97&98 (1987) 871-874.
4. I. Pereyra, M.P. Carreño and F. Alvarez, J. of Non-Cryst. Sol.,110(1989) 175-178.
5. M.P. Carreño, I. Pereyra, A. Kamazawa and A. Arasaki, J. of Non-Cryst. Sol., 114 (1989) 762-764.

Complex Refractive Index of a-Si:F Thin Films

M. García-Castañeda and A. Mariño-Camargo

Departamento de Física, Universidad Nacional de Colombia,
Bogotá, Colombia (S.A.)

Abstract. The effect of fluorine on the complex refractive index (n, k) of a-Si:F thin films produced by r.f. sputtering in argon atmospheres between 20 and 60 mTorr is studied. Samples produced at 20 mTorr show a significant decrease of n when the partial pressure of fluorine is increased. Furthermore the k values shift to shorter wavelengths, in the same way as the optical gap does.

1. Experimental

a-Si:F thin films were made by r.f. sputtering in a gas mixture of Ar and SiF_4 at two different argon partial pressures (20 and 60 mTorr). The substrate temperature was held at about 300°C. For each value of argon partial pressure, a set of samples were prepared in which the fluorine content was changed by increasing the partial pressure of SiF_4 from 0 up to about 10^{-3} Torr [1].

From spectrophotometric measurements of reflectance and transmittance of films prepared on sapphire substrates, the absorption coefficient α was deduced. The optical gap E_0 was determined by using the Tauc [2] relation. With the collected data, the effect of the fluorine content on the complex refractive index (n, k) was calculated for each film, using the consistent numeric scheme recently developed for such a case [3].

2. Results and Discussion

Figures 1a and 1b show the results, as a function of the incident radiation energy, of measurements of the absorption coefficient and the optical gap (E_0), respectively, of films fabricated in an argon atmosphere of 20 mTorr and various fluorine concentrations. With the increase of fluorine content, the absorption edge is shifted to higher energies. Such a behavior can be attributed, as happens with hydrogen [4], to the removal of dangling bonds by fluorine, eliminating the cause of the large absorption at low energies.

The increase of the optical gap from 1.4 eV to 1.6 eV approximately, on the other hand, is correlated with the increase of the activation energy E_a and the concomitant decrease of dark conductivity σ_D [1,5] observed in these films. The further increase of the optical gap for fluorine partial pressure $\geq 10^{-3}$ Torr has been attributed to the formation of SiF_4 clusters [6].

Figures 2a and 2b show the real part of the complex refraction index, n, as a function of incident photon energies for different fluorine concentra-

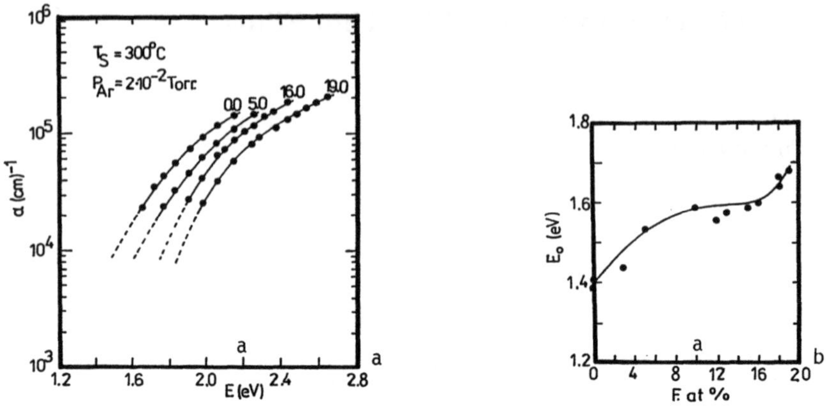

Fig. 1 (a) The absorption coefficient (α) as a function of the photon energy for different fluorine concentrations and at an argon pressure of 20 mTorr. (b) The optical gap (E_0) as a function of fluorine concentration

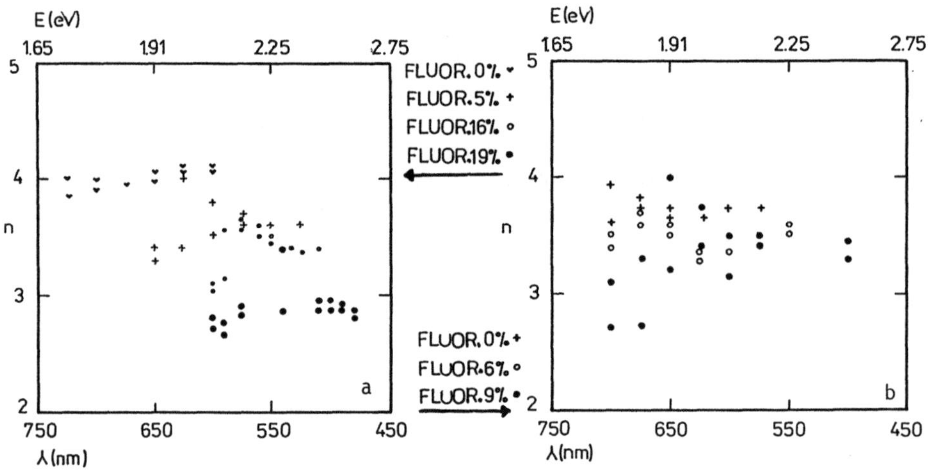

Fig. 2 The real part of the complex refractive index (n) as a function of photon energy for different fluorine concentrations and argon partial pressure of (a) 20 and (b) 60 mTorr

tions, for argon partial pressure of 20 and 60 mTorr, respectively. The general trend is a decrease of the n values when the fluorine content is increased. The decrease is even more marked in samples produced in argon atmospheres at about 20 mTorr, in which the oxygen content is not detectable by the Rutherford backscattering technique [6]. This is not the case in samples produced in higher argon pressures (60 mTorr). For those films the oxygen content is high (about 10%) and low fluorine incorporation does not diminish the density of the samples in a noticeable way.

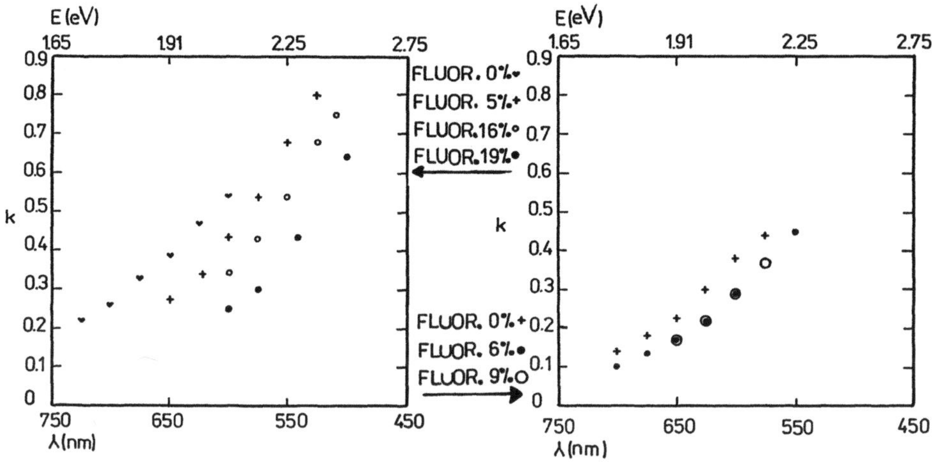

Fig. 3 The imaginary part of the complex refractive index (k) as a function of photon energy for different fluorine concentrations and two different deposition conditions, **(a)** 20 and **(b)** 60 mTorr argon pressure

Dispersion in n values is obeyed to a $\pm 5\%$ error assumption in film thickness measurement, in which a Dektak surface profile system was used.

The behavior of the imaginary part of the complex refractive index k is shown in Figs. 3a and 3b for the two different deposition conditions. A shift to the shorter wavelength region is observed for a partial argon pressure of 20 mTorr with different fluorine concentrations. This is well correlated with the optical gap shift and it has been attributed to the fact that both fluorine and oxygen act as efficient defect passivation agents.

References

1. A. Mariño, J.M. Díaz: In *Proc. of the 5th Latin American Symposium on Surface Physics.* Bogotá, Colombia, 1988, Ed. by M. Cardona, J. Giraldo (World Scientific, Singapore)
2. J. Tauc, R. Grigorovici, A. Vancu: Phys. Status Solidi **15**, 672 (1966)
3. M. García-Castañeda, H. Sánchez-Machet: Thin Solid Films **176**, 69 (1989)
4. F. Freeman, W. Paul: Phys. Rev. B **20**, 716 (1979)
5. W. Beyar, J. Chevalier, K. Reichelt: Solar Energy Mater. **9**, 229 (1983)
6. A. Mariño, K. Reichelt: Revista Colombiana de Física **20**, #3 (1988)

Bonding Structure of Amorphous SiN_x:H Films

M.M. Guraya[1], *H. Ascolani*[1], *G. Zampieri*[1], *J.I. Cisneros*[2], *J.H. Dias da Silva*[2], *and M.P. Cantão*[2]

[1]Centro Atómico Bariloche, 8400 Bariloche, Argentina
[2]UNICAMP, CP 6165, 13081 Campinas, SP, Brazil

We present a study of amorphous SiN_x:H thin films with $0 \le x < 1.5$, prepared by the RF reactive sputtering method. By combining results of x-ray photoemission spectroscopy (XPS), electron energy loss spectroscopy (EELS), and infrar ed (IR) absorption, we have determined all the atom and bond densities.

In Table 1 the results of the measurements are shown. The nitrogen content x was determined from the ratio of the intensities of the N-1s to Si-2p XPS peaks. The integrated absorption of any vibrational band in the IR region, I_{m-1} , is related to the corresponding bond density, (m-1), by

$$(m-1) = A_{m-1} \cdot I_{m-1} \quad . \tag{1}$$

Using constants A_{Si-H} and A_{N-H} from the literature, /1,2/ we obtained the densities of Si-H and N-H bonds. The sum of both densities is the hydrogen atom density [H] .

Table 1: Results of the measurements.

x	$\hbar\omega_p$ eV	I_{Si-H} (	I_{N-H} cm^{-1}	I_{Si-N})	(Si-H) (	(N-H) 10^{22}	(Si-N) cm^{-3}	(Si-Si))
0.0	17.14	84	-	-	1.2	-	-	9.46
0.36	19.02	190	2	2276	2.3	0.06	5.03	5.76
0.49	19.60	201	5	3760	2.4	0.2	6.61	4.76
0.87	20.57	200	11	6370	2.2	0.3	10.86	2.01
1.35	21.90	63	68	7380	0.7	1.9	12.99	1.17

From the EELS spectra we obtained the plasmon energies $\hbar\omega_p$, which are related to the densities of valences electrons, n_V, through

$$(\hbar\omega_p)^2 = \frac{4\pi\hbar^2 e^2}{m} n_V \tag{2}$$

where m and e are the mass and electric charge of the electron. The density of valence electrons can be written as

$$n_V = n_{Si}[Si] + n_N[N] + [H] \tag{3}$$

where n_{Si} and n_N are the numbers of valence electrons per Si and N atom. Since [H] and the ratio [N] / [Si] = x can be determined from the IR and XPS measurements, respectively, eqs. (2) and (3) allow us to evaluate [Si] and [N]. We have used $n_N = 3$ because the N-2s electrons do not participate in the plasma oscillation./3/

A detailed balance of the total numbers of atoms and bonds give the relations:

$$3[N] = (Si-N) + (N-H) + 2(N-N) + (N-) \tag{4}$$

Springer Proceedings in Physics, Volume 62
Surface Science Eds.: F.A. Ponce and M. Cardona

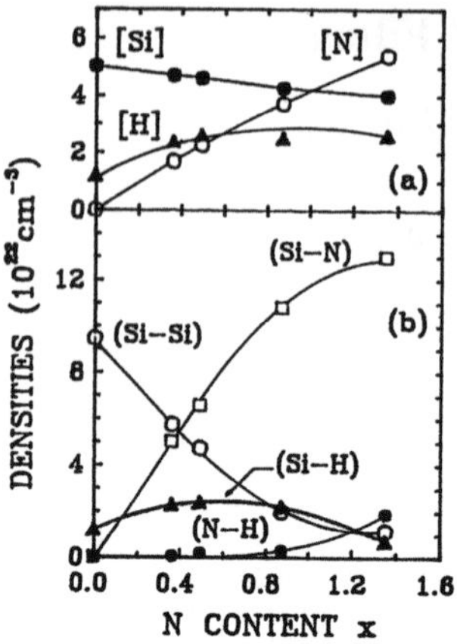

Figure 1: Atom (a) and bond (b) densities as a function of the N content x.

$$4\,[Si] = (Si\text{-}N) + (Si\text{-}H) + 2(Si\text{-}Si) + (Si\text{-}) \qquad (5)$$

where (N-) and (Si-) are the densities of N and Si dangling bonds. We will neglect in the following these densities because they are three orders of magnitude smaller than the other calculated densities. /2/ At low N contents the density of N-N bonds can also be neglected and eq. (4) be used to determine the (Si-N) term. The resulting values of (Si-N), when plotted as a function of the integrated absorption I_{Si-N}, show a linear relationship in agreement with eq. (1). From the slope we obtain $A_{Si-N} = 1.76x10^{19}$ cm^{-2}. Having determined A_{Si-N} we can use eqs. (1) and (4) to evaluate (Si-N) and (N-N) for the most nitrogenated sample. We obtain (N-N) = $0.64x10^{22}$ cm^{-3}. Knowing (Si-N) for all the samples, we determine the density of Si-Si bonds using eq. (5). In fig. 1 all the atom and bond densities are shown as a function of x. The most important conclusions we can extract are:

i) The H content doubles with the incorporation of N and then remains approximately constant.

ii) For x < 1, Si-N bonds increase at the expense of Si-Si bonds; N atoms are fully coordinated by three Si atoms and H binds exclusively to Si.

iii) For x > 1, Si-N bonds increase at the expense of both Si-Si and Si-H bonds; however the ideal threefold coordination of N with Si atoms is not reached and some N-H and N-N bonds begin to appear.

iv) Near stoichiometry Si-N bonds dominate but substantial amounts of Si-Si, N-H, and N-N bonds are observed.

References:

/1/ H. Shanks, C.J. Fang, L. Ley, M. Cardona, F.J. Demond, and S. Kalbitzer, Phys. Stat. Sol. 100, 43 (1980).
/2/ A. Morimoto, Y. Tsujimura, M. Kumeda. and T. Shimizu, Jpn. J. Appl. Phys. 24, 1394 (1985).
/3/ M.M. Guraya, H. Ascolani, G. Zampieri, J.I. Cisneros, J.H. Dias da Silva, M.P. Cantão, and F.C. Marques, Phys. Rev. B (to be published).

TFTs with an a-SiC_x:H Insulator Layer

I. Pereyra, M.P. Carreño, and A.M. de Andrade

Laboratorio de Microelectrônica, Escola Politécnica da Universidade de São Paulo, Caixa Postal 8174, CEP 01051, São Paulo, Brazil

Abstract. Amorphous hydrogenated silicon carbide (a-SiC_x:H) with very high resistivity (10^{15} ohm.cm) and optical gap (Eo~4 eV) is used as insulator layer in thin film field-effect transistors. The static and dynamic characteristics are studied and the possibility of applying these transistors to address a-Si based sensors is confirmed. Low values of the threshold voltage (V_{th} ~ 2.5 V) were obtained and instabilities did not pose any serious problem: the devices show little change after long term ac operation.

1. Introduction

Amorphous silicon thin film transistors have been already fabricated with satisfactory performance for applications in a-Si:H based image sensors and for addressing liquid crystal displays. Their application in thin film integrated circuit technology is limited by their intrinsic instability and speed. The speed is somewhat difficult to improve due to the fact that electron scattering in amorphous materials imposes a lower limit for the carrier mobility (μ). Attempts to increase the mobility require more complicated structures such as amorphous superlattice active layers.

On the other hand, the instability is related not only to bulk phenomena in the a-Si:H active layer but mainly with allowed states in the semiconductor/insulator interface. As a consequence, the major efforts have been in the direction of the improvement and the understanding of the insulator's role. Works with SiO_2 as insulator layer have been already reported. Silicon dioxide, however, has the disadvantage of not being obtainable by the same deposition technique as the a-Si:H active layer, which makes processing more difficult. This problem can be avoided utilizing hydrogenated amorphous silicon alloys and, in fact, recently, most of the works dealing with a-Si:H TFTs report devices with amorphous silicon nitride (a-SiN:H) insulator layer. Another possibility is to use glow discharge amorphous silicon carbide (a-SiC_x:H), which

Springer Proceedings in Physics, Volume 62
Surface Science Eds.: F.A. Ponce and M. Cardona

has the advantage of presenting a lower density of interface states with a-Si:H than the a-SiN:H. However, specimens with sufficiently high resistivity were not available for this application. Recently, Onmori et al. [1] reported plasma enhanced chemical vapor deposited a-SiC_x:H with more suitable properties. They reported resistivities of the order of 10^{15} ohm.cm and an optical gap of 4 eV. We think that this material is a good candidate for insulator gate applications and in this work we present the performance of a-Si:H TFTs produced with these a-SiC_x:H films as the insulator layer.

2. Experimental, Results and Discussion

The layers defining the thin film transistors were deposited successively onto Corning glass substrates covered with nickel-chromium metal gates, as it is shown in a cross-sectional view in Fig.1. The intrinsic a-Si:H active layer, the a-SiC_x:H insulator layer and the n^+ a-Si:H injecting contact were deposited in only one vacuum step by the Plasma Enhanced Chemical Vapor Deposition (PECVD) technique from appropriate mixtures of SiH_4, CH_4, PH_3 and H_2. The thickness of the intrinsic a-Si:H layer and n^+ a-Si:H contact are 150 nm and 35 nm respectively. The a-SiC_x:H insulator layer is 100 nm thick, with a resistivity of 2.10^{15} ohm.cm and an optical gap of 4 eV. Finally, a 1 μm thick aluminum layer is evaporated to define the source and drain contacts. Transistors with channel lengths (L) of 10, 15 and 20 μm and channel widths (W) of 1 and 2 mm were fabricated to check the scaling law.

The typical output characteristics of a transistor with L=10 μm and W=2 mm is shown in Fig.2a. The curves were obtained for Vg varying from 3 to 12 volts in 3 volts step and it can be observed a well defined saturation region. To obtain the threshold voltage (V_{th}) we use the relation:

$$I_{ds} = \mu o.C_{ox}.(W/L).V_{ds}.(Vg - V_{th}), \quad (1)$$

obtained from the basic expression of MOSFET theory for small V_{ds} values [2]. The plot of I_{ds} vs. Vg, for V_{ds}= 0.35 V, for

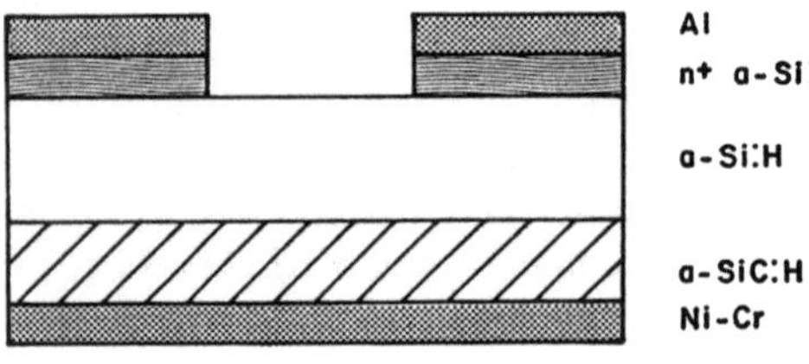

Fig.1 Cross-sectional view of the transistors.

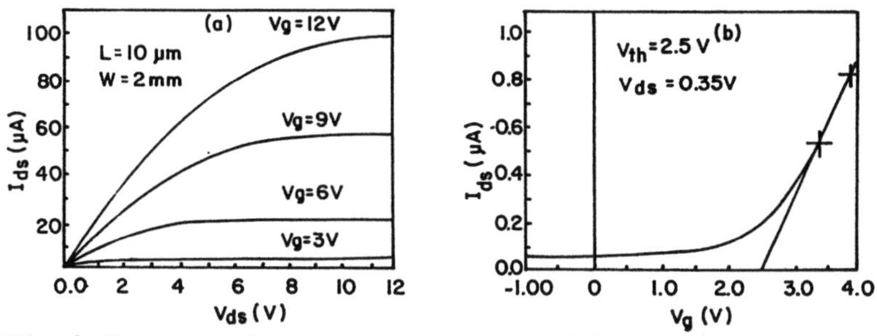

Fig.2 I_{ds} vs. V_{ds} output curves (a) and I_{ds} vs. Vg transfer curve (b) for a transistor with L = 10 µm and W = 2 mm for Vg=3,6,9 and 12 volts.

the transistor of Fig.2a, is shown in Fig.2b. The obtained V_{th} values are about 2.5 V.

Capacitance measurements indicated that the capacitance for static operation is about 25 nF/cm^2 and the C vs. Vg curve of most of the transistors showed the typical feature of MOS structures, with "accumulation" and "depletion" regions. With the obtained value of the capacitance and Eq.(1), the electron channel mobility can be estimated. Typical values were about 0.35 cm^2/V sec.

In order to support our method of extracting parameters, we made a numeric simulation substituting the obtained values of electron mobility, threshold voltage and insulator capacitance in the basic MOSFET equation [2]. The result is shown in Fig.3, where we can see a reasonable fit, particularly in the triode region. The saturation current is somewhat lower than the theoretical one, but the saturation condition (V_{ds}>Vg-V_{th}) is reasonably well satisfied. This result is in some way important because it justifies the parameters V_{th} and obtained but, it is somehow unexpected because the simulation does not take into account considerations about the amorphous material structure.

To evaluate the performance of the a-Si:H TFTs in switching circuits, we studied their response to a high frequency gate signal. The switching speed of the devices was sufficient to be operated with a 10 kHz clock. On the other hand, a slight diminution of the output current was oberved after some measurement cycles, but it was never larger than 20% and disappeared after few minutes of rest.

Summarizing, the typical values of threshold voltage and channel electron mobility obtained were in the range of 2.2 to 2.8 V and 0.30 to 0.40 cm^2/V sec respectively. These results prove the viability of a-SiC$_x$:H as insulator layer in a-Si:H

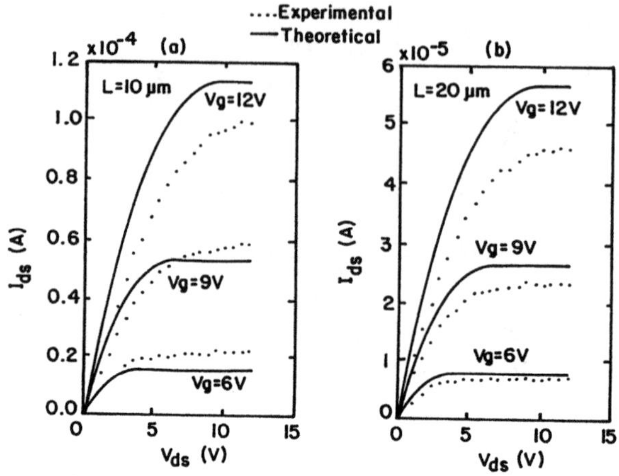

Fig.3 Computer simulation with the basic MOSFET model for two transistors: (a) the same as that of the Fig.2, with V_{th}=2.49 V and μ = 0.33 cm^2/V sec and (b) transistor with L = 20 μm and V_{th}=2.64 V and = 0.39 cm^2/V sec.

based TFTs. In fact, these transistors were used to address an a-Si based image sensor [3]. Some instability exists but it does not seem to be a serious problem.

3. References

1. R.K. Onmori, I. Pereyra, C.A. Sassaki and M.P. Carreño, 9th European Photovoltaic Solar Energy Conference, Freiburg, FRG. (1989),33-36.
2. Y. P. Tsividis, "Operation and Modeling of the MOS transistor", McGraw-Hill Ed., 1987, Cap.4.
3. C.A. Sassaki. A.T. Arasaki, M.P. Carreño, A. Komazawa and I. Pereyra, J.of Non-Cryst. Sol, 115(1989) 90-92.

Thermal Depth Profiling of Solar Cells by Acoustic Calorimetry

M. Fracastoro-Decker, E.A.M. Fagotto, and F. Decker

Instituto de Física, UNICAMP, 13081 Campinas, SP, Brazil

Abstract. The electroacoustic technique has been used to investigate thermal dissipation processes in p-n^+ Si solar cells. The acoustic signal has been measured as a function of the amplitude and frequency of the applied voltage modulation. Experiments can be understood using a vector model which takes into account the different locations of power generation and absorption in the cell.

1. Introduction

Thermal dissipation processes consume more than 80% of the incident radiation in a solar cell. Whereas the total amount of energy lost can be easily calculated from the measured cell efficiency, most interesting for the purpose of improving the cell performance is to know the spatial location of the power dissipated within the cell. Several thermal wave analytical techniques have been applied to characterize these dissipation processes in semiconductor devices. Between these there are methods based on photoacoustic phenomena [1-3]. The acoustic effect consists of the production and propagation of a pressure wave in a closed chamber (acoustic chamber) by means of a modulated illumination (photoacoustic effect) or of an alternating current (electroacoustic effect) in a solid in contact with the chamber. The aim of this work is to use the electroacoustic (EA) technique to identify the solar cell regions where the power is generated or absorbed, using a vector description of the power losses inside the cell which is an extension of the Cahen-Wolf model [4,5].

2. Results and Discussion

The EA technique was applied to two different p-n^+ Si solar cells with a very narrow barrier (~ 0.1 μm), of thicknesses 0.35 mm and 1 mm, respectively. The acoustic signal (modulus and phase) was measured with and without light, both as a function of the amplitude and of the frequency of the modulated applied voltage (sinusoidal or square wave). The signal was measured using two acoustic sensors (Sennheiser microphones mod. KE4) at the emitter and at the base side of the cell (the first one was removed when measuring the EA signal with light).

Springer Proceedings in Physics, Volume 62
Surface Science Eds.: F.A. Ponce and M. Cardona

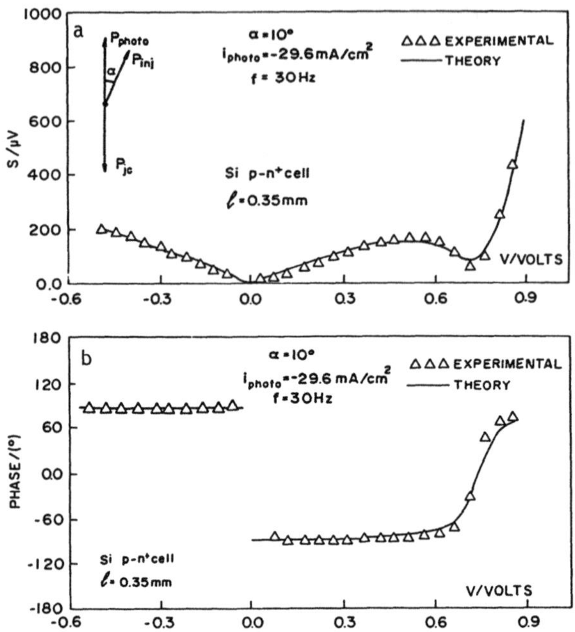

Figure 1. Modulus (a) and phase (b) of the EA signal from a Si solar cell vs. applied voltage amplitude. Insert shows the three vector contributions to the signal.

The model used to describe the thermal phenomena occurring into the cell takes into account the following mechanisms of power dissipation or absorption: a) heating of the junction by photogenerated carriers when sliding down along the barrier; b) junction cooling as hot carriers from the emitter pass the potential barrier and are injected into the base; c) recombination of injected carriers in the bulk. We suppose that only these mechanisms give rise to an acoustic effect when the potential barrier is modulated. Whereas the Cahen-Wolf description treats these contributions as scalar quantities, positive for heating and negative for cooling processes, here we assume that the power dissipation (or absorption) associated to each mechanism can be represented by a vector, whose phase angle depends on the spatial region of the cell where it is originated and on the modulation frequency. Relative to vectors corresponding to processes a) and b), which occur close to the emitter surface of the cell, the vector representing the power dissipated in process c), taking place inside the bulk, makes an angle α (see insert in Figure 1). The measured EA signal is proportional to the total power dissipated into the device, i.e., to the sum of the three vectors:

$$P = P_{photo} + P_{jc} + P_{inj} \tag{1}$$

where

$$|P_{photo}| = (i_{photo}/e)(E_{gap} - \Delta E_c - \Delta E_v - eV) \quad (2a)$$
$$|P_{jc}| = (i_{dark}/e)(E_{gap} - \Delta E_c - \Delta E_v - eV) \quad (2b)$$
$$|P_{inj}| = (i_{dark}/e)\ E_{gap} \quad (2c)$$

and

$$i_{dark} = i_0\ (\exp(eV/nkT) - 1). \quad (3)$$

In the above equations i_o and n are the diode saturation current and quality factor. i_{photo}, the current due to photogenerated carriers, is considered constant in the potential range of our experiments.

The modulus and phase of the EA signal, S, which is proportional to vector **P**, were calculated as a function of the amplitude of the potential modulation, V, using eqs. (1)-(3), and were fitted to the experimental data (corresponding to a fixed modulation frequency) leaving as free parameters i_{photo} and the proportionality factor between **S** and **P**. The angle α (which is a measure of how far from the emitter surface injected carriers recombine, on the average) was determined using experimental results not shown in these figures. The results of the comparison between the calculated and experimental signal (amplitude and phase) are shown in Figure 1. At negative voltages the EA signal is mainly due to photogenerated carriers; for $V > 0$ the contribution of injected carriers starts to grow until at high, positive bias their recombination dominates the signal. The agreement of the model to the data is excellent and shows that a vector picture is necessary to take into account correctly the different contributions to the power dissipation into the device.

Measurements of the EA signal as a function of the potential modulation frequency were performed with the aim of investigating in more detail the thermal power profile into the cell. Figures 2a and 2b show that the EA signal amplide decreases about 1.5 decades for each frequency decade for both the thinner and the thicker cell, and that the signal arising from the bulk recombination of injected carriers is higher when measured from the emitter side than from the base side of the cell, at frequencies $f > f_c$ (where the cut-off frequency is defined by $f_c = D/(\pi l^2)$, D being the thermal diffusion coefficient of the sample and l its thickness). This means that the power is generated closer to the emitter surface than to the cell back surface. For $f < f_c$ the two signals are almost equal, because the sample is "thermally thin", i.e., both microphones test the whole sample thickness at these frequencies.

We also measured the frequency dependence of the EA signal (detected from the base side) arising from injected and from photogenerated carriers (for equal dissipated electrical powers), in the thinner and in the thicker cell (Figure 3a and 3b, respectively). Having in mind an exponential distribution (starting from the emitter surface) of the power dissipated by injected carriers, and an almost δ-function

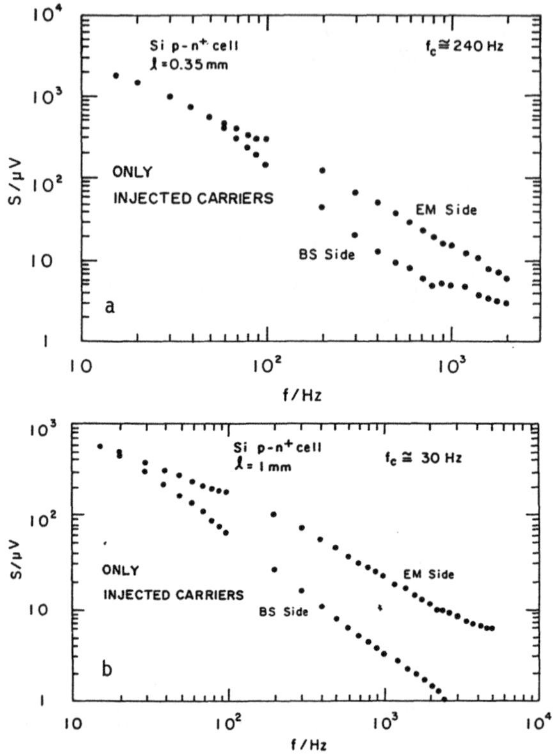

<u>Figure</u> <u>2</u>. EA signal amplitude vs. potential modulation frequency detected from the base or emitter side of a thinner (a) and a thicker (b) Si solar cell.

distribution of the power arising from photogenerated ones, one would expect a more rapid decay of the signal arising from photogenerated carriers at frequencies higher than f_c (at lower frequencies the whole sample thickness is monitored by the microphone). This is exactly what the experiments show in the case of the thinner cell, the two signals being almost equal up to about 200 Hz, and then separating in the expected way for $f > f_c$. For the thicker cell the same experiment shows no appreciable difference between the signal of the injected and of the photogenerated carriers. This is because the region where the power is dissipated is comparatively almost equally far from the microphone in the two cases.

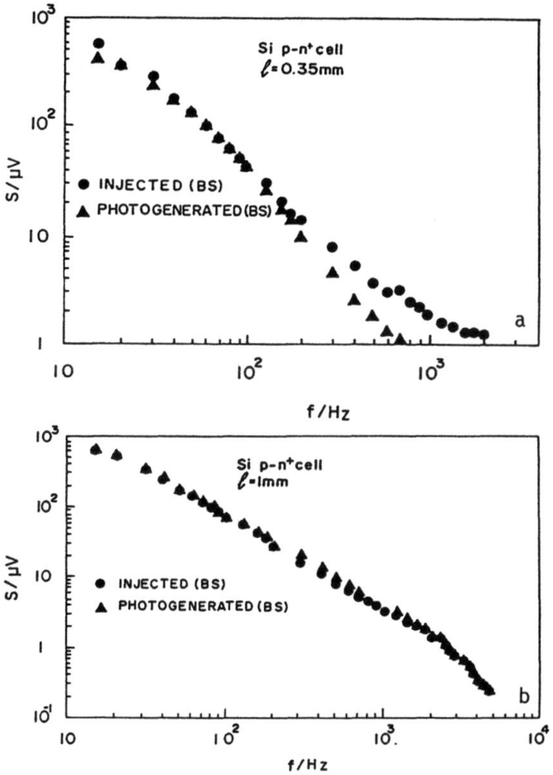

<u>Figure</u> <u>3</u>. EA signal amplitude vs. potential modulation frequency due to photogenerated or injected carriers. (a) thinner cell; (b) thicker cell.

3. Conclusions

Acoustic calorimetry allows the thermal depth profiling of power losses inside a photovoltaic solar cell. The experimental results described in this paper can be interpreted in terms of power generation and absovption processes taking place very close to the emitter side (i.e. within the space-charge region of the cell) and of a power dissipation mechanism due to the recombination of the injected carriers towards the base. The analysis of the resultant acoustic signal as a vector sum of these contributions allows a direct insight into the spatial distribution of the power loss mechanisms inside the cell.

References

1. <u>Photoacoustic and Thermal Wave Phenomena in Semiconductors</u>, A. Mandelis Ed., North Holland, N. York, 1987.

2. Photoacoustic and Photothermal Phenomena, P. Hess and J. Pelzl Eds., Springer, Berlin, 1988.
3. D. Cahen, B. Büchner, F. Decker and M. Wolf, IEEE Trans. on Electronic Devices, 37, 498 (1990).
4. M. Wolf, Energy Convers., 11, 63 (1971).
5. D. Cahen, H. Flaisher and M. Wolf in Ref. 2; p. 247.

Study of the Optical Properties of CdTe Thin Films Grown by rf Sputtering

M. García-Rocha, M. Meléndez-Lira, S. Jiménez-Sandoval, and I. Hernández-Calderón

Departamento de Física, Centro de Investigación y de Estudios Avanzados del IPN, Apdo. Postal 14-740, 07000 México, D.F., Mexico

Abstract. We present the results of photoreflectance spectroscopy and spectral photoresponse obtained from experiments on as-grown and heat treated microcrystalline CdTe thin films deposited by rf sputtering. After thermal annealing, the photoreflectance and photoconductivity experiments show the same trend observed in absorption measurements: a shift towards lower energies of the fundamental band gap. The shift is explained in terms of effects induced by structural modifications such as reduction of the hexagonal phase and quantum size effects.

1. Introduction

Sophisticated and expensive techniques are employed for the elaboration of CdTe based novel devices using superlattices and quantum well structures. However, large-scale application of CdTe in solar energy conversion devices requires the use of economic growth methods. In this respect, rf sputtering appears as a promising technique for the growth of large area CdTe solar cells. We present the results obtained from the optical characterization by means of photoreflectance and spectral photoresponse of CdTe thin films grown by rf sputtering.

2. Experimental Details

The standard diode sputtering technique was employed for the growth of CdTe thin films on top of Corning glass substrates at fixed temperatures in the range from 40 to 240 °C. Thermal annealing of the as-grown films was performed in a nitrogen flux for one hour at 400 °C. X-ray diffraction and electron microscopy revealed the presence of cubic (sphalerite) and hexagonal (wurzite) grains and columnar structures with average diameter within the 50-400 Å range [1,2]. The photoreflectance measurements were performed in the usual configuration in the 1 to 3 eV energy range with a chopped He-Ne laser beam as modulating agent. The photoconductivity measurements were done with the sample contained in a Faraday box with ohmic contacts in the lateral configuration and a typical polarization voltage of 20 eV. All experiments were performed at room temperature.

Springer Proceedings in Physics, Volume 62
Surface Science Eds.: F.A. Ponce and M. Cardona

3. Results and Discussion

Figure 1 shows the photoreflectance spectra of two CdTe samples. A shift of the fundamental band gap, E_o, toward lower energies after heat treatment was observed in all the samples. This shift has also been measured from the absorption spectra of the same films [3]. Ou et. al., noticed a similar shift in the absorption spectra of electrochemically deposited CdTe films[4]; they attributed this change to defect induced absorption. However, for our films, we obtained a very good agreement of the calculated absorption coefficient with a model of direct transitions between parabolic bands [3], discarding defect induced absorption and favorable to a reduction of the energy gap between the conduction and valence bands. The obtained values of E_o for the as-grown CdTe films from absorption [3] and photoreflectance measurements were in the 1.52 to 1.58 eV range; always larger than the typical value for a CdTe single crystal, $E_o \sim 1.50$ eV [5]; further discussion about E_o follows.

The typical spectral response of the CdTe thin films before and after thermal annealing is shown in Fig. 2. The films presented much wider spectra with máxima at higher energies than that expected for a single crystal [6-8]. The photoresponse of the films was always 3 or 4 orders of magnitude lower than that of a single crystal for the same measurement conditions.

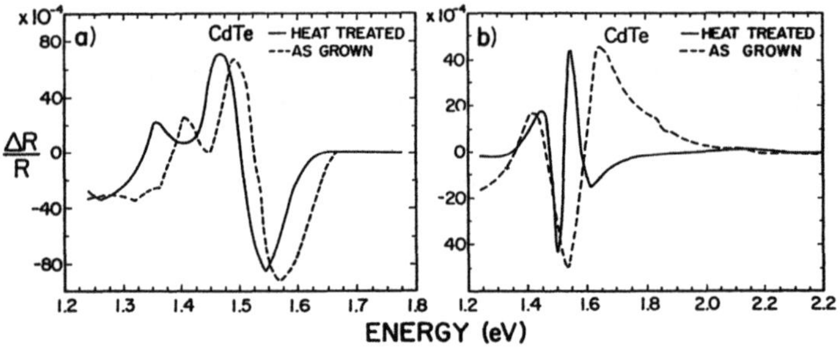

Figure 1. Photoreflectance spectra before and after thermal annealing. A shift of E_o toward lower energies after the heat treatment is clearly seen. The CdTe films were grown at (a)190 °C, and (b) 40 °C.

After heat treatment, the photocurrent peak sharpens and shifts toward lower energies; repeating the behavior already observed in absorption and photoreflectance experiments. Both photocurrent and dark current increased after annealing, indicating an improvement of the transport properties of the film.

The shift after thermal annealing of the Eo transition towards lower energies, approaching the value of the bulk single crystal, Eo=1.5 eV, is associated to two main effects: 1) Transformation of the hexagonal phase regions of the film to the cubic crystalline structure (zincblende). The band gap of hexagonal CdTe is not known, however, for other II-VI compounds which present both types of bulk crystalline structure, the hexagonal phase presents larger values of E_o [5]; this could also be the case for CdTe. Other

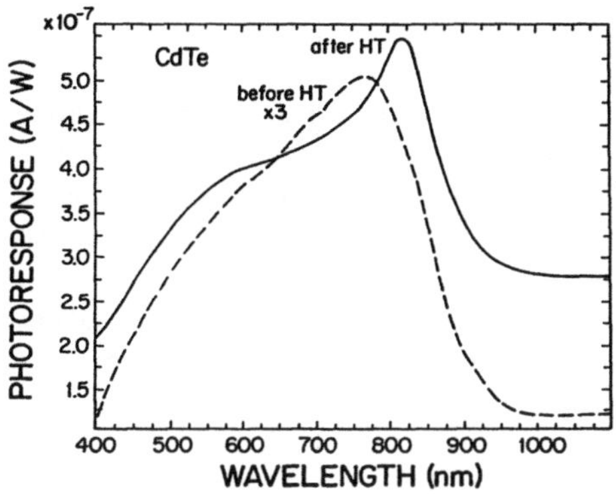

Figure 2. The spectral response of CdTe thin films before (dash curve) and after heat treatment (HT). The shift of the peak towards lower energies of the CdTe thin film after annealing is also observed in these experiments.

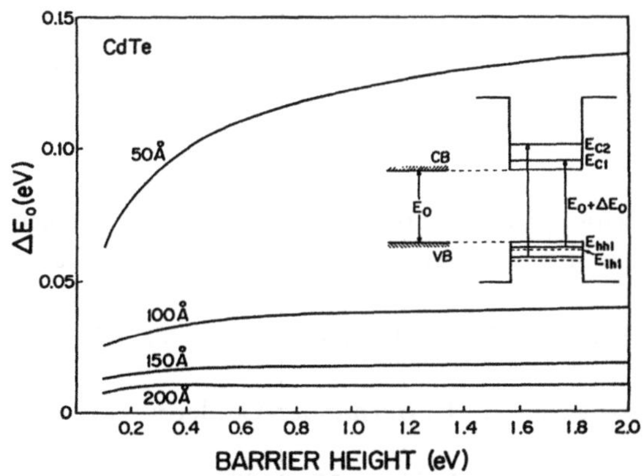

Figure 3. Calculated shift of the band gap of CdTe microcrystals due to quantum size effects. The calculation is based on a model of a spherical well with finite barrier heights and includes conduction and valence band shifts

origin of the shift can be due to quantum size effects. Since the film is composed of grains and columns with average dimensions in the range from 50 to 500 Å, electron and hole confinement can take place, with the grain boundaries acting as potential barriers. A simple calculation of the shift in E_o considering a three dimensional potential well with finite barriers, and taking into account the electron and hole parameters of CdTe, is shown in Fig. 3. Considering that most of the films exhibited shifts in the range from 15 to 60 meV, the values in Fig. 3 are consistent with the experiments, supporting the possibility of quantum size effects. At this moment, it is not clear which effect is the responsible of the shift, however, we believe that both effects are simultaneously contributing to the observed reduction of the band gap.

4. Conclusions

Microcrystalline CdTe thin films have been characterized by photoreflectance and photoconductivity techniques. The shift towards lower energies of the features associated to the fundamental band gap after the annealing process in the photoreflectance and photoconductivity experiments is associated to transformation of the hexagonal metastable phase to the cubic one and quantum size effects due to the microcrystalline nature of the films. Other possible contributions such as strain and intrinsic Franz-Keldysh effects are under study.

Acknowledgments

MGR, MML and SJS thank CONACyT of Mexico for partial funding. This work was partially supported by CONACyT through grants AE-77/88 and P228CCOX891688.

References

1. I. Hernández-Calderón, J.L. Peña, and S. Romero, in Lectures on Surface Science I, edited by G.R. Castro and M. Cardona, (Springer, Berlin, 1987), p. 56.
2. I. Hernández-Calderón, S. Jiménez-Sandoval, J. L. Peña, and V. Sailer, J. Cryst. Growth **86**, 396(1988).
3. M. Meléndez-Lira, S. Jiménez-Sandoval, I. Hernández-Calderón, J. Vac. Sci. Technol. **A7**, 1428(1989).
4. S. S. Ou, M. Stafsudd, B.M. Basol, J. Appl. Phys. **55**, 3769(1984).
5. D.L. Greenaway and G. Harbeke, in *Optical Properties and Band Structure of Semiconductors,* (Pergamon, New York, 1968).
6. M. García-Rocha, I. Hernández-Calderón, to be published.
7. R. H. Bube in *"Photoconductivity of Solids",* (R. E. Krieger, New York, 1978).
8. H. B. DeVore, Phys. Rev. **102**, 86(1956).

Part VIII

Superlattices and Quantum Effects

Metallic Superlattices: Structural and Elastic Properties

M. Grimsditch[1] *and I.K. Schuller*[2]

[1]Materials Science Division, Building 223,
Argonne National Laboratory, Argonne, IL 60439, USA
[2]Physics Department, Building 019,
University of California–San Diego, La Jolla, CA 92093, USA

Abstract. The fabrication and structural characterization of metallic superlattices is discussed. Methods to determine the elastic properties of these materials (which can only be prepared as thin films) are reviewed and the results obtained using various techniques are summarized. Contrary to theoretical expectation, the elastic properties are found to depend on the modulation wavelength.

From the vast literature that exists on thin film growth [1] it may be argued on energy grounds that if (in thermodynamic equilibrium) material A grows as a smooth layer on material B, then B will grow as islands on material A. This conclusion would preclude the possibility of preparing thinly layered superlattices. Experimentally however, it has been found that with modern technologies, like MBE and sputtering, it is indeed possible to fabricate such structures [2] and it must therefore be argued that kinetic effects are also important in determining thin film growth.

That layered structures can be produced is clearly shown in Fig. 1 which shows θ-2θ x-ray scans for Mo/Ni superlattices [3]. In samples with large modulation wavelengths (Λ) peaks are observed at the Mo[110] and Ni[111] positions: this is expected since materials usually grow normally to the densest planes; (110) for bcc, (111) for fcc and (001) for hcp. As Λ is reduced additional peaks appear indicating the existence of a superstructure. It can be easily shown that peaks are expected at

$$\frac{1}{\bar{a}} \pm \frac{n}{\Lambda} \tag{1}$$

where $\bar{a}$ is the average lattice constant and n an integer. The intensity of the peaks depends on many factors including the lattice parameters of the constituents, scattering factors, roughness, etc. We shall return to this point later; for the moment it is sufficient to note that from the x-ray spectra and Eq. 1, Λ and $\bar{a}$ may be obtained without resorting to any specific model.

Having produced these novel materials there is of course interest in determining their physical properties. Many of them have been discussed extensively in the literature [2,4], here we will concentrate only on the elastic properties of superlattices. In the continuum limit it is straightforward to calculate the elastic constants (C_{ij}) of a layered system [5-7]. Two main features arise from these calculations: the C_{ij} of the superlattice are an average of those of its constituents, i.e.,

$$C_{ij}(B) < C_{ij} < C_{ij}(A)\,, \tag{2}$$

and they are also independent of Λ. Figure 2 shows the results obtained for the biaxial modulus (Y_B) of Cu/Ni superlattices [8] in clear contradiction with theory. What is even more spectacular (and controversial) is that Y_B for Cu/Ni (Λ = 20 Å) is larger than Y_B for diamond! The reason for the controversy lies in the severe difficulties encountered in measuring elastic constants of materials which can only be prepared as thin films. There have been a number of techniques developed to deal with this problem [9-16] but most of them require the removal of thin film from the substrate and are consequently open to criticism. Brillouin scattering [11] is one of the few techniques which does not require the removal of the film from the substrate and it has played an important role in the study of metallic superlattices.

Springer Proceedings in Physics, Volume 62
Surface Science Eds.: F.A. Ponce and M. Cardona

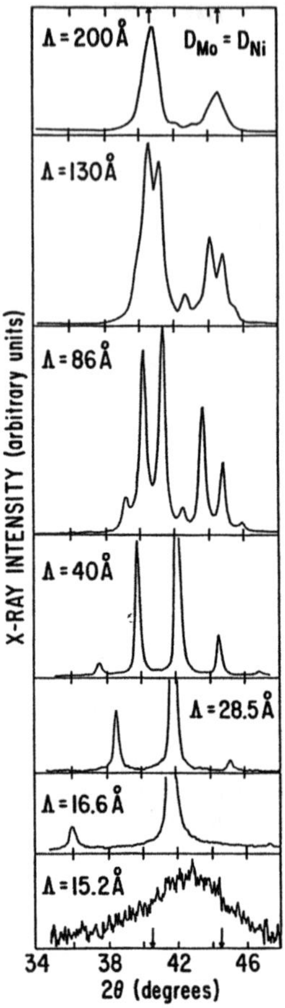

Fig. 1. X-ray spectra of Mo/Ni superlattices (Ref. 3).

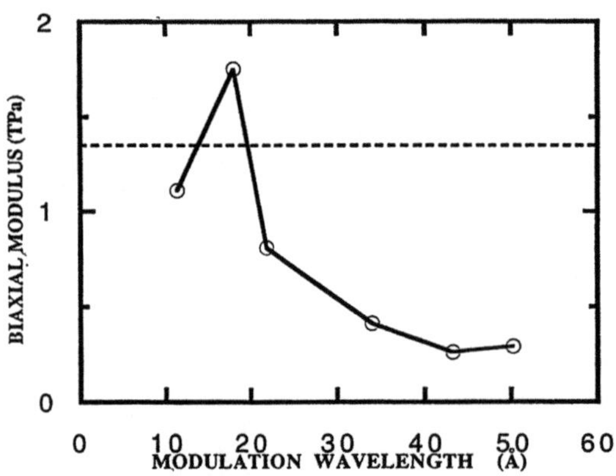

Fig. 2. Biaxial modulus of Cu/Ni supoerlattices (Ref. 8). The dashed line is the biaxial modulus of diamond.

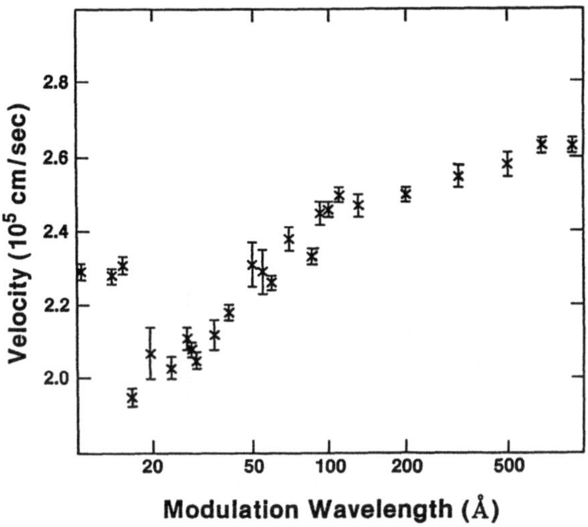

Fig. 3. Surface wave velocity in Mo/Ni superlattices (Ref. 3).

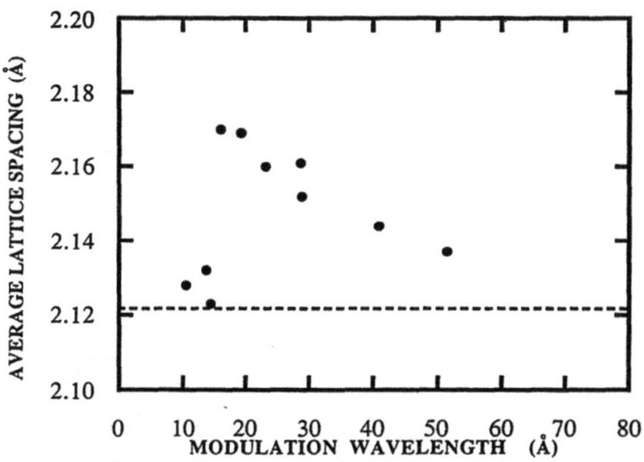

Fig. 4. Average lattice spacing perpendicular to the layering of Mo/Ni superlattices (Ref. 3).

Figure 3 shows the surface wave velocity in Mo/Ni superlattices determined by Brillouin scattering [3]. Since the velocity is related to the elastic constants its Λ dependence is also in contradiction with continuum theory [5-7]. Table I contains a summary of all the experimental determinations of elastic constants of superlattices performed to date: it contains the system studied, C_{ij} measured, existence of an anomaly, correlation with a structural change (to be discussed below) and the reference. Although there are some contradictions for some systems (notably Cu/Ni), the inescapable conclusion that can be drawn from Table I is that the elastic constants of superlattices are anomalous in the sense that they do not behave as expected from continuum theory.

Before diving into the possible origins of the effect it is interesting to note that the results given in Table I show no evident correlation with the crystal structure of the constituents (viz., bcc, fcc, hcp, etc.) nor with the fabrication method (viz., sputtering or evaporation). From the experimental standpoint however, it has been found that in all cases where the average lattice constant has been determined from x-rays, there is a strong correlation between changes in $\bar{a}$ with respect to the average of the bulk materials and anomalies in the elastic properties.

Table I. Superlattices in which elastic constants have been determined

System	Elastic Constant	Anomaly	Correl. Struc.	Ref.
Cu/Ni	Y_B	Yes		8
	Y,F	Yes		17
	Y_B, T, F, C_{66}	No		18
	C_{44}	No		19
Cu/Pd	Y_B	Yes		9
	Y_B, Y, F, C_{66}	No		13
	C_{44}	Yes		20
Mo/Ni	C_{44}	Yes	Yes	3
	C_{33}	Yes	Yes	12
Pt/Ni	C_{33}	Yes	Yes	12
Ti/Ni	C_{33}	Yes	Yes	12
Cu/Nb	C_{44}	Yes	Yes	21, 22
	C_{44}, C_{33}	Yes		23
	Y_B	Yes	Yes	24
	C_{12}	?	?	25
NbN/AlN	C_{44}	No	Yes	26
GaAs/AlAs	C_{33}, C_{44}	Yes	Yes	27
Nb/Si	C_{44}	Yes	Yes	28
Au/Cr	C_{44}	Yes	Yes	29
Ag/Pd	Y_B	Yes		30
Au/Ni	Y_B	Yes		9
Cu/Au	Y_B	No		30
Cu/Al	?	Yes		31
V/Ni	C_{44}	Yes	Yes	15
Fe/Pd	C_{44}, C_{11}?	Yes		32
Co/Ag	C_{44}	Yes	Yes	33,34
Mo/Ta	C_{44}	Yes	Yes	35
	C_{33}	No	No	35
Co/Cu	C_{44}	Yes	Yes	36

System	Elastic Constant	Anomaly	Correl. Struc.	Ref.
Fe/Cu	C_{44}	Yes	Yes	37
ZrN/AlN	C_{33}	Yes		38

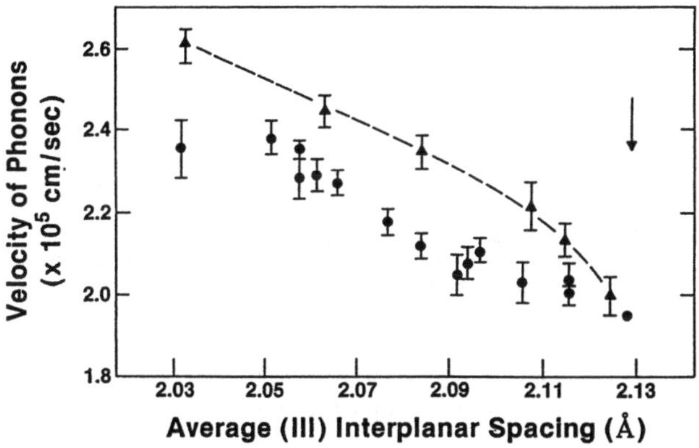

Fig. 5. Molecular dynamics calculations (Ref. 39) of a shear sound velocity as a function of strain in a Ni crystal.

This correlation is indicated in the fourth column of Table I: In all cases but one, whenever changes in $\bar{a}$ were detected by x-rays, anomalies were also observed in some elastic constant. The exception is a Λ independent C_{33} in Mo/Ta where changes in $\bar{a}$ were present; note however that C_{44} does indeed change in this superlattice. Figure 4 shows the experimentally determined $\bar{a}$ for Mo/Ni obtained from the spectra in Fig. 1; it clearly has very similar behavior to the velocity shown in Fig. 3. The correlation between $\bar{a}$ and C_{ij} was explained in Ref. 39, using Molecular Dynamics techniques where it was shown that the elastic softening is a direct consequence of the lattice expansion. These calculations are shown in Fig. 5 together with the experimental results of Fig. 3 plotted as a function of measured lattice expansion. The origin of the expansion remains however unexplained by these calculations.

There have been a number of models proposed to explain the observed behavior [40-51]. At present it has not been possible to determine which one of the models is correct. This is due to the fact that none of the models are capable of predicting *a priori* the behavior expected for a given system. Furthermore, for those models that have some predictive capabilities there is at least one experimental piece of evidence which contradicts the prediction. Notwithstanding the above comments, the model described in Refs. 44-47 is at present the most likely candidate.

In order to understand at a microscopic level the elastic anomalies in superlattices two avenues of research appear to be necessary: the theoretical models must be enhanced so that they acquire predictive capabilities and, more experimental work is required to determine the detailed atomic structure at interfaces. The latter should be aimed at obtaining structural information parallel to the layers as well as roughness, interdiffusion, etc. Figure 6 shows an x-ray spectrum from a Mo/Ni superlattice and two model fits: the upper fit was obtained using the bulk properties of Mo and Ni and "perfect" interfaces, the lower portion is a fit containing 7 fitting parameters (individual lattice constants of the constituents, nonuniform lattice expansion, roughness, etc. etc.) [52]. Although excellent fits can be obtained enormous care must be taken before the fit parameters can be assigned a reliable physical meaning.

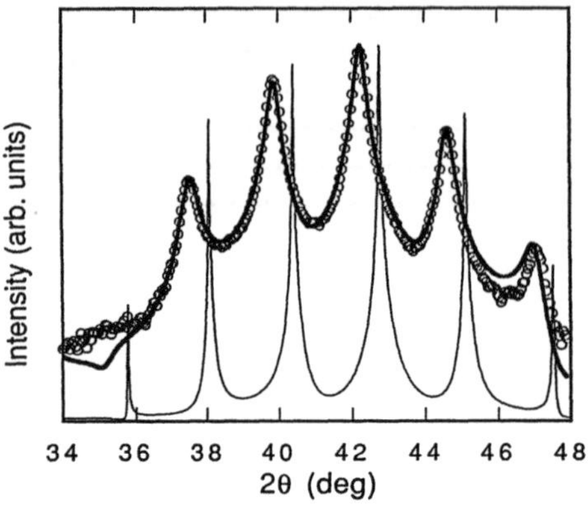

Fig. 6. Experimental x-ray spectrum from a Mo/Ni superlattice (dots) and fits to the data (full line). The light line is obtained using bulk values for the relevant parameters, the heavy line is a fit with 7 fitting parameters.

This work supported by the U.S. Department of Energy, BES-Materials Sciences, under contract #W-31-109-ENG-38 and the Office of Naval Research under grant #N00014-88K-0210.

References

1. See for instance Epitaxial Growth, J. W. Matthews ed., Academic Press Inc., NY (1975).
2. For an early review see Synthetically Modulated Structures, L. L. Chang and B. C. Giessen eds., Academic Press, New York 1985,
3. R.Khan, C. S. L. Chun, G. P. Felcher, M. Grimsditch, A. Kueny, C. M. Falco, and I. K. Schuller, Phys. Rev. B. 27, 7186 (1983).
4. See for instance Physics, Fabrication and Applications of Multilayered Structures, P. Dhez and C. Weisbuch eds., Plenum Publishing Co., NY (1988).
5. S. M. Rytov, Akust. Zh. 2 71 (1956) [Sov. Phys.-Acoust. 2 68 (1965)]
6. J. Sapriel, B. Djafari-Rouhani and L. Dobrzynski, Surf. Sci. 126, 197 (1983)
7. M. Grimsditch, Phys. Rev. B 31, 6818 (1985) and M. Grimsditch and F. Nizzoli, Phys. Rev. B 33, 5891 (1986)
8. T. Tsakalakos and J. E. Hilliard, J. Appl. Phys. 54, 734 (1982)
9. W. M. C. Yang, T. Tsakalakos and J. E. Hillard, J. Appl. Phys. 48, 876 (1977)
10. M. Barmatz, L. R. Testardi and F. J. DiSalvo, Phys. Rev. B. 12, 4367 (1975) and B. S. Berry and W. C. Pritchet, IBM J. Res. Develop. 19, 334 (1975).
11. J. Sandercock in "Topics in Applied Physics" Vol. 51 "Light Scattering in Solids III" eds. M. Cardona and G. Guntherodt, (Springer, N.Y., 1982)
12. B. Clemens and G. Eesley, Phys. Rev. Lett. 61, 2356 (1988).
13. A. Moreau, J. B. Ketterson, and B. D. Davis, J. Appl. Phys., to be published.
14. A. Fartash, I. K. Schuller, and M. Grimsditch, Appl. Phys. Lett. 55, 2614 (1990)
15. R. Danner, R. P. Huebener, C. S. L. Chun, M. Grimsditch, and I. K. Schuller, Phys. Rev. B. 33, 3696 (1986)
16. R. J. Bower, Appl. Phys. Lett. 23, 99 (1973) and C. Y. Ting and B. L. Crowder, J. Electrochem. Soc. 129, 2590 (1982)
17. L. R. Testardi, R. M. Willens, J. T. Krause, D. B. McWhan, and S. Nakahara, J. Appl. Phys. 52, 510 (1981)
18. A.Moreau, J. B. Ketterson, and J. Mattson, J. Appl. Phys. Lett., to be published.
19. J. Mattson, R. Bhadra, J. B. Ketterson, M. B. Brodsky, and M. Grimsditch, J. Appl. Phys., 67, 2973 (1990).

20. J. R. Dutcher, S. Lee, J. Kim, G. Stetgeman, and C. M. Falco, Proc. Mater. Res. Soc. 160, (in press), J. Mater. Sci. Eng. A. 127 (in press) and J. Mater. Sci. Eng. B. 6, (in press)
21. A. Kueny, M. Grimsditch, K. Miyano, I. Banerjee, C. M. Falco, and I. K. Schuller, Phys. Rev. Lett. 48, 166 (1982).
22. I. K. Schuller and M. Grimsditch, J. Vac. Sci. Technol. B4, 1444 (1986).
23. J. A. Bell, W. R. Bennett, R. Zanoni, G. I. Stegeman, C. M. Falco and C. T. Seaton, Sol. St. Comm. 64, 1339 (1987).
24. A. Fartash, I. K. Schuller, and M. Grimsditch (to be published)
25. J. A. Bell, R. J. Zanoni, C. T. Seaton, G. I. Stegeman, and C. M. Falco, Appl. Phys. Lett. 51, 652 (1987).
26. R. Bhadra, M. Grimsditch, J. Murduck, and I. K. Schuller, Appl. Phys. Lett. 54, 1409 (1989).
27. M. Grimsditch, R. Bhadra, I. K. Schuller, F. Chambers, and G. Devane, Phys. Rev. B. 42, (1990).
28. E. Fullerton, I. K. Schuller, and M. Grimsditch, (to be published).
29 P. Bisanti, M. B. Brodsky, G. P. Felcher, M. Grimsditch, and L. R. Sill, Phys. Rev. B. 35, 7813 (1987).
30. G. E. Henein and J. E. Hilliard, J. Appl. Phys. 54, 728 (1983).
31. V. S. Kopan and A. V. Lysenko, Fiz. Metal. Metalloved. 29, 183 (1970).
32. P. Baumgart, B. Hillebrands, R. Mock, and G. Guntherodt, Phys. Rev. B. 34, 9004 (1986).
33. S. M. Hues, R. Bhadra, M. Grimsditch, E. Fullerton, and I. K. Schuller, Phys. Rev. B. 39, 12966 (1989).
34. E. Fullerton, S. Hues, M. Grimsditch, and I. K. Schuller (to be published).
35. J. A. Bell, W. R. Bennett, R. Zanoni, G. I. Stegeman, C. M. Falco, and F. Nizzoli, Phys. Rev. B 35, 4127 (1987).
36. J. R. Dutcher, S. M. Lee, C. D. England, G. I. Stegeman, and C. M. Falco, Jour. Mater. Scie. Eng. A126, 13 (1990).
37. E. Fullerton, R. Bhadra, I. K. Schuller, and M. Grimsditch, (to be published).
38. W. J. Meng, G. L. Eesley and K. A. Svinarich (to be published).
39. I. K. Schuller and A. Rahman, Phys. Rev. Lett. 50, 1377 (1983).
40. T. B. Wu, J. Appl. Phys. 53, 5265 (1982).
41. W. E. Pickett, J. Phys. F. 12, 2195 (1982).
42. A. F. Jankowski and T. Tsakalakos, J. Phys. F. 15, 1279 (1985).
43. A. F. Jankowski, J. Phys. F. 18, 413 (1988).
44. D. Wolf and J. F. Lutsko, Phys. Rev. Lett. 60, 1170 (1988).
45. D. Wolf and J. F. Lutsko, J. Appl. Phys. 66, 1961 (1989).
46. D. Wolf and J. F. Lutsko, J. Mater. Res. 4, 1427 (1989).
47. D. Wolf, Surf. Sci. 225, 117 (1990).
48. M. L. Hueberman and M. Grimsditch, Phys. Rev. Lett. 66, 1403 (1989).
49. R. C. Cammarata and K. Sieradzki, Phys. Rev. Lett. 62, 2005 (1989).
50. A. Banerjea and J. R. Smith, Phys. Rev. B. 35, 5413 (1987).
51. B. W. Dodson, Phys. Rev. B. 37, 727 (1988).
52. E. Fullerton, I. K. Schuller, H. Vanderstraeten, and y. Bruynseraede (to be published).

Analysis of the Tight-Binding Description of the Structure of Metallic 2D Systems

R. Baquero

Departamento de Física, Centro de Investigación y de Estudios Avanzados del IPN, Apdo. Postal 14-740, 07000 México, D.F., Mexico

Abstract. Bidimensional metallic systems as interfaces, quantum wells and superlattices with sharp interfaces became recently available and their properties can now be experimentally studied in detail. To calculate the Local density of States (LDOS) for surfaces, interfaces, quantum wells and superlattices we use empirical tight-binding hamiltonians together with the Green function matching method (GFM). In this paper we show some examples of our results employing the method just outlined to describe metallic 2D systems. In particular, we refer briefly to the effect on the LDOS of the very recently established contraction of the first interatomic layer distance in the Ta(001) surface. We then discuss the Nb-V ideal (100) interface and conclude that under certain conditions the V-side of an interface can show magnetism as the V(001) surface does. As a last example, we present a calculation that relates the changes with gold coverage of the reaction rate of the catalytic reaction of cyclohexene into benzene on a Pt(001) surface to the changes on the LDOS of the outermost Pt atomic layer. We show that the behavior of the LDOS around the Fermi level is an important factor to the explanation of the behavior of this catalytic reaction. We conclude by stating that the empirical tight-binding method is a very simple and useful tool for the description of 2D metallic systems. The advantage is that the computational demands are low and all the ingredients to take full profit of this method are available (reliable tight-binding parameters and suitable methods for the calculation of the Green function.).

1. Introduction

2D-systems as interfaces, quantum wells and superlattices with very sharp interfaces became recently available. The detailed experimental study of their properties is therefore now possible.

Springer Proceedings in Physics, Volume 62
Surface Science Eds.: F.A. Ponce and M. Cardona

Ab initio calculations constitute the most accurate way of analyzing physical systems. Nevertheless, new insights and meaningful results can also be obtained for problems of current interest with much simpler methods that allow an easy and quick test of models, hypothesis, and geometrical situations with relatively low computational demands. In this paper, we make use of empirical tight-binding hamiltonians using the parameters that became available for many elements from the work of Papaconstantopulos.[1] To calculate the Green function we make use of the Green function matching method cast in the form suitable for a discrete description of the systems.[2] The method has been described widely in the literature and we therefore refer the interested reader to the literature (See refs. 1-5).

2. The Ta(001) surface

Ta is a bcc transition metal and its (001) surface has been studied previously experimentally by Bartynski and Gustafsson using inverse photoemission[6] and theoretically with a slab calculation by Krakauer.[7] More recently, Bartynski et al. succeeded in determining the geometry of the clean (001) tantalum surface more accurately by photo- electron diffraction spectroscopy.[8] The main result of this study was to establish that the first interlayer distance in the Ta(001) surface was shorter by about 10-15%. Very recently, Jensen, Bartynski and Weinert using Auger spectroscopy,[9] could establish that the just mentioned contraction was 13.5%. Most of the desagreement with experiment of the calculated LDOS desapears when one takes into account this contraction as shown in more detail in another paper in these Proceedings. The remaining desagreement might be attributable to a posible contraction of the second interatomic layer of less magnitude.

3. The influence of stress on the V/Ta(001) interface

Tight-binding analysis can give also good results in interfaces as well. Here we present our results for the V/Ta(001) interface. Since V has a lattice parameter of 3.03A and Ta 3.3A there is an important mismatch that will play a role for two semi-infinite systems forming the interface in a free standing configuration. This mismatch will cause stress at the interface and the purpose of our calculation is to show how important is this effect. The situation is different

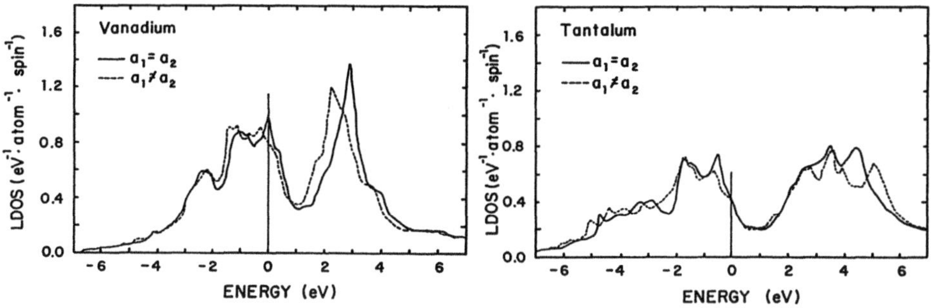

Fig. 1– The LDOS at the V– and Ta– side of the corresponding interface atomic layer are shown. Stress (....) plays an important role mainly in V. The origin is at the Fermi level, E_F.

to the one when an overlayer is grown on a substrate since the overlayer most of the time adopts the configuration of the substrate.

In Fig.1, we show the Local Density of States for the V- and Ta- interface atomic layer. The origin is at the Fermi level, E_F. In the case of V an important difference arises above E_F. The main one is the change in the position of the high-energy peak by more than 1 eV. There are differences below E_F as well. A very important point here is the change in the population at E_F. There is a 20% change. The density of states at E_F is very important in determining the magnetic properties of a transition metal surface. In the case of V(100) it has been predicted that magnetism would occur due to the very high density of states according to the criterion of Allan for magnetism[10] to occur in surfaces. When stress is not taken into account the population that appears at the interface is similar to the one at the surface. The conclusion that some V-interfaces might show magnetism for the same reason as it occurs in surfaces appears very naturally as we have found previously.[11] Our new calculation shows that magnetism will not occur in the V side of a V/Ta(001) interface due to stress. Stress is then a very important factor and should be included for the proper description of an interface.

Magnetism in the V-side of some interfaces might nevertheless occur even if it might not occur in V/Ta. The reason is that magnetism is due to the localization of d-states on the surface. Whenever the other element of the interface will not have d-states LDOS available near E_F, these electrons will behave similar as in the surface and magnetism might occur. This might be the case for interfaces of V with some dielectrics (this is important for tunneling experiments), some semiconductors or even for some non-transition metals and the noble metals.

In the bottom of Fig.1 the LDOS for the interface atomic layer of Ta appears. We see again that stress is important. The change at E_F is not very spectacular in this case but one can conclude certainly that one has to be very careful when neglecting stress in the description of an interface. A more complete version of this work will be published elsewhere.[12]

4. The Dehydrogenation of the Cyclohexene Molecule

The cyclohexene molecule has an hexagonal shape with a double bond between two carbon atoms. This bond forces four carbon atoms altogether to be in the same plane. The other two carbons in the molecule are in another plane at a certain angle with respect to the first. The molecule gets in contact with the metallic Pt(100) surface through the hydrogen atoms. The details of the models and of the chemical reaction are not important here. The only thing that we would like to point is that the reaction gets started by a transfer of charge from the metal surface to the molecule. This transfer destabilizes the double bond between the two carbon atoms and breaks it (See Fig.2). Equilibrium is reached when each of the carbon atoms next to the double bond ones, looses one of its hydrogen atoms. In this process also two new double bonds are formed. A further breaking of one of these double bonds by a similar process ends up with three double bonds and constitutes the formation of a benzene molecule.

Group-VIII metals are chemically active. The convertion of cyclohexene to benzene through dehydrogenation of the first takes place on Pt(001) surfaces at a certain rate and certain partial pressures.[13,14] When a monolayer of gold is deposited on the Pt(001) surface the reaction rate rises by a factor of four. If another layer is grown the reaction rate goes down again to a value of about the same one it had on the clean Pt surface. We study this reaction in more detail elsewhere.[15] Here we want only to show that the LDOS at the Pt(001) outermost atomic layer correlates very well with the observed behavior of the reaction rate.

We will use a detailed analysis of the changes in the local density of states (LDOS) of the surface layer as the

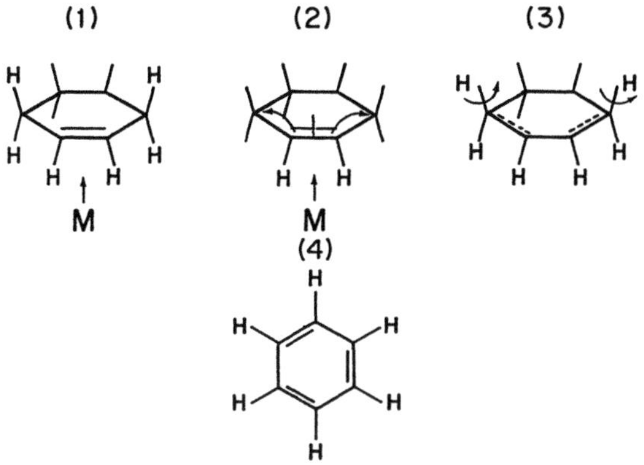

Fig. 2- We sketch here a possible model for the catalytic reaction: 1) charge is transfered to the molecule from the Pt(100) surface. 2) This charge destalizes the double bond and breaks it. 3) Two hydrogen atoms are lost and two double bonds are formed. 4) The further breaking of one double bond leads to a benzene molecule.

overlayer is grown on top of it and correlate the changes in the chemical reaction to physical parameters of the active layer. More precisely, to the changes in the LDOS around the Fermi level, E_F.

To describe the two metals we use again empirical tight-binding hamiltonians with an orthogonal s-p-d basis and an interatomic interaction up to third nearest neighbors.The Green function is calculated again using the Surface Green Function Matching[1-5] (SGFM) method. From the Green function of the system we can calculate the LDOS projected on any atomic layer and we look, in particular, at the changes around E_F at the Pt outermost atomic layer.

The possible causes for the observed reactivity enhancement[13] considered before are the following. First, gold on top of platinum could provide the active sites for the reaction. Second, the platinum atoms located below the gold layer could be the active centers for cyclohexene dehydrogenation. Their activity has to be governed by the influence of the gold layer on the Pt(001) first layer, in this case. A third given explanation is that the active sites are the platinum atoms that have not been covered by gold, i.e. defect sites in the gold layer. The bonding at these sites would be modified by the presence of gold.

Poisoning has been found to play an important role in the sense that the bonding to Pt directly is very strong and

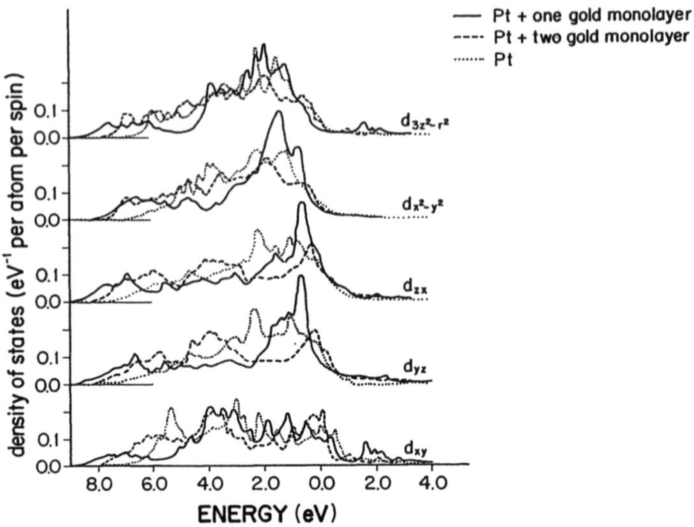

Fig. 3- The contribution of the d-states to the LDOS of the outermost Pt(100) layer is shown for the three cases considered: pure Pt, one and two gold monolayers. The origin is at the Fermi level, E_F. The intensity of the peak just below E_F for the important d_{zx} and d_{yz} states behaves approximately in the ratio 1/4/1 as the reaction rate does.

breaks the cyclohexene molecules leaving the surface covered by C atoms that would prevent further activity on that site. A second gold layer would inhibit the molecule to approach the Pt atomic layer and therefore would diminish the activity again. Here we show a complementary point of view.

In order to explore deeper these points of view we present here the changes that the LDOS suffers during the process of growing gold monolayers on an ideal clean Pt(001) surface.

An analysis of the symmetry of the wave functions of the states taking part in the reaction shows that only the electronic states of d-symmetry take place in the reaction. And to the extend in which proximity between the molecule and surface states is necessary, only the states that have a spatial projection in the direction perpendicular to the surface should be considered. That is we have to look at the contribution of the d_{zx}, d_{yz} and $d_{3z^2-r^2}$ states to the LDOS around the Fermi level for the three cases, i.e., pure Pt, one and two atomic layers of gold as overlayers.

In Fig. 3, we analyze in detail the contribution to LDOS of all the states of d-symmetry. We show three curves for each state contribution. They correspond again to the outermost platinum atomic layer in the clean surface case and the one - and two - monolayer systems. Here again, we see that the three curves differ sometimes even substantially. A striking characteristic of some of these curves is the pronounced peak that develops in the one-nonolayer case. This can be observed mainly in the $d_{x^2-y^2}$, d_{zx} and d_{yz} contributions. For our case of interest, the changes in d_{zx} and d_{yz} are the meaningful ones. It is to be observed that in these curves the peak shows at energies closer to the Fermi level. These d_{zx} and d_{yz} states are the ones that probably participate the most in the dehydrogenation of the cyclohexene molecule.

Let us recall first that the dehydrogenation of cyclohexene is increased about four times by the addition of a gold monolayer compared to the clean Pt(100) surface. The activity goes back to the one of the clean surface when a second monolayer is grown.

We can observe in Fig. 3, that there is a qualitatively similar behavior of the LDOS of the d_{zx} and d_{yz} states on the outermost platinum layer and the one of the reaction rate, around E_F. The changes in these curves are big and follow the behavior of the chemical reaction when gold in grown on the Pt(001) surface.

Let us try to be more quantitative. The basic idea is that in molecules as cyclohexene the Π-like bonding is chemically very active and when the molecule approaches the surface, the reaction very probably starts with a charge transfer from the surface to the molecule. The amount of charge available at the Fermi level of states of the right symmetry can be therefore essential for the reaction to get started.

Let's take as a measure of the activity the value of the LDOS at the peak that appears in all curves just below E_F. The contribution to LDOS of the d_{zx} and d_{yz} states at the selected energy behaves as the ratio 1-4-1 which is exactly the same behavior of the reaction rate.

On this basis. we propose that is the occupied electronic states of d_{zx} and d_{yz} symmetry around the Fermi level, one of the factors that control the enhancement of the activity of the dehydrogenation of cyclohexene to benzene.

5. Conclusions

2D-metallic systems as surfaces, interfaces, quantum wells and superlattices with sharp interfaces that avoid interdiffussion to a great extend are now available.

Tight-binding is a technique to calculate the electronic density of states for these systems that is very easy to implement and has very low computational demands. The necessary formalism and algorithms to calculate the Green are available.

We have shown in three very different examples, i.e., the description of the Ta(001) surface, the V/Ta(001) interface and the catalytic reaction of cyclohexene to benzene on the Pt(001) surface that this way of describing physical problems gives meaningful results even quantitatively. A more complete version of this work will be published elsewhere.[14]

6. References

1- D.A. Papaconstantopulos, *The Electronic Band Structure of Elementary Solids* (Plenum, New York, 1986).

2- R. Baquero, V.R. Velasco, and F. García-Moliner, Phys. Scr. **38**, 742 (1988).

3-F. García-Moliner and V.R. Velasco, Prog. Surf. Sci. **21**, 93 (1986).

4- R. Baquero and A. Noguera, Rev. Mex. Fís. **35**, 614 (1989).

5- J.C. Slater and G.F. Koster, Phys. Rev. **94**, 1498 (1954).

6- R.A. Bartynski and T. Gustafsson, Phys.Rev.B **35**, 939 (1987).

7- H. Krakauer, Phys. Rev. B **30**, 6834 (1990).

8- R.A. Bartynski et al.,Phys. Rev. B **40**, 5340 (1989).

9- E. Jensen, R.A. Bartynski and M. Weinert, Phys. Rev. B **41**, 12468 (1990).

10- G. Allan, Phys. Rev. **B19**,4774 (1979).

11- R. Baquero, A. Noguera, A. Camacho, and L. Quiroga, Phys. Rev. **B42** (1990), in press.

12- R. Baquero, A. Noguera, A. Camacho and L. Quiroga (to be published).

13- J.W.A. Sachtler, M.A. Van Hove, J.P. Bibérian and G.A. Samorjai, Phys. Rev. Lett.,**45**, 1601 (1980).

14- R. Baquero, D. Martínez, A. Noguera and J. Mendieta (to be published).

Magneto-Electro Optical Absorption of a Semiconductor Superlattice

Z. Barticevic[1], *M. Pacheco*[2], *and F. Claro*[3]

[1]Departamento de Física, Universidad Santa María, Casilla 110V, Valparaiso, Chile
[2]Departamento de Física, Universidad de Santiago, Casilla 5659, Santiago, Chile
[3]Facultad de Física, Universidad Católica de Chile, Casilla 6177, Santiago, Chile

The optical properties of semiconductor heterostructures are strongly affected by the presence of external fields. We investigate the optical spectrum of a superlattice in the presence of both an electric and a magnetic field parallel to the growth axis. The electric field combined with the superlattice potential introduces the so called Stark ladders in the spectrum [1], while the magnetic field quantizes the motion in the x-y plane. We show that the absorption spectrum for interband transitions contains only resonances and exhibits a remarkable structure that depends on the ratio between the fields. Our results are obtained within the envelope function approximation [2]. Simple parabolic bands for electrons and holes are considered. The envelope function of the superlattice subject to the electric field is constructed in a tight-binding scheme from a linear combination of isolated quantum well states in a magnetic field.

Consider an electron moving in a superlattice potential $V(z) = V(z+d)$ and external electric $\vec{F} = F\hat{z}$ and magnetic $\vec{B} = B\hat{z}$ fields. Here $\hat{z}$ is the growth direction of the superlattice. Using the Landau gauge, $\vec{A} = (0, Bx, 0)$ the z-dependent envelope function $\chi(z)$ is a solution of

$$\left(\frac{\hbar^2}{2m^*(z)}[-\frac{d^2}{dz^2} + \frac{1+2l}{r^2{}_B}] + V(z) + eFz\right)\chi(z) = \epsilon\chi(z) \ . \tag{1}$$

Here $m^*(z)$ is the electron or hole effective mass, l is a non-negative integer and $r_B = (\hbar c/eB)^{1/2}$. Using the tight-binding approximation to treat electrons and holes the envelope functions can be written as

$$\chi(z) = \sum_n C_n \phi(z - nd) \tag{2}$$

where $\phi(z - nd)$ is the lowest energy eigenfunction of an isolated quantum well in the presence of the magnetic field B, centered at $z = nd$. Replacing Eq.(2) in Eq.(1) we obtain

$$\sum_n C_n[\epsilon - \epsilon_p(l) - eFz - V(z) + V_p(z - nd)]\phi(z - nd) = 0. \tag{3}$$

Here $V_p(z - nd)$ is a single quantum well potential centered at $z = nd$. If $m^*_{w(b)}$

Springer Proceedings in Physics, Volume 62
Surface Science Eds.: F.A. Ponce and M. Cardona

is the effective mass in the well (barrier) region and ϵ_0 is the lowest allowed energy of the quantum well, then the energies of the well in a magnetic field will be given by $\epsilon_p = \epsilon_0 + \frac{\hbar^2}{r_B^2 m_w^*}(l+1/2)$ if $\frac{\hbar^2}{r_B^2}(l+1/2)\left(\frac{1}{m_w^*} - \frac{1}{m_b^*}\right)/V_0 \ll 1$. This last condition is valid for heterostructures of $GaAs/GaAlAs$ for example, where the superlattice potential V_0 is at least several hundred times the usual magnetic energies.

The coefficients C_n of the cell eigenfunctions obey the simple recursion formula

$$C_{n+1} + C_{n-1} + \frac{\Delta}{\lambda}(\xi - n)C_n = 0 \tag{4}$$

where $\Delta = eFd$ is the Stark cell energy, $\lambda = V_0 \int_{-L/2}^{L/2} \phi(z)\, \phi(z+d)\, dz$ is the transfer integral between nearest neighbors, and $\xi = [\epsilon - \epsilon_0 - \frac{\hbar^2}{r_b^2 m_w^*}(l + 1/2)]/\Delta$. Eq. (4) is obeyed by Bessel functions $J_{l-\xi}(2\lambda/\Delta)$. Proper behaviour for large index demands $\xi = m$ an integer. Thus the energy is quantized and given by

$$\epsilon_{l,m} = \epsilon_0 + \hbar\omega_B(l + 1/2) + meFd \tag{5}$$

with $\omega_B = eB/m_w^* c$. Note that the spectrum is entirely discrete, with two superimpossed ladders of levels whose spacing may be commensurate or incommensurate depending on the ratio of the natural frequencies involved.

Applying the previous solutions to both the conduction and valence band wave functions and using the properties of Bessel functions one obtains the absorption coefficient for a photon of frequency ω,

$$\alpha = \Gamma \frac{B}{F\omega} \sum_l J^2_{[\epsilon_B l - \epsilon_f]}(\eta)\, \delta(\epsilon_B l - \epsilon_f - [\epsilon_B l - \epsilon_f]), \tag{6}$$

where $\eta = 2(\lambda_v + \lambda_c)/\Delta$, ϵ_g is the energy gap, Γ is a constant and $[x]$ is the integral part of x . We have used the dimensionless reduced magnetic and photon energies $\epsilon_B = (1/m_e^* + 1/m_h^*)\hbar eB/c\Delta$ and $\epsilon_f = (\hbar\omega - \epsilon_g - \epsilon_e - \epsilon_h)/\Delta - \epsilon_B/2$, respectively, where $\epsilon_e(\epsilon_h)$, $m_e^*(m_h^*)$ are the isolated-well energy level and effective mass for electrons (holes).

Equation (6) shows that the spectrum consists of point resonances, each with a strength determined by the properties of Bessel functions. Assume first the reduced magnetic energy to be a rational $\epsilon_B = p/q$, where p and q are integers having no common divisors. A resonance occurs when

$$q\epsilon_f = lp - Mq$$

with M, l integers, the first unrestricted, the second non-negative. This is the Diophantine equation, which has an infinity of solutions l, M provided $\epsilon_f = N/q$ with N an integer [3]. One can show that [4]: (1) Resonances are equally spaced, with spacing p/q. (2) The strength of the resonances is determined by the function

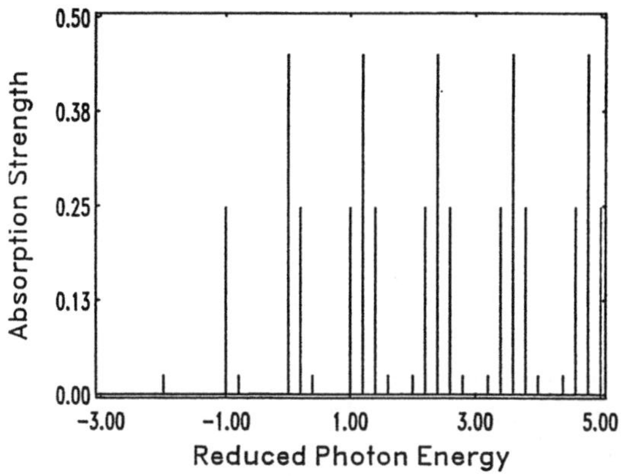

Fig. 1 Absorption strength S for $\epsilon_B = 6/5$,and $\eta = 1.2$. Energies are in units of the Stark period eEa.

$$S(p,q,N) = \sum_{s \geq -\frac{Nq_{r-1}}{q}} J^2_{Np_{r-1}+sp}(\eta). \tag{7}$$

Here p_{r-1}/q_{r-1} is the next to the last convergent in an even expansion of the rational p/q. (3) If the reduced photon energy ϵ_f increases by p/q only the limit in the sum (7) changes, becoming $s \geq -(Nq_{r-1}+1)/q$, and either a term is added or the sum remains unchanged. When the photon energy is large the terms added are small and the spectrum becomes periodic with a period p/q. (4) At zero reduced photon energy ($N = 0$) there is always a resonance since we measured energies from the effective band edge. Decreasing the energy by p units below the edge just causes the lowest index term in (7) to disappear and the strength is decreased. Absorption in the gap is thus finite but reduced as the energy is made smaller. As an illustration we show in Fig.1 the case $p/q = 6/5$.

The spectrum also has a period for fixed photon energy and varying magnetic field. To see this we redefine the reduced photon energy as $\bar{\epsilon}_f = \epsilon_f + \epsilon_B/2$, to make it independent of B. Given $\bar{\epsilon}_f = N/Q$ with N and Q prime to each other, the relevant Diophantine equation is $p(2n+1) - 2M_q = 2p\bar{\epsilon}_f$. We have shown earlier that [4] : (1) Resonances occur whenever $p/q = 2P/Q(2S+1)$, with P and S arbitrary integers. (2) The absorption strength function is invariant under the change $q/p \longrightarrow q/p + Q$, and $q/p \longrightarrow Q - q/p$. We note that the factor in Eq. (6) is proportional to the magnetic field, so that the absorption coefficient is not truly periodic in the inverse field, but contains a decaying factor B^{-1}.

In conclusion, we have shown within a two band-model that the absorption of a superlattice in the presence of electric and magnetic fields parallel to the axis is a sequence of resonances with interesting properties. The spectrum is periodic for large energies and exhibits a fine structure characteristic of the ratio between the fundamental magnetic and electric energies. There is absorption in the gap that acquires a Franz-Keldysh envelope in the limit of low electric fields. The absorption spectrum is proportional to a strength function S that for fixed photon energies exhibits a characteristic period in the inverse field.

This research was supported in part by Fondo Nacional de Ciencias, Grant 90/0375 and by Universidad Santa Maria, Grant 901101

References

[1] G.H. Wannier, Phys. Rev. Lett. 117, 432 (1960). The existence of this quantization has been the subject of controversy in the past but was recently verified experimentaly in semiconductor superlattices. See E.E. Mendez,F.Agulló-Rueda and J.M.Hong, Phys. Rev. Lett.60, 2426 (1988)

[2] M. Altarelli, " Heterojunctions and Semiconductor Superlattices" ed. by G. Allan, G. Bastard, N. Boccara, M. Lannoo, and M. Voos (Springer, Berlin, 1986)

[3] See for example C. D. Olds, "Continued Fractions" (Random House, New York, 1963) Ch. 2.

[4] F. Claro, M. Pacheco, Z. Barticevic, Phys. Rev. Lett. 64 3058 (1990)

A Tight-Binding Study of Interface States in Ultra-Thin Quantum Wells of HgTe in CdTe

F.J. Rodríguez, A. Camacho, and L. Quiroga

Departamento de Física, Universidad de los Andes,
A.A. 4976, Bogotá D.E., Colombia

Abstract. We present a calculation of the electronic structure of systems composed by a reduced number of HgTe layers between CdTe substrates. Special attention is drawn to the anisotropic dispersion relation through the entire two-dimensional Brillouin zone for a monolayer of HgTe in CdTe. Interface states depend sensitively on the valence-band-offset. The electronic properties of these heterostructures are studied for different small numbers of layers in the quantum well as a function of the valence-band-offset.

Heterostructures built up from II-VI semiconductors such as CdTe and HgTe have been increasingly proposed as promising candidates for devices because of their contrasting electronic properties [1]. Their high quantum efficiency and their ability to be taylored to peak performance at selected spectral regions, make these systems very important. However, there are still many open questions at the very fundamental understanding of their behavior. The biggest controversy is about the value of the valence-band-offset (ΔE_V) at the interfaces. Magneto-optical experiments in superlattices fix a value for this parameter in 40 meV [2], which is expected from the known common-anion rule. On the other hand photoemission experiments in interfaces give a value of 350 meV [3]. The bulk valence-band edge of HgTe is higher in energy than the CdTe bulk valence edge.

Springer Proceedings in Physics, Volume 62
Surface Science Eds.: F.A. Ponce and M. Cardona

In this work we study theoretically the interface state structure of ultrathin (001) quantum wells (QW's) of HgTe in CdTe. We use a realistic empirical tight-binding parametrization including interactions up to the second-neighbors and spin-orbit coupling. The Hamiltonian matrix elements have been taken from Bryant [4]. The interface energy levels, density of states and the electronic state projected structure (ESPS) are obtained through a layer Green function calculation [5]. We chose for ΔE_V the two extreme values proposed in Ref.[2, 3].

We report the projected electronic band structure and the dispersion relation of the interface states along the main directions of the two-dimensional Brillouin zone for a monolayer QW of HgTe in CdTe. In addition, results concerning the variation of the interface state energies as a function of the number of layers of HgTe are presented. We mean by layer a unity formed by one anionic plane (Te) and one cationic plane (Hg). Its width is 3.24 Å.

In the study of effects produced by the interfaces on the electronic spectrum of a heterostructure is very important to determine the ESPS. In Fig. 1a we show the ESPS of pure CdTe and the new features induced by a monolayer of HgTe (quantum well N=1) for ΔE_V = 40 meV. We observe two new states: a resonant state within the valence band for any **k** in the two-dimensional Brillouin zone and one state with energy in the gap of CdTe very near to the bottom of the conduction band. Our method of calculation allows the decomposition of the local density of states in the contributions coming from each atomic orbital. The resonant state is mainly of P_x and P_y anionic character and it is consequently a bridge-bond type state. It coresponds to a state localized at the monolayer of HgTe but extended in

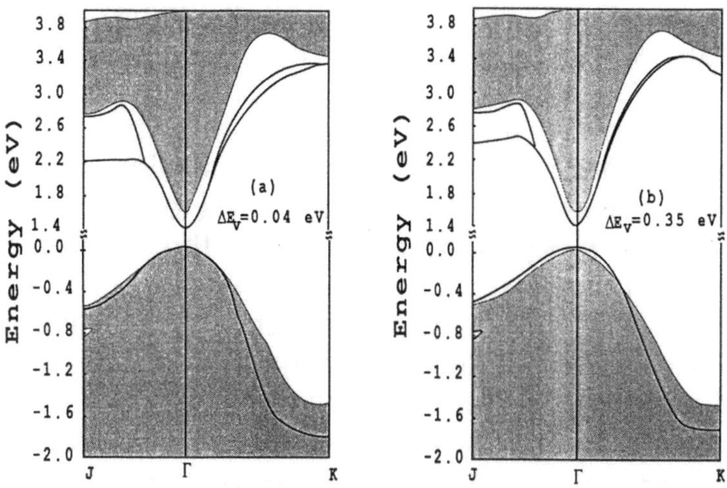

Figure 1. Electronic state projected structure of CdTe (shadow regions) and new features introduced by a monolayer of HgTe in CdTe (lines). The zero of energies is the top of the bulk CdTe valence band. (a) ΔE_V=40 meV and (b) ΔE_V=350 meV.

CdTe. The dispersion in Γ-K line is greater than the CdTe band edge itself. The state coming from the conduction band presents a strong weight of cation atomic s state as expected. We observe a splitting of this state along the Γ-K and Γ-J lines, which become very large near the J point. This behavior is very similar to the one reported for the cationic ideal surfaces of HgTe and CdTe [4].

In Fig. 1b is shown the ESPS for ΔE_V = 350 meV. The state coming from the conduction band presents a splitting as above, out of the zone center, but shows a larger dispersion along the Γ-K direction. By constrast the state coming from the valence band has a different behavior as for the smaller valence-band-offset case. It is now a true localised state at the monolayer plane, with energy in the CdTe gap for the whole Γ-J direction. In Γ-K direction it transforms from a true localised state for small **k** values to a resonant state within the valence band for **k** larger than (0.4,0.4)2π/a approximately. This fact is due to greater dispersion of this monolayer state than the upper valence band edge. This result is similar to the one reported in Ref. [6, 7] for wide QW's obtained in the framework of a **kp** method, where the isotropic approximation for the in plane dispersion was assumed and only small valence-band-offsets were considered. However, in the monolayer case we find that hybridization of the localised state with the heavy-hole band takes place along the Γ-K direction for the upper limit of ΔE_V's proposed. It is worthnoting that our approach allows us to consider in an exact way the in plane symmetry. Therefore, anisotropies in the dispersion relation are a direct product of our scheme of calculation.

The monolayer confined states evolve into interface states as the number of layers in the QW increases. As a consequence of the interface coupling appear one symmetric and one antisymmetric states with energies lying in the CdTe gap. The presence of these states has been explained in Ref.[6,7] as due to the coupling of the light-hole CdTe band with the electronic HgTe band, which have the same symmetry but inverted effective-masses.

In Fig. 2 we show, as a function of the HgTe number of layers, the transition energies at the two-dimensional Γ point, between the antisymmetric interface state (A, empty) and the following occupied states: the fundamental heavy-hole state (hh) and the symmetric interface state (S). The latter one lies at higher energy than the hh state. There has been shown for superlattices [8] that these transitions present very different oscillator strenghts according to the incident light polarization and therefore they could be used in optical absorption experiments to

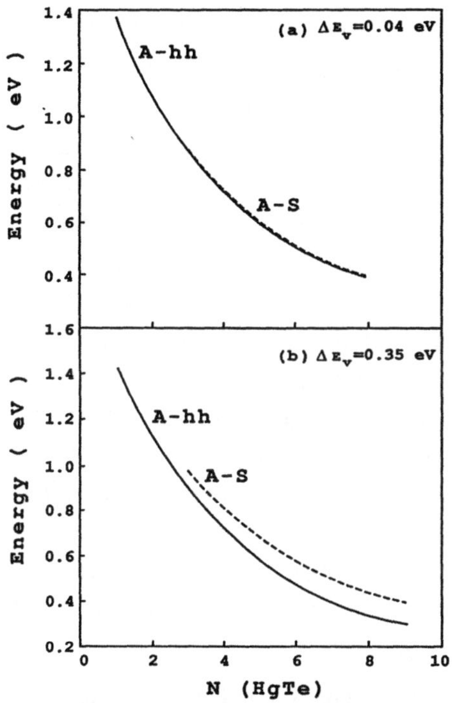

Figure 2. Transition energies, as a function of the HgTe layer number in the QW, between the interface antisymmetric state (A) and the two uppermost occupied states, the symmetric interface state (S) and the fundamental heavy-hole (hh) state. (a) ΔE_V=40 meV and (b) ΔE_V=350 meV.

discriminate the valence-band-offset value. In Fig. 2a, ΔE_V = 40 meV, it can be observed that the differences between A-hh and A-S transition energies remain below 10 meV for layer number up to 8. By contrast, if ΔE_V = 350 meV, as in Fig. 2b, the same differences can reach 100 meV for 8 layers. This can be understood if one realizes that the S interface state follows the bulk HgTe valence edge while the hh state is determined by the confinement properties of the QW.

In this work we reported the band structure of ultrathin QW's of HgTe in CdTe in the framework of a tight-binding calculation. We limit ourselves to ideal interfaces, i.e. lattice mismatch, relaxation and possible cationic interdifussion effects have been neglected. The dispersion relation of confined states in a monolayer shows a very anisotropic character along the main directions in the two-dimensional Brillouin zone. The monolayer confined states, as the interface states in QW's of thicker widths, are highly sensitive to the valence-band-offset. This feature provides a tool to decide experimentally the value of the valence-band-offset in HgTe-CdTe heterostructures. More elaborated interface models including some degree of relaxation and/or disorder are being undertaken at present by our group and results will be published elsewhere.

This work was partially supported by COLCIENCIAS (Colombia) through project No. 1204-05-003-90.

REFERENCES

[1]. J.P.Faurie, IEEE J. Quantum Electron. QE-22, 1656 (1986).

[2] Y.Guldner, G.Bastard, J.P. Vieren, M.Voos, J.P.Faurie and A.Million, Phys.Rev.Letts. 51, 907 (1983).

[3] S.P. Kowalczyk, J.T.Cheung, E.A.Kraut and R.W.Grant, Phys.Rev.Letts. 56, 1605 (1986).

[4] G. W. Bryant, Phys. Rev. B35, 5547 (1987).

[5] F. Rodríguez, A. Camacho, L. Quiroga and R. Baquero, Phys. Status Solidi (b) 160, 127 (1990).

[6] N.A. Cade, J.Phys. C: Solid State Physics, 18, 5135 (1985).

[7] Y.R.Lin-Liu and L.J.Sham, Phys.Rev. B32, 5561 (1985).

[8]. Y. -C. Chang, J.N. Schulman, G. Bastard, Y. Guldner and M. Voos, Phys. Rev. B31, 2557 (1985).

Part IX

Long Range Interaction: Magnetism and Superconductivity

Local Pair Phenomenological Approach to the Normal State Properties of High T_c Superconductors

B.R. Alascio[+], *R. Allub*[+], *C.R. Proetto, and C.I. Ventura*[++]

Centro Atómico Bariloche, C.N.E.A.,
8400 S.C. de Bariloche, Río Negro, Argentina

Abstract. A mixed Fermion-Boson Hamiltonian in which a band of paired states overlaps the Fermi level of a wider fermion band is studied. This model leads quite naturally to the one-particle form of the self-energy which explains some of the anomalous normal state properties of high temperature superconductors. This model also leads to the observed forms of the tunneling conductance and photoemission spectrum.

1. Introduction

Since the discovery by Bednorz and Müller [1] of high temperature superconductors (HTS) research has focused not only in the superconducting properties but also in the normal state properties.

Varma et al [2] have recognized distinctive anomalies common to most HTS in their normal phase. Some of these anomalies are common to BiO and CuO although it is still debatable if they belong to the same family. Furthermore it is not yet clear if other Mott insulators which metalize by doping and do not become superconducting may have similar properties.

Specifically, the authors of Ref.2 identify several physical properties: resistivity, tunneling conductance, nuclear relaxation rate, specific heat and thermal conductivity, Raman scattering and optical conductivity, and photoemission, that are anomalous as compared to normal metals and are common to most HTS. They propose that these anomalies can only be understood outside the Fermi liquid picture. They characterize the metallic state of HTS by the term "marginal Fermi liquid". As will be seen below the term "marginal" stems from the fact

---

+ Member of the Carrera del Investigador Científico del Consejo Nacional de Investigaciones Científicas y Técnicas (CONICET).
++ Fellow of the CONICET.

Springer Proceedings in Physics, Volume 62
Surface Science Eds.: F.A. Ponce and M. Cardona

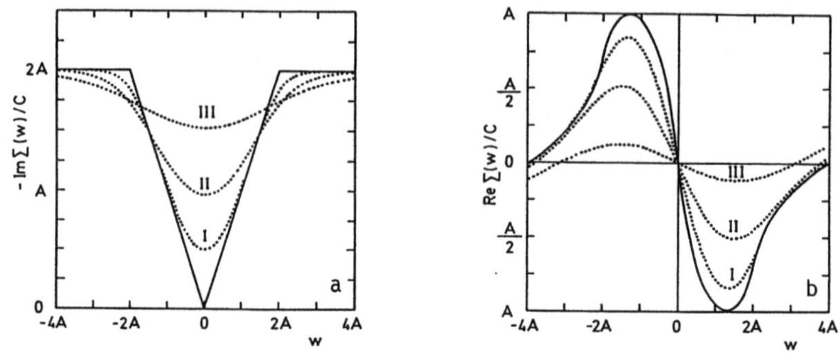

Fig.1) (a) Imaginary part of the self-energy as function of energy ω. Full line: T=0; pointed lines: I) T=A/4, II) T=A/2, III) T=A. (b) Real part of the self-energy as function of energy ω. Full line: T=0; pointed lines: I) T=A/4, II) T=A/2, III) T=A.

that the one particle self-energy vanishes at the Fermi level, as for a Fermi liquid, but with a linear energy dependence which prevents the definition of a quasiparticle.

In the phenomenological approach of Ref.2 they assume the existence of excitations which give rise to charge and spin polarizabilities $P(q,\omega)$ of the form :

$$\mathrm{Im}P(q,\omega) \;\alpha\; \begin{cases} -\dfrac{\omega}{T} & ,\text{for} \, |\omega|<T \\ -\,\mathrm{sgn}(\omega) & ,\text{for} \, |\omega|>T \; . \end{cases} \tag{1}$$

These ansatz polarizabilities and their Kramers-Kronig related real parts give rise to a rather particular one particle self-energy:

$$\sum(k,\omega) \;\alpha\; \omega \ln \frac{x}{\omega_c} - i\,\frac{\pi}{2}\,x \tag{2}$$

where $x=\max(|\omega|,T)$ and ω_c is a necessary cutoff. The real and imaginary parts of the self-energy (2) are represented by the full lines in Figs.1(b) and 1(a) respectively.

Varma et al. [2] claim that all the properties listed above can be understood on the basis of Eqs.(1) and (2). However, two of the most crucial experimental results, tunneling conductance and photoemission spectra, do not follow directly from their assumption.

In what follows we will show that the same form of the self-energy (Eq.(2)) can be derived from a model in which a band of paired states crosses the Fermi level of a normal Fermi liquid. Furthermore, the same picture allows a clear interpretation of the tunnel conductance and photoemission spectra which differs from the one proposed in Ref.2.

2. The Model

To derive the self-energy, photoemission and tunnel conductance we propose a Hamiltonian in which a band of paired states crosses the Fermi level of a metallic electron band. Our proposition is purely based on phenomenological grounds although the model has been originally proposed by Ranninger et al. [3] and extended and revisited by several authors [4-6] in connection with HTS. The Hamiltonian consists of: i) an uncorrelated electron band with a constant density of states, ii) a band of highly correlated states with attractive on-site interactions straddling the Fermi level, and iii) a mixing term between both kinds of states.

The Hamiltonian reads:

$$H = \sum_{k,\sigma} (\varepsilon_k - \mu)\, c^+_{k\sigma} c_{k\sigma} + \sum_{j,\sigma} (E_j - \mu)\, d^+_{j\sigma} d_{j\sigma}$$

$$- \frac{U}{2} \sum_{j,\sigma} n^d_{j\sigma} n^d_{j\bar{\sigma}} + \sum_{j,k,\sigma} \left(V_{kj} c^+_{k\sigma} d_{j\sigma} + V^*_{kj} d^+_{j\sigma} c_{k\sigma} \right), \quad (3)$$

where $c^+_{k\sigma}$ creates an electron with crystal momentum k, spin σ and energy ε_k in a broad uncorrelated band, $d^+_{j\sigma}$ is the electron creation operator in another set of states labeled by the quantum numbers ($j\sigma$) with energies E_j. To fix ideas in the following we will consider the latter as localized states, so that we can take $V_{kj} = W N^{-1/2} e^{-ikR_j}$. We assume that particles in d-states are strongly correlated by an attractive potential -U (U>0). $n^d_{j\sigma} = d^+_{j\sigma} d_{j\sigma}$ is the d-state number operator, N the total number of sites, μ denotes the chemical potential. The last term in (3) represents the hybridization between localized and extended states.

We assume U>W and reduce the Hilbert space eliminating the high energy singly occupied d-electron states. This is achieved by a generalized Schrieffer-Wolff transformation [3]. The transformed Hamiltonian reads:

$$\tilde{H} = \sum_{k,\sigma} (\varepsilon_k - \mu)\, c^+_{k\sigma} c_{k\sigma} + \sum_j 2(\Delta_j - \mu)\, A^+_j A_j +$$

$$\frac{W}{2N} \sum_{j,k,k'} g_{jk} \left[e^{-i(k+k').R_j} \left(c^+_{k'\downarrow} c^+_{k\uparrow} + \right.\right.$$

$$\left.\left. + \; c^+_{k\downarrow} c^+_{k'\uparrow} \right) A_j + \text{H.c.} \right], \qquad (4)$$

where $A^+_j = d^+_{j\uparrow} d^+_{j\downarrow}$ is the creation operator for the electron pair at site j with energy $2\Delta_j = (2*E_j - U)$, $g_{jk} = \frac{W\,U}{\left[(U/2)^2 + (\Delta_j - \varepsilon_k)^2\right]}$. The elimination of the electron hybridization term of Eq (3) gives rise to a new mixing term in which two electrons from the uncorrelated band simultaneously enter a paired state and vice versa. In what follows we will assume $(\Delta_j - \varepsilon_k) << U$, so that $g_{jk} = 4W/U = g$.

3. The One Particle Self-Energy

From the transformed Hamiltonian (4) we can calculate the self-energy as well as other thermodynamical properties. To obtain the one particle self-energy we begin by calculating its imaginary part using second order perturbation theory to include the effect of the mixing term. Assuming a continuous distribution $\rho_b(\Delta)$ for the energies of the paired states we obtain :

$$-\text{Im} \sum(\omega) = \frac{\hbar}{\tau(\omega)} =$$

$$\pi\, (W\,g)^2 \rho_c \int \rho_b(\Delta) \frac{\cosh(\beta\omega/2) \quad d\Delta}{\cosh(\beta\Delta)\, \cosh[\beta(\Delta - \omega/2)]} \qquad (5)$$

where $\tau(\omega)$ is the relaxation time and ρ_c is the density of band states per site. This result was first obtained by Eliashberg [6].

To show the similarity between the phenomenological $\text{Im}\sum(\omega)$, Eq.(2), and the result corresponding to Eq.(5) we evaluate (5) assuming a constant density of band

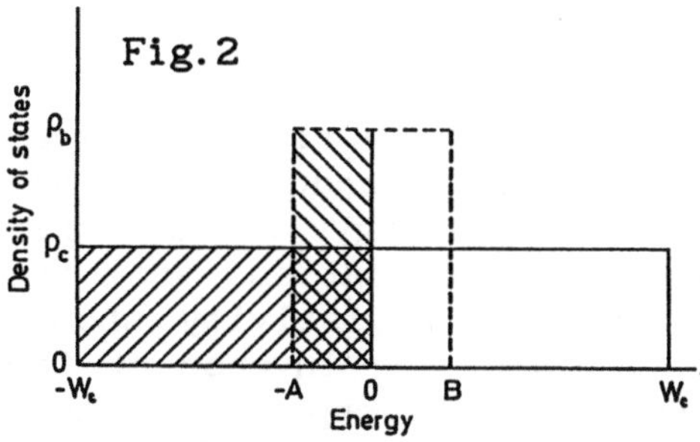

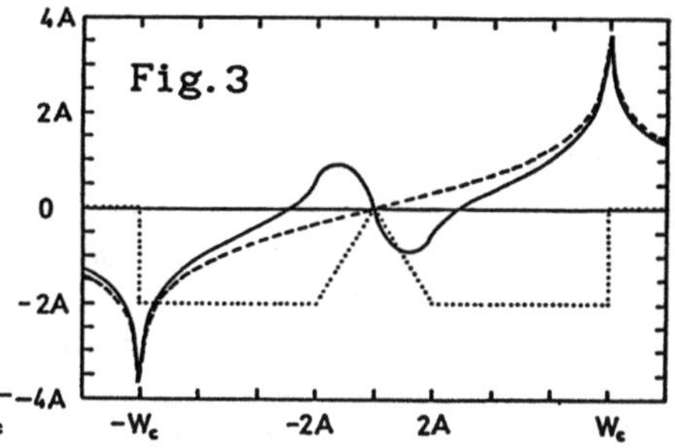

Fig.2) Schematic drawing of the band configurations used in this work. ρ_c corresponds to the density of states associated with the single electrons, while ρ_b corresponds to the density of paired states.

Fig.3) Pointed line: imaginary part of the self-energy given by Eq.(6) at T=0; full line: Kramers-Kronig related real part of the self-energy; dashed line: real part of the self-energy assuming a constant imaginary part between $-W_c$ and W_c.

states ρ_c extending from $-W_c$ to W_c and also a constant density of paired states from -A to B as shown in Fig.2. With these assumptions we obtain :

$$-\mathrm{Im}\sum(\omega) = \frac{\hbar}{\tau(\omega)} =$$

$$c\,T\,\coth(\beta\omega/2)\,\ln\left[\frac{\cosh(\beta B)\,\cosh[\beta(A+\omega/2)]}{\cosh(\beta A)\,\cosh[\beta(B-\omega/2)]}\right] \qquad (6)$$

where $c = \pi\,(W\,g)^2\rho_c\rho_b$.

To visualize the form of the self-energy obtained from these scattering processes we plot in Fig.3 the real and imaginary parts of the self-energy, and compare them with those resulting from an elastic scattering process with a constant relaxation time. We also show the imaginary part of $\sum(\omega)$ of Eq.(6) at different temperatures in Fig.1(a).

Through the Kramers-Kronig relations we calculate the real part of the self-energy from $\mathrm{Im}\sum(\omega)$. As pointed out in [2] the critical dependence of $\mathrm{Re}\sum(\omega)$ on ω at ω=0 is a direct consequence of the form of $\mathrm{Im}\sum(\omega)$ and leads to the "marginal Fermi liquid " concept. In

Fig.1(b) we display Re $\sum(\omega)$ as obtained from (6) for different temperatures.

The linear temperature dependence of the resistivity as well as the shape of the thermal conductivity are consequences of this form of the self-energy. We can infer from the results shown in Fig.1 that deviations from the linear temperature dependence of the resistivity should appear at high temperatures. In fact, the inverse of the relaxation time at $\omega=0$ as a function of temperature shows saturation for temperatures larger than the bandwidth of the paired states.

4. Tunneling Conductance and Photoemission Spectrum

These two experiments are of crucial importance to the phenomenological analysis of the normal phase as they probe directly into the density of states near the Fermi level.

Both experiments seem to indicate a non analytic V-shaped form of the density of states centered at the Fermi level (E_F) in several of the HTS (including the three dimensional compounds based on $BaBiO_3$) : $D(w) = D(E_F) + K * |w-E_F|$. The absolute values of the parameters $D(E_F)$ and K cannot be determined from these experiments. However, the ratio between them as obtained from tunneling [7] is quite consistent with that obtained from photoemission experiments [8].

The authors of Ref.[2] link this behaviour of the tunnel conductance and the photoemission spectra to the imaginary part of the one particle self-energy proposed in their paper.

However, as pointed out in Ref.[9], this form of the self-energy leads to an inverted V shape (Λ) or cusp in the density of states at the Fermi level (see Fig.4), contrary to what tunneling and photoemission experiments indicate.

In what follows we show that the same Fermion-Boson picture that gives rise to the V-shaped self-energy is consistent with the tunneling and photoemission results if one includes the possibility of tunneling into the paired (bosonic) states.

To analyze the tunneling experiment it is necessary to add to Hamiltonian (3) terms accounting for the interaction of the system with the tunnel electrons:

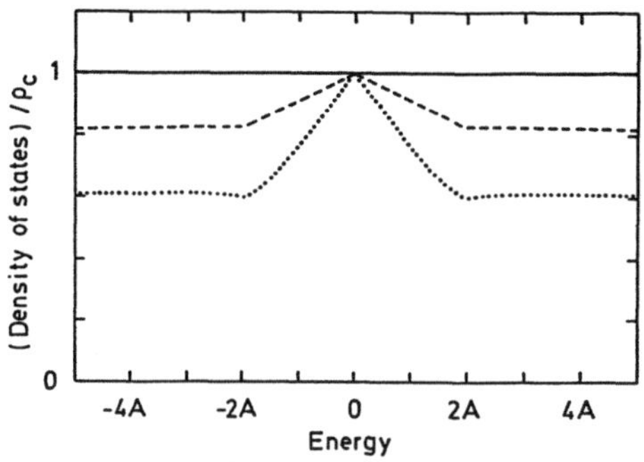

Fig.4) Renormalized fermionic density of states corresponding to the self-energy of Eq.(6). Full line: c=0, dashed line: c=2, dotted line: c=5, where $c=\pi\rho_b\rho_c(Wg)^2$.

$$H_T=\sum_{k,\sigma}(\varepsilon'_k-\mu)\ b^+_{k\sigma}b_{k\sigma}+\sum_{k,k',\sigma}(\ T_{kk'}\ c^+_{k\sigma}b_{k'\sigma}+T^*_{kk'}\ b^+_{k'\sigma}c_{k\sigma})$$

$$+\sum_{j,k,\sigma}(\ G_{kj}\ b^+_{k\sigma}d_{j\sigma}+G^*_{kj}\ d^+_{j\sigma}b_{k\sigma}\ ). \qquad (7)$$

Here $b^+_{k\sigma}$ denotes the creation operator for tunnel electrons with crystal momentum k , spin σ and energy ε'_k. The first term in (7) corresponds to the band of tunnel electrons, while the next two terms represent the interaction of that band with the HTS system.

To evaluate the tunneling conductance we again use the canonical transformation that reduces the dynamical space to paired d-states. The transformed tunneling Hamiltonian (7) reads :

$$\tilde{H}_T=\sum_{k,\sigma}(\varepsilon'_k-\mu)\ b_{k\sigma}^+b_{k\sigma}+\sum_{k,k',\sigma}(T_{kk'}c^+_{k\sigma}b_{k'\sigma}+T^*_{kk'}\ b^+_{k'\sigma}c_{k\sigma}\ )$$

$$+\sum_{j,k,k'}\frac{g_{jk}}{\sqrt{N}}\left[G_{k'j}(\ b^+_{k'\downarrow}c^+_{k\uparrow}+c^+_{k\downarrow}b^+_{k'\uparrow})\ A_j+\text{H.c.}\right], \qquad (8)$$

where the first term corresponds to a normal tunneling process, while the second gives rise to a new tunneling process in which an added electron can join one of the electrons in the fermion band to enter a paired state.

Using Hamiltonian (8), the Fermi Golden Rule allows us to calculate the probability per unit time for an

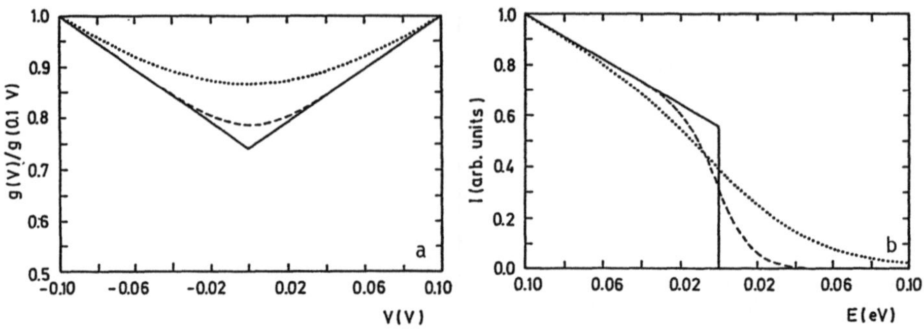

Fig.5) (a) The ratio of tunneling conductances $g(V)/g(V_0)$, V_0=0.1eV, full line: T=0, dashed line: T=100 K, dotted line: T=300 K, taking A=B=0.1eV. (b) Photoemission spectrum plotted as a function of initial-state energy, full line: T=0, dashed line: T=100 K, dotted line: T=300 K, taking A=B=0.1eV.

electron to tunnel through the two possible processes into the HTS. From it one can calculate the tunneling current J_t . As above, assuming constant densities of states for the localized and extended bands (see Fig.2), we obtain for the conductance ($g(V)=dJ_t/dV$) the following result :

$$\frac{g(V)}{g(0)} = 1 + \frac{\rho_b\, 8(W/U)^2}{\beta} \left[\coth(eV\beta/2)\right] \ln\left[\frac{\cosh(\beta B)\,\cosh[\beta(A+eV/2)]}{\cosh(\beta A)\,\cosh[\beta(B-eV/2)]}\right], \tag{9}$$

where the first term corresponds to the normal tunneling contribution, while the second originates from the new tunneling process. To obtain this result we have considered the coupling parameters $T_{kk'}$ and G_{kj} as constants of the same order of magnitude.

We display in Fig.5(a) the ratio $g(V)/g(V_0)$, V_0=0.1V, for the temperatures T=0,100,300 K. In order to fit this curve to the experimental results of Ref.7 we have taken $[\rho_b\ (W/U)^2]$= 0.4 states/eV/cell. The value of the conductance at V=0 and T=0 K corresponds entirely to the normal tunneling process, while the $|V|$ dependence of the conductance arises from the anomalous tunneling process.

The energy-width of this V-shaped dependence extends from (-2A) to (2B).

With a similar treatment for the photoemission process we obtain two contributions to the photocurrent. The first corresponds to the normal processes in which an electron from the uncorrelated band absorbs a photon to be ejected from the sample. In the second process the electron is ejected from a paired state leaving its partner in one of the conduction band states. The resulting photoemission spectrum is shown in Fig.5(b) for T=0,100 and 300 K. In order to fit the data of Ref.8 we take $[\rho_b(W/U)^2] = 1$ states/eV/cell. As in the tunneling result, the intensity at the Fermi level corresponds to the normal processes, while the linear energy dependence results entirely from the anomalous contribution.

5. Discussion and Conclusions

To summarize, we have considered a model in which the Fermi level of a band of uncorrelated states is straddled by a band of paired states. From it we have derived the form proposed on phenomenological grounds by Varma et al [2] for the self energy of the uncorrelated states. From this self-energy one can calculate the one-particle density of states, which shows a <u>cusp</u> at the Fermi level [9] instead of a minimum as some experiments seem to indicate. We have solved this contradiction calculating the tunneling conductance and photoemission spectrum which arise from new processes inherent to the proposed model. Our calculations lead to the observed form of these properties and are not based on the density of states obtained from the self-energy. In addition, recently Kirtley and Scalapino [10] argued that the shape of the tunneling conductance in HTS could be due to strong inelastic scattering rather than to a density of states effect. In their approach they relate the scattering mechanism to magnetic fluctuations.

We have made no attempt to justify the model though there have been approaches to this question by several authors (see for example Refs. 4,5 and 6). The existence of paired states is easier to justify in the "skipping valence" BiO superconductors than in the CuO based compounds (see [11], for example). However the inclusion of both compounds in the same family of HTS is still a matter of discussion.

Regarding our results we remark that to obtain the tunneling conductance and photoemission spectrum consistent with experimental values we have had to assume that the density of paired states has a rather large value ($\rho_b \simeq$ 10states/eV/cell). Since the width of this band has to be larger than $\simeq$ 0.15eV [7,8], this implies that the number of paired states is of the order of the number of unit cells. This fact indicates that the paired states should arise from intrinsic properties of the material. The density of paired states is also consistent with the magnitude of the linear term in the resistivity [12](assuming reasonable values for the parameters involved: $n_e \simeq 10^{22}/cm^3$, $m/m_e \simeq 5$, $W \simeq 0.6eV$, ρ_c= 1state/eV/cell).

In conclusion, we have shown that the peculiar phenomenological form for the one particle self-energy proposed by Varma et al [2] can be derived quite straightforwardly from a "mixed Fermion-Boson" model. The correct linear energy dependence of the tunnel conductance and photoemission spectrum also follow straightforwardly from the same model. However, these last results are not derived from the self-energy through the density of states but follow from new processes allowed by the model.

References

[1] G.Bednorz and A.Müller, Z.Physik B64, 189 (1986).

[2] C.M.Varma, P.B.Littlewood, S.Schmitt-Rink, E.Abrahams and A.Ruckenstein , Phys.Rev.Lett. 63, 1996 (1989); C.M.Varma, Int.J. of Mod.Phys. B3, 2083 (1989).

[3] J.Ranninger and S.Robaszkiewicz , Physica 135B, 468 (1985); R.Micnas, J.Ranninger, and S.Robaszkiewicz, Rev.Mod.Phys.62, 113 (1990).

[4] L.Joffe, A.I.Larkin, Yu N.Ovchinnikov, and Yu Lu, in "Strongly Correlated Electron Systems", ed. by G.Baskaran, A.E.Ruckenstein, E.Tossatti and Yu Lu, World Sci.Publ.Co., Singapore(1989).

[5] I.O.Kulik , Int.J.of Mod.Phys. B2, 851 (1988).

[6] G.M.Eliashberg , Pisma Zh.Eksp.Teor.Fiz. 46, 94 (1987).

[7] M.Gurvitch, J.M.Valles Jr., A.M.Cucolo, R.C.Dynes, J.P.Garno, and L.F.Schneemeyer , Phys.Rev.Lett. 63, 1008 (1989).

[8] J.M.Imer, F.Patthey, B.Dardel, W.D.Schneider, Y.Baer, Y.Petroff, and A.Zettl, Phys.Rev.Lett. 62, 336 (1989).

[9] P.Lee, in Proc. of the Symposium on High Temperature Superconductivity, Los Alamos (1989).
[10] J.R.Kirtley and D.J.Scalapino, Phys.Rev.Lett. 65, 798 (1990).
[11] C.M.Varma, Phys.Rev.Lett. 61, 2713 (1988).
[12] E.J.Osquiguil, L.Civale, R.Decca, and F.de la Cruz , Phys.Rev. B38, 2840 (1988) ; R.Decca, E.J.Osquiguil, F.de la Cruz, C.D'Ovidio, M.T.Malachevsky, and D.Esparza, Solid State Commun.69, 355 (1989).

Low Dimensional Magnetism

N. Majlis

Instituto de Física, Universidade Federal Fluminense, 24020, Niteroi, Brazil

Abstract. The magnetic behaviour of surfaces is to a great extent determined by the various kinds of anisotropy energies [1]. In section 1 I shall discuss examples of ferromagnetic surfaces in which the local values of the exchange couplings and anisotropy constants near and at the surface allow the existence of an ordered ferromagnetic phase at the surface at temperatures above the bulk transition temperature. We shall also see how the possibility of competing interactions at the surface can in principle explain some intriguing experimental results in Tb [2] and in Gd [3]. In section 1.2 I review the calculation of the power absorbed in a spin wave resonance experiment, in which the resonant frequency is determined by surface localized modes, and I review the possibilities offered by this technique to obtain information on the surface parameters. In section 2, I review some not so well known theoretical results on the effects of the dipole-dipole interactions in two dimensional systems, and sketch a spin-wave analysis of some recent experiments on epitaxial transition-metal monolayers. We shall see that the spin wave approximation gives a rather good description of the low temperature properties of these systems, if dipolar interactions and the appropriate anisotropy energy terms are included in the Hamiltonian besides exchange interactions.

1. Magnetism at Surfaces

1.1 First Order Phase Transition at Ferromagnetic Surface

Terbium and Gadolinium are among the metals which exhibit a surface ferromagnetic phase (SFM) at temperatures above the bulk transition temperature T_c to the paramagnetic phase [1]. Besides, in the same series of measurements where this interesting phenomenon was observed, some intriguing abrupt variations were found of the magnetization as a function of temperature, both in Tb [2] and in Gd [3], within a small range of T between the bulk and the surface transition temperatures. In both systems, the magnetization M_s (T) of the surface has a deep minimum followed by a steep maximum, and then decreases rapidly to vanish at the surface transition temperature T_s. This suggests that a first order phase transition may be occurring at the surface in this interval of temperatures [1]. In this respect, some relevant information was obtained through photoemission measurements with spin polarization analysis of emitted electrons, which led to the conclusion that in Gd the surface spins are aligned predominantly antiparallel to the bulk spins [3]. It seems reasonable to consider, in consequence, models in which some effective exchange surface-bulk interactions are antiferromagnetic [4].

We have adopted a Heisenberg semi-infinite model, as a convenient representation of Gd or Tb surfaces. The spins on the first plane are assumed to couple antiferromagnetically with those of the second plane, while all other interactions, intra or interplane, are FM.

Springer Proceedings in Physics, Volume 62
Surface Science Eds.: F.A. Ponce and M. Cardona

The presence of an easy axis anisotropy is enough, when combined with the enhancement of the exchange interaction at the surface, to maintain the surface ferromagnetic at $T > T_c$.

In order to induce some kind of helical local arrangement of the magnetization in some range of T, we need some competition between FM and AFM couplings, so we added a second nearest-neighbour interaction, assumed FM, between the first and the third plane. The model incorporates also two other important elements: 1) enhancement of the exchange couplings at the surface; 2) an easy-axis, surface single-ion anisotropy.

With these elements the Hamiltonian is:

$$
\begin{aligned}
H = & -\frac{1}{2} J_{\parallel} \sum_{\langle i,j\rangle} S_{oi}\cdot S_{oj} + \frac{1}{2} J_{\perp}^{(1)} \sum_{\langle i,j\rangle} S_{oi}\cdot S_{1j} \\
& - \frac{1}{2} J_{\perp}^{(2)} \sum_{[i,j]} S_{oi}\cdot S_{2j} - D \sum_{i} (S^{z}_{oi})^{2} \\
& - \frac{1}{2} J \sum_{(p,q)\geq 2,\ \langle i,m\rangle} S_{pi}\cdot S_{qm}
\end{aligned}
\qquad (1.1)
$$

where the first lower index of the spin operators denotes the plane, and the second the point on the two dimensional plane lattice. Angular (square) brackets denote pairs of first (second) nearest neighbours, respectively.

A mean field approximation calculation based on this Hamiltonian leads to the following results [5]: for some choices of the parameters of the model, the most stable phase at low T is a helical (HEL) arrangement at the surface, with the bulk ferromagnetically ordered. The spins are assumed to be aligned parallel to the surface in all planes. The rotation angles α_0 and α_1 of the average magnetization of the first two planes relative to the bulk spin quantization axis z, vary with T. The magnetization of the third plane is maintained equal to the bulk one at the same temperature, which is calculated also in the mean field approximation. At some temperature T^* another phase, which we call AFM (+), minimizes the Helmholtz free energy F. In this phase, $\alpha_0 = \pi$ and $\alpha_1 = 0$. The change from HEL to AFM (+) occurs with a jump in the magnetization of the first two planes. The discontinuity of the order parameter and the equality of F for both competing phases at the transition temperature, characterize a first order phase transition.

As T increases above T^* the angles of phase HEL continue varying until at T^{**} they reach the values $\alpha_0 = 0$ and $\alpha_1 = \pi$, of the phase we call AFM(-). At a higher temperature T^{***}, below but close to T_c, the free energy of AFM(-) equals that of AFM(+) and at any higher T it has the lowest value of F, all along up to T_c, so that at T^{***} we have another first order phase transition, similar to that at T^*. At temperatures $T > T_c$, however, there is no distinction between AFM(+) and AFM(-), since the bulk is paramagnetic, in first approximation.

Another phase, which we call FLOP exists at temperatures between T_c and T_s, but it is metastable. In the FLOP phase, the first two planes are perpendicular to the z axis, and mutually antiparallel.

Finally, at a temperature T_s the surface also turns paramagnetic. Fig.1 shows $m_0(T)$.

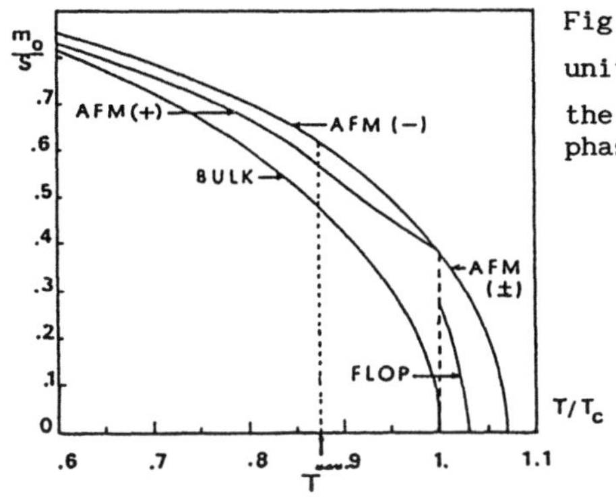

Fig.1.Surface magnetization $m_o(T)$ in unit of S, as a function of T/T_c, near the transition to the paramagnetic phase.

As indicated above, we have a possibility of the coexistence of two competing phases which become more and more nearly equally stable as temperature increases, in the interval $Tc < T < Ts$. This fact could explain the experimental results for Gd and Tb [1-3], if one assumes the existence of 180° domains at the surface, corresponding to an admixture of the AFM(+), (-) phases, which is a metastable configuration around T_c. The consideration of the dipolar field tends to strengthen this argument, since antiparallel domains on the surface have a lower dipolar energy than a single domain. Clearly, we do not have a quantitative description of the domain structure, but this picture seems to be coherent with hysteresis effects which were reported for the Gd surface [6].

The measured magnetization, in this model, would be an average over several antiparallel domains, and therefore smaller than that of each domain. At some temperature very close to T_c, where the residual magnetization of the planes near the surface gets sufficiently small, the domains are freer to rotate, and the apparent magnetization increases if they become mainly parallel. This process would explain the sudden growth of the average measured surface magnetization $\bar{M}_s(T)$ above T_c and close to T_s. Since this happens at T very close to T_s, though, both the local and the average magnetization $\bar{M}_s(T)$ ultimately decrease very rapidly as T grows and approaches T, as found experimentally.

1.2 Ferromagnetic resonance of surface modes

The technique of ferromagnetic resonance has been widely used in the study of ferromagnetic films and layered structures [7-9]. We have extended previous calculations at low T of the power absorption of a semi-infinite ferromagnetic [10], to finite T, by application of the Layer Random Phase approximation (LRPA) [1]. The hamiltonian we chose is:

$$H = - \sum \frac{I_{ij}}{z} (S_i^+ S_j^- + S_i^- S_j^+ + 2\eta_{ij} S_i^z S_j^z) \qquad (1.2)$$

where a uniaxial exchange anisotropy η_{ij} was introduced. We assume that the easy axis z is along the surface, that the magnetization is parallel to the surface on all planes and that the exchange anisotropy constant η is = 1 for all but the surface plane where $\eta_s > 1$. We considered the possibility of different isotropic exchange couplings on the surface $J_\parallel$ and between the surface and the bulk $J_\perp$. As is well known, the existence of surface modes with energies detached from the bulk continuum depends on the values of the surface parameters [1]. We shall consider a case in which two surface modes exist as discrete energy states (isolated poles of the Green's function) in certain intervals of T, with energies lying above the bulk continuum. These modes are called upper and lower "optic" surface magnons in analogy with the phonon case. These particular calculations were performed for a (111) surface plane in an fcc lattice. A uniform constant magnetic field H_o in the z direction, and a perpendicular radio frequency field along the y axis are applied to the film. The incident rf field is the plane wave

$$\vec{h}_{rf}(x,t) = \frac{h_y}{\sqrt{2}}\left(u^+ e^{-i(\Omega t - k_x x)} - u^- e^{i(\Omega t - k_x x)}\right), \tag{1.3}$$

where $u^{\pm}$ are the circular polarization versors around the z axis, and x is perpendicular to the film.

The rate of power absorption by the system can be expressed as [10]:

$$P = -\pi(g\mu_B)^2 \int \omega \, d\omega \; \mathrm{Im} \sum_{\ell,m} \left[G_{\ell m}(\omega,k_\parallel) + G_{\ell m}(-\omega,k_\parallel)\right] \cdot$$

$$\cdot \; |h_y|^2 \cdot N_s \cdot \delta_{\vec{k}_\parallel,0} \; . \tag{1.4}$$

The Green's functions in Eq.(1.4) are defined as:

$$G_{\ell m}(\omega,\vec{K}_\parallel) = \sum_{\Delta_\parallel} e^{i\vec{K}_\parallel \cdot \vec{\Delta}_\parallel} \cdot \langle\langle S^+_{\vec{\ell}_\parallel,\ell}(t);\, S^-_{\vec{\ell}_\parallel+\vec{\Delta}_\parallel,m}(0)\rangle\rangle_\omega \tag{1.5}$$

and

$$\langle\langle A(t);\, B(0)\rangle\rangle_\omega = \mathbb{F}_\omega\Big[-\, i\theta(t)\langle[\, A(t),\, B(0)]\rangle\Big] \quad , \tag{1.6}$$

where $\langle X \rangle = T_r(\rho X)/\, T_r\rho$, $\rho = e^{-\beta H}$,

and $\mathbb{F}_\omega[\, f(t)]$ symbolizes the Fourier transform of f. The spectral representation of the Green's function is:

$$G_{nm}(K_\parallel,\nu) = \frac{1}{6I\pi} \sum_\alpha \frac{X^\alpha_n(k_\parallel)^* \cdot X^\alpha_m(k_\parallel)\, \sigma_m}{\nu - \nu_\alpha(\vec{k}_\parallel)} \quad , \tag{1.7}$$

where $6I\nu_\alpha(\vec{k}_\parallel)$ is the eigenenergy of the α mode, and $X^\alpha_n(\vec{k}_\parallel)$ is the amplitude of its wave-function at plane n. The parameters σ_m are defined as

$$\sigma_m = \langle S^z_m \rangle / \langle S^z_b \rangle , \tag{1.8}$$

where $\langle S^z_b \rangle$ is the thermodynamic average of the z component of spin in the bulk, and they must be calculated self-consistently [1], for each T and H_o. The surface modes depend on the distance to the surface, which is proportional to the plane index n, through an attenuation factor $\xi = e^{-a/L}$ which is the transfer matrix, or the "state ratio" for that particular mode [1], for which the length L can be defined then as the localization length. We

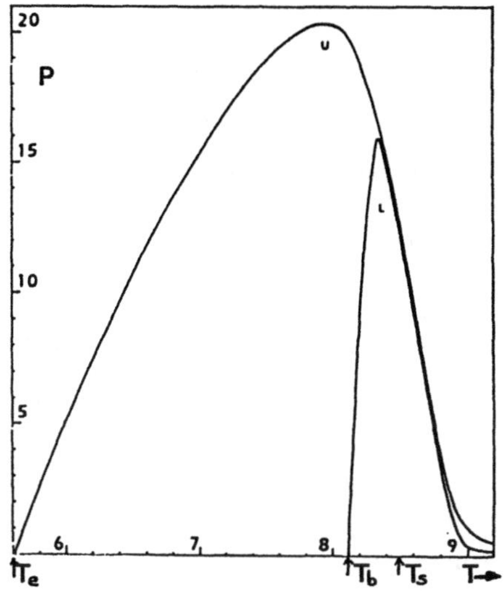

Fig. 2. Absorption intensity at the frequencies ν^U_{opt} and ν^L_{opt}, vs. reduced temperature T, for $g\mu_B B_o/6I=0.01$. T_e = edge of absorption; T_b = bulk Curie temperature; T_s = surface Curie temperature.

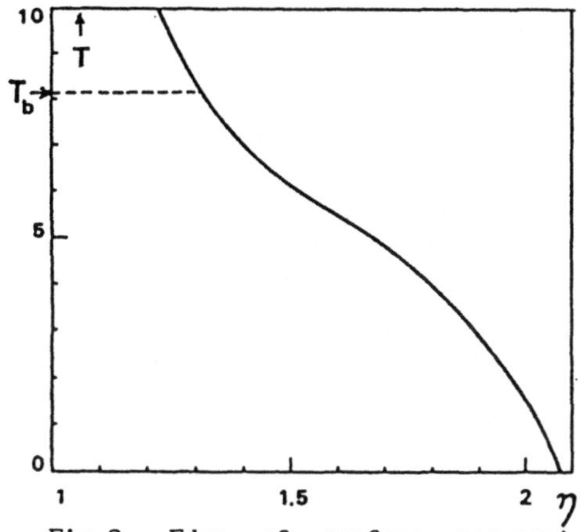

Fig.3. Edge of surface resonant absorption T_e as a function of anisotropy η.

have assumed that the r.f. field is constant inside the sample, which is justified, as a good approximation, only if the skin depth of the radiation is appreciably longer than L. With the former assumptions the resonance intensity for the particular mode α is proportional to:

$$P_\alpha \propto \langle S_b^z \rangle^2 \cdot \sum_{m,n} X_n^{\alpha *}(0)\, X_m^\alpha(0)\, \sigma_m\, \nu_\alpha(0) \quad , \tag{1.9}$$

which depends on T and the applied field H_o. For the parameters chosen in the present example, the system has an SFM, so that $T_s > T_c$, and it displays two optic surface modes. The lower energy one only separates from the bulk continuum spectrum (at $\vec{k}_\parallel = 0$) above T_c. Fig.2 shows the absorption intensity for the lower and upper optic surface magnons at $\vec{k}_\parallel = 0$ as function of the dimensionless reduced temperature. We only need the amplitudes X_n^α and the energies ν_α for $\vec{k}_\parallel = 0$, due to wave vector conservation. The upper optic mode exhibits an absorption threshold temperature T_e, which depends on the anisotropy parameter, for constant exchange integrals, as shown in Fig.3. For such a case, measuring T_e would give valuable information on η.

2. Two-dimensional magnetism and the dipole-dipole interaction

Mermim-Wagner theorem establishes the absence of long range order at finite T in 2D systems with isotropic short range interactions. However, monolayers of several transition [12] and rare earth [11] metals have been found in the last five years to exhibit ferromagnetic long range order, which persists up to fairly large T, of the order of half the bulk T_c. It is possible to reconcile these findings with the former theoretical results, by appealing to anisotropy in the exchange interactions, or local anisotropy energy terms, or by invoking the presence of magnetic dipolar interaction, which is both anisotropic and of long range. The possibility of experimentally realizing systems of strictly two dimensions was provided by the relatively recent epitaxial techniques, which have allowed to produce very good monolayers of many ferromagnetic elements and alloys. Although in general one must assume that these systems are probably composed of a series of islands, the structural analysis and the magnetic measurements lead to the conclusion that these islands are very large, have a well defined crystalline structure, and exhibit long range ferromagnetic order over distances of hundreds of lattice constants [13]. Several of these 2D systems exhibit a magnetic ordered phase up to temperatures of hundreds of K. Their magnetization has been measured with great accuracy by the use of spin-polarized secondary electron emission spectroscopy (SPSEE) [13,14], spin-polarized low energy electron diffraction (SPLEED) [14] and magneto-optic Kerr effect [15]. In the case of Co/Cu(100) the magnetic field dependence of the magnon energy gap was measured with light scattering techniques [15].

We shall discuss briefly in what follows the role of different factors, including several types of anisotropy and the dipole-dipole interaction, in stabilizing the long range order in such 2D systems at finite temperatures. Application of this analysis has been made to the case of Co/Cu(100) monolayers. From comparison of spin wave calculations with the experimental results, we are able to establish the appropriate models for the anisotropies, amd also to propose some numerical estimates for their values. This is particularly important in view of further studies of the critical exponents of such 2D system [15,16]. We assume the following localized-spin model Hamiltonian:

$$\mathcal{H} = - J \sum_{\langle \ell,m \rangle} \vec{S}_\ell \cdot \vec{S}_m - g\mu_B H \sum_\ell S_\ell^y + \frac{(g\mu_B)^2}{2a^3} \sum_{\ell,m} \left| \frac{a}{\vec{r}_{\ell m}} \right|^3 \cdot \left[\vec{S}_\ell \cdot \vec{S}_m - \frac{3(\vec{S}_\ell \cdot \vec{r}_{\ell m})(\vec{S}_m \cdot \vec{r}_{\ell m})}{|\vec{r}_{\ell m}|^2} \right] +$$

$$+ D \sum_\ell \left(S_\ell^x \right)^2 + \frac{K_4}{2} \sum_{\langle \ell,m \rangle} \left[\left(S_\ell^y \right)^2 \left(S_m^y \right)^2 + \left(S_\ell^z \right)^2 \left(S_m^z \right)^2 \right] . \qquad (2.1)$$

As mentioned above, the dipolar interaction is capable of stabilizing ferromagnetic long-range order in a 2D system [17-18]. Its main effect is that of changing the dispersion relation for small k, from the isotropic exchange form $E_k \approx Dk^2$, to $E_k \approx A.\sqrt{k}$ [17]. In order to interpret the experimental results for the Co monolayers, we also assume an easy plane (D > 0) single-ion anisotropy, and an easy axis in-plane anisotropy. This is suggested by the experimental dependence of the coercive field on the orientation of an applied field in the plane relative to the crystal axes [15,19]. In Co/Cu(100) the easy directions are the diagonals of the square Bravais cells in 2D. Before performing the standard Holstein-Primakoff

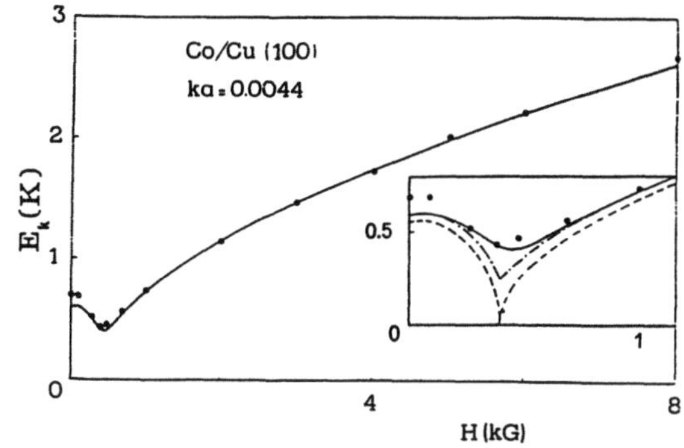

Fig.4. Energy of magnons with λ=363 nm (ka=0.0044), and $\theta_k = \pi/2$, versus magnetic field for a monolayer of Co/Cu(100). Full line: spin-wave calculation with field applied at an angle α=0.01 π with the y axis. [Insert: full line refers to ka=0.0044, α=0.01 π; dashed-dotted line: ka=0.0044, α=0; dashed line: ka=0, α=0]. Dots: experimental data [12].

transformation upon the spin operators, we must determine the ground state configuration. According to the experimental results [14,15], which show only in-plane-remanescence, we take as quantization axis a direction in the plane of the layer (which we choose as the yz plane) which makes an angle ϕ with the y = (010) axis along which the external magnetic field is applied. The ground state configuration is obtained by demanding that the linear terms in the boson operators a_k and a_k^+ vanish. This determines univocally the angle ϕ. It can be easily shown that this condition coincides with that obtained minimizing the energy of the system obtained in the classical limit $S \longrightarrow \infty$. The equation which determines ϕ is:

$$\sin\phi \cdot \left[g\mu_B H_y - 2K_4 S^3 z \cos\phi \cdot \cos 2\phi \right] = 0 \quad , \qquad (2.2)$$

where z = number of nearest neighbours, which is 4 for the square lattice.

The explicit expression for the field dependence of the magnon gap for Co, in the limit ka << 1, is:

$$E_k \approx \left[g\mu_B H \cos\phi - 2\, K_4 S^3 (\sin^4\phi + \cos^4\phi) + 2DS + \right.$$
$$\left. + 4\pi w\bar{e} - 2\pi wka + 2JS(ka)^2 \right]^{1/2} \cdot$$
$$\left[g\mu_B H \cos\phi - 2\, K_4 S^3 \cos 4\phi + 2\pi wka \cdot \sin^2(\theta_k - \phi) + 2JS(ka)^2 \right]^{1/2} , \qquad (2.3)$$

where $w = g^2\mu_B^2 S^2/a^3$ measures the strength of the dipolar interactions, $\bar{e}$ is a geometric factor of order 1, and θ_k is the angle between $\vec{k}$ and the y(≡ (010)) axis. This expression has a minimum for $H = H_o = 2K_4S^3z/(g\mu_B)$ (Fig.4). H_o is the minimum field capable of aligning the magnetization along its own direction. For k = 0 the minimum gap is zero (see insert in Fig.4) within the present approximation (harmonic) to the boson Hamiltonian. The angle ϕ decreases monotonically with increasing field, starting from the easy axis orientation $\phi = \pi/4$, down to $\phi = 0$ for $H = H_o$. The cusp gets rounded, and the minimum is not zero, when k > 0, and/or when the field

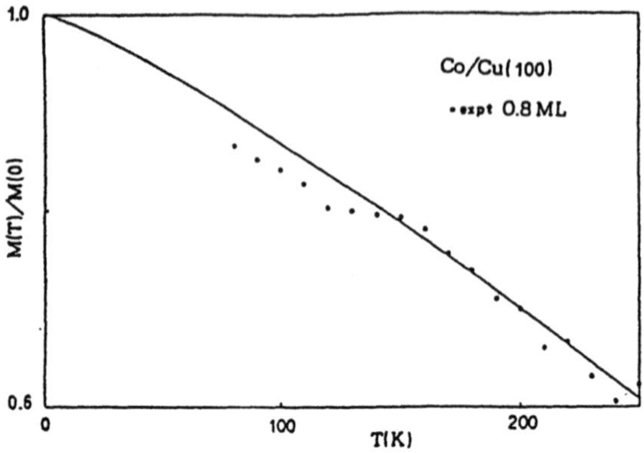

Fig.5. Temperature dependence of the spin-wave magnetization of a Co/Cu (100) monolayer (full line) compared with the experimental results (dots) for 0.8 ML [12].

makes a finite angle α with the (010) axis. In this last case, Eq.(2.2) is substituted by

$$g\mu_B H \sin(\phi - \alpha) - \frac{1}{2} K_4 S^3 z \sin 4\phi = 0 \qquad (2.4)$$

and ϕ decreases monotonically from $\pi/4$ to α as H grows from 0 to ∞.

As to the low temperature dependence of the magnetization, the experimental results suggest an approximately linear behaviour (see Fig.5). The calculation of the magnetization as function of T within the spin wave approximation, yields in fact a quasi-linear behaviour:

$$(M_s(0) - M_s(T))/ M_s(0) = A.T^s$$

where the theoretical exponents $s \cong 1.12$ coincides within 10% with the exponent obtained from the experimental results [15].

This quasi linear behaviour is also qualitatively consistent with the theoretical results previously obtained by Maleev [17], although the present calculations differ from Maleev's in that the k - space integrations which yield M(T) are performed now over the whole first Brillouin zone of the 2D reciprocal lattice. This requires calculating the dispersion relation numerically, including the contribution thereto of the long range dipolar sums [20].

In conclusion, with two adjustable anisotropy parameters, we find quantitative agreement by assuming that the monolayers of Co are modelled by local spins interacting through short range exchange interactions similar to those in the bulk, plus dipole-dipole interactions and small single-ion and two-ion anisotropy terms. At low temperatures well defined magnons exist, and they determine the temperature decrease of the magnetization.

However, the large values of the critical temperatures, around 300 K for Co monolayers on Cu, cannot be explained apparently without invoking an enhancement of the effective exchange interaction relative to the bulk value. Also, the very abrupt decrease of the magnetization as T approaches T_c is characteristic of 2D systems and requires a theoretical treatment outside the spin-wave approximation [21].

REFERENCES

1. N.Majlis, "Thin Films and Small Particles", Proc. 5th Latin American Symposium on Surface Physics, Bogotá, Colombia, 1988. Editors: M.Cardona, J.Grijaldo (World Scientific, 1989) pags.64-83.
2. C.Rau and C.Jin, J.Appl.Phys. 54 (1988) 3667.
3. D.Weller, S.F.Alvarado, W.Gudat, K.Schröder, M.Campagna, Phys.Rev.Lett. 63 (1985) 155.
4. J.M.Sánchez and J.L.Morán-Lopez, Phys.Rev.Lett.58 (1987) 1120.
5. N.Majlis and S.Salinas, Sol.State Comm. 73 (1990) 385.
6. D.Weller and S.F.Alvarado, Phys.Rev.B37 (1988) 9911.
7. Z.Frait, D.Fraitová and L.Pust, J.Magn.Mater. 54 (1986) 1307.
8. J.F.Cochran, B.Heinrich and A.S.Arrot, Phys.Rev.B34 (1986) 7788.
9. R.W.Teale and J.Pelegrini, J.Phys.F16 (1986) 621.
10. A.Duda and H.Puszkarski, Acta Physica Polonica A46 (1974) 97.
11. C.Rau and C.Jin, J.de Phys. C8 (1989) 1627.
12. O.Paul, M.Taborelli and M.Landolt, Surf.Sci. 211/212 (1989) 724.
13. Z.Q.Qiu, S.H.Mayer, C.J.Gutiérrez, H.Tang and J.C.Walker, PR. Lett. 63 (1989) 1649.
14. W.Dürr, M.Taborelli, O.Paul, R.Germar, W.Gudat, D.Pescia and M.Landolt, Phys. Rev. Lett. 62 (1989) 206.
15. D.Kerkmann, J.A.Wolff, D.Pescia, Th.Woike and P.Grünberg, Sol.State Comm. 72 (1989) 963; D.Kerkmann, Appl.Phys.A49 (1989) 523.
16. C.Rau, Appl. Phys. A49 (1989) 579.
17. S.V.Maleev, Sov.Phys. JETP 43 (1976) 1240.
18. Y.Yafet, J.Kwo and E.M.Gyorgy, Phys.Rev.B33 (1988) 6519.
19. D.Pescia and P.Grünberg, IFF Bulletin 33/1988, KWA Jülich.
20. M.G.Pini, A.Rettori, D.Pescia, N.Majlis, S.Selzer, Nato Workshop on "Microscopic Aspects of non Linearity in Condensed Matter", Editors: V. Pokrovsky and A.R.Bishop, Plenum Press, in press.
21. V.L.Pokrovski and M.V.Feigelman, Sov.Phys.JETP 45 (1977) 291.

Phase Transitions in Ultrathin Films

F. Aguilera-Granja and J.L. Morán-López

Instituto de Física, "Manuel Sandoval Vallarta",
Universidad Autónoma de San Luis Potosí,
78000 San Luis Potosí, S.L.P., Mexico

Abstract. The phase transitions in ultrathin Ising films are studied within the mean field approximation. Analytic expressions for the thickness dependence of the Curie temperature are given. The magnetization profiles along the films are calculated within this approximation for various temperatures. The model is applied to estimate the magnetic interactions in Fe and Gd ultrathin films. The experimental results in the case of Fe films on Au and Ag substrates can be very well reproduced.

1. Introduction

Although much is known about phase transitions in $3d$–systems, many aspects remain to be understood in systems of low dimensionality. Very often one finds unexpected and interesting properties at surfaces, thin films, etc. For example, recent experimental studies [1–4] on the magnetic properties of surfaces of Gd, Cr, and Tb have shown that a surface ordered magnetically can coexist with a magnetically disordered bulk phase.

From the theoretical point of view, one of the models more widely used to study the magnetic properties of surfaces is the Ising model. Within that model one can take into account in a straightforward manner the presence of the surface. The environmental effects produced by the surface can be simulated by assuming a location dependent coupling constant J_{mn}; m and n denote the position of the magnetic atoms in the lattice. The simplest case corresponds to a situation in which only the surface coupling constant J_S is assumed to be different from the rest. In that context, the coexitence of a paramagnetic bulk with a magnetic surface can be obtained if J_S exceeds a critical value J_{SC}. The estimation of J_{SC} has been the subject of various publications [5–10]. In general, it is found that it depends sensitively on the surface geometry, bulk structure and range of interactions.

On the other hand, with the development of modern vacuum science it is now possible to study experimentally the magnetic properties of low dimensional systems. In particular, by depositing magntic atoms on the top of non–magnetic substrates, the thickness dependence of the Curie temperature of ultrathin films of Gd on W(110) [11] and of Fe on Au(100) and Ag(100) [12–14], has been measured.

Here, we study some of these new experimental results by modelling the ferromagnetic thin film within the framework of the spin–$\frac{1}{2}$ Ising Hamiltonian. This model, solved in the mean field approximation, allows us to study the thickness dependence of the magnetization. As a result, we obtain an analytic expression of the thickness dependence of the Curie temperature T_C. Our model is applied to the cases of Fe on Au and Ag [12–14] and of Gd on W [11]. Despite the simplicity of the model, the first set of experiments can be very well reproduced.

2. Model

The model used to describe the magnetic properties of the thin film is shown in Fig. 1. The crystal is subdivided into planes parallel to the surface and we assume that the two surfaces are equivalent. The coordination numbers on the parallel planes and inter–planes are denoted by Z_0 and Z_1, respectively ($Z = Z_0 + 2Z_1$).

To follow the magnetic transitions, we define a magnetic long–range–order parameter η_j at each plane, as the difference of the spin–up and spin–down probabilities at the j–th plane. Within the mean field approximation and taking only first nearest neighbor interactions into account, the free energy of the system can be written in a very simple way [9,15]. For a temperature near and below the temperature T_C, it is possible to

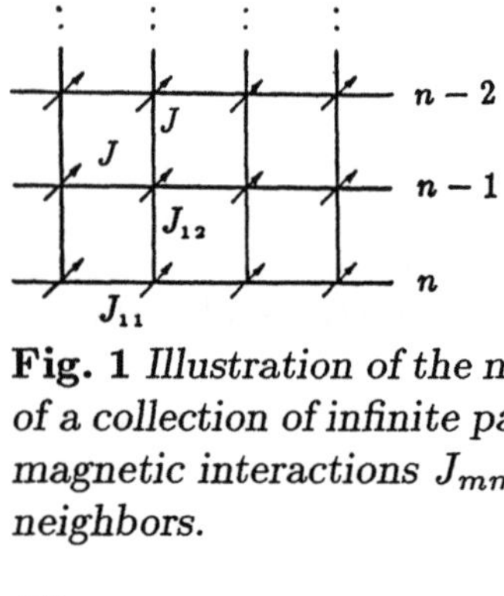

Fig. 1 *Illustration of the model used for the thin film. The model consists of a collection of infinite parallel planes with two equivalent surfaces. The magnetic interactions J_{mm} are assumed to be only between first nearest neighbors.*

perform a linearization on the free energy of the system. In that range of temperatures, it is easy to show [9] that the free energy reduces to the following set of homogeneous linear equations:

$$\hat{A}\bar{\eta} = 0, \tag{1}$$

where $\bar{\eta}$ is a vector with the η_n components, and $\hat{A}$ is a square $n \times n$ tridiagonal matrix with elements:

$$A_{mn} = (kT_C - Z_0 J_{mm})\delta_{m,n} - Z_1 J_{mn}(\delta_{m+1,n} + \delta_{m,n+1}). \tag{2}$$

The existence of the two symmetric surfaces in the system impose the additional symmetries on $\hat{A}$: $A_{kk} = A_{n+1-k,n+1-k}$ for $1 \leq k \leq n$, and $A_{l-1,l} = A_{n-l+2,n-l+1}$ for $2 \leq l \leq n$. Here, n is the thickness of the film. The same properties are obviously satisfied by J_{mn}. For semi–infinite systems, it is well known [5–10] that depending on the J_{mn}, the surface Curie temperature T_{CS} is equal or higher than the bulk Curie temperature $T_{CB} = ZJ/k$. The set of J_{mn} that makes $T_{CS} = T_{CB}$ defines the critical surface $f(J_{mn,c})$ and is obtained from the equation:

$$det\hat{A} = 0. \tag{3}$$

In principle, one can take any number of J_{mn} different from the bulk coupling constant J. Here, just for simplicity, we take only J_{11} and $J_{12} \neq J$.

In a thin film there is strictly only one transition temperature, the one that occurs at the surface of the thin film. We call that temperature T_C. In the monolayer and two–layer films (n=1,2) $kT_C = Z_0 J_{11}$ and $kT_C = Z_0 J_{11} + Z_1 J_{12}$, respectively. The transition temperature for $n = 2$ is given by:

$$x = \frac{1}{2}(a + \sqrt{a^2 + 8c^2}), \tag{4}$$

and in general, for $n \geq 3$:

$$(x-a)^2 \sin n\theta - 2(x-a)c^2 \sin(n-1)\theta + c^4 \sin(n-2)\theta = 0, \tag{5}$$

where

$$x = \frac{(kT_{CS} - Z_0 J)}{Z_1 J}, \quad a = \frac{Z_0(J_{11} - J)}{Z_1 J}, \quad c = (\frac{J_{12}}{J}), \quad \tan\theta = \sqrt{4x^{-2} - 1}. \tag{6}$$

The equation (5) was obtained by a cofactor expansion of the equation

(2) [15]. This equation is valid for values of the parameters such that $T_C \leq T_{CB}$. In the particular case of J_{11} and $J_{12} \neq J$, the critical surface in the parameter space of the J's that separates the two behaviors for $n \geq 3$ is given by the relation:

$$(2-a_c)^2 - 2(2-a_c){c_c}^2(\frac{n-2}{n-3}) + {c_c}^4(\frac{n-3}{n-1})=0. \qquad (7)$$

Here, the subindice c denotes the critical value of the quantities defined in Eq. (6). For situations where the Curie temperature is higher than the one in the infinite solid, the equation that holds for $n \geq 3$ is:

$$x^2(x-a)^2L(n) - 4x(x-a)c^2L(n-1) + 4c^4L(n-2) = 0, \qquad (8)$$

where

$$L(n) = (1+\sqrt{|4x^{-2}-1|})^n - (1-\sqrt{|4x^{-2}-1|})^n. \qquad (9)$$

3. Results

We applied our simple model to the recent experimental results on the thickness dependence of T_C. The magnetic properties of Fe films deposited on Au(100) and Ag(100), were investigated by means of spin–polarized secondary electron emission spectroscopy (SPSEE), spin–polarized low–energy electron diffraction (SPLEED) [12–13] and by the surface magneto–optic Kerr effect (SMOKE) [14]. The various experimental results of the thickness dependence of the Curie temperature are shown in Fig. 2. Open circles, triangles, and open squares correspond to the experiments reported in Ref. 12, 13, and 14, respectively. Based on these experimental results, which show that no induced magnetic mo-

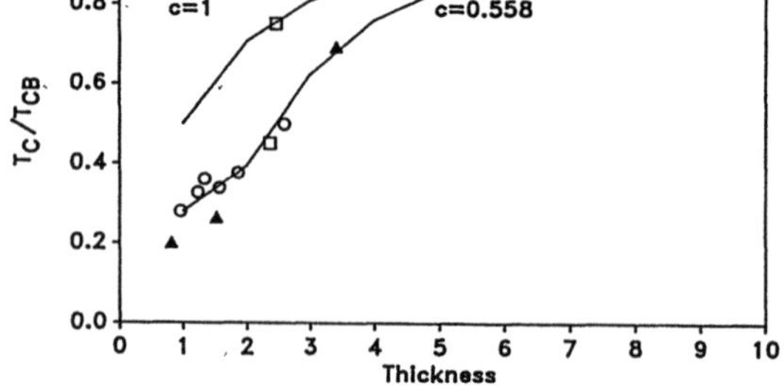

Fig. 2 *Dependence of (T_C/T_{CB}) on sample thickness in Fe thin films deposited on nonmagnetic substrates. The circles, triangles, and squares correspond to various experimental results* [12,13,14].

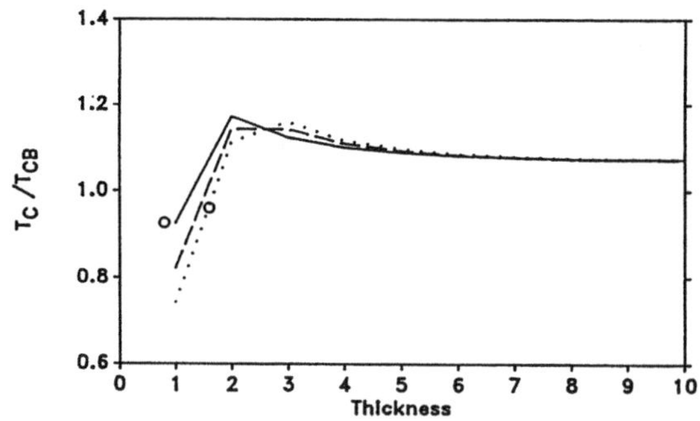

Fig. 3 *Dependence of the* (T_C/T_{CB}) *on sample thickness in Gd thin films deposited on W. The continuous, the dashed and the dotted line, correspond to the three models discussed in the text. The circles correspond to the experimental results* [11].

ment occurs in the subsurface, one can infer that the Fe films are isolated magnetically. One can then assume that the interfaces Fe–vacuum and the Fe–sustrate are magnetically similar. The parameters that define the geometry of the bcc Fe–thin film are $Z_0 = 0$ and $Z_1 = 4$. Furthermore, the two parameters, J_{12} and J, are estimated from the experimental values of the Curie temperatures of the monolayer and the bulk. Choosing the parameter J_{12} such that the experimental results for the monolayer are fitted [12], we obtained for $c = J_{12}/J$ the value 0.558. The results for the thickness dependence of the Curie temperature are shown in Fig. 2. It is remarkable that with such a small number of parameters one can reproduce fairly well Dürr's experimental results, as well as some other reported in the literature [13,14]. We show also the results for $c = 1$, which correspond to a situation in which the coupling constants are the same in all the system.

In the case of Gd, it has been observed by means of SPLEED and electron capture spectroscopy (ECS) that the surface Curie temperature of macroscopic samples is higher that the one in the bulk [1,2] by about 22 K. This means that the coupling constant at the surface is larger than J_{SC}. More recently, electron–spin resonance experiments in ultrathin films, show that the Curie temperature is a monotonic function of the thickness (see experimental points in Fig. 3). Even for samples as thick as 80 Å ($\sim$ 27 layers) the T_C is 5K smaller than T_{CB}.

To model this system, we take a fcc thin film with (111) surfaces. The coordination numbers are $Z_0 = 6$, $Z_1 = 3$ and the coupling constants are estimated from the bulk Curie temperature and the surface Curie temperature of the macroscopic sample. In order to reduce the number of parameters we assume the following three different relations: *i)* $J_{11} = \alpha J$

and $J_{12} = J$, *ii)* $J_{11} = \alpha^2 J$ and $J_{12} = \alpha J$, *iii)* $J_{11} = J_{12} = \alpha J$. In all the cases the parameter α is choosen to fit the T_{CS} of a macroscopic Gd sample [2]. Those values are 1.845, 1.282 and 1.485, respectively. The results of our calculation for the three different relations are shown in Fig. 3. The solid line corresponds to case *i)*, in which one assumes that only the J_{11} differs from the rest. The dashed line shows the results in which one assumes a smoother variation of the J's (case *ii)*). Results assuming that the coupling constants J_{11} and J_{12} are affected to the same extent (case *iii)*), are shown by the dotted line.

For the three cases considered, T_C for the monolayer is smaller than T_{CB}, and model *i)* agrees better with the experimental value. In the two–layer film, as a consequence of the larger coordination number, $T_C > T_{CB}$. A common characteristic of the three models, is that T_C reaches a maximum, higher than T_{CB}, then it decreases as a function of thickness and goes to the surface Curie temperature of the semi–infinite system. This behavior is not in accordance with the experimental findings [11]. Other fact that our model can not reproduce is that $T_C < T_{CB}$ in the 80 Å film. To get that kind of behavior it may be necessary to assume surface perturbations over very large distances. This problem has been studied within a more accurate method [16] and by assuming anisotropic interactions. The improvements on the model do not remove the disagreement between theory and experiment.

We calculated also the dependence of the magnetization on the film thickness and the temperature. In Figs. 4 and 5 we show two examples that illustrate the change of the magnetization for Fe and Gd thin films as a function of the thickness and of temperature. In these figures we only plot the magnetization in half of the film (there is a mirror symmetry in the film). The interaction parameters in Fig. 4 and Fig. 5 are the same as those used in the results presented in Fig. 2 and Fig. 3, respectively. In Fig. 4(a) we present the dependence of the magnetization in a Fe film made up of three layers. The critical temperature is $T_C = 0.63 T_{CB}$. As the film gets thicker, T_C increases, and in the case of a ten layer film of Fe it is $0.95 T_{CB}$. This result agrees with the experimental observation which shows that T_C is almost one for films thicker than five layers [13]. From the comparison of Fig. 4(a) and Fig. 4(b) one can see how the T_C increases with thickness.

In Fig. 5(a) we show the temperature dependence of the magnetization for a Gd film with three layers. The critical temperature in this case is $T_C = 1.14 T_{CB}$. We present only the magnetization at the surface; in the central layer, the magnetization is almost the same. As the film gets thicker, T_C decreases and takes the value of $1.075 T_{CB}$ in the case of a ten layer film. Our results do not agree with the experimental ones. It has

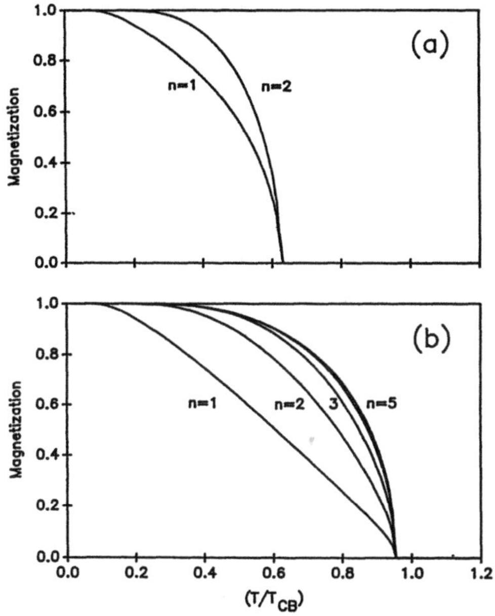

Fig. 4 *Temperature dependence of the magnetization in Fe thin films. Figures (a) and (b) correspond to a three and a ten layer film, respectively. The films have a mirror symmetry.*

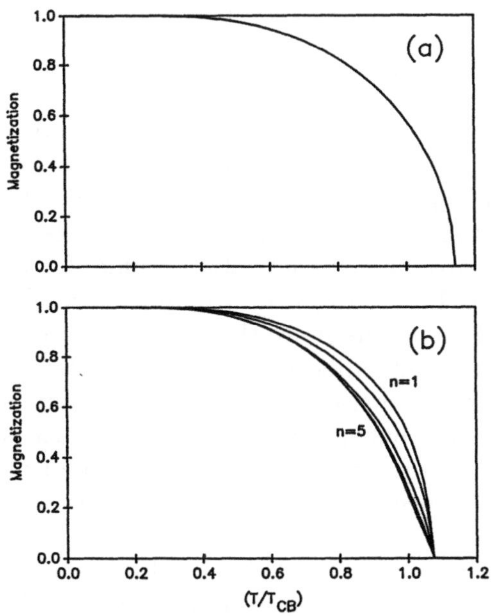

Fig. 5 *Temperature dependence of the magnetization in Gd thin films. Figures (a) and (b) correspond to a three and a ten layer film, respectively. The films have mirror symmetry.*

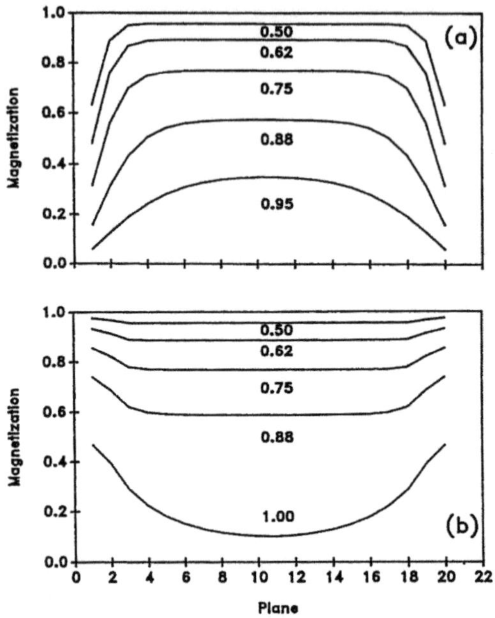

Fig. 6 *Magnetization profile across Fe (a) and Gd (b) films for various temperatures (normalized to T_{CB}). The numbers indicate the various T/T_{CB} values.*

been reported [11], that the Curie temperature of a 80 Åfilm is 5 degrees smaller than the bulk T_C. In our model we obtain the opposite behavior.

For a more complete study of the magnetic properties, we investigated the dependence of the magnetization as a function of the layer position across the film. We calculated the magnetization profiles for 20 layer films. The results are shown in Fig. 6. Fig. 6(a) contains the results for a Fe film. The interaction parameters used are the same as those in Fig. 2. One clearly sees that the magnetization is larger in the internal planes than at the surfaces. As the temperature increase, in materials like this the magnetization decreases, but always the magnetization is larger in the internal planes than at the surface. However, as the system get closer to T_C, both the surface and the inside planes become paramagnetic at the same temperature (ordinary transition [6]).

In Fig. 6(b), the magnetization profile for a Gd thin film is shown. The interactions parameters are the same as those used in Fig. 3. In this case the magnetic order is larger at the surface than in the internal planes. As the temperature increase, the magnetic order decrease but the magnetization in the internal planes is much smaller that at the surface.

In macroscopic systems the bulk magnetization goes to zero (extraordinary magnetic transition [6]).

In conclusion, by using a very simple Ising film with two free surfaces, we can obtain analytic expressions for the thickness dependence of the Curie temeperature. This allows us to estimate the magnetic interactions in these low dimensional systems. We can reproduce fairly well the recent experimental results of Fe on Au and Ag. In contrast, our model fails to explain the experimental findings of Gd on W, where interactions of larger range may be important. The method allows also to calculate the temperature dependence of the magnetization and the magnetization profiles across the films.

Acknowledgements. This work was partially supported by Dirección General de Investigación Científica y Superación Académica de la Secretaría de Educación Pública through Grant C89–08–0081.

REFERENCES

[1] C. Rau and S. Eichner, in *Nuclear Methods in Materials Research*, edited by K. Bethege, H. Bauman, H. Hex, and F. Rauch (Vieweg, Braunschweig, 1980), p.354.
[2] D. Weller, S.F. Alvarado, W. Gudat, K. Schroder, and M. Canpagna, *Phys. Rev. Lett.* **54**, 1555 (1985).
[3] C. Rau and S. Eichner, *Phys. Rev. Lett.* **47**, 939 (1981).
[4] C. Rau, C. Jin, and M. Robert, *J. Appl. Phys.* **63**, 3667 (1988).
[5] D.L. Mills, *Phys. Rev. B* **3**, 3887 (1971).
[6] T.W. Burkhart, and E. Eisenriegler, *Phys. Rev. B* **16**, 3213 (1977).
[7] K. Binder and D.P. Landau *Phys. Rev. Lett.* **52**, 318 (1984).
[8] E.F. Sarmento and C. Tsallis, *J. Phys. C* **18**, 2777 (1985).
[9] F. Aguilera–Granja and J. L. Morán–López, *Phys. Rev. B* **31**, 7146 (1985).
[10] J.M. Sanchez and J.L. Morán–López, in *Magnetic Properties of Low Dimensional Systems*, edited by L.M. Falicov and J.L. Morán–López, Springer Proceedings in Physics, Vol. 14 (Springer Verlag, Berlin, 1986).
[11] M. Farle and K. Baberschke, *Phys. Rev. Lett.* **58**, 511 (1987).
[12] W. Dürr, M. Taborelli, O. Paul, R. Germar, W. Gudat, D. Pescia, and M. Landot, *Phys. Rev. Lett.* **62**, 206 (1989).
[13] M. Stampanoni, A. Vaterlaus, M. Aeschlimann, and F Meier, *Phys. Rev. Lett.* **59**, 2483 (1987).
[14] J. Araya–Pochet, C.A Ballentine, and J.L. Erskine, *Phys. Rev. B* **38**, 7846 (1988).

[15] F. Aguilera–Granja and J. L. Morán–López, *Solid State Commun.* **74**, 155 (1990).
[16] J.M. Sanchez and J.L. Morán–López, in *Magnetic Properties of Low Dimensional Systems II: New Developments*, edited by L.M. Falicov, F. Mejía–Lira and J.L. Morán–López, Springer Proceedings in Physics, Vol. 50, (Springer Verlag, Berlin, 1990).

Magnetotunneling Current Through Semiconductor Microstructures

G. Platero[1] *and C. Tejedor*[2]

[1]Instituto de Ciencia de Materiales (CSIC)
[2]Departamento de Fisica de la Materia Condensada, Universidad Autonoma, Cantoblanco, E-28049 Madrid, Spain

1. INTRODUCTION

Tunneling is a quantum mechanical effect which can be analyzed both by transport and spectroscopic experiments.The developement of growth techniques of semiconductor microstructures allowing the design of the required potential profiles has allowed the study of this quantum process.

The aim of this work is the analysis of the effect of a magnetic field on the electronic tunneling current through semiconductor single and double barriers. We have considered a magnetic field applied in the plane of the barriers.In this configuration, the magnetic and barrier potentials are superimposed in the growth direction and the electronic spectrum consists of dispersive bands rather than degenerate bulk Landau levels, whose associated eigenstates are localized in the growth direction.As an external bias is applied, electronic current flux takes place and we will show below that in this configuration, there is a discrete set of tunneling channels available to produce the current [1,2].

We have calculated the tunneling current in the scheme of the Transfer Hamiltonian techniques[3]. We start in section 2 describing briefly the theoretical formalism [1,3,4] .In section 3 the main features of the current density as a function of B and of the external bias V for the case of a single barrier are discussed. Two main features have been observed : negative differential resistance regions in the I/V characteristic curve with non resonant origin and two superimposed periodicities for the current as a function of the inverse of the magnetic field. We will explain these quantum features in terms of the discret character of the tunneling channels.In section 4 we will discuss the resonant magnetotunneling through a double barrier system. Both, coherent and sequential contributions to the current are analyzed. In order to describe the coherent process a first-order perturbation approach is not able to account for the virtual transitions with the well states.For this purpose we have extended the Transfer Hamiltonian method up to infinite order in perturbation theory (Generalized Transfer Hamiltonian)[4]. The sequential tunneling calculated includes possible scattering processes in the well and is obtained invoking current conservation through the whole system. The characteristic curve I/V is qualitatively very different for coherent and sequential magnetotunneling and their relative intensity can be controlled by changing the external field and sample characteristics.

2. THEORETICAL METHOD

The Schrödinger equation for a barrier potential V(z) separating two semiinfinite crystals in the presence of a transverse magnetic field is given , in the effective mass approximation by

$$[-\frac{\hbar^2}{2m^{\times}m_0}\frac{\partial^2}{\partial z^2}+\frac{e^2B^2}{2m^{\times}m_0}(z+z_0)^2+V(z)-E_n(k_y)]\phi(z)=0 \qquad [1]$$

where the Landau gauge is considered: $\vec{A}=(0,-Bz,0)$,
$m^{\times}$ and m_0 are the effective and free electron mass respectively and z_0 is the magnetic orbit

Springer Proceedings in Physics, Volume 62
Surface Science Eds.: F.A. Ponce and M. Cardona

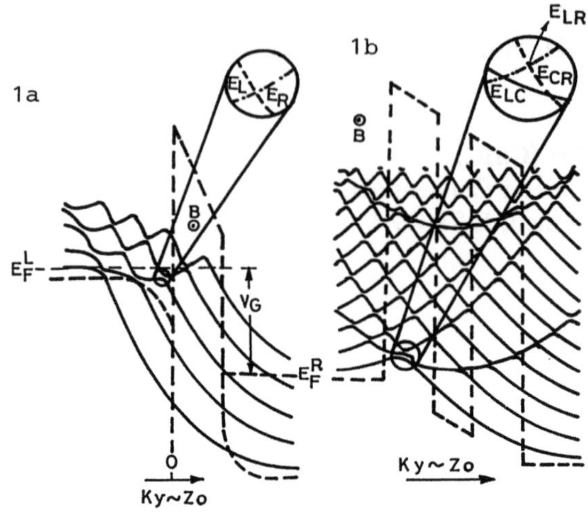

Fig. 1: Dispersion relation of the magnetic levels of simple (1-a) and double (1-b) barriers with B parallel to the interfaces.

center:

$$z_0 = \frac{\hbar k_y}{eB} = l_m^2 k_y \quad . \qquad [2]$$

The eigenstates have the form $e^{ik_x x + ik_y y}\phi(z)$ and the energy of the magnetic level $E_n(k_y)$ is related to the total energy of the eigenstate by means of

$$E = \frac{\hbar^2 k_x^2}{2m^{\times} m_0} + E_n(k_y) \quad . \qquad [3]$$

For a given value of k_y (or z_0) the one-dimensional equation [1] can be numerically solved by a finite elements method. Figure 1 shows the band structures of single and double barriers with an applied bias and in the presence of a magnetic field. The main result is that states far from anticrossings have wave functions localyzed in only one side of the barriers, while just states close to the anticrossings have wave functions with significant weigth in more than one region. Therefore, the latter are the channels allowing the carrier flow between different parts of the system. This is the origin of the tunneling current theoretically analyzed below as a phenomenon out of the equilibrium.

The Generalized transfer Hamiltonian method (GTHM)[4] is an extension of the usual transfer Hamiltonian formalism[3] for including the possibility of resonant processes. The total Hamiltonian H is separated in left and right terms by means of two auxiliary Hamiltonians $H_L \equiv H$ in the left side and $H_R \equiv H$ To recover H, the required perturbations written as:

$$H \equiv H_L + V_L e^{\eta t} = H_R + V_R e^{\eta t} \qquad [4]$$

are switched on adiabatically ($\eta \rightarrow 0$). One starts with a time dependent wave function for the total system

$$|\Psi(t)> = f(t) e^{-i\omega_L t} | L > + \sum_R a_R(t) e^{-i\omega_R t} | R > \qquad [5]$$

where $\omega_L = E_L/\hbar$ and $\omega_R = E_R/\hbar$. At the initial time, $|\Psi(-\infty)>$ must describe a particle on the left side. This is fulfilled by imposing

$$f(-\infty)=1;\ a_R(-\infty)=0 \quad . \qquad [6]$$

After a while the particle in a precise left state $|L>$ can in principle evolve to any state $|R>$ to the right. The time dependent coefficients are determined from the Schrödinger equation by an expansion in a perturbation series. The transition probability from left to right per unit time is given by

$$P_{LR} = \lim_{\eta\to 0} \frac{d}{dt} \,|a_R(t)|^2 = \frac{2\pi}{\hbar}\,\delta(E_L - E_R)\,|t_{LR}|^2 \ , \qquad [7]$$

where the transmission t_{LR} has the next expression for discrete problems[4]:

$$t^d_{LR} = < L\,|\,V_L + V_L(E_L - H + \mathbb{P}_R V_R)^{-1}\mathbb{P} V_R\,|\,R> \quad . \qquad [8]$$

$\mathbb{P}_R$ is the projection operator on the state $|R>$ and $\mathbb{P} = 1 - \mathbb{P}_R$. As we see, from [7] the transmission probability is only different from zero for $E_L = E_R$ (see fig. 1) where $E_L(E_R)$ are the eigenvalues of the auxiliar hamiltonian $H_L(H_R)$. Furthermore, as the potentials in eq. 1 depend only on z, k_x and k_y are good quantum numbers and have to be conserved in the tunneling process. It means that only the crossings $E_L = E_R$ (anticrossings of the total hamiltonian H spectra) are the possible tunneling channels. In fact looking at the wavefunctions corresponding to an energy and k_y of the spectrum, only those associated with the anticrossings have weight on both sides of the barriers, and therefore can contribute to the tunneling current. Once t_{LR} is known the current is evaluated adding the transmission probabilities between occupied states to the right and empty states to the left.

3. MAGNETOTUNNELING THROUGH A SINGLE BARRIER

In the case of a single barrier the transmission probability is given by the first term of [8], which is corresponding to first order in perturbation theory. Fig. 2 shows the current density as a function of the magnetic field for an AlGaAs 100 Å barrier between two semiinfinite GaAs media doped with n= 10^{18} cm^{-3} and an external bias of 0.2 V. There are two main features observed in the tunneling current as a function of the inverse of magnetic field and as a function of the external bias. In the first case (fig. 2) the current presents two different quantum oscillations which are periodic with the inverse of the field. The first type of oscillations comes from the bulk Fermi level oscillations as a function of magnetic field (SdH) , the second ones are specific from the tunneling process and comes from the fact that the number of available channels changes with field : for small fields there are many channels and to lose or gain one does not practically affect the total current whose oscillatory behavior is mainly due to the bulk Fermi level. However, for higher fields very few channels contribute to the current and to lose a channel is critical for the total current. This behaviour has been experimentally observed[5]. The second interesting feature is the fact that the characteristic curve I/V for a single barrier presents regions of NDR. This interesting fact can also be explained in terms of tunneling channels. For a fixed magnetic field , as the bias increases, a channel crosses the left Fermi energy and current of electrons takes place. For higher bias the channels move down in energy more rapidly than the lowering of the barrier and the effective barrier seen by the electrons is higher giving a NDR region with no resonant origen. Once the bias increases enough to allow the entrance of the next channel, the current increases sharply again. This effect is well observed in samples with a non degenerated semiconductor to the left because in this case the next crossing crosses the Fermi level for higher bias and its contribution to the current is not superimposed to that corresponding to the preceeding one[6]. In fig. 3 the I/V characteristic curve is shown for the sample described above. In this case the NDR regions are not appreciated because the emitter is highly doped and there is a superposition of the current contribution coming from several channels.

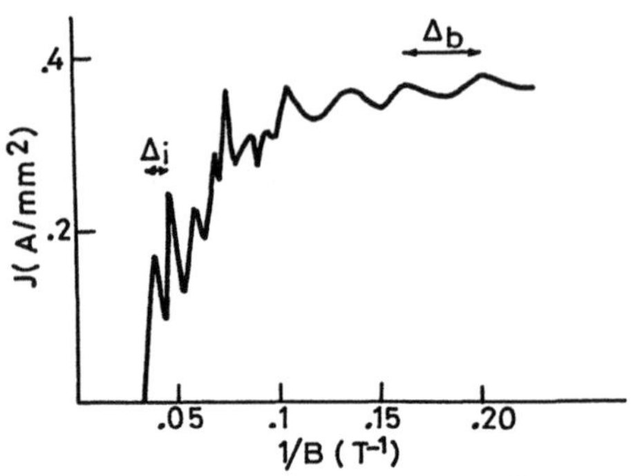

Fig.2: Current density (A/mm^2) through a single barrier (B⊥J) as a function of 1/B (T^{-1}). Δb (Δi) are the bulk (interfacial) period of the oscillations.

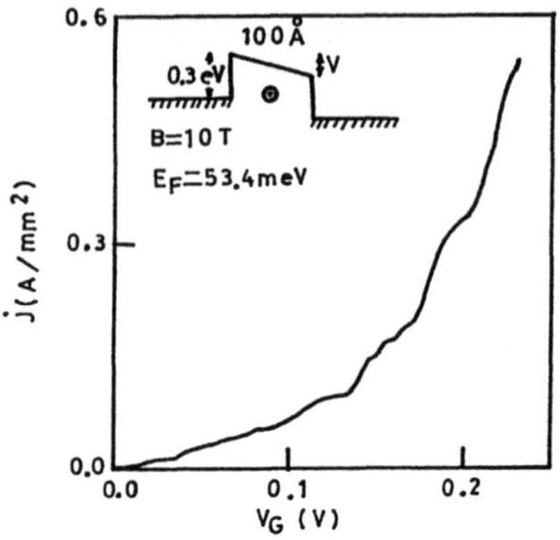

Fig.3: Current density (A/mm^2) through a single barrier as a function of the external bias (v).

4. RESONANT MAGNETOTUNNELING THROUGH A DOUBLE BARRIER

In order to analyze the coherent contribution to the current a first order perturbation term (which allows an accurate description of the transmission probability coefficient for tunneling through a simple barrier) is not able to describe the virtual transitions with the well states . The transmission probability per unit of time from left to right through a double barrier system is given by [7] where t_{LR} is defined in [8] , in terms of the Green function of the system. We see from [7] that the available coherent tunneling channels are those corresponding to the crossings E_{LR} as shown in figure 1-b. The transmission coming from the virtual processes involved in the coherent tunneling is represented by energy differences (E_L-E_C), where E_C is the energy of a state localized in the well, appearing in the denominators of the Green's function[4]. These denominators give very important contributions to the transmission so that when a state localized in the well with energy E_C is close to E_L a peak appears in the current. From figure 1-b it is very simple to visualize that such a fact is going to occur several times when varying the bias. This gives a structure of narrow peaks of j as a function of the bias.

The sequential process is a three step process : the electrons cross the first barrier, then spend some time in the well where scattering processes can take place,and finaly cross the second barrier. There is a pair of tunneling channels associated with each sequential process: lc (left-center) and cr (center-right) which are not at the same energy nor at the same momentum k_y (see fig. 1-b). In order to analyze the sequential contribution to the current we consider a macroscopic model which includes all the possible scattering processes in the well , which one would expect to be important when the well width is large, i. e. when the barriers are well separated.

In order to account also for these processes, a microscopic model for possible elastic or inelastic scattering processes is outside the scope of this work. However it is possible to calculate all the contributions to the sequential tunneling current in a general way by imposing electronic current conservation [7]. The double barrier system can be seen as two resistors in series, so that the current crossing the first barrier should be the same as the current crossing the second one [7]. It means that there is charge accumulated within the well in equilibrium, and the Fermi level in the well can be calculated from the condition described above : $J_L = J_R$ where J_L and J_R are the current through the first and second barriers,respectively.

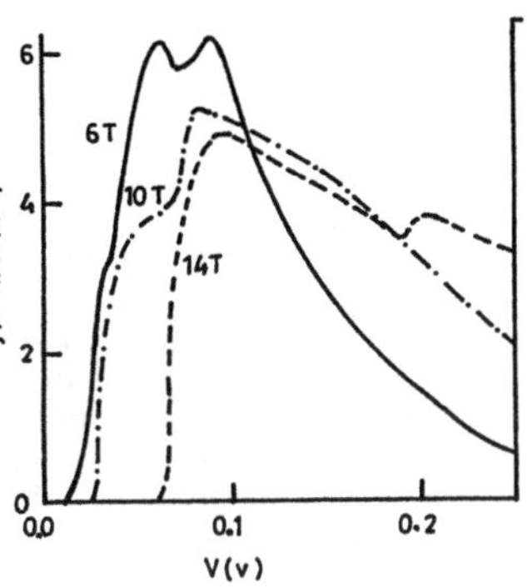

Fig.4

Sequential tunneling current for a 100A AlGaAs DB and a 70A GaAs well as a function of the bias for B=14,10 and 6T.

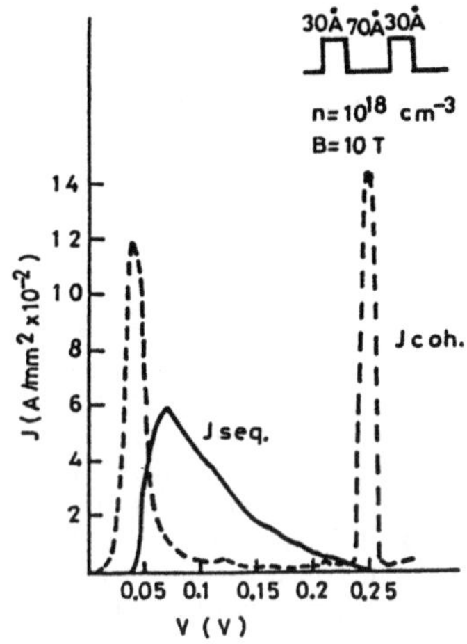

Fig.5

Coherent (dashes line) and sequential (solid line) tunneling current for a 30 A AlGaAs DB and a 70 A GaAs well and a B = 10 T, as a function of the bias.

A double barrier where both coherent and sequential processes are possible, behaves like two channels in parallel. The resistance of each channel is proportional to the inverse of the corresponding transmission probabilities T_C or T_S respectively [7]. This gives for the total transmission probability: $T = T_C + T_S$. From the total transmission coefficient the total current can be obtained. Fig. 4 shows the total sequential contribution to the magnetotunneling current for a 100 Å AlGaAs double barrier separated by a 70 Å GaAs quantum well. For this sample, composed of thick barriers, the coherent contribution to the current is less than two orders of magnitude smaller than the sequential one which determines in this case the current features as a function of the external bias. Three different cases are calculated corresponding to three magnetic fields : B = 6T, B = 10T and B = 14T. The current behaviour is similar to that observed experimentally : there is a threshold bias for the current which moves to higher bias as the magnetic field increases, the current peak broadens and also decreases with the field. As the barriers become thinner, the coherent tunneling current increases faster than the sequential one and is the first one which controls the total current. As it has been described before, narrow peaks appear in the characteristic curve for the coherent contribution to the current. An example of this behaviour for narrow barriers is shown in fig. 5 where the I/V characteristic curve for a 30 Å AlGaAs double barrier separated by a 70 Å GaAs quantum well is represented, the Al barrier concentration is 40%. Here, the sequential contribution to the current is very small compared with the coherent one. The strong increase of the coherent tunneling contribution for thinner barriers comes from the fact that now the wavefunctions corresponding to the well states are not localized anymore in the well only but have non-negligible weight outside of it, therefore the number of available coherent tunneling channels increases, giving a high contribution to the current. The experimental work [8] [12] has been done for thick barrier samples where the sequential tunneling predominates. Unfortunately there is no experimental information to compare with for thin barriers in this configuration of the field.

5. SUMMARY

The Transfer Hamiltonian techniques are suitable to describe tunneling processes for systems with localized eigenstates in the current direction as in the case of transverse magnetotunneling ($B \perp J$) . For this configuration we conclude that there is a discrete set of tunneling channels which correspond to the anticrossings in the dispersion relation . We have analyzed the magnetocurrent through a single barrier and some interesting quantum features have been observed : The current presents two different quantum oscillations as a function of the magnetic field which correspond to two periodicities of the magnetocurrent as a function of 1/B. One kind of these oscillations is a bulk effect and comes from the bulk Fermi level oscillations with the field.The other one is explained in terms of the variation of available channels with B.Other interesting result is the appearence of NDR in the characteristic curve I/V with non resonant origin.The magnetotunneling current through a double barrier has been calculated by means of the GTH formalism and the coherent and sequential contributions are analyzed and compared each other for different samples and fields. We have calculated the sequential process imposing current conservation through the whole system. The relative intensity of coherent and sequential processes can be controlled varying the field intensity and the sample characteristics.

We are indebted to CASA for the CRAY computing facilities. This work has been supported in part by the Comision Interministerial de Ciencia y Tecnologia of Spain under contract MAT88-0116-C02-01/02

REFERENCES

1. L.Brey, G.Platero and C.Tejedor, Phys. Rev. B, 38, 9649 (1988)

2. G.Platero, L.Brey and C.Tejedor, Phys. Rev. B,40,8548 (1989)

3. C.B.Duke, "Tunneling in solids" Solid State Physics. Supplement 10. (Academic Press. New York, 1969)

4. L.Brey, G.Platero and C.Tejedor, Phys. Rev. B, 38, 10507 (1988)

5. T.W.Hickmott,Solid State Commun., 63, 371 (1987)

6. P.Schulz and C.Tejedor,Phys. Rev. B, 39,11187 (1989)

7. M.Büttiker, IBM J. Res. Dev. 32, 63 (1988)

8. M.L.Leadbeater, L.Eaves, P.E.Simmonds, G.A.Toombs, F.W.Sheard, P.A.Claxton, G.Hill and M.A.Pate, Solid State Electronics, 31, 707 (1988)

9. M.L.Leadbeater, E.S.Alves, L.Eaves, M.Henini, O.H.Hughes, A.Celeste, J.C.Portal, G.Hill and M.A.Pate, Phys. Rev. B, 39, 3438 (1989)

10. S.Ben Amor, K.P.Martin, J.J.L.Rascol, R.J.Higgins, A.Torabi, H.M.Harris and C.J.Summers, Appl. Phys. Lett. 53, 2540 (1988)

11. P.Gueret, C.Rossel, E.Marclay and H.Meier, J.Appl.Phys,66 ,278 (1989)

12. A.Zaslavsky,Y.P.Li,D.C.Tsui,M.Santos and M.Shayegan, Phys. Rev. B, 42, 1374 (1990)

Preparation and Properties of High-T_c Superconducting Bi(Pb)-Sr-Ca-Cu-O Thick Films by a Melting-Quenching-Annealing Method

M.E. Gómez, L.F. Castro, G. Bolaños, O. Morán, and P. Prieto

Departamento de Física, Universidad del Valle,
Apartado Aéreo 25360, Cali., Colombia

We have prepared superconducting Bi(Pb)-Sr-Ca-Cu-O thick films (30-100 µm) on MgO (100) substrates using a rapid melting-quenching-annealing method. A layer of $Bi_{1.6}Pb_{.4}Sr_2Ca_2$- $Cu_{3.6}O_x$ mixture on MgO (100) substrates was melted at 1050 °C for typically 2 min. The samples were then directly cooled to room temperature. After an annealing at 860 °C for about 10h in air the films exhibited superconductivity with T_c(zero)=108 K. The critical current densities at 77 K and zero magnetic field were about 10^3 A/cm^2. Morphology and composition of the films were investigated by SEM, EDX and X-ray diffraction measurements.

1. INTRODUCTION

The superconducting Bi-Sr-Ca-Cu-O system has been of great interest since it was first reported that it was possible to prepare oriented materials /1/, which were much less sensitive to atmospheric conditions than $YBa_2Cu_3O_x$ type materials. For practical applications the preparation of oriented superconducting films is important because the transport properties of the H-T_c superconductors are highly anisotropic and the critical currents in oriented films are much higher than those in unaligned materials. Highly oriented Bi-Sr-Ca-Cu- O thick films can be prepared by a Melting-Quenching-Annealing (MQA) method on MgO substrates /2/. The MQA technique is based on the controlled crystallization of the Bi-Sr-Ca-Cu-O precursors, which have been melted at elevated temperatures. Quenching and annealing steps are used to control the crystallization process. Melting, specially in thick films, seems to allow the formation of denser materials than bulk samples prepared by solid state reaction and sinterization.

The Bi-system has a number of superconducting phases with different stoichiometries and crystal structures. The phases can be represented by an homologous series given $Bi_2Sr_2Ca_{n-1}Cu_nO_{2n+4}$. The known superconducting phases in this system are: i) a single perovskite-like layered phase consisting of one CuO layer sandwiched between Bi-O layers with an ideal chemical composition of $Bi_2Sr_2CuO_6$ (abbreviated as 2201) with a T_c of about 10 K /3/. ii) a double perovskite layered phase with an ideal chemical composition of $Bi_2Sr_2CaCu_2O_8$ (abbreviated 2212) with a T_c of 85 K /4/ and iii) a triple CuO layered structure with a chemical composition of $Bi_2Sr_2Ca_2Cu_3O_{10}$ (abbreviated 2223) with a T_c near 110 K /5/. Only the last two phases are of possible technological interest. Despite extensive efforts 2223-phase has not been prepared as a single phase, one reason is that the 2223 phase has a defective structure with numerous intergrowths of the lower T_c phases (i.e. 2201 and 2212) in addition to other insulating phases such CuO, $(Ca,Sr)_2Cu_2O_3$. Substitution of Bi by Pb and Sb seems to facilitate the preparation of material with a high T_c up to 105 K /6/, but the formation of 2223 phase is a very slow process even then.

In this communication we report the production and characterization of the superconducting Bi(Pb)-Sr-Ca-Cu-O thick films prepared on MgO substrates with T_c above 100K and critical current densities of about 10^3 A/cm^2 at 77 K using a MQA process.

2. EXPERIMENTAL PROCEDURE

Bi(Pb)-Sr-Ca-Cu-O thick films were prepared on MgO single crystals with (001) orientation using the MQA process. Bi_2O_3, PbO, $SrCO_3$, $CaCO_3$ and CuO in powder were mixed in mole fractions which are equivalent to $Bi_{1.6}Pb_{.4}Sr_2Ca_2Cu_{3.6}O_{10-x}$, then the mixture was placed as a thin layer on top of MgO substrate and heated at 1050 °C for typically 2 minutes for a completed melting. The melt was then quenched to room temperature at a very high cooling rate by putting the coated substrate on a stainless-steel plate maintained at room temperature. Subsequently the films were heat treated at a temperature of 860 °C in air for typically 10 hours for recrystallization. This simple procedure gave dense, uniform and well crystallized films with thicknesses between 30 and 150 µm, measured by examining the cross section using light microscopy. The samples were characterized with X-ray diffraction, optical microscopy, scanning electron microscopy (SEM) and energy dispersive analysys (EDX), resistivity and critical current measurements. Resistivity measurements with different current densities were made using the four probe technique. The transport critical currents of the films were measured using the four probe method across a 2 mm long and 100 µm wide bridge obtained by an etching process using dilute HCl. The 1 µV/cm criterion was used for the determination of J_c at 77 K. No external magnetic field was applied.

3. RESULTS AND DISCUSSION

A cross sectioned view of the interface between the substrate and film is shown in the SEM micrograph in Fig. 1. Platelet shaped grains with grain size above 10 µm are observed in

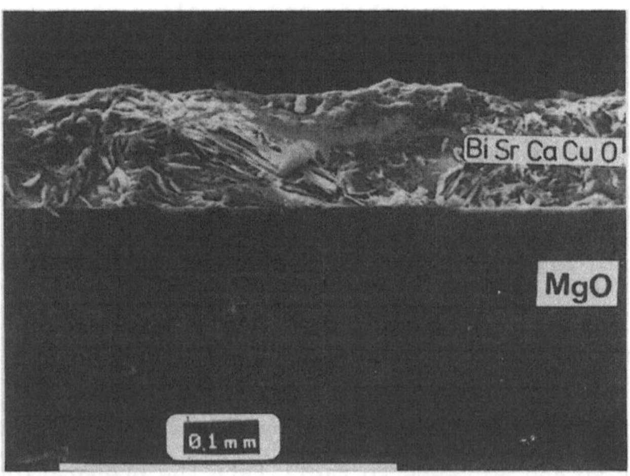

Fig. 1 Cross section SEM micrograph of a textured thick film on a MgO substrate.

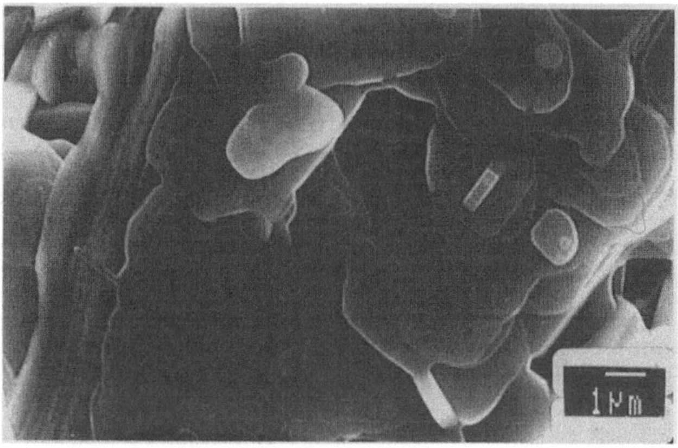

Fig. 2 Top view SEM micrograph Bi(Pb)-Sr-Ca-Cu-O thick film on MgO.

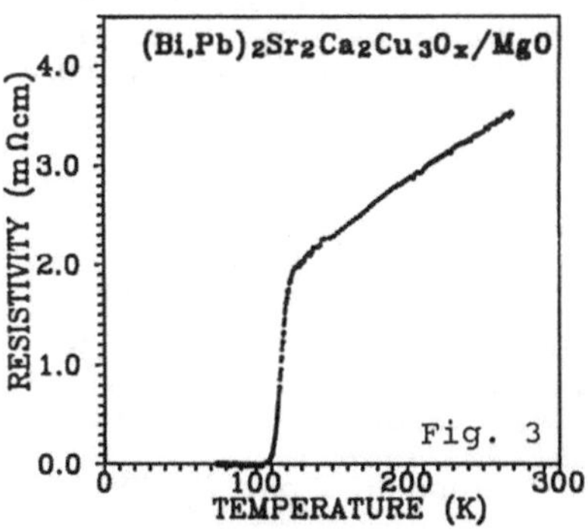

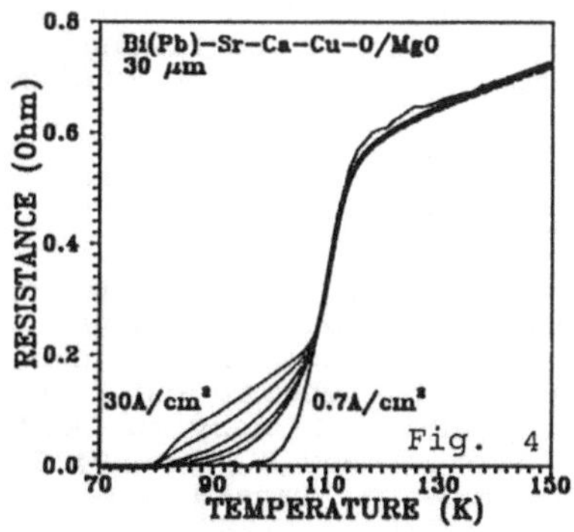

Fig. 3 Temperature dependence of the electrical resistivity for a $Bi_{1.6}Pb_{.4}Sr_2Ca_2Cu_{3.6}O_x$ with a thickness of 90 μm.

Fig. 4 Resistivity vs temperature curves obtained at various current densities for a 30 μm thick film prepared by the MQA process.

Fig.2. The morphology suggests that the films have textured structure as it was also found by X-ray diffraction measurements. The average composition based on EDX microanalysis corresponded approximately to the 2223 and 2212 phases. CuO and Ca_2PbO_x were also detected. A lower concentration of Pb after the crystalization process was measured, indicating that Pb acted only as a flux favoring the formation of the high T_c phase. Fig. 3 shows the temperature dependence of the resistivity for a film with a thickness of about 90 μm. A current density of about 1 A/cm² was used for this measurement. With decreasing temperature the resistivity decreases linearly above 120 K. The film shows a T_c near 106 K. The critical temperature $T_c(R=0)$ is strongly affected by the current used during its determination. Fig. 4 shows the resistivity vs temperature curve near the normal- superconducting transition in a film with a thickness of about 30 μm. As the current density increases from .7 to 30 A/cm² the zero resistance temperature shifts from 100 to 80 K.

The broadening of the transition has been observed in Pb-doped Bi-Sr-Ca-Cu-O bulk samples by applicaton of a magnetic field /7/. This behavior is probably due to the smaller overall volume fraction of the superconducting phase in this sample. The critical current density at 77 K, determined using the average thickness of the film, was about 800 A/cm^2. This current density is higher than the J_c reported for sintered bulk samples. An impediment for higher J_c values are the defects and microcracks that are always present in these samples.

In summary, superconducting thick films with T_c above 100 K were produced using a MQA process. The T_c for this film is strongly dependent on the measuring current density. Such marked dependence of T_c on J suggests that the high T_c phase is weakly linked in forming complex Josephson networks in these samples. A systematic study of the superconducting properties with the thickness is in progress.

Acknowledgment

The authors wish to thank W. Sybertz from IFF-Research Centre Jülich, FRG, for perfoming the SEM and EDX microprobe analysis. Financial support from COLCIENCIAS is gratefully acknowledgment.

REFERENCES

1 Y. Akamatzu, M. Tatsumisago, N. Tohge, S. Tsuboi and M. Minami; Jpn J. Appl. Lett. 27 L1696 (1988).
2 P. Prieto, G. Zorn, R. Arons, S. Thierfeldt. M. E. Gómez, B. Kabius,. W. Sybertz and K. Urban; Sol.St.Comm. 69 235 (1989)
3 C. Michel;, M. Hervieu, M.M. Borel, A. Granden, F. Deslardes, J. Provost and B. Raveau; Z. Phys. B68 421 (1987).
4 H. Maeda, Y. Tanaka, M. Fukutami and T. Asano; Jpn. J. Appl. Phys. 27 L209 (1988).
5 J. M. Tarascon, Y. Le Page, P. Barboux, B.G. Bagly, L.H. Green, W.R. Mehinnon, G.W. Hull, M. Giroud and P.M. Hwang; Phys. Rev. B37 9382 (1988).
6 B. Kabius, M. E. Gómez, P. Prieto, S. Thierfeldt and R. R. Arons; Physica C 162-164 635 (1989).
7 H. Kamahura, K. Togano, M. Vehara, H. Maeda, K. Takahashi and E. Yanagisawa; Jpn. J. Appl. Phys. 27 L1514 (1988).

Theoretical Analysis of Surface States in Ta(100)

R. Baquero[1], *R. De Coss*[2], *and A. Noguera*[2]

[1]Departamento de Física, Centro de Investigación y de Estudios Avanzados del IPN, Apdo. Postal 14-740, 07000 México, D.F., Mexico
[2]Instituto de Física, Universidad Autónoma de Puebla, Apdo. Postal J-48, Puebla, Pue., Mexico

Abstract. This is the first calculation on the influence of the very recently found first interatomic layer contraction on the electronic structure of the Ta(100) surface.

1. Introduction

Ta is a bcc transition metal and its (001) surface has been studied previously experimentally by Bartynski and Gustafsson using inverse photoemission[1] and theoretically with a slab calculation by Krakauer[2]. This surface was assumed to be ideal. A reasonable agreement between theory and experiment was reported although a detailed comparison showed clear differences.

More recently, Bartynski et al. succeeded in determining the geometry of the clean (001) tantalum surface more accurately by photo- electron diffraction spectroscopy[3]. The main result of this study was to establish that the fist interlayer distance in the Ta(001) surface was shorter by about 10-15%.

Very recently, Jensen, Bartynski and Weinert[4], using Auger spectroscopy, could establish that the just mentioned contraction was 13.5%. We study now the effect of this contraction on the surface states of the Ta(001) surface.

2. Results

In Fig. 1, we present the lenses diagrams for the Γ-M region of the 2D-Brillouin zone. We have used an s-p-d orthogonal basis and the tight-binding parameters of Papaconstantopulos[5]. The Green function for the surface has been calculated with the Surface Green Function Matching method[6](SGFM) which allow the use of the bulk tight-binding parameters.

The black squares in Fig. 1, are the inverse photoemission experimental points of reference 1. The dots are the surface states in the bulk gaps. The lines show the

Springer Proceedings in Physics, Volume 62
Surface Science Eds.: F.A. Ponce and M. Cardona

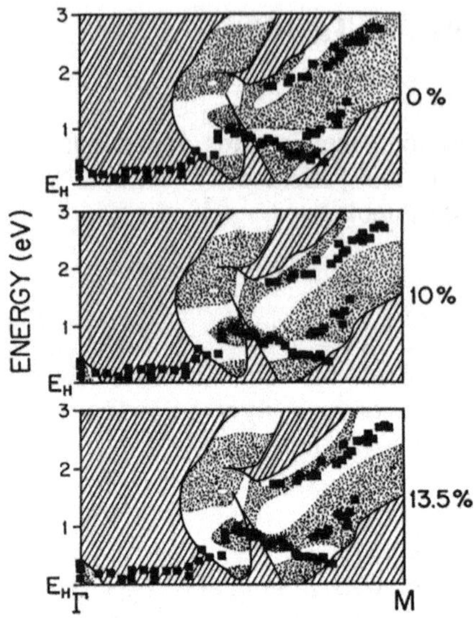

Fig. 1. The Γ–M interval. Theory compared to inverse photoemission experiments. The calculated surface states in the gaps appear as dots. The black squares are the experimental points of ref. 1. The percentage contraction of the first interlayer distance is indicated. E_H is the Fermi level.

zones where the bulk states lie. The top figure is the ideal Ta(001) surface with no contraction of the first interlayer distance. Notice the small gap around Γ just above the Fermi level. The surface states detected experimentally do not appear in this calculation. The band of resonances that starts near Γ ends in a region where there are no surface states. A careful examination of the states further to the M point shows clear differences between theory and experiment also in this region.

Let us look now at the influence of a 10% contraction. Notice that the states in the gap move to lower energies. The most important fact might be that surface states appear in the Γ point near the Fermi level as detected experimentally.

When the 13.5% contraction is taken into account there is also a better agreement in the higher part of the spectrum near the M point. The agreement is not yet total when one looks at it in detail, but the dispersion curves of the surface states seem to agree better. This might be due also to the fact that the tight-binding parameters in the contracted case have been calculated using the Harrison Scaling law which is thought to be valid only up to about 10% deformation. We have here a 13.5% one.

3. Conclusions

In conclusion, we can say that the surface states that have been experimentally detected at the Γ point just above the Fermi level seem to be directly related to the contraction of the first interlayer distance in the Ta(001) clean surface. This contraction also influences the exact position in energy of the surface states in the Γ-M interval of the 2D-First Brillouin zone. These states have been related to the reconstruction of the W(001) surface. This is the first study of the consequences of this very recently found contraction of the first interatomic distance on the electronic structure of the Ta(100) surface.

References

1.R.A. Bartynski and T. Gustafsson, Phys.Rev.B **35**, 939 (1987).

2.H. Krakauer, Phys. Rev. B **30**, 6834 (1984)

3.R.A. Bartynski et al.,Phys. Rev. B **40**, 5340 (1989).

4.E. Jensen, R.A. Bartynski and M. Weinert, Phys. Rev. B **41**, 12468 (1990).

5.D.A. Papaconstantopoulos, *The Electronic Band Structure of Elementary Solids* (Plenum, New York, 1986)

6.F. García-Moliner and V. Velasco, Prog. Surf. Sci. **21**,93 (1986).

3. Conclusions

Magneto-Optical Studies of Ultrathin Fe/W(100) Films

J. Araya-Pochet[1], *C.A. Ballentine*[2], *and J.L. Erskine*[2]

[1]Escuela de Fisica, Universidad de Costa Rica, San José, Costa Rica
[2]Department of Physics, University of Texas, Austin, TX 78712, USA

Abstract. The magnetization of Fe on W(100) as a function of applied field was determined using magneto-optical techniques. A monolayer (ML) is found to be ferromagnetic with in-plane easy axis for samples grown above room temperature. Remanence is found for monolayers grown below 200 K with magnetic field applied either along the surface or perpendicular to it. Ultrathin films 2 to 6 ML exhibit a peculiar hysteresis loop.

1. Introduction

The magnetic properties of low-dimensional systems have been explored theoretically and experimentally[1]. Our understanding of the novel properties associated with thin film structures is limited by the ability to stabilize the ideal structures postulated in theoretical calculations. The growth of nearly perfect crystalline structures can be achieved, to some extent, by Molecular Beam Epitaxy, but in most cases of practical interest, defects will play an important, if not dominant role in thin film magnetic properties. For this reason, it is of importance to understand how defects can alter the magnetic properties.

It seems reasonable to assume that defects become important when they limit the range of the exchange interaction[2] which is basically measured by the correlation length ξ, a quantity that diverges at the Curie temperature. Defects can also induce surface anisotropies due to dipole-dipole interactions and to magnetostriction caused by lattice misfit[3]. It is also important to note that domain wall motion can be influenced by the presence of defects (Barkhausen effect).

2. Fe on W System

In our present study, we have stabilized Fe layers on the (100) face of clean W by electron beam evaporation. Earlier studies[4,5] have shown that a monolayer (ML) of Fe grows pseudomorphically on W(100) despite a large lattice misfit (10%). This system is thermodynamically stable up to the Fe melting point. Below 4 ML, Fe grows layer-by-layer, but annealing can induce the formation of three-dimensional islands. Evidence of a reduction of the thin film magnetic moment per atom due to a strong hybridization between the W 4d and Fe 3d states has been reported[6]. Such an effect could be compensated by a negative pressure on the Fe overlayer.

3. Experimental Details

Tungsten crystals were spark cut and mechanically polished to an accuracy of 1° and cleaned by repeated *in situ* annealing of the crystals to 1200 K in an O_2

Springer Proceedings in Physics, Volume 62
Surface Science Eds.: F.A. Ponce and M. Cardona

atmosphere followed by flashing to 2500 K to remove the surface oxide. Surface chemical composition was established by Auger spectroscopy which indicated oxygen levels below 2% and carbon levels below 4%.

The multilayer Fe films were grown at a substrate temperature of about 1000 K for the first layer; subsequent layers were grown below 200 K. Monolayer films were grown at 900, 300, and 110 K. Vacuum chamber pressure was maintained in the middle 10^{-10} Torr range during evaporation and measurements. Film thickness was determined by timed Auger analysis and by a quartz microbalance. With these techniques, an absolute thickness accuracy of 0.3 monolayer and a relative thickness accuracy of 0.1 monolayer can be achieved.

The crystalline structure was determined by a video LEED system with a coherence length of about 100$Å^{-1}$. This system permits spot profile analysis as well as beam intensity analysis as a function of electron energy. Our video LEED studies show features characteristic of a crystal having a step density corresponding to average terrace widths of 100Å and a preferred orientation along a 01 axis.

The long range magnetic order was established by *in situ* surface magneto-optic Kerr effect (SMOKE) experiments. Our system, described elsewhere[7], is capable of detecting the magnetization of samples magnetized either along the surface (longitudinal configuration) or perpendicular to it (polar configuration).

4. Results

The monolayer Fe films were found to be ferromagnetic below room temperature. Those grown above room temperature exhibited an in-plane easy axis and remanence with the applied field along either the (01) or (11) direction. The low temperature grown monolayer films yielded polar and longitudinal SMOKE signals with square hysteresis loops in both cases. The studied multilayer films (2 to 6 layers) exhibit peculiar hysteresis loops (see Figures 1 and 2) as long as the applied field was along the surface and perpendicular to the steps. Rotation of the crystal by 90° restores the normal behavior.

Square loops were also recovered after a small (about 0.25 Langmuir) oxygen dose or by annealing. We also observed that the Fe to W Auger ratios decreased by annealing suggesting island formation. Island formation can also be induced by oxygen dosage[8].

It is clear that 2 ML films exhibit a nearly zero magnetization up to a critical applied field (about 80 Oersted). A similar behavior was reported by de Waard *et al.*[9] on polycrystalline and single Fe crystals with applied field along the (100) direction. In this experiment a zero SMOKE signal (consistent with zero magnetization) was found below a critical applied field, but normal bulk behavior was determined by measuring the (bulk) magnetization using an induction method. This lag of the surface magnetization was attributed to observed "pinning" of domain walls by dislocations at the surface.

The sudden "jump" in the magnetization as a function of applied field observed in our 4 layer film (Fig. 2) resembles the observed "jump" on a Fe(110) (thick) film grown on GaAs. This peculiar behavior was attributed by Prinz[10] to a predicted first-order transition in magnetization as a function of applied field along a [111] crystallographic direction. Such a phase transition was observed along the [110] direction, in this case probably due to the presence of a uniaxial anisotropy.

In summary, we have found a monolayer of Fe on W(100) to be ferromagnetic with easy axis that depends on the method of preparation. Peculiar behavior of the hysteresis loops can be either attributed to surface morphology (defects) or

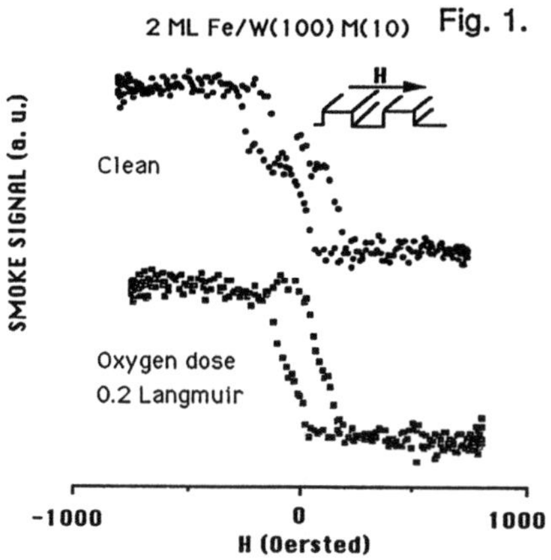

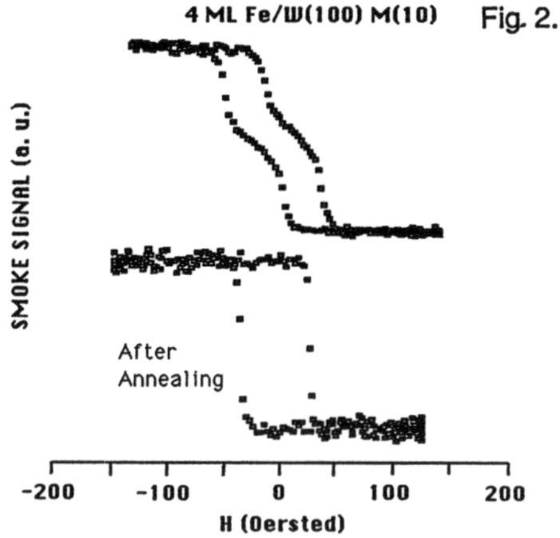

Fig. 1. Magnetization as a function of applied field for a 2 ML film. Oxygen dosage causes the peculiar hysteresis loop to revert back to a normal rectangular loop.

Figure 2. Magnetization as a function of applied field. The sudden "jump" in the magnetization is quenched by annealing. This procedure probably induces the formation of three-dimensional islands.

to a first-order phase transition induced by surface anisotropies. Further experiments on this system are being pursued at this time.

Acknowledgements. We would like to thank M. Drakaki and J. Chen for sharing some of their data. This work was supported by NSF DMR89-22359 and NSF INT90-00058.

References.

1. Magnetic Properties of Low Dimensional Systems II, (Eds.) L.M. Falicov, F. Mejíra-Lira, and J.L. Morán-López, Springer Proceedings in Physics, Vol. 50 (Springer Verlag, Berlin, 1990); Magnetic Properties of Low Dimensional Systems, (Eds.) L.M. Falicov and J.L. Morán-López, Springer Proceedings in Physics, Vol. 14 (Springer Verlag, Berlin, 1986); Magnetism in Ultrathin Films (feature issue of) *Appl. Phys. A* **49** [5,6] (1989).
2. W. Dürr, M. Tamborelli, O. Paul, R. Germer, W. Gudat, D. Pescia, and M. Landolt, *Phys. Rev. Lett.* **62**, 206 (1989).
3. C. Cahppert and P. Bruno, *J. Appl. Phys.* **64**, 5736 (1988).
4. X.L. Zhou, C. Yoon, and J.M. White, *Surf. Sci.* **203**,**53** (1988).
5. P.J. Berlowitz, J.W. He, and D.W. Goodman, private communication.
6. Soon C. Hong, A.J. Freeman, and C.L. Fu, *Phys. Rev. B* **38**, 12156 (1989).
7. C.A. Ballentine R.L. Fink, J. Araya-Pochet, and J.L. Erskine, *Appl. Phys. A* **49**, 459 (1989).
8. A.A. Chernov, private communication.
9. H. de Waard, E. Uggerhøj, and G.L. Miller, *J. Appl. Phys.* **46**, 264 (1975).
10. G.A. Prinz, *Phys. Rev. Lett.* **47**, 1761 (1981).
11. D. Mukamel, M.E. Fisher, and E. Domany, *Phys. Rev. Lett.* **37**, 565 (1976).

Part X

Applications:
Catalysis, Corrosion, Absorbates

Chemisorption Studies of Catalytic Reactions

M.H. Farías

Instituto de Física, Universidad Nacional Autónoma de México,
Laboratorio de Ensenada, Apdo. Postal 2681, 22800 Ensenada,
Baja California, Mexico

Abstract. We review the surface science point of view of catalysts used for hydrodesulphurization (HDS) processes. Surface science has given insight on the characterization and subsequent understanding of HDS catalysts. Adsorption of thiophene on single crystals is one of the most studied systems from which information like C-S bond breaking precedes C-C bond breaking, and hydrogenation of the sulphur atom takes place before the final C-S hydrogenolysis step, is obtained. Most models can explain the experimental results on the effect of the promoter on HDS catalysts.

1. Introduction

Crude oil has to be purified because of several reasons. The catalysts that are used for processing of oil products deteriorate with contaminants like sulphur or metals. Also, there is a need to reduce emissions from engines which contain sulphur oxide and nitrogen oxides and contribute to the acid rain.

The reducing treatments, hydrotreating, commonly used in industry, utilize catalysts made of small MoS_2 or WS_2 particles inside pores of an alumina support. They contain minor amounts of a promoter of catalytic activity like cobalt or nickel. The location of the ions of these elements in the catalyst has been the subject of many studies.

In the last decade a large amount of information has appeared in the literature from careful surface science studies, solid state chemistry and organometallic chemistry about the structure and function of the catalyst and the promoter in these materials. In this work we will review only the surface science point of view of catalysts used for hydrodesulphurization (HDS) processes. For more general review articles we refer the reader to references [1-6].

2. The Mo-S/alumina catalyst

The catalysts used for HDS are made by pore impregnation of alumina with some aqueous solutions. The resulting precursor is submited to a sulphiding procedure, resulting in a completely sulphided compound where the Mo is in the form of MoS_2 and the promoter ions are in a sulphidic environment.

The structure of MoS_2 is hexagonal, forming layers of Mo between two layers of S and these groups of layers interact with others weakely through S-S Van der Waals forces.

Crystals grow in the form of platelets, much larger parallel to the basal planes than perpendicular to them. In the real catalysts, MoS_2 crystals are very small and they look somewhat wrinkled as measured by high resolution transmission electron microscopy (HRTEM) [7]. They grow either parallel or edged to the alumina surface [8]. The addition of cobalt in the sulphidic form results in Co_9S_8 crystallites on the surface of the support, in Co ions adsorbed on the surface of MoS_2 crystallites, and in tetrahedral sites in the alumina lattice [9]. The structure of Co adsorbed on MoS_2 has been called Co-Mo-S phase [10]. Most of the Co is present in this form on the edges of the MoS_2 crystallites and as Co_9S_8.

3. Surface Science approach

Surface scientists have been interested, among other things related to HDS catalysts, in finding the exact location of the promoter. However, it is very difficult to prepare a MoS_2 single crystal with a reasonably large and flat edge surface. Molybdenum single crystals with adsorbed sulphur have been used by several researchers as a model catalyst for HDS in an attempt to study catalytic reactions on metal single crystals at elevated pressures, and the structure of adsorbates on those single crystal surfaces and of their desorption and decomposition characteristics.

4. Thiophene Adsorption

By using thermal desorption spectroscopy (TDS) of several molecules, among them thiophene, from clean, sulphided and carbided Mo(100) surfaces, it was shown by the Somorjai group [11-15] that thiophene decomposes upon heating, leaving a surface with sulphur and carbon impurities. However, sulphur or carbon preadsorbed reduced the decomposition of thiophene, and for more than half monolayer of preadsorbed S or C most of the decomposition of thiophene was blocked and only associatively chemisorbed thiophene was observed.

The dissociatively chemisorption of thiophene on clean single crystals is not characteristic of molybdenum only, as was demonstrated on Mo, Pt, Ni and Cu surfaces using near-edge X-ray absorption fine structure (NEXAFS) [16,17] and high resolution electron energy loss spectroscopy (HREELS) [18-21]. NEXAFS measurements showed that thiophene, at low temperatures and at monolayer coverage, adsorbs flat on the surface of Pt(111) and at 290 K C-S bond starts breaking and the molecule breaks completely by 470 K [16]. HREELS spectra presented evidence of a Pt-S bond around 350 K, indicating also C-S bond breaking. Also, gave evidence that the C-C and C-H bonds remained after the appearance of the C-S bond breaking, suggesting that C_4H_4 was bonded to the Pt surface. More evidence that sulphur breaks from thiophene at around 350 K and that only desorption of thiophene is obtained after heating a fully sulphided surface was given from X-ray photoelectron spectroscopy (XPS) and TDS measurements [18].

Careful TDS and HREELS measurements of thiophene adsorption on Mo(100) by Zaera et al. [17] showed that at low thiophene coverage there is only one H_2 TDS peak at around 340 K and that a C-H bond disappears by 310 K, while at high coverage, most thiophene adsorbs perpendicular to the surface, with the sulphur atom down. For high thiophene coverage at 500 K there is still a C_4H_2S fragment, which loses sulphur after 600 K.

5. Adsorption of Other Molecules

Roberts and Friend, studied by TDS the adsorption of tetrahydrothiophene and butanethiol [22]. They found the same intermediate molecule, which gave at the end either butane or butene. Also using TDS, Benziger and Preston studied the adsorption of methanethiol and methanol on W(211), either clean, oxidized, carbided or sulphided [23]. They observed also decomposition of the molecule on a clean surface and a decrease of it on the oxidized, carbided and sulphided surfaces. Preadsorption of oxygen reduced the decomposition of both adsorbates, while preadsorption of sulphur or carbon only reduced the decomposition of methanol.

6. Thiophene Desulphurization

A high pressure-low pressure cell was combined with surface science techniques on Mo(100) single crystal surfaces to study thiophene desulphurization by Somorjai and coworkers [24-26]. Similar properties were found between Mo(100) and unsupported MoS_2. Also, similar product distribution of butadiene, butene and butane was found [24]. Deactivation by H_2S was observed, while butene had no influence on catalytic activity [26]. They found that hydrogenation of preadsorbed sulphur has a rate two orders of magnitude smaller than the turnover frequency of thiophene hydrodesulphurization [25], which proves that reduction of sulphur which is bound to the metal surface is not a step in the thiophene hydrodesulphurization on Mo(100), and that the sulphur removed from thiophene is not deposited in four-fold hollow or bridge sites. Carbon atoms went to four-fold hollow sites under the conditions of this study, which did not block the active sites for HDS of thiophene [26].

7. The Role of Carbon

Carbon-supported metal sulphides were observed to have higher HDS activity than those supported by alumina [27-29]. The authors consider that metal cations in contact with carbon (a soft ligand) would keep their catalytic activity, while in contact with oxygen (a hard base) would be inhibited. Also, the small relative size of carbon would not block the adsorption of molecules as much as oxygen or sulphur. Lee and Boudart showed that Mo_2C is about as good a HDS catalyst as MoS_2/alumina [30].

8. Sulphided Metal vs Metal Sulphide

Because of the great similarities between the HDS of Mo(100) and MoS_2, Somorjai et al. concluded that also in commercial HDS the reaction does not take place on bare metal sites. Besides, the slow hydrogenation of sulphur adsorbed on Mo(100) allowed to propose that hydrogenation of the sulphur atom in thiophene takes place before the final C-S hydrogenolysis step [26].

There are still questions about the differences between sulphided metals and metal sulphides. Tatarchuck et al. in a study of supported ruthenium catalysts compared mildly sulphided Ru particles on alumina with fully sulphided ones. They found differences in the process of HDS of thiophene, and the RuS_2 catalysts were much more active than the mildly sulphided ones.

9. The Effect of the Promoter

The effect of the addition of Co or Ni as promoters of catalytic activity for HDS is not well undertood at present. Several models have been proposed and although much work has been done more experimental evidence is needed in order to differentiate among them.

One model proposes that Co or Ni increase the number of active sites at the catalyst surface in what is called a textural promoter, in which the promoter is not being involved itself in the catalysis. The textural effect was proposed by Voorhoeve [31,32] with Co or Ni atoms intercalated between MoS_2 or WS_2 layers, which would donate electrons to surface Mo or W ions in the 4+ state (see Fig.1). However, because Ni ions in the 2+ state instead of Ni atoms are introduced, and the formation of ternary sulphides has not been possible, Farragher and Cossee proposed that the Co or Ni is located at the edges of the MoS_2 crystals between alternate layers in a decoration or pseudo-intercalation [33] (see Fig.1). On the other hand, from infrared (IR) studies of NO molecules adsorbed on Co promoted MoS_2 catalysts, Topsoe and Topsoe concluded that the promoter ions are located in the plane of the Mo cations of the MoS_2 layers [34] (see Fig.1). Although there is some doubt in the interpretation of their experiment, their model is widely accepted. Changes in the MoS_2 crystal size of the catalyst by the addition of a promoter have been observed by Delmon [35] and by Candia et al. [36]

Chianelli et al. proposed that the promoter ion influences a neighbouring Mo site and creates a much more active site, in what is called the electronic effect model [37,38].

The determination of the number of active sites in the HDS catalysts has been an important subject of research. Adsorption of oxygen was measured by Tauster et al. [39], Bachelier et al. [40], Zmierczak et al. [41], and others. However, it cannot be used for the quantitative determination of active sites because it reflects the general state of dispersion rather than specific sites [41]. Also, adsorption

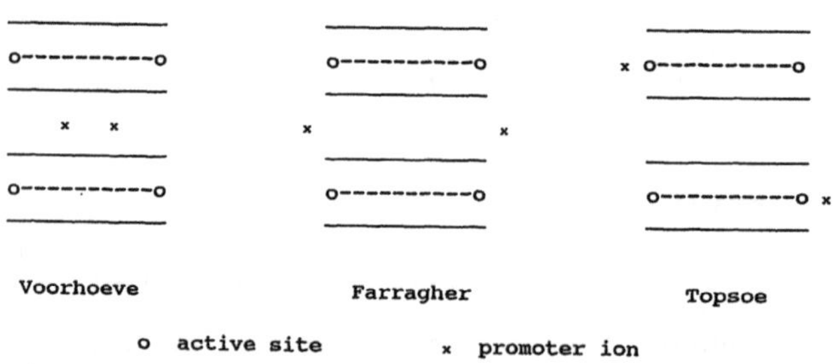

Fig.1 Scheme of the location of the active sites and the Co promoter ions in MoS_2.

of CO has been attempted, however, the determination of the active surface area of an HDS catalyst remaines unsolved.

De Beer et al. suggested that cobalt sulphide and nickel sulphide might be the catalysts instead of the promoter [27,42]. They found that the activities of cobalt sulphide and nickel sulphide on carbon for HDS of thiophene were higher than that of MoS_2/C. This is not an isolated case because many transition sulphides have a higher HDS activity than MoS_2 or WS_2 [43]. More work has been done in this area but no clear picture has emerged yet.

We may say that most models proposed can explain the experimental results. More experimental evidence is necessary to differentiate between the textural, the modified Mo site, and the new catalytic Co site models.

10. Conclusions

Surface science measurements give insight into the characterization and subsequent understanding of HDS catalysts.

From adsorption and decomposition of thiophene on several metal single crystals we can conclude: (1) C-S bond breaking precedes C-C bond breaking, (2) a C_4H_xS fragment is formed on the surface, (3) at low coverage thiophene adsorbs flat on the metal surface, and at high coverage, it adsorbs either perpendicular or tilted.

Preadsorption of oxygen reduces the decomposition of tetrahydrothiophene and butanethiol, while preadsorption of carbon or sulphur only reduce the decomposition of methanol.

Sulphur removed from thiophene is not deposited in four-fold hollow or bridge sites on Mo(100). Carbon atoms do not block the active sites for HDS of thiophene on Mo(100).

Probably HDS does not take place on bare metal sites. Hydrogenation of the sulphur atom in thiophene takes place before the final C-S hydrogenolysis step.

Metal sulphides are more active for HDS than sulphided metals.

Most models proposed can explain the experimental results on the effect of the promoter (the textural model, the modified Mo site model, and the new catalytic site model). Even more work has to be done in this area.

11. References

1. V.H.J. de Beer and C.G.A. Schuit, Preparation of Catalysts (B. Delmon, P.A. Jacobs and G. Poncelet, Eds.), Elsevier, Amsterdam, 1976, p. 343.
2. F.E. Massoth, Adv. Catal., 27, 265 (1978).
3. P. Grange, Catal. Rev.-Sci. Eng., 21, 135 (1980).
4. P. Ratnasamy and S. Sivasanker, Catal. Rev.-Sci. Eng., 22, 401 (1980).
5. R.R. Chianelli, Surface Properties and Catalysis by Non-Metals (J.P. Bonnelle, B. Delmon and E. Derovane, Eds.), Reidel, 1983, p. 361.
6. R. Prins, V.H.J. de Beer and G.A. Somorjai, Catal. Rev.-Sci. Eng. 31, 1 (1989).
7. H. Topsoe and B.S. Clausen, Catal. Rev. -Sci. Eng., 26, 395 (1984).

8. T.F. Hayden and J.A. Dumesic, J. Catal., 103, 366 (1987).
9. H. Topsoe, B.S. Clausen , R. Candia, C. Wivel and S. Morup, J. Catal., 68, 433 (1981).
10. C. Wivel, R. Candia, B.S. Clausen, S. Morup and H. Topsoe, J. Catal., 68, 453 (1981).
11. M. Salmeron, G.A. Somorjai, A. Wold, R.R. Chianelli and K.S. Liang, Chem. Phys. Lett., 90, 105 (1982).
12. M.H. Farias, A.J. Gellman, G.A. Somorjai, R.R. Chianelli and K.S. Liang, Surf. Sci., 140, 181 (1984).
13. M. Salmeron and G.A. Somorjai, Surf. Sci., 126, 410 (1983).
14. A.J. Gellman, M.H. Farias, M. Salmeron and G.A. Somorjai, Surf. Sci., 136, 217 (1984).
15. D.G. Kelly, M. Salmeron and G.A. Somorjai, Surf. Sci., 175, 465 (1986).
16. J. Stohr, J.L. Gland, E.B. Kollin, R.J. Koestner, A.L. Johnson, E.L. Muetterties and F. Sette, Phys. Rev. Lett., 53, 2161 (1984).
17. F. Zaera, E.B. Kollin and J.L. Gland, Surf. Sci., 184, 75 (1987).
18. J.F. Lang and R.I. Masel, Surf. Sci., 183, 44 (1987).
19. J. Stohr, E.B. Kollin, D.A. Fischer, J.B. Hastings, F. Zaera and F. Sette, Phys. Rev. Lett., 55, 1468 (1985).
20. F. Zaera, E.B. Kollin and J.L. Gland, Langmuir, 3, 555 (1987).
21. B.A. Sexton, Surf. Sci., 163, 99 (1985).
22. J.T. Roberts and C.M. Friend, J. Am. Chem. Soc., 108, 7204 (1986).
23. J.B. Benziger and R.E. Preston, J. Phys. Chem., 89, 5002 (1985).
24. A.J. Gellman, D. Neiman and G.A. Somorjai, J. Catal., 107, 92 (1987).
25. A.J. Gellman, M.E. Bussell and G.A. Somorjai, J. Catal., 107, 103 (1987).
26. M.E. Bussell and G.A. Somorjai, J. Catal., 106, 93 (1987).
27. J.C. Duchet, E.M. van Oers, V.H.J. de Beer and R. Prins, J. Catal., 80, 386 (1983).
28. J.P.R. Vissers, J. Bachelier, H.J.M. Ten Doeschate, J.C. Duchet, V.H.J. de Beer and R. Prins, Proceedings 8th Int. Congress on Catalysis, Berlin, (Verlag Chemie, Weinheim), 1984, p. II-387.
29. J.P.R. Vissers, B. Scheffer, V.H.J. de Beer, J.A. Moulijn and R. Prins, J. Catal., 105, 277 (1987).
30. J.S. Lee and M. Boudart, Appl. Catal., 19, 207 (1985).
31. R.J.H. Voorhoeve and J.C.M. Stuiver, J. Catal., 23, 228, 243 (1971).
32. R.J.H. Voorhoeve, J. Catal., 23, 236 (1971).
33. A.L. Farragher and P. Cossee, Proc. 5th Int. Congress Catal., Palm Beach, 1972, (North-Holland, Amsterdam, 1973), p. 1301.
34. N.-Y Topsoe and H. Topsoe, J. Catal., 84, 386 (1983).
35. G. Hagenbach, Ph. Courty and B. Delmon, J. Catal., 31, 264 (1973).
36. R. Candia, B.S. Clausen and H. Topsoe, Bull. Soc. Chim. Belg., 90, 1225 (1981).
37. R.R. Chianelli, T.A. Pecoraro, T.R. Halbert, W.-H. Pan and E.I. Stiefel, J. Catal., 86, 226 (1984).
38. S. Harris and R.R. Chianelli, J. Catal., 98, 17 (1986).
39. S.J. Tauster, T.A. Pecoraro and R.R. Chianelli, J. Catal., 63, 515 (1980).

40. J. Bechelier, J.C. Duchet and D. Cornet, J. Phys. Chem., 84, 1925 (1980).
41. W. Zmierczak, G. MuraliDhar and F.E. Massoth, J. Catal., 77, 432 (1982).
42. V.H.J. de Beer, J.C. Duchet and R. Prins, J. Catal., 72, 369 (1981).
43. M.J. Ledoux, O. Michaux, G. Agostini and P. Panissod, J. Catal., 102, 275 (1986).

Catalytic Behavior of Perovskite-Type Oxides

E.A. Lombardo and J.O. Petunchi

Instituto de Investigaciones en Catálisis y Petroquímica,
INCAPE, Santiago del Estero 2654, 3000 Santa Fe, Argentina

Abstract. $LaCoO_3$ and two partially substituted cobaltates containing either Sr or Th were selected to study the effect of partial substitution of the La and the hydrogen reduction treatment upon the catalytic behavior. In order to find the correlation between surface chemistry and catalytic properties a battery of tools were employed such as XRD, XPS, hydrogen chemisorption and two well characterized test reactions. A reduction model is proposed which also interprets the deep effect of the accompanying cations (La, Sr, Th) in modifying the catalytic behavior of the transition metal.

1. Introduction

Supported group VIII metal catalysts, are of widespread use in industrial processes and atmospheric pollution control. Due to both their practical importance and theoretical interest they have attracted much attention from academia and industry.

One of the key questions concerning the performance of these systems is how the support and certain additives or promoters influence the dispersion and the stability of such metals which in turn affect the catalytic behavior. Any endeavor to address this question entails a knowledge of the relation between the catalytic and solid state properties of catalysts.

Perovskite-type oxides appear as a family of promising compounds to gain insigth into the relationships connecting solid and catalytic properties. These isomorphic solids (general formula ABO_3) are highly versatile materials due to the possibility of accomodating a large variety of elements within the same crystalline structure. Multicomponent perovskites can also be attained by partial substitution of cations A and B. Besides, after treatment under controlled reducing atmosphere it is possible to modify the oxidation state of the transition metal cation located at the B site which in many cases can be reduced to the metallic state to obtain a well characterized and highly dispersed "metal on oxide solid" /1/.

In what follows it will be shown how the combination of a suitable surface spectroscopic technique (XPS), chemisorption measurements and the use of well characterized test reactions to probe the surface may yield a detailed picture of the solid transformations which occur when the starting oxide is progressively reduced. $LaCoO_3$, $La_{0.6}Sr_{0.4}CoO_3$ and $La_{0.8}Th_{0.2}CoO_3$ perovskites have been selected to illustrate this procedure.

2. Experimental

Mixed oxide preparation. $LaCoO_3$ was obtained by precipitation of $La(NO_3)_2$ with $K_3Co(CN)_6$. The lanthanum cobaltates partially substituted by either Sr or Th were prepared by freeze-drying. In all cases the dry solids were then fired in oxygen at 950°C for 16 h to develop the perovskite structure /2/.

Springer Proceedings in Physics, Volume 62
Surface Science Eds.: F.A. Ponce and M. Cardona

2.1 Reduction Procedure and Hydrogen Chemisorption Measurements

The hydrogen reduction of the oxides was followed using a standard gas recirculation system /3/ which included a liquid air trap. The extent of reduction was calculated from the hydrogen uptake. Hydrogen chemisorption was performed in a similar setup. The adsorption isotherms were obtained at 298 K. The amount of chemisorbed hydrogen on a given sample is reported as the difference between two successive isotherms. Between them the solid was evacuated at 298 K until a vacuum better than 10^{-3} Pa was reached.

To study the surface modifications effected by hydrogen reduction a pretreatment chamber with a lateral cold trap, having a volume of 1 liter, was connected directly to an ESCA 750 Shimadzu instrument. This spectrometer is driven by a computer system (ESCAPAC 760) which allows both the accumulation of data and their processing.

The oxide powder was pelleted to form a tablet and then subjected to the following treatment: a) evacuation at 673 K, final pressure 10^{-5} Pa, b) reduction at the desired temperature, c) evacuation of the hydrogen after the probe has been cooled down at room temperature, d) evacuation at 573 K in the analyzer chamber. The spectra were always taken at 573 K. More details are given elsewhere /4/.

2.2 Reaction Procedures

The catalytic experiments were done in the same recirculation system used for reduction into which a bulb, containing most of the reaction volume (430 cm^3), was incorporated through a bypass. The exit stream from the bulb could be sampled for gas chromatographic analysis /5/.

2.3 Treatment of the XPS Data

The atomic fraction of the elements on the surface was calculated using the area under the peaks, the Scofield photoionization cross sections, the mean free paths and the instrumental function given by the ESCA manufacturer /6/.

To estimate the contribution of Co^o, Co^{2+} and Co^{3+} to the Co2p peaks the curve resolver Model CR-6B included in the ESCAPAC unit, was used. A similar procedure was applied to deconvolute the O1s peak /7/.

3. Results and Discussion

3.1 Bulk Reduction

$LaCoO_3$, $La_{.8}Th_{.2}CoO_3$ and $La_{.6}Sr_{.4}CoO_3$ were reduced at temperatures between 523 and 773 K. At the latter complete reduction of the cobalt was achieved in the three cases, e.g.

$$2\ LaCoO_3 + 3\ H_2 \xrightarrow{773\ K} La_2O_3 + 2\ Co^o + 3\ H_2O \quad .$$

The extent of reduction vs. time plots showed typical saturation curves at each temperature. No autocatalytic effects were observed in any case. At intermediate temperatures the Sr containing oxide was easier to reduce than both the La and La-Th solids.

3.2 Surface Reduction

The surface Co^o concentration and the Co/A cation ratios were calculated from the XPS data also obtained at reduction temperatures between 523 and 773 K.

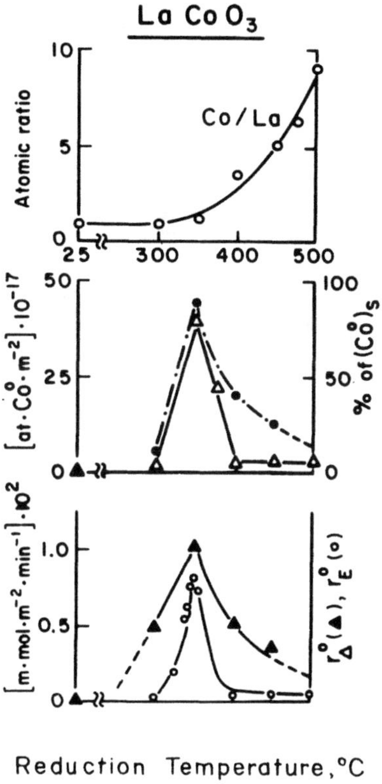

a) Relative migration of the elements after hydrogen reduction (XPS)

b) H_2 chemisorption (Δ) and Co^o surface concentration % of $(Co^o)_s$ (●) (XPS)

c) Cyclopropane (Δ) + H_2
r_A^o initial rate
H_2/Δ-1, 170 Torr, 250°C
Ethene + H_2
r_E^o, initial rate
H_2/E-2, 270 Torr, -20°C

Figure 1. Surface composition of $LaCoO_3$ for different temperature and surface treatments

In $LaCoO_3$ the width of the Co2p peaks goes through a maximum at 573 K while it moves downwards in binding energies from the starting value of 780.6 eV. Fig. 1a shows the onset of cobalt segregation at 623 K when the bulk average oxidation state of the cation goes below Co^{2+}. On the surface, however, the maximum percentage of metallic cobalt is detected at this temperature (Fig. 1b).

A similar pattern is observed on Figs. 2a and 2b for $La_{.6}Sr_{.4}CoO_3$ but the onset of cobalt segregation and the maximum in Co^o concentration moves down to 523 K. This correlates, however, with the more facile reduction of this oxide, i.e. the bulk average oxidation state of cobalt becomes now lower than two at 523 K. Note, that in this case the strontium (oxide) segregates to the surface together with the cobalt (Fig. 2a).

Fig. 3a shows that Co segregates out of the oxide matrix in the $La_{.8}Th_{.2}CoO_3$, while the surface % Co^o for this oxide shows a completely different pattern to those observed for the other two perovskites.

The hydrogen chemisorption data shown in Figs. 1b, 2b and 3b confirm the trends of the XPS data.

A puzzling question arises at this point: Why does the surface concentration of metallic cobalt go through a maximum in the first two cases while this is not the case for the Th containing oxides? An important clue to solve this puzzle is given by the evolution of the O1s signal along the reduction process which is shown in Table 1 for the three solids.

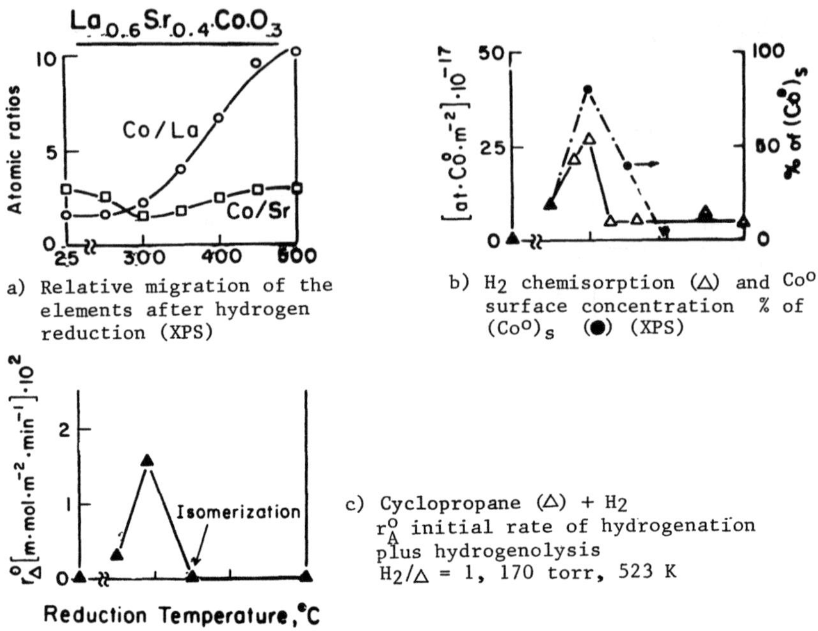

a) Relative migration of the elements after hydrogen reduction (XPS)

b) H_2 chemisorption (△) and Co^o surface concentration % of $(Co^o)_s$ (●) (XPS)

c) Cyclopropane (△) + H_2
r^o_Δ initial rate of hydrogenation plus hydrogenolysis
H_2/Δ = 1, 170 torr, 523 K

Figure 2. Surface composition of $La_{.6}Sr_{.4}CoO_3$ for different temperatures and surface treatments

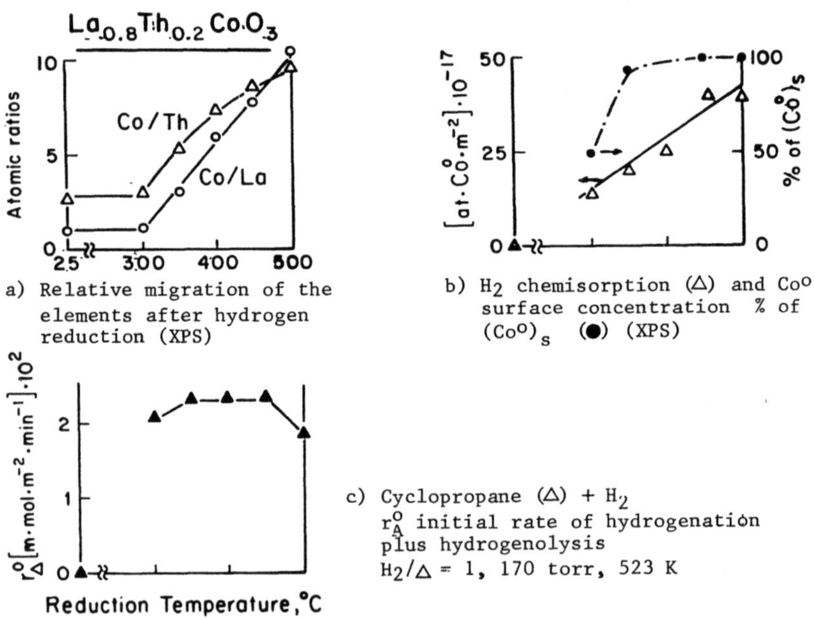

a) Relative migration of the elements after hydrogen reduction (XPS)

b) H_2 chemisorption (△) and Co^o surface concentration % of $(Co^o)_s$ (●) (XPS)

c) Cyclopropane (△) + H_2
r^o_Δ initial rate of hydrogenation plus hydrogenolysis
H_2/Δ = 1, 170 torr, 523 K

Figure 3. Surface composition of $La_{.8}Th_{.2}CoO_3$ for different temperatures and surface treatments

Table 1. The hydroxylation of the oxides upon the H_2 reduction is revealed by the O1s spectra. Binding energies (BE) and full width at half maximum (FWHM) are given in eV

Reduction temperature	$LaCoO_3$ BE	$LaCoO_3$ FWHM	$La_{.6}Sr_{.4}CoO_3$ BE	$La_{.6}Sr_{.4}CoO_3$ FWHM	$La_{.8}Th_{.2}CoO_3$ BE	$La_{.8}Th_{.2}CoO_3$ FWHM
Oxide	529.7	1.8	529.7	2.2	529.8	2.1
573 K	530.1	1.8	532.8	2.2	529.8	2.2
673 K	530.1	4.1	532.8	2.3	530.2	2.0
773 K	532.2	3.8	532.6	2.4	530.2	2.0

The O1s spectrum is in most cases a composite of two peaks, one located at 529.7 ± 0.2 eV due to the lattice oxygen and another at 532.2 ± 0.2 eV assigned to the hydroxyls, in agreement with Fleisch et al. /8/.

The unreduced oxides show a single peak corresponding to the lattice oxygen. The $LaCoO_3$ after reduction at or around 673 K shows a large contribution of hydroxyls which definitely overtakes the lattice signal at the highest reduction temperature. The OH signal develops at much lower temperatures in the Sr containing perovskite. Note that this signal has already blocked the lattice oxygen peak at 573 K. This indicates that the hydroxylated crúst is at least a few atomic layers thick. On the other hand, the reduced Th containing oxide only shows the presence of lattice oxygen even after reduction at 773 K.

These observations indicate that there is a correlation between the decrease of Co^o exposed on the surface and the increase in OH surface population. The catalytic test will help to further clarify the surface transformations which occur upon reduction and evacuation.

3.3 Test Reactions

Two well characterized types of reactions were selected to probe the surface.

Ethylene may react with hydrogen to produce only ethane, whereas cyclopropane may yield: propylene (isomerization) which typically occurs over basic oxides, propane (hydrogenation) over transition metals and their oxides, methane and ethane (hydrogenolysis) over transition metals. Then, the overall reaction pattern may be correlated with the surface transformations which occur during reduction.

Fig. 1c shows a close agreement between the rates of hydrogenation and hydrogenolysis of ethylene and cyclopropane and the concentration of exposed metallic cobalt. No isomerization product (propylene) was detected in the latter case.

These results, together with tracer experiments reported elsewhere /3/, clearly indicate that the loci of catalytic activity are small clusters of Co^o.

The Sr containing perovskite produces an activity pattern which mimics the Co^o surface concentration. Besides, at 623 K when the hydrogenation-hydrogenolysis activity decays to zero, propylene was the only reaction product detected. This also correlates well with the XPS results which indicate that the SrO migrates to the surface of the reduced solid (Fig. 2a).

The $La_{.8}Th_{.2}CoO_3$ oxide, the most active of this series, is also less affected by reduction and only yields hydrogenation and hydrogenolysis products from cyclopropane (Fig. 3d). Note that this oxide is the only one that was not covered by hydroxyl groups after reduction at high temperatures (Table 1).

3.4 Reduction Model

Through the combination of a surface spectroscopic technique, chemisorption measurements and the appropriate use of test reactions the reduction process can now be modelled.

Two different reduction regimes occur in these solids. At low temperatures the bulk and the surface are homogeneously reduced up to Co^{2+}, but at around 523 K for the Sr containing oxide and 573 K for the other two oxides the surface becomes more reduced than the bulk. In other words, the reduction process becomes diffusion limited at higher temperatures.

However, despite the segregation of metallic cobalt to the grain boundaries both the % Co^{o} calculated from the XPS data and the amount of hydrogen chemisorbed goes through a maximum in the $LaCoO_3$ and the Sr containing oxide. The catalytic reaction undoubtedly probes the upper most layer of the solid and also confirms these results (Figs. 1 and 2).

The only possible explanation of these observations is that the surface cobalt has been partially reoxidized upon hydrogen evacuation. This reoxidation may have occurred through reaction with neighbouring OH groups covering the oxide surface. As a matter of fact, this redox process is known to occur in other solids such as $Mo(CO)_6/\gamma\text{-}Al_2O_3$ /9/, upon evacuation at temperatures as low as 573 K.

This hypothesis is further substantiated by the fact that the Th containing oxide which do not show the Ols peak of the hydroxyls (Table 1) do not exhibit the sharp maximum in % Co^{o} as the other oxides do (Fig. 3).

This study has also served the purpose of illustrating the drastic changes produced in the catalytic behavior by partial substitution of the A cation and/or an adequate hydrogen pretreatment which can now be better understood in terms of the proposed model.

Acknowledgements

This work was supported by a grant from CONICET.

We are indebted to the Japan International Cooperation Agency for the donation of an ESCA 750 Shimadzu spectrometer.

References

1 Lombardo, E.A. and Petunchi, J.O., in "Perovskite Catalysts", Catal. Today, N° 8 August (1990).
2 Miró, E., Flesia, M. and Lombardo, E.A., Rev. Fac. Ing. Quím. Santa Fe 44, 15 (1980).
3 Petunchi, J.O., Ulla, M.A., Marcos, J.A. and Lombardo, E.A., J. Catal. 70, 356 (1981).
4 Lombardo, E.A., Tanaka, K. and Toyoshima, I., J. Catal. 80, 340 (1983).
5 Ulla, M.A., Migone, R.A., Petunchi, J.O. and Lombardo, E.A., J. Catal. 105, 107 (1987).
6 Wagner, C.D., Riggs, W.M., Davis, L.E., Moulder, J.F. and Muilemberg, I. (eds.), in Handbook of X-ray Photoelectron Spectroscopy (Perkin-Elmer Corp., Palo Alto, CA 1987).
7 Marcos, J.A., Buitrago, R.H. and Lombardo, E.A., J. Catal. 105, 95 (1987).
8 Fleisch, T.N., Hick, R.F. and Bell, A.T., J. Catal. 87, 398 (1984).
9 Brenner, A. and Burwell, L., Jr., J. Catal. 52, 353 (1973).

Implanted Tin Oxide Thin Films for Selective Gas Sensing*

F.C. Stedile[1], *C.V. Barros Leite*[2], *W.H. Schreiner*[1], *and I.J.R. Baumvol*[1]

[1]Laboratório de Filmes Finos, Instituto de Física,
Universidade Federal do Rio Grande do Sul,
P.O. Box 15051, 91500 Porto Alegre, RS, Brazil
[2]Departamento de Física, PUCRJ, 22452 Rio de Janeiro, Brazil

Abstract. Tin oxide thin films obtained by reactive sputtering were submitted to different thermal annealings and then implanted with Fe^+, Cu^+, Zn^+, Ga^+, As^+. A subsequent air annealing was performed after implantation. The system was characterized by Sheet Resistance measurements, Rutherford Backscattering Spectroscopy and Nuclear Resonant Scattering analyses. Some correlations between the depth distribution of the implanted species and the O/Sn ratios as a function of depth were found. The results indicate the possibility of modulating the O/Sn ratio depth profile with implantation which is useful for the production of stable and selective gas sensors.

1. Introduction

Tin oxide is a material with high transparency in the visible region of the spectrum and with a modifiable electrical conductivity [1,2]. Oxygen deficiency, good electrical conductivity and transparency are not normally compatible. Other oxides absorb visible light strongly when rendered O deficient, but corresponding O loss from SnO_2 does not result in a light absorbing material, the metal-oxide remaining transparent. All technological applications of SnO_x thin films are consequences of their unusual electrical and optical properties. These applications range from contacts in liquid crystal displays and trimmed resistors in hybrid circuits to gas sensors. The gas sensing characteristics of the tin oxide are closely related to its electrical conductivity. Most of the research efforts on the gas sensor properties of tin oxide thin films have concentrated on the properties of undoped films deposited by several different methods. Although undoped, oxygen deficient SnO_x films have a high electrical conductivity, the sensing selectivity to different gaseous species can be further enhanced by doping with appropriate impurities. A rather complete work upon the consequences of doping SnO_x films was made by Astaf'eva and Skornyakov [8]. The films obtained by spray-pyrolysis of $SnCl_4$ had the impurities added to the original solution mainly as chlorides in amount of 1 at %.

Another doping technique that has been considered is ion implantation, commonly used for doping Si and GaAs semiconductor devices as well as other materials when controlled concentration of dopants at a chosen depth is desirable. Several authors have studied the effect of doping SnO_x by ion implantation [9-14].

In the present work we grew tin oxide films by reactive magnetron sputtering. The deposition conditions were as follows: substrate: oxidized silicon wafers; P_{base}: 10^{-5} Pa; P_{Ar}: $2x10^{-2}$ Pa; P_{O2}: $1x10^{-2}$ Pa; P_{dc}: 500 V; I_{dc}: 40 mA; target: 99.999% Sn; $d_{target-substrate}$: 6 cm; $T_{subst} < 100\ ^oC$; t_{depos}: 30 min; thickness: 2000 Å. The samples were submitted to different thermal annealings as well as to implantation with different ions (Fe^+, Cu^+, Zn^+, Ga^+, As^+) with the same dose ($1x10^{16}$ cm^{-2}) and energy (70

*Supported in part by CNPq and FINEP, Brazilian agencies

keV). After each processing step we characterized the films by means of Sheet Resistance measurements (R_S), Rutherford Backscattering Spectroscopy (RBS) and Nuclear Resonant Scattering (NRS). By doing so, we followed the modifications which occured in the film structure and composition, such as the concentration of impurities and the O/Sn ratio as a function of depth in the film, besides the electrical conductivity.

2. Experimental Results

2.1 R_S Measurements

The data obtained before and after ion implantation are summarized in Table I. The implantation in samples submitted to pre-implantation annealing in vacuum led to an increase in their R_S, while the implantation in samples submitted to pre-implantation annealing in air diminished the R_S(Ga^+ is an exception).

From Table II we see that after the last performed air annealing, the three unimplanted samples displayed the same R_S values.

Table I: Sheet Resistance (R_S) values from samples submitted to a thermal treatment and then implanted with different ionic species. The entries are the R_S in Ω/□. The upper limit of sensitivity of the four-point probe was $R_S \sim 10^7$ Ω/□.

Implanted Species	Pre-implantation treatment R_S(Ω/□)		
	As-deposited	Annealed in Vacuum	Annealed in air
unimplanted	5.7×10^6	3.6×10^5	3.6×10^5
Fe^+	$> 10^7$	1.8×10^6	2.5×10^4
Cu^+	$> 10^7$	9.1×10^5	5.5×10^4
Zn^+	1.4×10^6	1.4×10^6	9.1×10^4
Ga^+	4.1×10^6	2.2×10^6	$> 10^7$
As^+	$> 10^7$	4.5×10^5	7.0×10^4

Table II: Post-implantation Sheet Resistance (R_S) values from the samples submitted to different pre-implantation annealings, implanted with the indicated atomic species and post-implantation annealed in air. The entries are the R_S in Ω/□. The upper limit of sensitivity of the four-point probe was $R_S \sim 10^7$ Ω/□.

Implanted Species	Thermal treatments R_S(Ω/□)		
	As-deposited + Annealed in air	Annealed in vacuum + Annealed in air	Annealed in air + Annealed in air
unimplanted	2.3×10^5	9.1×10^4	2.5×10^5
Fe^+	$> 10^7$	4.8×10^5	1.2×10^4
Cu^+	4.5×10^3	7.7×10^4	4.5×10^3
Zn^+	2.7×10^4	3.2×10^3	9.1×10^3
Ga^+	3.6×10^3	6.9×10^4	3.6×10^3
As^+	1.4×10^4	2.3×10^3	3.2×10^4

Details on the specific effects of each implanted species and the subsequent thermal treatments are also given in Table II. With one exception (Fe^+ implantation) all post-implantation annealings led to a broadly ranged but comparable Sheet Resistance (from 4.8×10^5 to 2.3×10^8 Ω/□).

2.2 RBS Measurements

All implantations led to impurity distribution profiles with maximum concentration at about one third of the depth from the outermost surface (as predicted by the TRIM calculations) and that despite the thermal treatments the distributions of the implanted atoms remained within the SnO_x film, i.e., diffusion of the implanted species into the substrate and/or loss of impurities by evaporation at the outermost surface were negligible.

In our analyses we dealt with three main variables: i) the kind of thermal pre-treatment; ii) the implanted ionic species (Fe^+, Cu^+, Zn^+, Ga^+, As^+); and iii) whether or not the sample was submitted to post-implantation air annealing.

Comparing the spectra obtained from samples before and after the post-implantation air annealing we noticed that in most cases the implanted species depth profile showed little or no modifications caused by this air annealing (see for example, the RBS spectra of a film pre-annealed in air implanted with Zn^+ before and after the air annealing in Fig. 1). However,

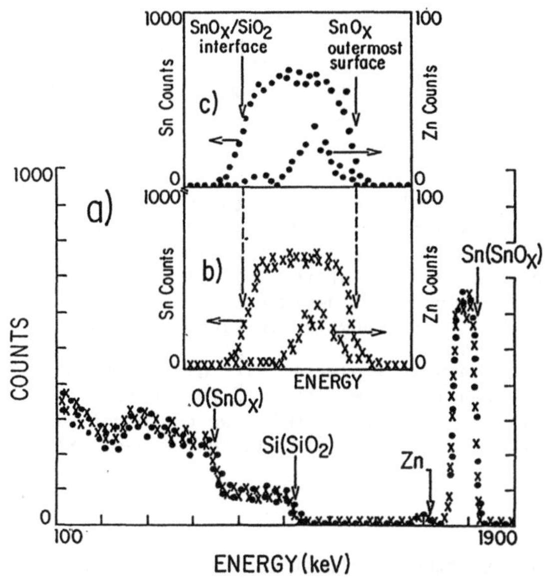

Fig. 1 - (a): RBS spectra from the SnO_x film submitted to a pre-implantation annealing in air and implanted with Zn^+ before (X) and after (*) the post-implantation air annealing; (b): Detail of the spectrum in (a) before the post-implantation air annealing. The energy scale for the Zn signal has been shifted, such that the energies of the α-particles scattered from Sn and Zn atoms at the SnO_x outermost surface appear in the same region of the spectrum. So the figure displays the superposition of the Sn and Zn signals giving the profiles at the same depth region of the film, which corresponds to an energy interval of 200 keV for both signals; (c): Detail of the spectrum in (a) after the post-implantation air annealing displaying the superposition of the Sn and Zn signals.

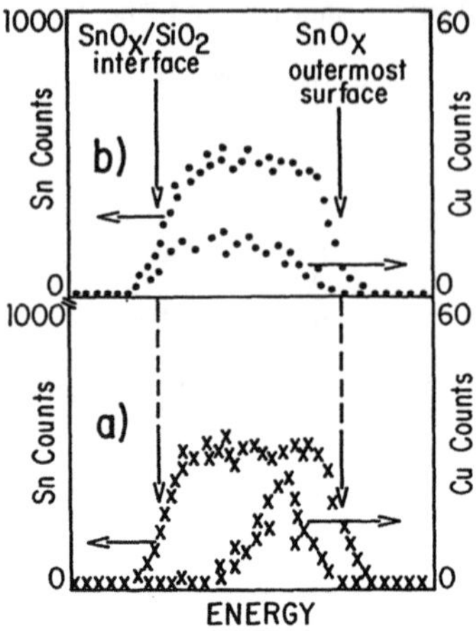

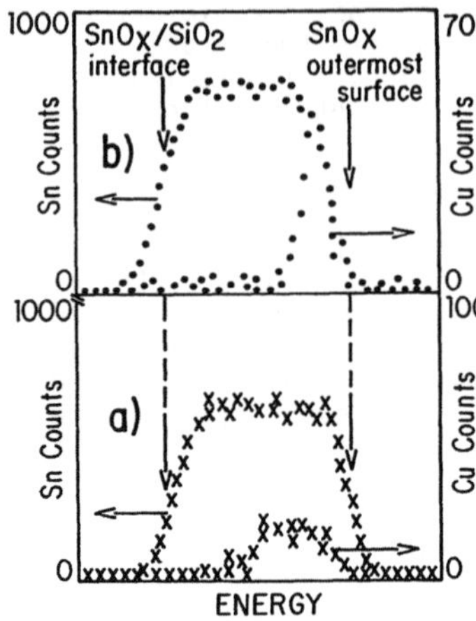

Fig. 2 - (a): Detail of the RBS spectrum from an as-deposited SnO_x film implanted with Cu^+ before the post-implantation air annealing displaying the superposition of the Sn and Cu signals giving the profiles at the same depth region of the film; (b): Detail of the RBS spectrum after the post-implantation air annealing displaying the superposition of the Sn and Cu signals.

Fig. 3 - (a): Detail of the RBS spectrum from a SnO_x, film pre-implantation annealed in air before the post-implantation air annealing displaying the superposition of the Sn and Cu signals giving the profiles at the same depth region of the film; (b): Detail of the spectrum after the post-implantation air annealing displaying the superposition of the Sn and Cu signals.

there were four exceptions, namely the implanted Zn^+ in the as-deposited sample and the implanted Cu^+ in all samples, which diffused during the air annealing. The Zn^+ and Cu^+ ions implanted in as-deposited samples migrated from their original positions towards the back of the SnO_x film, leading to a more or less homogeneous concentration versus depth profile (see Fig. 2), while the Cu^+ implanted ions in samples pre-annealed in vacuum and in air segregated at the outermost surface (see Fig. 3).

Comparing the RBS spectra from samples implanted with different ions and submitted to the same thermal treatments, we observed that before the post-implantation air annealing of all implanted samples we could not distinguish among the profiles of the implanted species within the depth resolution of the RBS analyses.

2.3 NRS Measurements [15]

The as-deposited tin oxide films presented an almost constant stoichiometry as a function of depth, with an O/Sn ratio about 1.5, whereas the sample annealed in air after reactive sputtering deposition reached the stoichiometry of SnO_2 at the outermost surface (see Fig. 4) with the O/Sn ratio decreasing somewhat at deeper regions in the film, but remaining greater than 1.5. The details of the analysis of tin oxide films with nuclear resonant scattering are given in Reference 7.

After the post-implantation air annealing the implanted samples corresponding to the three different pre-annealing conditions displayed different results. This is illustrated for the case of the Fe^+ implantation in Fig. 5.

A further illustration is shown in Fig. 6a, where the sample submitted to a pre-implantation annealing in air, implanted with Cu^+ and again an-

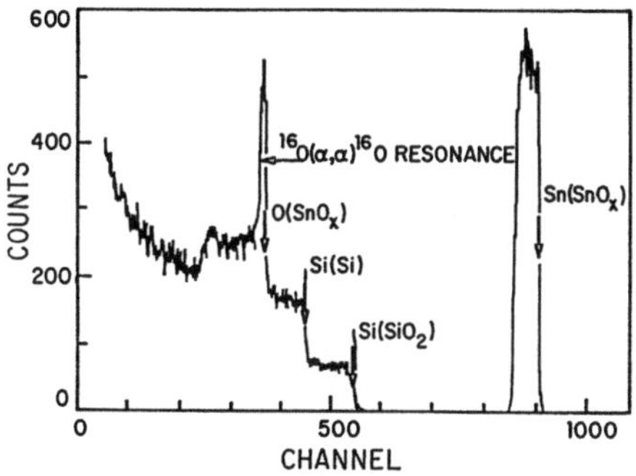

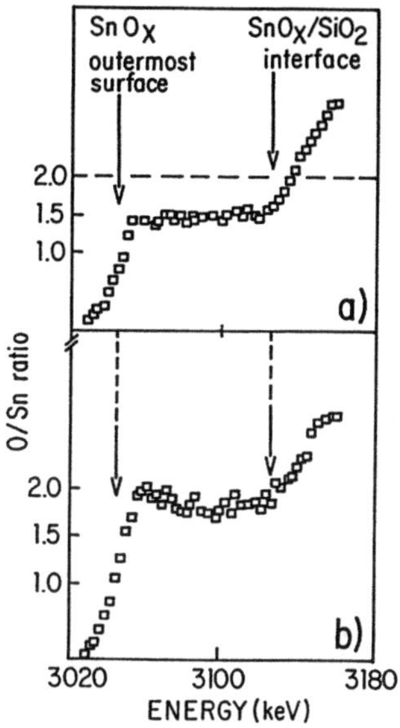

Fig. 4 - Left: $^{16}O(\alpha,\alpha)^{16}O$ Nuclear Resonant Scattering spectrum at an incident energy of the alpha particles of 3.088 MeV; Right: O/Sn ratio as a function of the depth in the SnO_x film as obtained from the $^{16}O(\alpha,\alpha)^{16}O$ Nuclear Resonant Scattering: (a) as-deposited sample; (b) sample annealed in air during 4 hours at 400 °C.

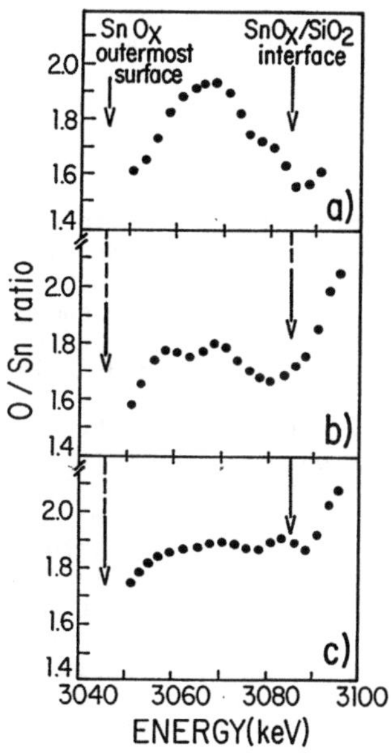

Fig.5 - O/Sn ratio as a function of the depth in the film as obtained from the $^{16}O(\alpha,\alpha)^{16}O$ Nuclear Resonant Scattering: (a) as deposited sample implanted with Fe^+ (10^{16} cm^{-2}; 70 keV) and submitted to a post-implantation annealing in air during 1 1/2 hour at 400°C; (b) sample pre-implantation annealed in vacuum (4 hours; 400°C; P = 5.10^{-5} Pa) implanted with Fe^+ (10^{16} cm^{-2}; 70 keV) and submitted to post-implantation annealing in air during 1 1/2 hour at 400°C; (c) sample pre-implantation annealed in air (4 hours; 400°C) implanted with Fe^+ (10^{16} cm^{-2}; 70 keV) and submitted to a post-implantation air annealing during 1 1/2 hour at 400°C.

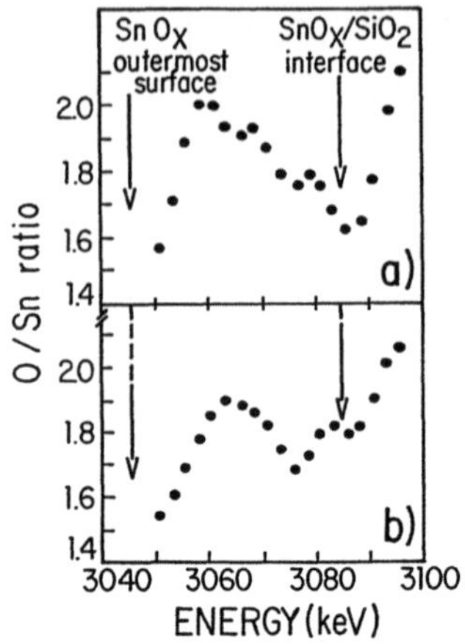

Fig. 6 - O/Sn ratio as a function of the depth in the film as obtained from the $^{16}O(\alpha,\alpha)^{16}O$ Nuclear Resonant Scattering: (a) sample pre-implantation annealed in air (4 hours; 400 °C) implanted with Cu^+ (10^{16} cm^{-2}; 70 keV) and submitted to a post-implantation annealing in air during 1 1/2 hour at 400 °C; (b) sample pre-implantation annealed in air (4 hours; 400 °C) implanted with Zn^+ (10^{16} cm^{-2}; 70 keV) and submitted to a post-implantation air annealing during 1 1/2 hour at 400 °C.

nealed in air reaches the SnO_2 stoichiometry at the outermost surface of the tin oxide film; the O/Sn ratio decreases going deeper into the film, until a value of 1.6 at the SnO_x/SiO_2 interface. The stoichiometry of the film submitted to a pre-implantation annealing in air, implanted with Zn^+ and post-annealed in air, however, reaches its maximum (1.9) at about one third from the outermost surface, decays to 1.7 and again increases as we go deeper into the tin oxide film (see Fig. 6b).

3. Discussion and Conclusions

This work shows the effects of the ion implantation of different atomic species on the electrical characteristics, the depth profiles of the O/Sn ratio and of the implanted species of tin oxide thin films. The films were deposited by Reactive Sputtering and all dopants were implanted at the same energy and dose. These conditions led to films with a high degree of purity and reproducibility, apart from the very good control of the deposition and doping parameters.

The implanted species were chosen based on the results reported by Astaf'eva and Skornyakov [8] in order to have dopants which induce either a decrease, an increase, or even those which do not alter the R_s of the film.

Some features of the observed effects of ion implantation upon the electrical conductivity of the SnO_x films can be explained by taking into account the modifications on the crystalline structure and stoichiometry of the oxide that ion implantation induces: damaging of the crystalline structure (which leads to an increase of the electrical resistivity), preferential sputtering of the O or Sn atoms of the tin oxide and/or the vacuum annealing during implantation (both leading to a decrease in the R_s values). Concerning specifically the results of the present work, the competition among the above processes as well as the electrical activity of the impurity itself led to a decrease in the R_s values of the films submitted to pre-implantation annealing in air, with the exception of Ga^+. The main factor acting in films pre-annealed in vacuum was the amorphization of their polycrystalline structure by ion implantation, leading to an increase in the film electrical resistivity.

The post-implantation air annealing also induced many modifications. It could lead to tin oxidation, creating a more complete oxide (SnO_X where $X \rightarrow 2$), rising the R_S, and/or to a further organization of the tin oxide crystalline structure and to the activation of the implanted impurities, diminishing the R_S. In the case of films pre-annealed in vacuum, implanted and then annealed in air, the competition among these trends led to smaller or unaltered values of R_S (compare results of Table I and II).

We also observed the effect of the electrical activation of the implanted species, by air annealing, which reduced the R_S of the implanted films when compared to the unimplanted ones and to the implanted SnO_X films before the post-implantation annealing. The effect of post-implantation air annealing of the films implanted as-deposited was the lowering by many orders of magnitude of the R_S (Fe^+ is an exception). This was the expected result, for this annealing was the first performed thermal treatment of these films which allowed the film structure organization and the activation of the impurities. The post-implantation annealing in air of the films implanted after a pre-annealing in air led also to a decrease in the R_S, although smaller. In these films we can again clearly see the effect of impurity activation, when comparing the R_S of unimplanted and implanted films. Observing the first line in Table II, we notice similar R_S for the three unimplanted films. This is a reasonable result since, after this air annealing, all unimplanted films tend to have similar compositions. The results also agree with the one found in Reference 7 from air annealed films and the one of Table I from the unimplanted film annealed in air.

Excepting Fe^+, all other species induce , after post-implantation annealings in air (no matter what kind of pre-implantation annealing), a decrease in the R_S which is a desirable effect in SnO_X thin film technology. In our films, deposited and doped under controlled conditions, we did not find the same influences of the ionic species upon the conductivity as predicted by Astaf'eva and Skornyakov [8]. Neither did we find correlations among the ionic characteristics of the impurities and the constituents of the film (Sn^{4-} and O^{2-}).

The post-implantation annealing also induced different diffusion behaviour of the implanted species in the film, depending on the implanted species and the pre-implantation thermal treatment. Some of them exhibited unaltered profiles (see Fig. 1); in some other films migration of the impurities from their original positions towards the back of the film occurs forming a more or less homogeneous concentration versus depth profile or segregation at the outermost surface (see Figs. 2 and 3).

At present it is not clear the in - or out - diffusion of Cu dependence on sample preparation, the in-diffusion of Zn in as-deposited SnO_X samples, and the relative inertness of the other impurities upon thermal treatment. One has to take into account several mechanisms. For as-deposited SnO_X films implanted with any impurity we have an amorphous, oxygen-deficient thin film to start with. Thermal annealing originates crystallization, oxidation and diffusion mechanisms within the sample. For pre-annealed in air samples the story is different. Implantation at the doses used in this work amorphizes the film which was crystalline and well oxidized. Further thermal annealing in air leads to a new crystallization without major oxidation effects. We think these effects are not only interesting but also of theoretical and practical value.

Some correlations between the implanted impurities and the O/Sn ratio profiles of the same film were found. The Cu profile that in Fig. 3b has a maximum concentration at the outermost surface and then rapidly decreases is accompanied by the same general trend in the O/Sn profile shown in Fig. 6a: a maximum at the external surface followed by a decrease as we go deeper into the film. The Zn profile in Fig. 1c has a large peak at about one third of the SnO_X depth and a second one near the SnO_X/SiO_2 interface. In Fig. 6b these two peaks appear in the corresponding O/Sn profile at about the same regions. However Fe, another impurity, leads to different O/Sn ratio profiles without noticeable Fe redistribution.

On the basis of the present findings, we conclude that the depth profile of the O/Sn ratio in the SnO_x films can be strongly modulated by implanting different dopant species. Two main factors influence the depth profile of the O/Sn ratio: the kind of pre-implantation annealing and the nature of the implanted species (observing Figs. 5 and 6 we notice that Fe^+, Cu^+ and Zn^+ submitted to the same thermal treatments led to very different O/Sn ratio profiles). An interesting fact is that in none of these implanted films does the O/Sn ratio resemble that of the unimplanted film submitted to air annealing.

These findings are extremely important having in mind the use of SnO_x films for gas sensing devices. The usual operating temperature of these devices is around 400 °C where we find strong oxidizing and reducing effects on the films, depending on the gas ambient to which the sensor is exposed. Now we find that the O/Sn ratio profile can be modulated by impurity implantation. We expect that further work with correct O/Sn ratio profile tailoring can lead to stable (profile O/Sn constant in time) and selective (different O/Sn depth profiles) gas sensing devices.

References

[1] J. Robertson, J. Phys. C, 12 (1979) 4767.
[2] J. Robertson, Phys. Rev. B, 30 (1984) 3520.
[3] H. Windischmann and P. Mark, J. Electrochem. Soc., 126 (1979) 627.
[4] T.W. Capehart and S.C. Chang, J. Vac. Sci. Technol., 18 (1981) 393.
[5] T. Yamazaki, U. Mizutani and Y. Iwama, Jpn. J. Appl. Phys., 22 (1983) 454.
[6] Jay N. Zemel, Thin Solid Films, 163 (1988) 189.
[7] F.C. Stedile, B.A.S. de Barros Jr., C.V. Barros Leite, F.L. Freire Jr., I.J.R. Baumvol and W.H. Schreiner, Thin Solid Films, 170 (1989) 285.
[8] L.V. Astaf'eva and G.P. Skornyakov, Inorg. Mater., 17 (1981) 1208.
[9] A. Licciardello, O. Puglisi and S. Pignataro, J. Chem. Soc., Faraday Trans. 2, 81 (1985) 985.
[10] R. Kelly and E. Giani, Nucl. Instrum. and Meth., 209 (1983) 531.
[11] Shih-Chia Chang, J. Vac. Technol. A, 1 (1983) 524.
[12] D.D. Casey, Extended Abstract No. 336, Fall Meeting of Electrochemical Society, Atlanta, Georgia, 1977.
[13] J.C. Lou and M.S. Lin, Thin Solid Films, 110 (1983) 21.
[14] J.P. Biersack and L.G. Hagmark, Nucl. Instrum. and Meth., 174 (1980) 257.
[15] B.K. Patnaik, C.V. Barros Leite, G.B. Baptista, E.A. Schweikert, D.L. Cocke, L. Quiñones and N. Magnussen, Nucl. Instrum. and Meth., 35 (1988) 159.

Detection of Elementary Particles Using Superconducting Transition-Edge Phonon Sensors on Silicon Crystals

B. Cabrera

Physics Department, Stanford University, Stanford, CA 94305, USA

Abstract. Our group at Stanford, is developing a phonon-mediated radiation detector based on the collection of ballistic phonons generated by particle interactions within a silicon crystal. The phonons are sensed using titanium superconducting thin films on opposite sides of the crystal. The titanium films are 40 nm thick and have a transition temperature of about 400 mK. They are patterned into a 2 μm wide meander. The detector is operated at a temperature slightly below T_c and at a bias current of about 75 nA. Self-extinguishing voltage pulses of several microsecond duration are observed from alpha particle and x-ray interactions in the silicon substrate.

1. Introduction

For the past several years a number of groups in the United States and in Europe [1] have been developing new classes of detectors based on phonons in insulating crystals and quasiparticles in superconductors. These detectors would have lower thresholds and higher resolution than state-of-the-art semiconductor diodes, and are being developed for laboratory searches for dark matter and for neutrino physics.

Semiconductor diode particle detectors now provide the highest energy resolution (≈ 3 keV FWHM for 1 kg) and the lowest thresholds available (≈ 4 keV) for large mass detectors [2]. In the keV range, less than 30% of the deposition energy is converted directly into the electron-hole pair signal, the rest forming phonons. The characteristic energy of these phonons is ≈ 1 meV, 10^3 less than the excitation energy for an electron-hole pair in a semiconductor (≈ 1 eV). Thus in principle, energy resolutions over an order of magnitude better are possible if the phonon signal is used.

The desire for higher resolution and lower threshold detectors is motivated by several important experiments in weak interaction physics. These include a search for a hypothetical flux of weakly interacting massive particles (WIMPs) which may make up the dark matter around our galaxy, a first measurement at a reactor of the predicted coherent neutrino-nucleus elastic scattering process, a self-normalizing reactor neutrino oscillation measurement for detecting a finite neutrino mass, and eventually a solar neutrino observatory capable of measuring the flux and energy spectrum of solar neutrinos.

Springer Proceedings in Physics, Volume 62
Surface Science Eds.: F.A. Ponce and M. Cardona

We are investigating detectors [3] in which the high energy phonons which travel ballistically from the particle interaction region within the crystal are detected before they down-convert to the quasi-thermal phonons used in the bolometers. The advantage of such a detector over the simple bolometer is that imaging of each event is possible. The additional information substantially improves the background suppression. In this paper, we discuss this new type of detector which we call a SiCAD (silicon crystal acoustic detector).

2. Phonons Processes in Silicon Crystals

Within a single crystal of silicon, energy depositions of a few keV are contained within a sphere a few μm in diameter for electron recoils and a few tens of nm in diameter for nuclear recoils. A roughly thermal-like spectral phonon distribution is generated with a characteristic temperature of 10-20 K. This distribution arises from the rapid decay of electron-hole excitations to the bottom of the conduction band, first generating very short wavelength phonons, which quickly relax to longer wavelength phonons within less than ≈ 1 ns. The decay rates are very strongly dependent on phonon energy ($\propto \nu^5$) and for wavelengths greater than several hundred lattice spacings further decays are negligible. These longer wavelength phonons propagate throughout the crystal with little scattering and no dispersion. This mode has come to be called the the ballistic phonon mode [4]. Fig 1 shows the energy density arriving at a [100] silicon surface from a point source of purely ballistic phonons.

The effects of the anharmonic decay process are shown in Fig 2. We have run a simple Monte Carlo model where the lifetime for decay is assumed to be of the form $\tau_A = (25\ \mu s)\ (1\ \mathrm{THz} / \nu)^5$ for silicon crystals [5]. We use a model with only one phonon mode with a decay rate equal to the value obtained by averaging the real crystal over all phonon modes as well as all wavevector directions within each mode. In fact, the lifetime depends on the mode and the magnitude and direction of **k**, nevertheless, the results are qualitatively correct. For the calculation, the anharmonic lifetime τ_A is used to determine whether a phonon decays during each

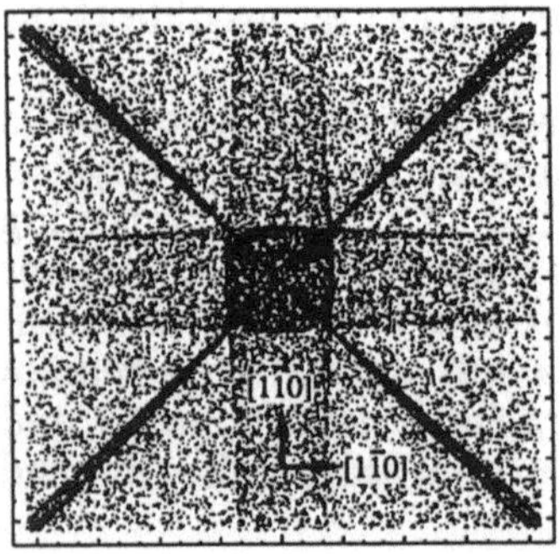

Fig 1. Calculated ballistic phonon energy density incident on surface of [100] silicon crystal.

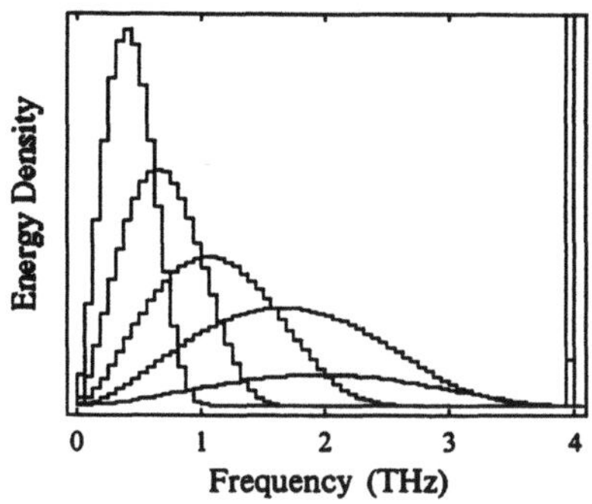

Fig 2. Anharmonic down-conversion of phonon energy distribution.

time step and a simple spherical density of states argument is used to determine the partitioning of the phonon energy into two phonons when a decay does occur. The probability peaks for decay phonons with half of the original energy and is proportional to $E_1^2 E_2^2$ where $E_1 + E_2 = E$ is the energy of the parent phonon. We have assumed a linear dispersion relation so that E is proportional to k and assume the density of states to be uniform in k-space. In Fig 2, we begin the calculation with a delta function of phonons at 4 THz. The five curves follow the time evolution of the Monte Carlo from 10 ns to 100 μs with each curve from right to left representing a factor of ten increase in the time. If we proceed exactly the same as in Fig 2 except that we begin the calculation with the initial energy uniformly distributed over the frequency an interval (e. g., 3-4 THz), remarkably, after several μsecs the resulting phonon energy distribution is the same as in Fig 2. Thus, the decay process forgets the initial distribution and depends only on the phonon physics below about 1 THz where the approximations that we have made are most applicable. These phonons have traveled several cms and the distribution has the same average phonon energy as a Planck distribution at 10-20 K, although the latter is skewed to lower energy with a high energy tail.

As the propagation distance increases through an intrinsic silicon crystal, a smaller fraction of the energy arrives ballistically and a larger fraction arrives at delayed times. Again in the spirit of our qualitative tour, we can estimate the fraction of ballistic energy transport as a function of crystal thickness with another simple Monte Carlo. We take a sphere of silicon and model it as an isotropic medium with both anharmonic decay and isotope scattering, but no focussing. We begin each of thirty thousand 8 THz phonons at the center of the sphere and follow each out towards the surface keeping track of anharmonic decay events and isotope scattering events. For the anharmonic decay events we use the same lifetime as above and assume that both of the daughter phonons continue along the same path as the parent phonon. For the isotope scattering, we assume that the scattering is isotropic and that the lifetime is given by $\tau_I = (0.4\ \mu s)\ (1\ THz / \nu)^4$ for silicon crystals, which for isotope scattering is the same in the real crystal for all phonon modes and for all wavevector directions [5]. Typical results of the calculation are shown in Fig 3. In the upper left we

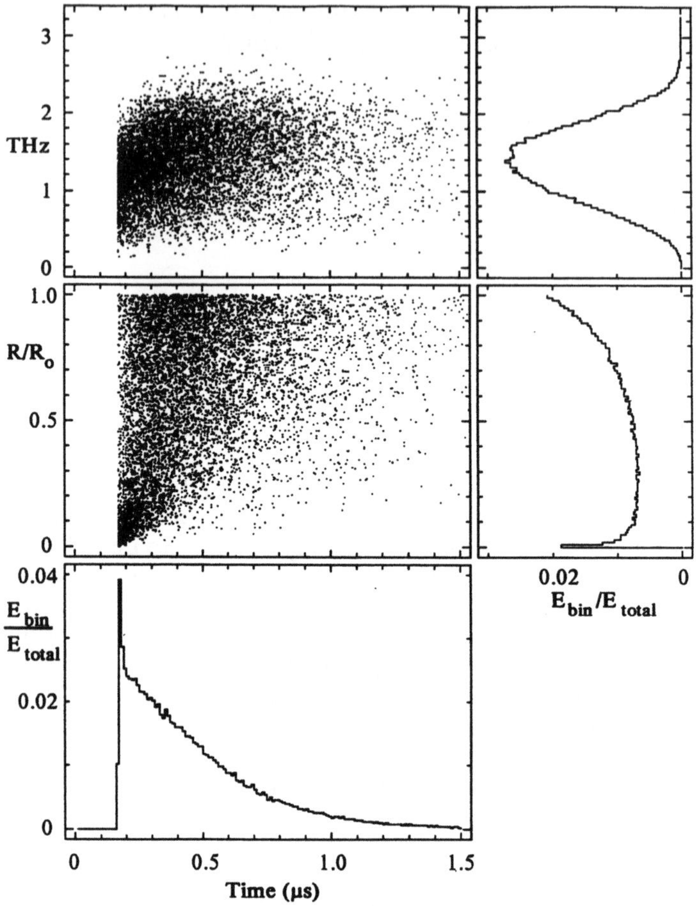

Fig 3. Results of Monte Carlo calculation with anharmonic decay and isotope scattering (see text for details).

record the arrival time and frequency of each phonon, where each point represents an equal amount of energy. The upper right histograms the phonon energy versus phonon frequency. The middle left plot records the arrival time versus the distance form the source for the point of last scattering before reaching the surface. The middle right histograms the phonon energy versus distances of last scattering from the source and the lower right histograms the phonon energy versus arrival time. The lower left portion of the middle density plot corresponds to ballistic phonons which have traveled from very near the source to the surface without scattering with a transit time very near the ballistic time. This ballistic component contains about 15% of the total energy for a 1 mm radius and about 11% for a 10 mm radius. The remaining phonon energy arrives after further scattering and is called quasidiffuse because of the wide range of diffusion lengths.

3. Superconducting Transition Edge Sensors

A superconducting transition-edge sensor consists of a thin-film patterned into a series meandering circuit and biased with a constant current. A phonon flux incident on the film will drive those portions normal where the phonon energy density exceeds a critical value. Then a voltage is seen across the circuit providing a signal which is proportional to the length of the circuit driven normal. If the current is below a latching critical current, then self-extinguishing pulses are seen, otherwise the resistive heating of the normal state is sufficient to expand the normal region across the whole circuit. Such a system is straightforward to manufacture over large crystal surfaces with photolithographic techniques, but the physics governing the response is non-linear. These types of devices, often called superconducting bolometers in the phonon physics literature, have been used for many years [6]. However, in the phonon experiments the devices are biased in temperature at the midpoint of the resistive transition and have a roughly linear response for small changes about the bias point. On the other hand, for our application better performance is obtained by biasing at the foot of the transition on the superconducting side, and our signal comes from those portions of the film that are driven fully into the normal state. To indicate this distinction, we prefer the term transition-edge sensor.

We estimate the threshold energy density E_ρ necessary for this superconducting to normal state transition, by integrating the heat capacity $C_{es}(T)$ of the superconductor from the bias temperature T_b to the critical temperature T_c. The time constants within the superconductor are sufficiently short to allow quasi-thermal equilibrium. For a film thickness d, the asymptotic form near T_c of the surface energy density E_σ given by E_ρ d is $E_\sigma \approx 5.0\ N(0)\ \Delta_0^2\ d\ (1-T/T_c)$, where Δ_0 is the gap at $T = 0$ and N(0) is the density of states at the Fermi surface in the normal metal. In Fig 4 we plot the surface critical energy density for 40 nm thick films in eV/μm^2 as a function of temperature for the elemental superconductors Al, Ti, Ir, & W.

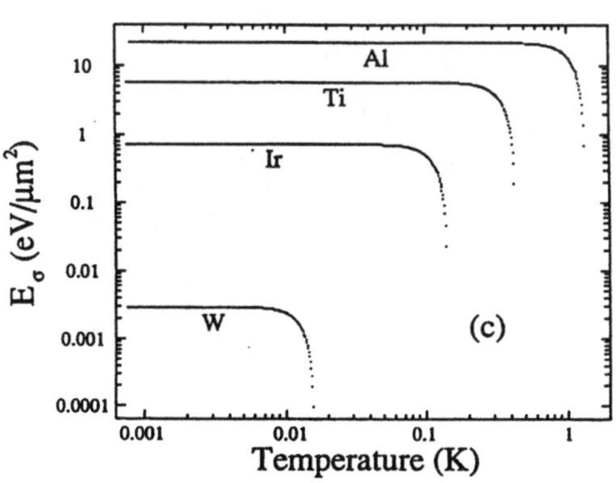

Fig 4. Critical surface energy density for 40 nm films.

4. Experiments

As examples of the current status of the research, we discuss two of our recent experiments using Ti transition-edge sensors in more detail. The first uses alpha particle interactions on a 1 mm thick silicon crystal and the second uses x-rays down to several keV.

4.1 Detection of Alpha Particles

During this past year we have performed a series of detailed experiments with alpha particles which have demonstrated the ballistic phonon detection concept [8]. These measurements use 5.5 MeV alpha particles form a 24 μCi source of ^{241}Am as the energy source. The alpha particles are stopped by the silicon in a surface layer about 20 μm thick and generate phonons which travel through 1 mm thick silicon crystals. We have demonstrated that about 1/3 of the phonon energy flux reaching the back face is ballistic both in timing experiments and in spatial distribution experiments. Fig 5a shows the timing difference between the front (t=0) incident alpha face and the back face as a function of the temperature. When the temperature is sufficiently close to T_c so that the ballistic component alone can exceed the threshold of the film, there is excellent agreement between the theoretical ballistic time for ST and FT phonons (dotted lines) and the experimental data. When the detector is cooled sufficiently below T_c, then the ballistic phonons alone are not sufficient to exceed threshold and longer times are needed to accumulate quasidiffuse phonons, so that the timing difference is seen to lengthen. The Ti films have a 2 μsec time constant for thermal recovery from a pulse. In Fig 5b, the spatial distribution of the ballistic focussing patterns are seen by plotting the distribution of events partitioned between two sensors which are located side by side with no gap. For diffuse propagation there would be little structure so that the size of the

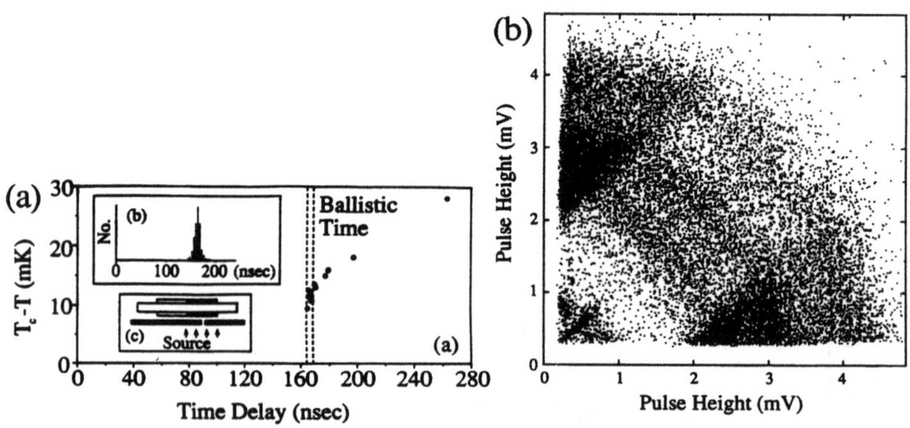

Fig 5. Alpha particle experiment showing (a) ballistic time-of-flight and (b) ballistic spatial distribution.

signal in channel A would smoothly vary with respect to its size in channel B. However, instead we find significant structure which can be nicely explained by the known ballistic focussing anisotropy shown in Fig 1 [8].

4.2 Detection of X-Rays with Transition Edge Sensors

We have performed several x-ray experiments using our Ti transition-edge sensors. These also used the 24 μC source of ^{241}Am. In addition to the several alpha lines around ≈ 5.5 MeV, the decay spectrum of ^{241}Am contains a nuclear gamma at 60 keV and two atomic x-rays at 14 and 18 keV. Foils of Pb or Sn which are 125 μm thick are placed between the source and detector and stop all but the 60 keV gamma rays which are attenuated in number by ≈ 0.5. As shown in Fig 4, our Ti films which are biased at $T_b/T_c \approx 0.95$ have a critical surface energy density $E_\sigma \approx 1$ eV/μm². We use cryogenic GaAs MESFET voltage-sensitive amplifiers with $\Delta V_{rms} \approx 1$ nV/√Hz at 1 MHz [9].

The 60 keV gamma rays have a 30 mm absorption length in the Si substrate, much longer than the crystal thickness, so that they interact at a nearly uniform rate throughout the interior of the crystal. Fig 6b is a pulse height spectrum obtained using the 125 μm Sn absorber. The prominent

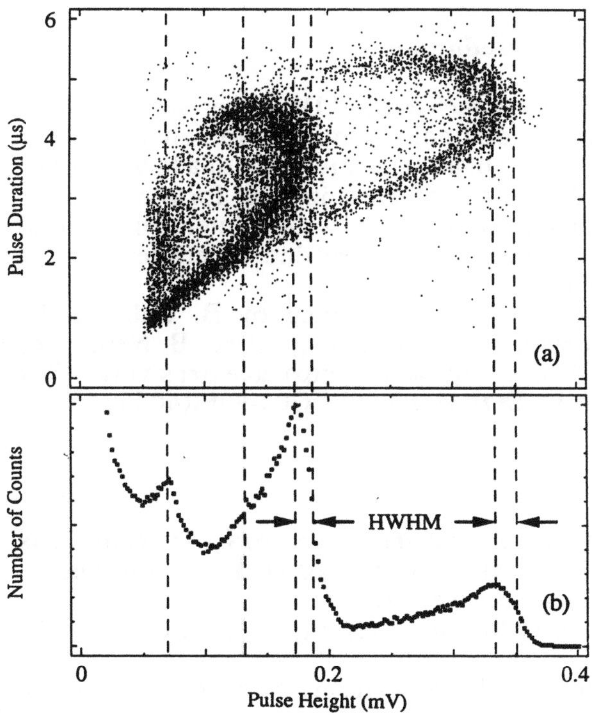

Fig 6. (a) Plot of pulse height vs. pulse duration. (b) Pulse height spectrum of 60 keV, 25 keV and 8 keV.

peak at the upper end of the spectrum is the 60 keV photopeak, the sharp feature at the lower end is consistent with secondary 8 keV x-rays produced in the surrounding Cu and then striking the front side of the detector, and the central peak around 25 keV is due to secondary emission of K_α x-rays from the Sn. The central peak disappears when the Sn absorber is replaced by Pb which has its K-edge above 60 keV.

Our first direct data on the distribution of signal amplitudes as a function of distance into the silicon is shown in Fig 6a. The graph shows a two dimensional plot of the height of each pulse versus its duration. The 60 keV and 25 keV branches are clearly visible and we estimate an energy resolution of ~ 2.5 keV (FWHM). This technique is effective in separating two branches for each energy, because events close to the titanium do not drive a large area normal, but do greatly exceed threshold in the smaller normal region. This area takes a longer time to return to equilibrium since the excess phonon energy must leave the film. On the other hand, events with the same peak height which are further from the film, just exceed threshold, so that the relaxation to equilibrium is faster. Thus the upper branches in Fig 6b correspond to events near the titanium film and the lower branches to events further from the films. At the 60 keV peak, we estimate that the event locations are about 100 μm from the crystal surface.

5. Conclusions and Acknowledgements

Imaging phonon-mediated detectors such as SiCADs are becoming a reality in the laboratory. The prospects look promising for obtaining the additional improvements necessary to achieve thresholds of a keV or better in large crystals. Such a detector is of great interest for a number of experiments including dark matter searches for weakly interacting neutral particle candidates and low energy reactor neutrino experiments to set better limits on the neutrino mass.

The work at Stanford has been performed by B. Cabrera, B. Dougherty, A. T. Lee, K. Irwin, and B. A. Young. Also, B. Neuhauser now at San Francisco State University has participated extensively. This work has been funded in part by DOE Grant DE-FG03-90ER40569.

References

[1] See for example: **Low Temperature Detectors for Neutrinos and Dark Matter III**, eds. L. Brogiato, D. V. Camin, and E. Fiorini (Editions Frontiere, France, 1990).

[2] See for example: D. O. Caldwell, et al, Phys. Rev. Lett. **61**, 510 (1988).

[3] B. Cabrera, J. Martoff and B. Neuhauser, Nucl. Instr. & Meth. **A275**, 97(1989).

[4] See for example: G. A. Northrop and J. P. Wolfe, Phys. Rev. **B22**, 6196 (1980).

[5] S. Tamura, Phys. Rev. **B31**, 2574 (1985); H. J. Maris, Phys. Rev. **B41**, 9736 (1990).
[6] See for example: **Nonequilibrium Phonon Dynamics**, ed. W.E. Bron, NATO ASI Series **B124**, Plenum Press, N.Y., 1985.
[7] See for example: M. Tinkham, **Introduction to Superconductivity**, (Krieger Publishing, 1975).
[8] B. A. Young, B. Cabrera and A. T. Lee, Phys. Rev. Lett. **64**, 2795 (1990).
[9] A. T. Lee, Rev. Sci. Instrum. **60**, 3315 (1989).

Characterization of Corrosion Film in Galvanized Steel Exposed to Atmospheric Corrosion

C. Beltrán, L. Cota, and M. Avalos-Borja

Instituto de Física, Universidad Nacional Autónoma de México,
Apdo. Postal 2681, 22800 Ensenada, B.C., Mexico

ABSTRACT. Specimens of commercial galvanized steel wire were exposed to the atmospheric environment for periods up to six months. The corrosion film formed was characterized by Auger Electron Spectroscopy (AES), Scanning Electron Microscopy (SEM), and X-Ray Diffraction (XRD). With these techniques we analyse morphology and composition as a function of time.

INTRODUCTION

The main aspects of corrosion studies are the identification of reaction products and their influence in the corrosion rate. It is clear that the behavior of metals in the environment depends on the composition, structure, solubility, thickness, adhesion, etc., of the solid compound that forms on the surface during reaction. The importance of this film is that it frequently forms a 'barrier' that isolates the metal from the environment and, therefore, controls the corrosion rate.

The corrosion of galvanized steel has been studied before [1,2], but since the results depend to some extent on the particular environment, we were interested in studying this phenomena in the coastal zone of northwest Mexico.

EXPERIMENTAL

We used commercial grade galvanized steel (like the one used for 'barbed wire' fences), in two forms: a) straight sections of 80 mm in lenght and b) segments of 700-750 mm in length rolled in spirals of 60 mm diameter by 150 mm height, both of them supported on PVC frames. The first type of samples were used for SEM and AES studies, and the second ones, for the collection of enough reaction film for the X-Ray studies. The pre-cleaned (ultrasonically immersed in acetone and dried in air) wires were exposed to the atmosphere for periods of 4, 8, and 15 days, and 1, 3, and 6 months.

At the above indicated times, samples type (a) were directly observed in the scanning microscope (Jeol JSM-5300) and in the Auger system (PHI-595). From samples type (b) we obained powder to be analysed by X-Rays (General Electric GEXRD6).

RESULTS

Visual inspection of fresh unexposed sample shows a gray surface, with metallic shine and smooth. SEM also shows that there are no important features aside from some scratches most likely produced during manufacturing. Auger spectroscopy from the as-received sample, shows that the elements present at the surface are zinc, oxygen, and minor contaminants as carbon, chlorine, sulfur, calcium and silicon. After argon sputtering cleaning, the surface consists mainly of zinc (Fig. 1).

After being exposed to atmospheric corrosion, visual changes were observed, for example, the surface was not shiny, but granular, and the color was dark gray. In the samples exposed for 4 days, SEM shows regions like 'sponges' surrounded by an amorphous material (Fig. 2a). Auger spectroscopy (Fig. 2b) indicates the presence of Zn and O in a

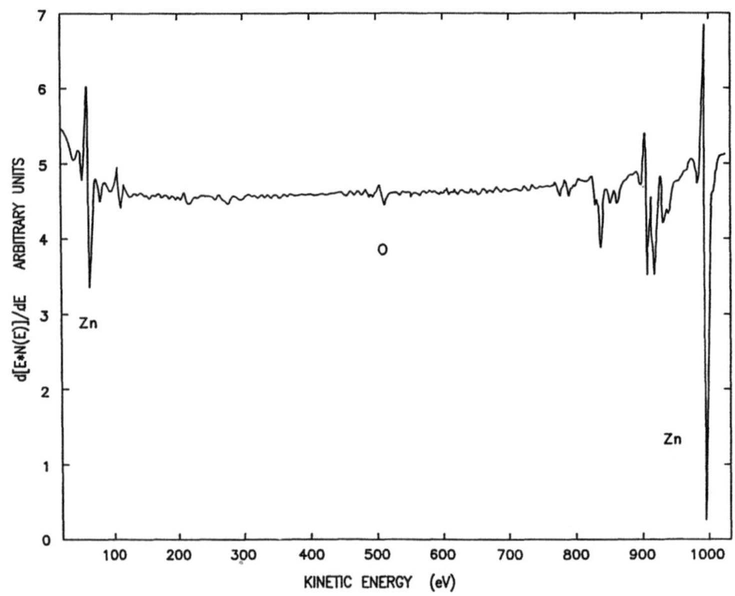

Fig. 1. Auger spectrum from as-received sample, after Ar cleaning.

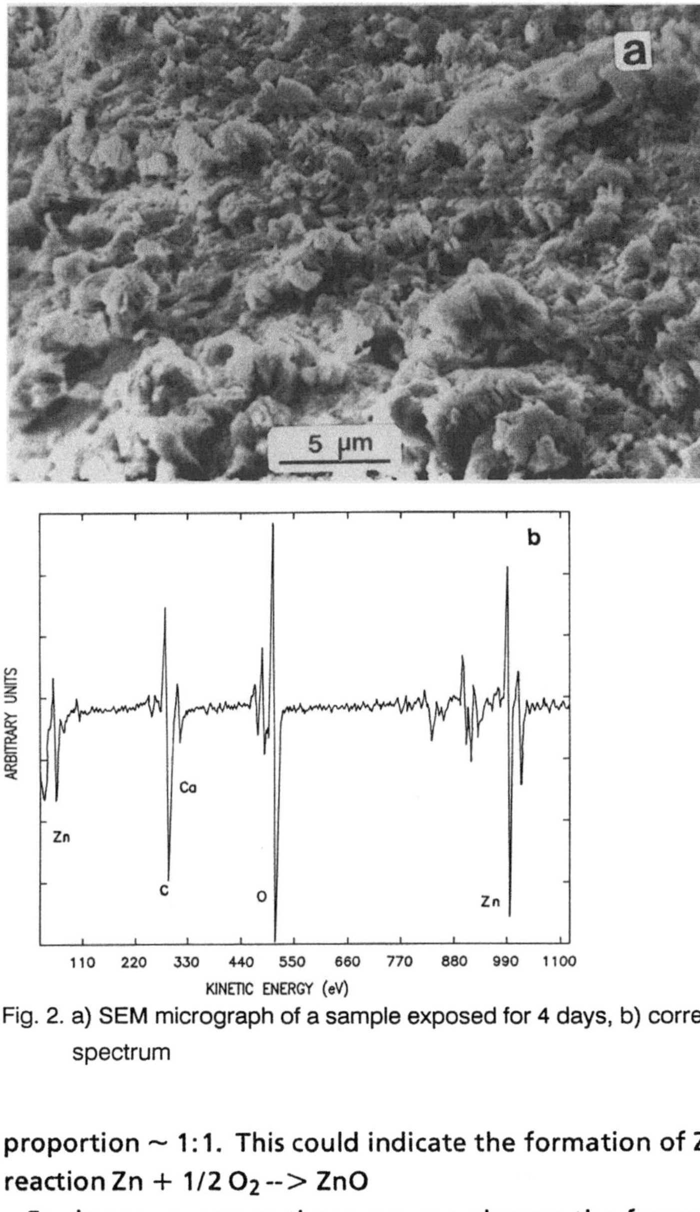

Fig. 2. a) SEM micrograph of a sample exposed for 4 days, b) corresponding Auger spectrum

proportion ~ 1:1. This could indicate the formation of ZnO through the reaction $Zn + 1/2\ O_2 \rightarrow ZnO$

For longer exposure times, we can observe the formation of products with a 'needle'-like morphology, as shown in Fig. 3a. Auger spectroscopy (Fig. 3b) indicates a proportion of Zn to O of ~ 1:2 which might indicate the existence of $Zn(OH)_2$ (since Auger is not sensitive to H, we can not determine this possibility with better accuracy). The chemical reaction leading to this product will be $ZnO + H_2O \rightarrow Zn(OH)_2$.

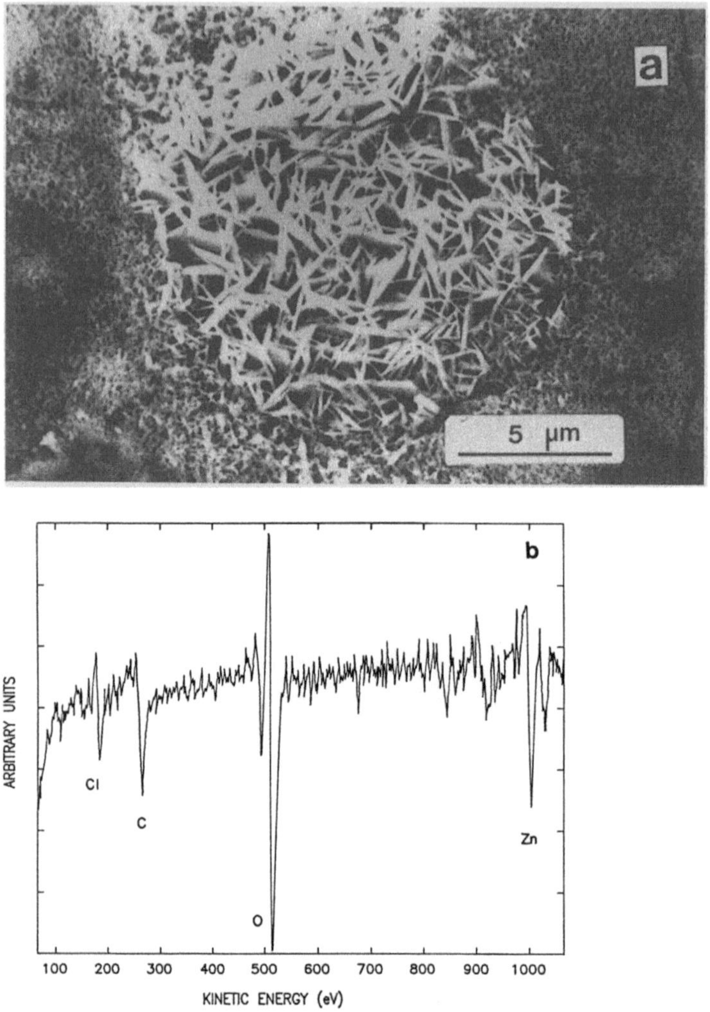

Fig. 3. a) SEM micrograph of a sample exposed for 8 days, b) corresponding Auger spectrum.

We also observe (for periods between 3 and 6 months) structures with cubic shapes, as shown in Fig. 4a. Auger spectroscopy (Fig. 4b) indicates that this structures are made of NaCl, certainly coming from the sea breeze.

In samples exposed 6 months, the needle-like features associated with hydroxides exhibit a greater concentration of Cl as shown in the Auger spectrum in Fig. 5. The formation of $Zn_5(OH)_8Cl_2$ was proved by

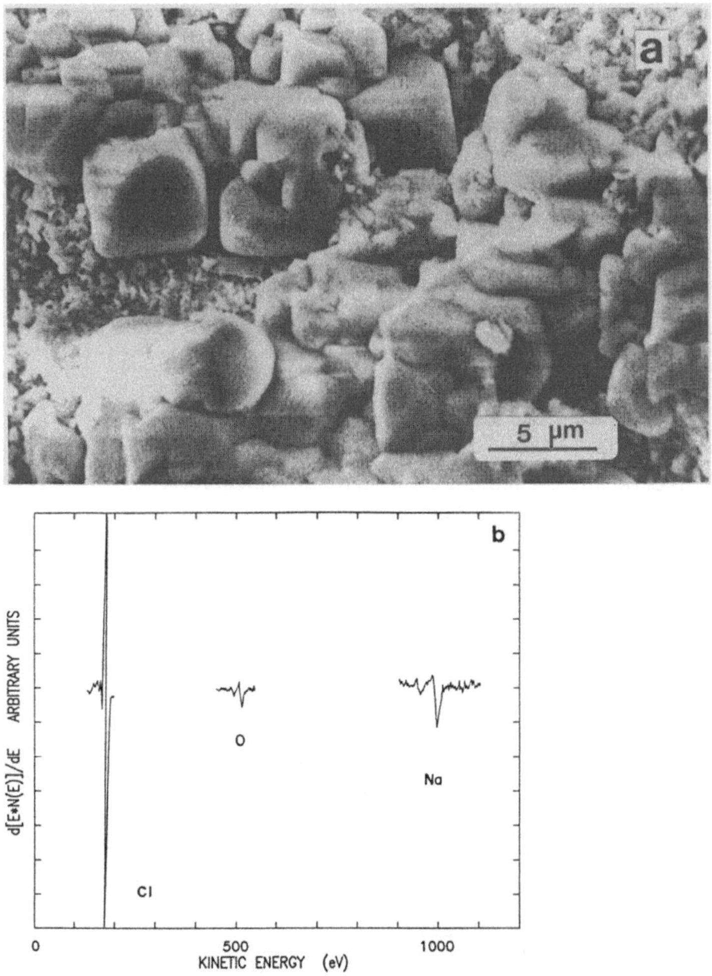

Fig. 4. a) SEM micrograph of a sample exposed for 3 months, b) corresponding Auger spectrum.

X-rays analysis by comparing the spectrum form the sample (Fig. 6a) with the corresponding spectrum obtained from X-ray card No. 7-155 from JCPDS (Fig. 6b). The formation of that compound is explained by the following reaction: $5Zn(OH)_2 + 2NaCl \rightarrow Zn_5(OH)_8Cl_2$.

It is worth mentioning that other workers (1-4) have found the formation of zinc carbonates for samples exposed to similar conditions. However we did not detect the presence of those compounds.

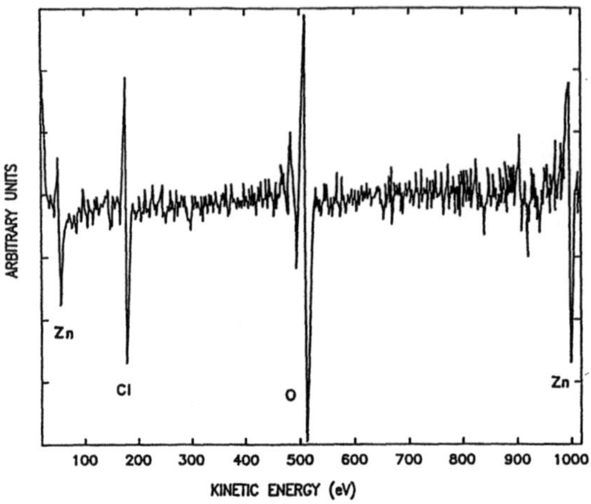

Fig. 5. Auger spectrum of needle-like features, after 6 months exposure.

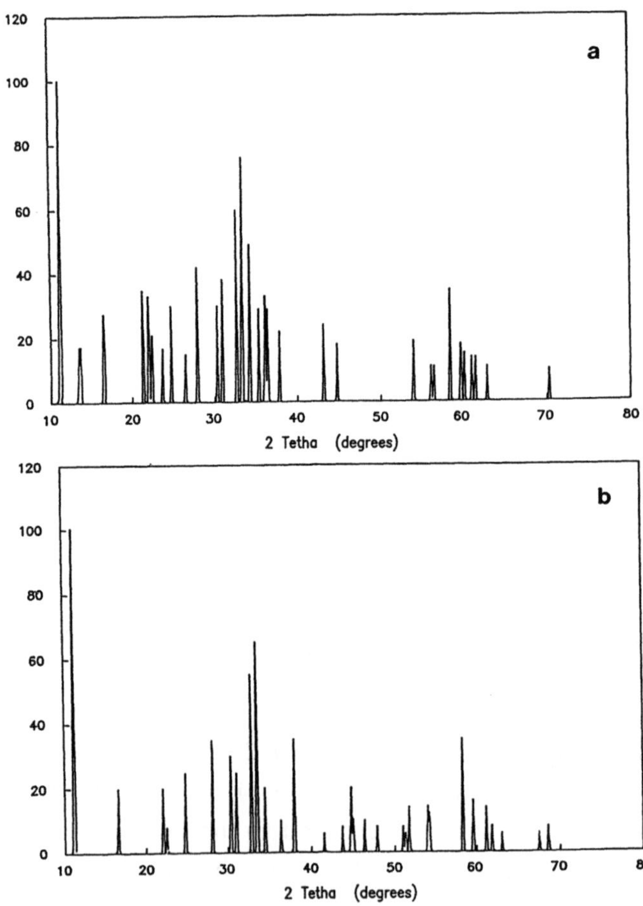

CONCLUSIONS

The initial stages of corrosion on galvanized steel reveals the presence of zinc oxide layers, and a later formation of zinc hydroxides with a peculiar 'needle' or laminar morphology. Samples exposed for more than 3 months show large areas of this product, identified by Auger and X-Rays as $Zn_5(OH)_8Cl_2$. We found no evidence of zinc carbonates.

ACKNOWLEDGEMENTS

We thank I. Gradilla for technical assistance. One of us (CB) thanks the Universidad 'Juarez' Autonoma de Tabasco and CICESE for support during this work.

REFERENCES

[1] Flinn et al. in "Degradation of Materials Due to Acid Rain", R. Baboian, Ed. Am. Chem. Soc., 1986, p.119.

[2] Haynie et al., in "Atmospheric Factors Affecting the Corrosion of Engineering Materials", ASTM STP 648, S.K. Coburn, Ed. Am. Soc. for Testing and Mat., 1978 P.30.

[3] S.R. Dunbar and W. Showak in "Atmospheric Corrosion", W.H. Ailor, Ed. Wiley, 1982, p. 529.

[4] R.A. Legault in "Atmospheric Corrosion", W.H. Ailor, Ed. Wiley, 1982, p.607.

Fig. 6. a) Experimental x-ray spectrum from powder collected from a sample exposed by 6 months, b) "theoretical" x-ray spectrum for $Zn_5(OH)_8Cl_2$.

Index of Contributors

List of Participants

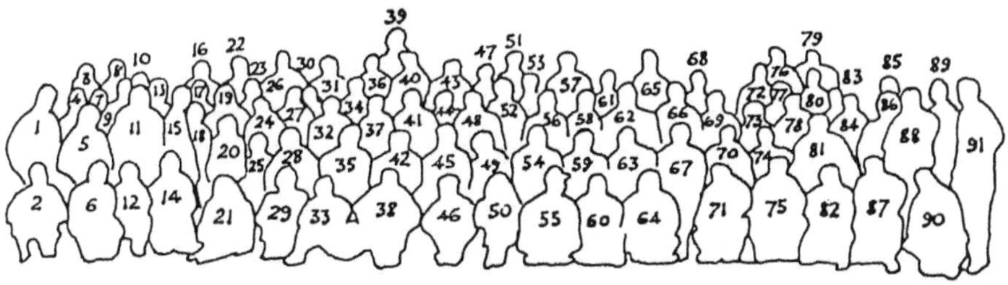

SIXTH LATIN AMERICAN SYMPOSIUM ON SURFACE PHYSICS, CUSCO, PERU. 2-7 September 1990

(1) L. Delgado, (2) E. Llachua, (3) J. Giraldo, (4) N. Majlis, (5) F. Briones, (6) F. Umeres, (8) G. Monsivais, (10) L. Bolarte, (11) W. Estrada, (13) A. Plata, (14) M. Miki, (15) A. Chernov, (16) V. Fuenzalida, (17) G. Gordillo, (18) G. Platero, (19) Z. Basticevic, (20) R. Barrio, (21) A. Camacho, (22) J. Barzola, (23) P. Feibelman, (24) V. Acosta, (25) C. Ocal, (26) E. Artacho, (27) D. Acosta, (28) P. Echenique, (29) P.Orozco, (30) P.Taylor, (31) J. Morán-Lopez, (32) M. E. Gómez, (33) H. Sánchez, (34) E. Lerner, (35) A. Eguiluz, (36) M. Farías, (37) R. Huacoto, (38) M. Grimsditch, (40) S. Camargo, (41) A. Talledo, (42) F. Sánchez-Sinencio, (43) C. Rivasplata, (44) J. Hernández, (45) I. Hernández-Calderón, (46) L. Catalán, (47) M. Rocca, (48) M. P. Carreño, (49) M. Cardona, (50) G. Patroni, (51) L. Mochán, (52) M. Del Castillo, (53) H. Lotsch, (54) F. Ponce, (55) M. Horn, (56) C. Oviedo, (58) S. Ferrer, (59) M. Salmerón, (60) V. Latorre, (62) H. Galindo, (63) E. Anda, (64) P. Schabes, (65) R. Caldera, (66) E. López, (67) M. José-Yacamán, (68) Y. Noguchi, (69) B. Alascio, (70) B. Cabrera, (71) A. Valera, (73) L. Viscido, (74) C. Cabrera, (75) C. Polo, (76) J. Luyo, (77) J. M. Heras, (78) I. Schuller, (79) W. More, (80) R. Nicolsky, (81) F. Decker, (82) J. Agreda, (83) H. Nowak, (84) J. Ramírez, (85) C.Pinto de Melo, (87) R. Asomoza, (88) R. Baquero, (89) C. Granqvist, (90) S. Castañeda, (91) W. Schreiner.

MIX
Papier aus verantwortungsvollen Quellen
Paper from responsible sources
FSC® C105338

If you have any concerns about our products,
you can contact us on
ProductSafety@springernature.com

In case Publisher is established outside the EU,
the EU authorized representative is:
Springer Nature Customer Service Center GmbH
Europaplatz 3, 69115 Heidelberg, Germany

Printed by Libri Plureos GmbH
in Hamburg, Germany